Operator Theory
Advances and Applications
Vol. 78

Editor
I. Gohberg

Partial Differential Operators and Mathematical Physics

International Conference in Holzhau, Germany
July 3–9, 1994

Edited by

M. Demuth
B.-W. Schulze

Birkhäuser Verlag
Basel · Boston · Berlin

Volume Editorial Office:

Prof. Michael Demuth
Technische Universität Clausthal
Institut für Mathematik
Erzstrasse 1
38678 Clausthal-Zellerfeld
Germany

Prof. Dr. Bert-Wolfgang Schulze
Max Planck-Arbeitsgruppe
"Partielle Differentialgleichungen und Komplexe Analysis"
Universität Potsdam
Institut für Mathematik
Postfach 60 15 53
14415 Potsdam
Germany

A CIP catalogue record for this book is available from the Library of Congress, Washington D.C., USA

Deutsche Bibliothek Cataloging-in-Publication Data
Partial differential operators and mathematical physics :
international conference in Holzhau (Germany), July 3–9, 1994
/ ed. by M. Demuth ; B.-W. Schulze. - Basel ; Boston ; Berlin:
Birkhäuser, 1995
(Operator theory ; Vol. 78)

NE: Demuth, Michael [Hrsg.]; GT

Softcover reprint of the hardcover 1st edition 1995

Printed on acid-free paper produced from chlorine-free pulp
Cover design: Heinz Hiltbrunner, Basel

ISBN 978-3-0348-9903-1 ISBN 978-3-0348-9092-2 (eBook)
DOI 10.1007/978-3-0348-9092-2

9 8 7 6 5 4 3 2 1

Contents

Preface ix

S. Albeverio, F. Ru-Zong, M. Röckner, W. Stannat:
A remark on coercive forms and associated semigroups 1

W. Arendt, S. Monniaux:
Domain perturbation for the first eigenvalue of the Dirichlet Schrödinger operator 9

E. Bernardi, A. Bove:
Geometric transition for a class of hyperbolic operators with double characteristics 21

P. Boggiatto, E. Buzano, L. Rodino:
Multi-quasi-elliptic operators in $\mathbb{R}^n$ 31

L. Boutet de Monvel:
Real analogue of the Bergman kernel 43

V.V. Chepyzhov, M.I.Vishik:
Attractors of non-autonomous evolution equations with translation-compact symbols 49

J. M. Combes, P. D. Hislop:
Localization for 2–dimensional random Schrödinger operators with magnetic fields 61

Yu. L. Daletskiĭ, V. R. Steblovskaya:
Some problems of calculus of variations in infinite dimensions 77

M. Demuth, F. Gesztesy, J. van Casteren, Z. Zhao:
Finite capacities in spectral theory 89

J. Dereziński:
Classical N-body scattering 99

P. Duclos, P. Šťovíček:
Quantum Fermi accelerators with pure-point quasi-spectrum 109

Yu.V. Egorov, V.A. Kondratiev:
On moments of negative eigenvalues of an elliptic operator 119

L. Erdős:
Magnetic Lieb-Thirring inequalities and stochastic oscillatory integrals 127

B. V. Fedosov:
On the trace density in deformation quantization 133

D. Fujiwara:
The stationary phase method with remainder estimate as dimension of the space goes to infinity 135

P. B. Gilkey, B. Botvinnik:
The eta invariant, equivariant spin bordism, and metrics of positive scalar curvature 141

J. Ginibre, G. Velo:
Generalized Strichartz inequalities for the wave equation 153

B. Helffer:
Around the transfer operator and the Trotter-Kato formula 161

R. Hempel, I. Herbst:
Bands and gaps for periodic magnetic Hamiltonians 175

M. Hieber:
H^∞-calculus for second order elliptic operators in divergence form 185

T. Ichinose:
Path integral for the relativistic Schrödinger semigroup 191

V. Ivrii:
Semiclassical spectral asymptotics and multiparticle quantum theory 199

S. T. Kuroda, K. Kurata:
Product formulas and error estimates 213

A. Laptev:
On inequalities for the bound states of Schrödinger operators 221

M. Levitin, D. Vassiliev:
Some examples of two-term spectral asymptotics for sets with fractal boundary 227

O. Liess:
Estimates for Fourier transforms of surface-carried densities on surfaces with singular points 235

A. Martinez, G. Nenciu:
On adiabatic reduction theory 243

I. McGillivray:
Recurrence for fractional powers of diffusion operators in terms of volume-growth 253

S. Nakamura:
Band spectrum for Schrödinger operators with strong periodic magnetic fields 261

V. S. Rabinovich:
Mellin pseudodifferential operators with operator symbols and its applications 271

M. Reissig, K. Yagdjian:
Hypoellipticity of certain differential operators with degeneration of infinite order 281

I. Roitberg:
On approximation of solutions of elliptic boundary value problems for Petrovskii elliptic systems by linear combinations of fundamental solutions 285

Y. Saitō:
The reduced wave operator with two unbounded media 291

E. Schrohe, B.-W. Schulze:
Mellin quantization in the cone calculus for Boutet de Monvel's algebra 299

B.-W. Schulze:
Transmission algebras on singular spaces with components of different dimensions 321

Z. G. Sheftel:
On approximation by solutions of non-local elliptic problems 343

A. Shlapunov, N. Tarkhanov:
A stability set in the Cauchy problem for elliptic systems 353

A. V. Sobolev:
Discrete spectrum asymptotics for the Schrödinger operator 357

P. Stollmann:
Convergence of Schrödinger operators on varying domains 369

G. Stolz:
Localization for the Poisson model 375

C. A. Tracy, H. Widom:
Systems of partial differential equations for a class of operator determinants 381

J. Voigt:
Absorption semigroups, Feller property, and Kato class 389

L. Weis:
Gaussian estimates and analytic semigroups 397

D. Yafaev:
New channels of scattering for long-range potentials 405

Participants 421

Talks 427

Preface

This volume contains the proceedings of the International Conference on "Partial Differential Equations" held in Holzhau/Erzgebirge, Germany, July 3–9, 1994. The conference was sponsored by the Max-Planck-Gesellschaft, the Deutsche Forschungsgemeinschaft, the Land Brandenburg and the Freistaat Sachsen.

It was initiated by the Max-Planck-Research Group "Partielle Differentialgleichungen und Komplexe Analysis" at the University of Potsdam as one of the annual meetings of the research group. This conference is part of a series begun by the former Karl-Weierstraß-Institute of Mathematics in Berlin, with the conferences in Ludwigsfelde 1976, Reinhardsbrunn 1985, Holzhau 1988 (proceedings in the Teubner Texte zur Mathematik **112**, Teubner-Verlag 1989), Breitenbrunn 1990 (proceedings in the Teubner Texte zur Mathematik **131**, Teubner-Verlag 1992), and Lambrecht 1991 (proceedings in Operator Theory: Advances and Applications, Vol. 57, Birkhäuser Verlag 1992); subsequent conferences took place in Potsdam in 1992 and 1993 under the auspices of the Max-Planck-Research Group "Partielle Differentialgleichungen und Komplexe Analysis" at the University of Potsdam.

It was the intention of the organizers to bring together specialists from different areas of modern analysis, geometry and mathematical physics to discuss not only recent progress in the respective disciplines but also to encourage interaction between these fields.

The scientific advisory board of the Holzhau conference consisted of S. Albeverio (Bochum), L. Boutet de Monvel (Paris), M. Demuth (Clausthal), P. Gilkey (Eugene), B. Gramsch (Mainz), B. Helffer (Paris), S.T. Kuroda (Tokyo), B.-W. Schulze (Potsdam).

The organizers would like to thank Frau Chr. Gottschalkson for executing a great part of the administrative work as well as Frau Albrecht from the Max-Planck-Institute of Mathematics in Bonn.

M. Demuth
Clausthal

B.-W. Schulze
Potsdam

Operator Theory:
Advances and Applications, Vol. 78

A Remark on Coercive Forms and Associated Semigroups

S. Albeverio, F. Ru-Zong, M. Röckner, W. Stannat

Abstract

It is shown how to determine a coercive closed form on a real Hilbert space directly from its associated semigroup.

AMS Subject Classification (1991)
Primary: 31C25
Secondary: 47D06

1 Introduction

The abstract theory of (real valued) coercive closed forms on real Hilbert spaces and associated semigroups is presented in [1], with the aim of application in the theory of Dirichlet forms. In particular, the relation between such forms, semigroups, associated resolvents and generators is studied. However, in contrast to the symmetric case, a direct connection between semigroups and coercive closed forms is missing. The purpose of this paper is to deliver this relation, complementing Diagram 3 in [1] this way. In particular, we show how one can construct the coercive closed form (including its domain) directly from the associated semigroup. In fact, one part of this result (i.e., Theorem 3.4(ii) below) in the case of complex-valued coercive closed forms can already be found in [3]. But for completeness we also give a proof of this in the real case, in particular adding the details of the underlying results from [5] which are not spelt out in the latter reference. We use the characterization of $(\mathcal{E},D(\mathcal{E}))$ in terms of its associated resolvent (cf. Proposition 2.5 below) in an essential way. We also discuss some consequences. Let us indicate shortly our results.

Let $(\mathcal{E},\mathcal{D}(\mathcal{E}))$ be a coercive closed form on a real Hilbert space $\mathcal{H}$ with associated semigroup $(T_t)_{t>0}$. For $t \in (0,\infty)$ let ${}^{(t)}\mathcal{E}(u,v) := \frac{1}{t}(u - T_t u, v)$; $u,v \in \mathcal{H}$. We shall prove in Theorem 3.4:

$$u \in \mathcal{D}(\mathcal{E}) \Longleftrightarrow \sup_{t>0} {}^{(t)}\mathcal{E}(u,u) < \infty$$

and

$$\mathcal{E}(u,v) = \lim_{t \downarrow 0} {}^{(t)}\mathcal{E}(u,v) \text{ for all } u,v \in \mathcal{D}(\mathcal{E}).$$

Moreover, the restriction of $(T_t)_{t>0}$ to $\mathcal{D}(\mathcal{E})$ is a strongly continuous semigroup on $\mathcal{D}(\mathcal{E})$. In particular

$$\lim_{t\downarrow 0} \mathcal{E}_1(u - T_t u, u - T_t u) = 0 \text{ for all } u \in \mathcal{D}(\mathcal{E}).$$

In Theorem 4.1 we shall give another characterization of $\mathcal{D}(\mathcal{E})$ by using a Tauberian theorem. For $u \in \mathcal{H}$ define $\alpha_u(t) := \int_0^t (u - T_s u, u)\, ds$; $t \in (0,\infty)$. Then

$$u \in \mathcal{D}(\mathcal{E}) \Longleftrightarrow \exists \lim_{t\downarrow 0} \frac{1}{t^2}\alpha_u(t) \in \mathrm{R}$$

and

$$\lim_{t\downarrow 0} \frac{1}{t^2}\alpha_u(t) = \frac{1}{2}\mathcal{E}(u,u) \text{ for all } u \in \mathcal{D}(\mathcal{E}).$$

2 Definitions and Preliminaries

Let $\mathcal{H}$ be a real Hilbert space with inner product $(\cdot,\cdot)$ and norm $||\cdot|| = (\cdot,\cdot)^{\frac{1}{2}}$. Let $\mathcal{D}(\mathcal{E})$ be a linear subspace of $\mathcal{H}$ and $\mathcal{E} : \mathcal{D}(\mathcal{E}) \times \mathcal{D}(\mathcal{E}) \longrightarrow \mathrm{R}$ a non-negative definite bilinear map.

For $\alpha \geq 0$ we set

$$\mathcal{E}_\alpha(u, v) := \mathcal{E}(u, v) + \alpha(u,v);\ u,v \in \mathcal{D}(\mathcal{E}). \tag{2.1}$$

Definition 2.1. $(\mathcal{E}, \mathcal{D}(\mathcal{E}))$ is said to satisfy the *weak sector condition* iff there exists a constant $K > 0$ such that

$$|\mathcal{E}_1(u, v)| \leq K\mathcal{E}_1(u, u)^{1/2}\mathcal{E}_1(v, v)^{1/2} \text{ for all } u, v \in \mathcal{D}(\mathcal{E}). \tag{2.2}$$

The constant K in Definition 2.1 is called *continuity constant.* It is easy to see that the weak sector condition is equivalent to the following: there exists a constant $K' > 0$ such that

$$|\mathcal{E}(u, v)| \leq K'\mathcal{E}_1(u, u)^{1/2}\mathcal{E}_1(v, v)^{1/2} \text{ for all } u, v \in \mathcal{D}(\mathcal{E}). \tag{2.3}$$

Definition 2.2. $(\mathcal{E}, \mathcal{D}(\mathcal{E}))$ is said to be *closed* if $\mathcal{D}(\mathcal{E})$ is complete with respect to the norm $\tilde{\mathcal{E}}_1^{1/2}$, where $\tilde{\mathcal{E}}(u,v) := 1/2\Big(\mathcal{E}(u,v) + \mathcal{E}(v,u)\Big)$; $u,v \in \mathcal{D}(\mathcal{E})$.

Definition 2.3. $(\mathcal{E}, \mathcal{D}(\mathcal{E}))$ is said to be a *coercive closed form* iff $\mathcal{D}(\mathcal{E})$ is a dense subspace of $\mathcal{H}$, and $(\mathcal{E}, \mathcal{D}(\mathcal{E}))$ is closed and satisfies the weak sector condition (2.2).

Proposition 2.4. Let $(\mathcal{E}, \mathcal{D}(\mathcal{E}))$ be a coercive closed form on $\mathcal{H}$. Then there exists unique strongly continuous contraction resolvents $(G_\alpha)_{\alpha>0}$ and $(\hat{G}_\alpha)_{\alpha>0}$ on $\mathcal{H}$ such that

$$G_\alpha(\mathcal{H}), \hat{G}_\alpha(\mathcal{H}) \subset \mathcal{D}(\mathcal{E}), \forall \alpha > 0 \text{ and } \mathcal{E}_\alpha(G_\alpha f, u) = (f, u) = \mathcal{E}_\alpha(u, \hat{G}_\alpha f) \tag{2.4}$$

for all $f \in \mathcal{H}, u \in \mathcal{D}(\mathcal{E}), \alpha > 0$. In particular, we have

$$(G_\alpha f, g) = (f, \hat{G}_\alpha g) \text{ for all } f, g \in \mathcal{H}, \tag{2.5}$$

i.e. $\hat{G}_\alpha$ is the adjoint of G_α for all $\alpha > 0$.

Proof. cf. e.g. [MR 92, Theorem I.2.8]. □

We assume from now on that $(\mathcal{E}, \mathcal{D}(\mathcal{E}))$ is a coercive closed form on $\mathcal{H}$. Denote by $(L, \mathcal{D}(L))$ and $(\hat{L}, \mathcal{D}(\hat{L}))$ the generators corresponding to $(G_\alpha)_{\alpha>0}$ and $(\hat{G}_\alpha)_{\alpha>0}$, respectively. Then we have (cf. e.g. [MR 92, Corollary I.2.10]),

$$\begin{aligned} G_\alpha(\mathcal{H}) &= \mathcal{D}(L) \subset \mathcal{D}(\mathcal{E}) \text{ and } \mathcal{E}(u, v) = (-Lu, v) \\ &\quad \text{for all } u \in \mathcal{D}(L), v \in \mathcal{D}(\mathcal{E}) \end{aligned} \tag{2.6}$$

and

$$\begin{aligned} \hat{G}_\alpha(\mathcal{H}) &= \mathcal{D}(\hat{L}) \subset \mathcal{D}(\mathcal{E}) \text{ and } \mathcal{E}(v,u) = (-\hat{L}u, v) \\ &\quad \text{for all } u \in \mathcal{D}(\hat{L}), v \in \mathcal{D}(\mathcal{E}). \end{aligned} \tag{2.7}$$

Define for $\alpha > 0$

$$\mathcal{E}^{(\alpha)}(u,v) = \alpha(u - \alpha G_\alpha u, v);\ u,v \in \mathcal{H}. \tag{2.8}$$

Proposition 2.5. Let $(\mathcal{E}, \mathcal{D}(\mathcal{E}))$ be a coercive closed form on $\mathcal{H}$.

(i) Let $u \in \mathcal{H}$. Then $u \in \mathcal{D}(\mathcal{E})$ if and only if $\sup_{\alpha>0} \mathcal{E}^{(\alpha)}(u, u) < \infty$.

(ii) $\mathcal{D}(L) = G_\alpha(\mathcal{H})$ is dense in $\mathcal{D}(\mathcal{E})$ and moreover, for all $u \in \mathcal{D}(\mathcal{E})$

$$\lim_{\alpha \longrightarrow \infty} \mathcal{E}_1(\alpha G_\alpha u - u, \alpha G_\alpha u - u) = 0. \tag{2.9}$$

In particular, $(\mathcal{E}, \mathcal{D}(\mathcal{E}))$ is uniquely determined by $(G_\alpha)_{\alpha>0}$ via (2.4).

(iii) $\lim_{\alpha \longrightarrow \infty} \mathcal{E}^{(\alpha)}(u, v) = \mathcal{E}(u, v)$ for all $u,v \in \mathcal{D}(\mathcal{E})$.

Proof. cf. [MR 92, Theorem I.2.13]. □

3 The main result

In this section, we shall assume that $(\mathcal{E},\mathcal{D}(\mathcal{E}))$ is a coercive closed form on $\mathcal{H}$ with continuity constant K. Denote the strongly continuous contraction semigroups corresponding to $(G_\alpha)_{\alpha>0}$ and $(\hat{G}_\alpha)_{\alpha>0}$ by $(T_t)_{t>0}$ and $(\hat{T}_t)_{t>0}$, respectively.

For $t \in (0,\infty)$ we define

$$^{(t)}\mathcal{E}(u,v) := \frac{1}{t}(u - T_t u, v);\ u,v \in \mathcal{H}. \tag{3.1}$$

$\mathcal{D}(\mathcal{E})$ is a real Hilbert space with respect to the scalar product $\tilde{\mathcal{E}}_1(u,v)(=\tilde{\mathcal{E}}(u,v)+(u,v))$. Denote by $\mathcal{D}(\mathcal{E})'$ the dual of $\mathcal{D}(\mathcal{E})$. Identifying $\mathcal{H}$ with its dual we have that $\mathcal{D}(\mathcal{E}) \subset \mathcal{H} \subset \mathcal{D}(\mathcal{E})'$ densely and continuously.

Since $\mathcal{E}$ is continuous on $\mathcal{D}(\mathcal{E})$ by (2.4), there exists a linear operator A on $\mathcal{D}(\mathcal{E})'$ with domain $\mathcal{D}(A) = \mathcal{D}(\mathcal{E})$ such that $\langle -Au,v\rangle = \mathcal{E}(u,v)$ for all $u,v \in \mathcal{D}(\mathcal{E})$, where $\langle\cdot,\cdot\rangle$ denotes the dualization between $\mathcal{D}(\mathcal{E})'$ and $\mathcal{D}(\mathcal{E})$. Note that (e.g. by [MR 92, Proposition I.2.16]) $(L,\mathcal{D}(L))$ is the part of $(A,\mathcal{D}(A))$ on $\mathcal{H}$, i.e., $\mathcal{D}(L) = \{u \in \mathcal{D}(A) | Au \in \mathcal{H}\}$ and $Au = Lu$ for $u \in \mathcal{D}(L)$.

Denote by $\mathcal{H}_\mathrm{C}$ the complexification of $\mathcal{H}$, i.e. $\mathcal{H}_\mathrm{C} = \mathcal{H}\times\mathcal{H}$ with addition given by $[u_1,v_1]+[u_2,v_2] := [u_1+u_2, v_1+v_2]$, scalar multiplication given by $(a+ib)[u,v] := [au-bv, av+bu]$ and inner product given by $([u_1,v_1],[u_2,v_2]) := (u_1,u_2)+i(u_1,v_2)-i(v_1,u_2)+(v_1,v_2)$. The bilinear form $(\mathcal{E},\mathcal{D}(\mathcal{E}))$ can be extended to a sesquilinear form $(\mathcal{E}^\mathrm{C},\mathcal{D}(\mathcal{E}^\mathrm{C}))$ on $\mathcal{H}_\mathrm{C}$ by

$$\mathcal{E}^\mathrm{C}([u_1,v_1],[u_2,v_2]) := \mathcal{E}(u_1,u_2) + i\mathcal{E}(u_1,v_2) - i\mathcal{E}(v_1,u_2) + \mathcal{E}(v_1,v_2) \tag{3.2}$$

for $[u_1,v_1],[u_2,v_2] \in \mathcal{D}(\mathcal{E}^\mathrm{C}) := \Big\{[u_1,u_2] \in \mathcal{H}_\mathrm{C} \Big| u_1,u_2 \in \mathcal{D}(\mathcal{E})\Big\}$. For $\lambda \in \mathrm{C}$ we set

$$\mathcal{E}^\mathrm{C}_\lambda([u_1,v_1],[u_2,v_2]) := \mathcal{E}^\mathrm{C}([u_1,v_1],[u_2,v_2]) + \lambda([u_1,v_1],[u_2,v_2]).$$

$\mathcal{D}(\mathcal{E}^\mathrm{C})$ is a complex Hilbert space with respect to the scalar product

$$\tilde{\mathcal{E}}^\mathrm{C}_1([u_1,v_1],[u_2,v_2]) := \tilde{\mathcal{E}}_1(u_1,u_2) + i\tilde{\mathcal{E}}_1(u_1,v_2) - i\tilde{\mathcal{E}}_1(v_1,u_2) + \tilde{\mathcal{E}}_1(v_1,v_2)$$

with corresponding norm $|[u,v]| := \tilde{\mathcal{E}}^\mathrm{C}_1([u,v],[u,v])^{1/2} = (\mathcal{E}_1(u,u)+\mathcal{E}_1(v,v))^{1/2}$. The reader should note that this is not the symmetric part of $(\mathcal{E}^\mathrm{C},\mathcal{D}(\mathcal{E}))$.

It is easy to check that

$$\left|\mathcal{E}^\mathrm{C}_1([u_1,v_1],[u_2,v_2])\right| \le \sqrt{2}K\left|[u_1,v_1]\right|\left|[u_2,v_2]\right| \tag{3.3}$$

for $[u_1,v_1],[u_2,v_2] \in \mathcal{D}(\mathcal{E}^\mathrm{C})$ and

$$\left|\mathcal{E}^\mathrm{C}_{1+\lambda}([u,v],[u,v])\right| \ge \left|[u,v]\right|^2 \text{ for } \lambda \in \mathrm{C} \text{ with } \mathrm{Re}(\lambda) \ge 0. \tag{3.4}$$

Clearly there exists an operator A^{C} on $\mathcal{D}(\mathcal{E}^{\mathrm{C}})'$ with domain $\mathcal{D}(A^{\mathrm{C}}) = \mathcal{D}(\mathcal{E}^{\mathrm{C}})$ and

$$\langle -A^{\mathrm{C}}[u_1,v_1],[u_2,v_2]\rangle = \mathcal{E}^{\mathrm{C}}([u_1,v_1],[u_2,v_2]);\ [u_1,v_1],[u_2,v_2] \in \mathcal{D}(\mathcal{E}^{\mathrm{C}}).$$

The following proposition is an immediate consequence of [Ta 79, Theorem 3.6.1] (cf. also [Ou 92, page 10]). For the reader's convenience we will present a short proof here. For Banach spaces E_1, E_2 we denote the space of all bounded linear operators from E_1 to E_2 by $\mathcal{L}(E_1,E_2)$ and the corresponding norm by $\|\ \|_{\mathcal{L}(E_1,E_2)}$. We set $\mathcal{L}(E_1) := \mathcal{L}(E_1,E_1)$. For the remaining terminology we refer to [MR 92, Section I.2].

Proposition 3.1. The operator $(A^{\mathrm{C}},\mathcal{D}(A^{\mathrm{C}}))$ generates a holomorphic semigroup $(S_z^{\mathrm{C}})_{z\in S(C)}$ om $\mathcal{D}(\mathcal{E}^{\mathrm{C}})'$ on some sector $S(C) = \left\{z \in \mathrm{C}\,\middle|\,|\mathrm{Im}(z)| \leq C\,\mathrm{Re}(z)\right\}$ for some $C \in (0,\infty)$.

Proof. By the Lax-Milgram Theorem (cf. [Ta 79, Lemma 2.2.1]) and (3.4) the inverse $(\lambda+1-A^{\mathrm{C}})^{-1}$ exists and is bounded on $\mathcal{D}(\mathcal{E}^{\mathrm{C}})'$ for $\lambda \in \mathrm{C}$ with $\mathrm{Re}(\lambda) \geq 0$. In particular, A^{C} is closed. The following inequalities are easy to verify using (3.3) and (3.4).

$$\left\|(\lambda+1-A^{\mathrm{C}})^{-1}\right\|_{\mathcal{L}(\mathcal{D}(\mathcal{E}^{\mathrm{C}})',\mathcal{D}(\mathcal{E}^{\mathrm{C}}))} \leq 1\,, \tag{3.5}$$

$$\|\lambda(\lambda+1-A^{\mathrm{C}})^{-1}\|_{\mathcal{L}(\mathcal{D}(\mathcal{E}^{\mathrm{C}})')} \leq \sqrt{2}K+1\,. \tag{3.6}$$

It follows that $A^{\mathrm{C}}-1$, and hence A^{C}, generates a holomorphic semigroup (cf. [Na 86, Theorem 1.14, A-II]).

□

Proposition 3.2. The operator $(A,\mathcal{D}(A))$ generates a strongly continuous semigroup $(S_t)_{t>0}$ on $\mathcal{D}(\mathcal{E})'$. The strongly continuous semigroup $(T_t)_{t>0}$ on $\mathcal{H}$ is the restriction of $(S_t)_{t>0}$ to $\mathcal{H}$.

Proof. Since $(A^{\mathrm{C}},\mathcal{D}(A^{\mathrm{C}}))$ is the generator of a holomorphic semigroup $(S_z^{\mathrm{C}})_{z\in S(C)}$, it follows from the well-known Hille-Yosida-Phillips Theorem (cf. [Pa 83, Theorem 1.5.3]) that $\|(\lambda-A^{\mathrm{C}})^{-n}\| \leq \frac{C}{(\lambda-1)^n}$ for some constant $C \in (0,\infty)$ and $\lambda > 1$. From $[Au,0] = A^{\mathrm{C}}[u,0]$ for all $u \in \mathcal{D}(\mathcal{E})$ and $(0,\infty) \subset \rho(A)$, it follows that for all $u \in \mathcal{H}$

$$\begin{aligned}\|(\lambda-A)^{-n}u\|_{\mathcal{D}(\mathcal{E})'} &= \|(\lambda-A^{\mathrm{C}})^{-n}[u,0]\|_{\mathcal{D}(\mathcal{E}^{\mathrm{C}})'} \leq \frac{C}{(\lambda-1)^n}\|[u,0]\|_{\mathcal{D}(\mathcal{E}^{\mathrm{C}})'}\\ &\leq \frac{2^{3/2}C}{(\lambda-1)^n}\|u\|_{\mathcal{D}(\mathcal{E})'}\ \forall\lambda>1.\end{aligned}$$

Hence this inequality holds for all $u \in \mathcal{D}(\mathcal{E})'$. Again by the Hille-Yosida-Phillips Theorem A generates a strongly continuous semigroup $(S_t)_{t>0}$ on $\mathcal{D}(\mathcal{E})'$. For $u \in \mathcal{H}$

and $\lambda > 1$ we have that $(\lambda - L)^{-1}u = (\lambda - A)^{-1}u$. Thus $S_t u = T_t u$ for all $t \geq 0$ by the uniqueness of the Laplace transform. □

Remark 3.3.

(i) Applying Proposition 3.2 to $(\hat{\mathcal{E}},\mathcal{D}(\mathcal{E}))$ with $\hat{\mathcal{E}}(u,v) := \mathcal{E}(v,u)$ we see that the dual statement of the above proposition is also true. Thus the operator $(\hat{A},\mathcal{D}(\hat{A}))$, where $\mathcal{D}(\hat{A}) = \mathcal{D}(\mathcal{E})$ and $\langle -\hat{A}u,v\rangle = \mathcal{E}(v,u)$, generates a strongly continuous semigroup $(\hat{S}_t)_{t>0}$ on $\mathcal{D}(\mathcal{E})'$ such that its restriction to $\mathcal{H}$ is just the strongly continuous contraction semigroup $(\hat{T}_t)_{t>0}$ on $\mathcal{H}$.

(ii) Let $(S_t^*)_{t>0}$ be the semigroup on $\mathcal{D}(\mathcal{E})$ dual to $(S_t)_{t>0}$. Then $(S_t^*)_{t>0}$ is strongly continuous again (cf. e.g. [Y 78, p.233]). It is easy to check that S_t^* is just the restriction of the operator $\hat{T}_t$ to $\mathcal{D}(\mathcal{E})$. Since the dual statement is also true we see that the restriction of $(T_t)_{t>0}$ on $\mathcal{D}(\mathcal{E})$ is a strongly continuous semigroup on $\mathcal{D}(\mathcal{E})$.

Theorem 3.4. Let $(\mathcal{E},\mathcal{D}(\mathcal{E}))$ be a coercive closed form.

(i) Let $u \in \mathcal{H}$. Then $u \in \mathcal{D}(\mathcal{E})$ if and only if $\sup_{t>0} {}^{(t)}\mathcal{E}(u,u) < \infty$.

(ii) For all $u,v \in \mathcal{D}(\mathcal{E})$ we have that $\lim_{t\downarrow 0} {}^{(t)}\mathcal{E}(u,v) = \mathcal{E}(u,v)$.

(iii) $\lim_{t\downarrow 0} \mathcal{E}_1(u - T_t u, u - T_t u) = 0$ for all $u \in \mathcal{D}(\mathcal{E})$.

Proof. Let us show (ii) first. For $u,v \in \mathcal{D}(\mathcal{E})$ it follows from Proposition 3.2 that

$$\lim_{t\downarrow 0} {}^{(t)}\mathcal{E}(u,v) = \lim_{t\downarrow 0} \frac{1}{t}(u - T_t u, v) = \lim_{t\downarrow 0} \frac{1}{t}\langle ,u - S_t u\rangle v = \langle -Au, v\rangle = \mathcal{E}(u,v).$$

(i): Suppose $u \in \mathcal{D}(\mathcal{E})$. By (ii) there exist a $\delta \in (0,\infty)$ such that

$${}^{(t)}\mathcal{E}(u,u) \leq \mathcal{E}(u,u) + 1 \quad \forall t \in (0,\delta).$$

Clearly, ${}^{(t)}\mathcal{E}(u,u) \leq (2/\delta)\|\mathcal{H}\|u^2$ for all $t \in [\delta,\infty)$. Hence $\sup_{t>0} {}^{(t)}\mathcal{E}(u,u) < \infty$. On the other hand, suppose that $u \in \mathcal{H}$ with $\sup_{t>0} {}^{(t)}\mathcal{E}(u,u) < \infty$. Then

$$\begin{aligned}
\mathcal{E}^{(\alpha)}(u,u) &= \int_0^\infty \alpha^2 e^{-\alpha t}(u - T_t u,u)\,dt = \int_0^\infty \alpha e^{-t}(u - T_{\frac{t}{\alpha}} u,u)\,dt \\
&\leq \sup_{t>0} {}^{(t)}\mathcal{E}(u,u)\Big(\int_0^\infty te^{-t}\,dt\Big) \\
&= \sup_{t>0} {}^{(t)}\mathcal{E}(u,u) \quad \forall \alpha > 0.
\end{aligned}$$

Therefore, $\sup_{\alpha>0} \mathcal{E}^{(\alpha)}(u,u) < \infty$ and, consequently, $u \in \mathcal{D}(\mathcal{E})$ by Proposition 2.5.
(iii): This is a consequence of Remark 3.3 (ii). □

4 An alternative characterization of $\mathcal{D}(\mathcal{E})$

Let $(\mathcal{E},\mathcal{D}(\mathcal{E}))$ be a coercive closed form. We shall give another characterization of $\mathcal{D}(\mathcal{E})$ in terms of its associated semigroup $(T_t)_{t>0}$. For $u \in \mathcal{H}$ we define $\alpha_u(t) := \int_0^t (u - T_s u,u)\, ds$; $t \in (0,\infty)$.

Theorem 4.1. Let $(\mathcal{E},\mathcal{D}(\mathcal{E}))$ be a coercive closed form.

(i) For all $u \in \mathcal{D}(\mathcal{E})$ we have that $\lim_{t\downarrow 0} \alpha_u(t)/t^2 = \mathcal{E}(u,u)/2$.

(ii) Let $u \in \mathcal{H}$. Then $u \in \mathcal{D}(\mathcal{E})$ if and only if the limit $\alpha_u(t)/t^2$ for $t \downarrow 0$ exists in R.

Proof. (i): Observe that $\mathcal{E}^{(\beta)}(u,u) = \beta^2 \int_0^\infty e^{-\beta t}\, d\alpha_u(t)$. Since $\lim_{\beta\to\infty} \mathcal{E}^{(\beta)}(u,u) = \mathcal{E}(u,u)$, it follows from [W 46, p.192, Theorem V.4.3] that $\lim_{t\downarrow 0} \alpha_u(t)/t^2 = \mathcal{E}(u,u)/2$.

For the proof of (ii) it now suffices to show that any $u \in \mathcal{H}$ such that $A := \lim_{t\downarrow 0} \alpha_u(t)/t^2$ exists in R, is in $\mathcal{D}(\mathcal{E})$. From [W 46, p.181, Theorem V.1] we have that

$$\limsup_{\beta\to\infty} \left|\beta^2 \int_0^\infty e^{-\beta t}(u - T_t u,u)\, dt - 2A\right| = 0.$$

Consequently, $\sup_{\beta>0} \mathcal{E}^{(\beta)}(u,u) < \infty$ and, therefore, $u \in \mathcal{D}(\mathcal{E})$ by Proposition 2.5. □

Acknowledgements
Two of the authors, S. A. and M. R., would like to thank the organizers for a very pleasant and stimulating conference in Holzhau. Financial support from the Sonderforschungsbereiche 237 (Essen-Bochum-Düsseldorf), 256 (Bonn), and 343 (Bielefeld) as well as from EC-Science Project SC1*CT92-0784 is gratefully acknowledged. The second named author, F. R.-Z., would also like to thank the Institute of Mathematics at the Ruhr-University Bochum and the Institute of Applied Mathematics at the University of Bonn for their hospitality.

References

[1] Ma, Z.-M. and Röckner, M. (1992). Introduction to the Theory of Non-Symmetric Dirichlet forms, Universitext, Springer-Verlag.

[2] Nagel, R. (1986). One-parameter Semigroups of Positive Operators, Lect. Notes in Math. 1184, Springer-Verlag.

[3] Ouhabaz, E.M. (1992). Propriétés d'ordre et de contractivité des semigroupes avec applications aux opérateurs elliptiques, Ph.D thesis, Besancon.

[4] Pazy, A. (1983). Semigroups of Linear Operators and Applications to Partial Differential Equations, Springer-Verlag.

[5] Tanabe, H. (1979). Equations of evolution, Pitman, London.

[6] Widder, D.V. (1946). The Laplace transform, Princeton University Press.

[7] Yosida, K. (1978). Functional Analysis, Springer-Verlag.

Sergio Albeverio
Institut für Mathematik, Ruhr-Universität Bochum,
Postfach 102148, 44780 Bochum, Germany.
BiBoS; SFB 237 Bochum-Essen-Düsseldorf; CERFIM, Locarno.

Fan Ru-Zong
Department of Probability and Statistics, Peking University,
Beijing 100871, P.R.China.
and Institut für Mathematik, Ruhr-Universität Bochum,
Postfach 102148, 44780 Bochum, Germany.

Michael Röckner and Wilhelm Stannat
Fakultät für Mathematik, Universität Bielefeld,
Postfach 100131, 33501 Bielefeld, Germany.

Operator Theory:
Advances and Applications, Vol. 78

Domain Perturbation for the first Eigenvalue of the Dirichlet Schrödinger Operator

W. Arendt, S. Monniaux

Introduction.

Let $\Omega \subset \mathbb{R}^N$ be an open connected set. We consider the Dirichlet-Schrödinger operator $H = -\Delta_\Omega^d + V$ on $L^2(\Omega)$ (where Δ_Ω^d denotes the Laplacian with Dirichlet boundary conditions and V is a suitable potential).

In a recent paper, F. Gesztesy and Z. Zhao [15] showed that the first eigenvalue $\lambda(H)$ of H is a strictly monotonic function with respect to the domain Ω (up to capacity, see below for the precise statement). Their proof is given with help of probabilistic methods.

The purpose of this article is to give an analytic proof of this result. In fact, we prove a generalization, allowing the potential to vary as well.

Our proof is based on a domination argument for positive irreducible semigroups (Section 2). In the main theorem (Theorem 3.1), the difference of two open sets is measured by capacity. Somé results concerning this notion are established in Section 1. In particular, we give a short proof of the fact that

$$H_0^1(\Omega) = \{u \in H^1(\mathbb{R}^N) \ : \tilde{u} = 0 \ q.e. \text{ on } \Omega^c\}$$

using the characterization of closed order ideals in $H^1(\Omega)$ which has been given recently by Stollmann [24]. This seems to be of independent interest.

Acknoledgement : The motivation for this work comes from a stimulating talk by Michael Demuth during the conference "Partial Differential Equations" held in Holzhau in July 1994.

1. A characterization of $H_0^1(\Omega)$ by capacity.

Let $\Omega \subset \mathbb{R}^N$ be an open set. Let

$$H^1(\Omega) = \left\{ u \in L^2(\Omega) : \frac{\partial u}{\partial x_j} \in L^2(\Omega), j = 1, \ldots, N \right\}$$

the first Sobolev space, which is a Hilbert space under the norm

$$\|u\|_{H^1(\Omega)} = \left(\int_\Omega |\nabla u|^2 dx + \int_\Omega |u|^2 dx \right)^{1/2}$$

(see [9, Chapter IX]). We need some basic properties of capacity and refer to Bouleau & Hirsch [8], Fukushima [14] or Ma & Röckner [17] for details.

The **capacity** of a subset A of $\mathbb{R}^N$ is defined by

$$\operatorname{cap}(A) = \inf\left\{\int_{\mathbb{R}^N} |\nabla\varphi|^2 dx + \int_{\mathbb{R}^N} |\varphi|^2 dx : \varphi \in H^1(\mathbb{R}^N), \varphi \geq 1_O \text{ a.e., } A \subset O\right\}.$$

Here 1_O denotes the characteristic function of O, open subset of $\mathbb{R}^N$.

One says a property is true **quasi everywhere** in $\mathbb{R}^N$ (*q.e.*) if there exists a set $A \subset \mathbb{R}^N$ of capacity 0 such that the property is true for all $x \in \mathbb{R}^N \setminus A$.

A function $u : \mathbb{R}^N \to \mathbb{R}$ is called **quasi-continuous** if for every $\varepsilon > 0$ there exists an open set $O_\varepsilon \subset \mathbb{R}^N$ such that $\operatorname{cap}(O_\varepsilon) < \varepsilon$ and u is continuous on $\mathbb{R}^N \setminus O_\varepsilon$.

It is well known that every $u \in H^1(\mathbb{R}^N)$ has a quasi-continuous representative, i.e. there exists a quasi-continuous function $\tilde{u} \in H^1(\mathbb{R}^N)$ such that $u(x) = \tilde{u}(x)$ *a.e.* The function $\tilde{u}$ is unique *q.e.*

By $H_0^1(\Omega)$ we denote the closure of the space of all test functions $C_c^\infty(\Omega)$ in $H^1(\Omega)$ (see [9, Chapitre IX]). It can now be characterized as follows.

Theorem 1.1. One has

$$H_0^1(\Omega) = \{u \in H^1(\mathbb{R}^N) : \ \tilde{u}(x) = 0 \text{ q.e. on } \mathbb{R}^N \setminus \Omega\}.$$

This characterization is well-known to potential analysts. A proof is given by Deny [13, Theorem 2, p. 143] (another is contained in [16, Theorem 3.1, p. 241], or [14, Example 3.3.2, p.81]). Here we give a short proof based on a recent result of Stollmann [24] characterizing closed ideals in $H^1(\mathbb{R}^N)$. Recall that $H^1(\mathbb{R}^N)$ is stable under the operation of taking the absolute value, i.e.

$$u \in H^1(\mathbb{R}^N) \text{ implies } |u| \in H^1(\mathbb{R}^N) \text{ and } \frac{\partial|u|}{\partial x_j} = \operatorname{sign}(u)\frac{\partial u}{\partial x_j}, \ j = 1, \ldots, N \quad (1.1)$$

(see [11, Chap.IV, §7, p.934] ; a proof by semigroup theory is given in [1, Section 2]).

A subspace J of $H^1(\mathbb{R}^N)$ is called an **ideal** if for $u \in J$, $v \in H^1(\mathbb{R}^N)$,

$$|v| \leq |u| \text{ a.e. implies } v \in J.$$

Theorem 1.2. (Stollmann [24]). A subspace J of $H^1(\mathbb{R}^N)$ is a closed ideal of $H^1(\mathbb{R}^N)$ if and only if there exists a Borel set Y in $\mathbb{R}^N$ such that

$$J = \{u \in H^1(\mathbb{R}^N) : \ \tilde{u} = 0 \text{ q.e. on } Y^c\}$$

Remark. The ideal property is important for the characterization of domination for semigroups defined by forms, see Stollmann and Voigt [25] for a special case and Ouhabaz [19], [20] for a general investigation. We refer to Schaefer [22] and Batty & Robinson [6] for basic properties of ordered Banach spaces.

In order to prove Theorem 1.1, we first show

Lemma 1.3. $H_0^1(\Omega)$ is a closed ideal in $H^1(\mathbb{R}^N)$.

Proof. *(a)* By [9, Lemme IX.5, p.111], one has $u \in H_0^1(\Omega)$ whenever $u \in H^1(\mathbb{R}^N)$ such that $\mathrm{supp}(u)$ is compact and included in Ω.

(b) It is easy to see (and also follows from an abstract result by Borwein & Yost [7, Corollary 1]) that $u \mapsto |u|$ is a continuous mapping on $H^1(\Omega)$. Consequently, for $u \in H^1(\mathbb{R}^N)$ also $v \mapsto u \wedge v$, $v \mapsto u \vee v$ are continuous mappings.

(c) Let $u \in H_0^1(\Omega)$, $v \in H^1(\mathbb{R}^N)$, $0 \leq |v| \leq |u|$. Let $\varphi_n \in C_c^\infty(\Omega)$ such that $\varphi_n \to u$ in $H_0^1(\Omega)$. Then $v_n = (v \vee -|\varphi_n|) \wedge |\varphi_n| \in H_0^1(\Omega)$ by *(a)* and $v_n \to v$ by *(b)*. Thus $v \in H_0^1(\Omega)$. □

We need the following well-known properties of capacity.

$$\mathrm{cap}(A \cup B) \leq \mathrm{cap}(A) + \mathrm{cap}(B) \text{ for all Borel sets } A,\, B \subset \mathbb{R}^N\,; \tag{1.2}$$

$$\lim_{n\to\infty} \mathrm{cap}(A_n) = \mathrm{cap}(\bigcup_{n\in\mathbb{N}} A_n)$$
whenever $(A_n)_{n\in\mathbb{N}}$ is an increasing sequence of Borel sets. (1.3)

If $Y \subset \mathbb{R}^N$ is a Borel set, we let

$$D_0(Y) = \{u \in H^1(\mathbb{R}^N) : \ \tilde{u} = 0 \ q.e. \text{ on } Y^c\}.$$

Proposition 1.4. If $C_c^\infty(\Omega) \subset D_0(Y)$ then $\mathrm{cap}(\Omega \setminus Y) = 0$.

Proof. Assume that $\mathrm{cap}(\Omega \setminus Y) > 0$. Let $(K_n)_{n\in\mathbb{N}}$ be an increasing set of compacts such that

$$\bigcup_{n\in\mathbb{N}} K_n = \Omega.$$

It follows from (1.3) that $\mathrm{cap}(K_n \setminus Y) > 0$ for some $n \in \mathbb{N}$.

Let $0 \leq 1_{K_n} \leq \varphi \in C_c^\infty(\Omega)$. Then $\varphi \notin D_0(Y)$. □

Proof of Theorem 1.1. By Theorem 1.2, there exists a Borel set Y such that $H_0^1(\Omega) = D_0(Y)$. It follows from Proposition 1.4 that $\mathrm{cap}(\Omega \setminus Y) = 0$.

This implies that $D_0(\Omega) \subset D_0(Y)$. In fact, let $u \in D_0(\Omega)$.

Then $N = \{x \in \Omega^c : \tilde{u}(x) \neq 0\}$ is of capacity 0.

Since $\{x \in Y^c : \tilde{u}(x) \neq 0\} \subset (\Omega \setminus Y) \cup N$, one has

$$\mathrm{cap}\{x \in Y^c : \tilde{u}(x) \neq 0\} \leq \mathrm{cap}(\Omega \setminus Y) + \mathrm{cap}(N) = 0.$$

Thus $u \in D_0(Y)$.

Conversely, since $C_c^\infty(\Omega) \subset D_0(\Omega)$ and $D_0(\Omega)$ is closed in $H^1(\mathbb{R}^N)$, it follows that $D_0(Y) = H_0^1(\Omega) \subseteq D_0(\Omega)$. □

Corollary 1.5. Let Λ, $\Omega \subset \mathbb{R}^N$ be open sets. Then $H_0^1(\Omega) = H_0^1(\Lambda)$ if and only if $\mathrm{cap}(\Omega \triangle \Lambda) = 0$.

Proof. Assume that $\operatorname{cap}(\Omega \triangle \Lambda) \neq 0$. Then $\operatorname{cap}(\Omega \setminus \Lambda) \neq 0$ or $\operatorname{cap}(\Lambda \setminus \Omega) \neq 0$.

If $\operatorname{cap}(\Omega \setminus \Lambda) \neq 0$, then by Proposition 1.4 there exists $\varphi \in C_c^\infty(\Omega)$, such that $\varphi \notin D_0(\Lambda) = H_0^1(\Lambda)$. In the other case, $C_c^\infty(\Lambda) \not\subset H_0^1(\Omega)$.

The converse implication follows directly from Theorem 1.1. □

2. Domination and eigenvalues for positive irreducible semigroups.

Let (Ω,ν) be a σ-finite measure space, let $\Lambda \subset \Omega$ be a measurable subset and let $p \in [1,\infty)$. We identify $L^p(\Lambda) = L^p(\Lambda,\nu)$ with a subspace of $L^p(\Omega) = L^p(\Omega,\nu)$, extending functions in $L^p(\Lambda,\nu)$ by 0 on $\Omega \setminus \Lambda$.

The following theorem is a generalization of [2, Theorem 1.3] (where it is assumed that $\Lambda = \Omega$). The argument is similar, but for the sake of completeness we include the proof.

Theorem 2.1. Let T be a bounded irreducible positive C_0-semigroup on $L^p(\Omega)$ with generator A, and let S be an irreducible C_0-semigroup on $L^p(\Lambda)$ with generator B. Assume that

(a) $0 \leq S(t)f \leq T(t)f \;\; (0 \leq f \in L^p(\Lambda),\, t \geq 0)$

(b) $\ker B \neq 0$

Then $\nu(\Omega \setminus \Lambda) = 0$ (so that $L^p(\Omega) = L^p(\Lambda)$) and $S(t) = T(t)$ $(t \geq 0)$.

We clarify some notations. If $f : \Omega \to \mathbb{R}$ is measurable, we write

$f \geq 0$ if $f(x) \geq 0$ $\nu - a.e.$;
$f > 0$ if $f \geq 0$ and $\nu(\{x : f(x) \neq 0\}) > 0$;
$f \gg 0$ if $f(x) > 0$ $\nu - a.e.$

If $Q \in \mathcal{L}(L^p(\Omega))$, we write $Q \gg 0$ if $Qf \gg 0$ whenever $0 < f \in L^p(\Omega)$.

The semigroup T is **irreducible** if $R(\mu, A) := (\mu - A)^{-1} \gg 0$ for all (equivalently one) $\mu > 0$ (see [18, p.306]).

By A' we denote the adjoint of A on $L^{p'}(\Omega)$, $\dfrac{1}{p} + \dfrac{1}{p'} = 1$.

Lemma 2.2. Let $0 < u \in L^p(\Omega)$ such that $\lambda R(\lambda, A)u \geq u$ for all $\lambda > 0$. Then $u \in \ker A$, $u \gg 0$ and there exists $\varphi \in \ker(A')$, $\varphi \gg 0$.

Proof. Let $0 \leq \varphi_0 \in L^{p'}(\Omega)$ such that $< u, \varphi_0 >> 0$. Since T is bounded, one has $\sup\{\|\lambda R(\lambda, A)\|,\, \lambda > 0\} < \infty$. Thus $\lambda R(\lambda, A)'\varphi_0$ has a ω^*-limit point $\varphi \in L^p(\Omega, \nu)'$ as $\lambda \searrow 0$.

Clearly, $\varphi \geq 0$ and $< u, \varphi > \geq < u,\varphi_0 >> 0$. Thus $\varphi > 0$.

Since for $0 < \mu \leq 1$, $0 < \lambda \leq 1$,

$$\mu R(\mu, A)'\lambda R(\lambda, A)'\varphi_0 = \frac{\mu\lambda}{\mu - \lambda}(R(\lambda,A)'\varphi_0 - R(\mu,A)'\varphi_0),$$

it follows that $\varphi \in \ker(A')$.

If $0 < f \in L^p(\Omega)$, then $< f,\varphi >=< \mu R(\mu,A)f,\varphi >> 0$ since $\varphi > 0$ and $\mu R(\mu,A)f \gg 0$. Hence $\varphi \gg 0$.

Finally, $\lambda R(\lambda,A)u - u \geq 0$ and $< \lambda R(\lambda,A)u - u,\varphi >= 0$. Since $\varphi \gg 0$, it follows that $\lambda R(\lambda,A)u - u = 0$, $(\lambda > 0)$; i.e. $u \in \ker(A)$. □

Proof of Theorem 2.1. Let $0 \neq v \in \ker(B) \subset L^p(\Lambda)$, $u = |v|$.

Then $u = |\lambda R(\lambda,B)v| \leq \lambda R(\lambda,B)u \leq \lambda R(\lambda,A)u$, $(\lambda > 0)$. It follows from Lemma 2.2 that $u \in \ker(B) \cap \ker(A)$ and $u \gg 0$ on Ω. Since $u \in L^p(\Lambda)$ so that $u(x) = 0$ *a.e.* on $\Omega \setminus \Lambda$, it follows that $\nu(\Omega \setminus \Lambda) = 0$ and so $L^p(\Omega) = L^p(\Lambda) =: E$.

Let $0 \leq \psi \in E'$. Then

$$(\lambda R(\lambda,A)' - \lambda R(\lambda,B)')\psi \geq 0 \text{ and } < u,(\lambda R(\lambda,A)' - \lambda R(\lambda,B)')\psi >= 0$$

since $u \in \ker(A) \cap \ker(B)$. Thus $(\lambda R(\lambda,A)' - \lambda R(\lambda,B)')\psi = 0$ since $u \gg 0$. Since $\text{span}E'_+ = E'$, it follows that $R(\lambda,A)' = R(\lambda,B)'$ $(\lambda > 0)$ and so $A = B$. □

If B generates a bounded C_0-semigroup and $\ker(B) \neq 0$, it is easy to see that also $\ker(B') \neq 0$. The converse is true if $1 < p < \infty$, but not for $p = 1$. So it is natural to ask whether in Theorem 2.1 it suffices to assume that $\ker(B') \neq 0$. The following example shows that this is not the case.

Example 2.3. Let $N \geq 3$, $0 < m \in C_c(\mathbb{R}^N)$. Denote by Δ the Laplacian on $L^1(\mathbb{R}^N) = E$. Then there exists $\mu_2 > 0$ such that

$$\sup\{\|e^{t(\Delta+\mu_2 m)}\|_{\mathcal{L}(L^1)},\ t \geq 0\} < \infty$$

(see [23, B5.2]). Let $0 < \mu_1 < \mu_2$, $S(t) = e^{t(\Delta+\mu_1 m)}$, $T(t) = e^{t(\Delta+\mu_2 m)}$. Then $0 \leq e^{t\Delta} \leq S(t) \leq T(t)$. In particular, both semigroups are irreducible. By [5, Remark 3.9], there exists $0 \ll \varphi \in \ker(B')$, $B = \Delta + \mu_1 m$. However, $S \neq T$. It follows from Theorem 2.1 that $\ker(B) = 0$. □

3. Strict monotonicity of the bottom of the spectrum.

Let $\Omega \subset \mathbb{R}^N$ be an open set and denote by Δ_Ω^d the Dirichlet Laplacian on $L^2(\Omega)$, i.e. $-\Delta_\Omega^d$ is associated with the Dirichlet form

$$a(u,v) = \int_\Omega \nabla u \nabla v\ dx,\ D(a) = H_0^1(\Omega).$$

We consider a potential $V \in L^1_{loc}(\mathbb{R}^N)$ such that V^- is relatively bounded with respect to the form a with form bound less than 1 ; i.e. we assume that there exists $0 \leq \alpha < 1$, $\beta \geq 0$ such that

$$\int_\Omega V^- u^2 dx \leq \alpha \int_\Omega |\nabla u|^2 dx + \beta \int_\Omega u^2 dx \tag{3.1}$$

for all $u \in H_0^1(\Omega)$.

Then we can define the self-adjoint operator $H = -\Delta_\Omega^d + V$ as usual to be the operator associated with the closed symmetric lower bounded form b on $L^2(\Omega)$ given by

$$D(b) = \{u \in H_0^1(\Omega) : \int_\Omega V^- u^2 dx < \infty\},$$

$$b(u,v) = \int_\Omega \nabla u \nabla v \, dx + \int_\Omega V uv \, dx.$$

By $\lambda(H) = \inf\{\sigma(H)\}$, we denote the bottom of the spectrum of H.

Now let $\Lambda \subset \mathbb{R}^N$ be another open set and $U \in L^1_{loc}(\mathbb{R}^N)$ another potential satisfying (3.1). Let $\tilde{H} = -\Delta_\Lambda^d + U$. Assume that

$$\Lambda \subset \Omega \text{ and } V \leq U. \tag{3.2}$$

Then it is not difficult to see that

$$\lambda(\tilde{H}) \geq \lambda(H), \tag{3.3}$$

(see the remark following Proposition 3.3 below).

Our aim is to prove the following result on strict monotonicity in (3.3).

Theorem 3.1. Assume in addition to (3.2) that

(a) Λ, Ω are connected;

(b) $\lambda(\tilde{H})$ is an eigenvalue of $\tilde{H}$.

Then $\lambda(\tilde{H}) = \lambda(H)$ if and only if $\text{cap}(\Omega \setminus \Lambda) = 0$ and $U = V$ *a.e.* on Ω.

This theorem has been proved by Gesztesy and Zhao [15] by probabilistic methods in the case where $U = V$. We give an analytic proof based on domination (see Section 2).

Before giving the proof we mention that in Theorem 3.1, one cannot replace condition *(b)* by the condition that $\lambda(H)$ is an eigenvalue of H. This can be seen by the following example.

Example 3.2. Let $\Lambda = \Omega = \mathbb{R}^N$, $N \geq 5$. Let $0 < m \in C_c(\mathbb{R}^N)$. Then $\mu_0 = \sup\{\mu > 0 : \lambda(-\Delta - \mu m) = 0\} \in (0,\infty)$ (see e.g. [5, Remark 2.11 or Section 4]). Thus $\lambda(-\Delta - \mu_0 m) = 0$. Moreover, by [10, Proposition 4.1], 0 is an eigenvalue of $-\Delta - \mu_0 m$. Letting

$$U = -\mu_0 m, \; V = -\frac{\mu_0}{2} m, \text{ and } A = B = \Delta,$$

one sees that the conclusion of Theorem 3.1 is false in this situation. Moreover, Theorem 3.1 implies that 0 is not an eigenalue of $-\Delta - \mu m$ for any $0 \leq \mu < \mu_0$.

For the proof of Theorem 3.1 we identify again $L^2(\Lambda)$ with a subspace of $\Lambda^2(\Omega)$ extending functions by 0, and $\Lambda^2(\Omega)$ with a subspace of $\Lambda^2(\mathbb{R}^N)$. Note that H generates a C_0-semigroup e^{-tH} on $\Lambda^2(\Omega)$. By the spectral theorem one has $\|e^{-tH}\| = e^{-t\lambda(H)} \quad (t \geq 0)$. (3.4)

Proposition 3.3. One has

$$0 \leq e^{-t\tilde{H}} f \leq e^{-tH} f \quad (0 \leq f \in L^2(\Lambda),\, t \geq 0). \tag{3.5}$$

Note that $e^{-t\tilde{H}} f$ and $e^{-tH} f$ are both functions defined on $\mathbb{R}^N$ by our convention and the inequality (3.5) holds *a.e.* As a consequence one has $\|e^{-t\tilde{H}}\| \leq \|e^{-tH}\|$ $(t \geq 0)$ and so (3.3) follows with help of (3.4).

As in [3] and [4] we denote by Δ_Ω the pseudo-Dirichlet Laplacian on $\Lambda^2(\Omega)$; i.e., $-\Delta_\Omega$ is associated with the form a_Ω given by

$$D(a_\Omega) = \{u_{|\Omega} : u \in H^1(\mathbb{R}^N); u(x) = 0 \text{ } a.e. \text{ on } \Omega^c\},$$

$$a_\Omega(u,v) = \int_\Omega \nabla u \nabla v \, dx.$$

Then by [3, Section 7] or [4],

$$e^{t\Delta_\Omega} f = \lim_{n\to\infty} e^{t(\Delta - n 1_{\Omega^c})} f \ (t \geq 0) \tag{3.6}$$

for all $f \in L^2(\Omega)$, where Δ denotes the Laplacian on $\Lambda^2(\mathbb{R}^N)$. The Dirichlet Laplacian is obtained by a second approximation. Let $\Omega_n \subset \Omega$ be open such that $\overline{\Omega}_n$ is compact, $\overline{\Omega}_n \subset \Omega_{n+1}$ $(n \in \mathbb{N})$ and $\Omega = \bigcup_{n\in\mathbb{N}} \Omega_n$. Then

$$e^{t\Delta_\Omega^d} f = \lim_{n\to\infty} e^{t\Delta_{\Omega_n}} f \quad (f \in L^2(\Omega),\, t \geq 0). \tag{3.7}$$

Remark. One has $e^{t\Delta_\Omega^d} = e^{t\Delta_\Omega}$ if Ω is of class C^1 (see [9, Chapitre IX]).

Proof of Proposition 3.3.

First step : Domination for the Dirichlet Laplacian .

It follows from the Trotter product formula that

$$e^{t(\Delta - n 1_{\Lambda^c})} f \leq e^{t(\Delta - n 1_{\Omega^c})} f \quad (0 \leq f \in L^2(\mathbb{R}^N) \ t \leq 0).$$

Thus it follows from (3.6) (and the same formula with Ω replaced by Λ) that

$$e^{t\Delta_\Lambda} f \leq e^{t\Delta_\Omega} f \quad (0 \leq f \in L^2(\Lambda),\ t > 0). \tag{3.8}$$

Let Λ_n be open sets such that $\overline{\Lambda}_n$ is compact, $\overline{\Lambda}_n \subset \Lambda_{n+1} \subset \Lambda$ $(n \in \mathbb{N})$ and $\bigcup_{n\in\mathbb{N}} \Lambda_n = \Lambda$. Choose Ω_n open such that $\Lambda_n \subset \Omega_n \subset \overline{\Omega}_n \subset \Omega_{n+1} \subset \Omega$ and $\bigcup_{n\in\mathbb{N}} \Omega_n = \Omega$. Then it follows from (3.7) and (3.8) that

$$e^{t\Delta_\Lambda^d} f = \lim_{n\to\infty} e^{t\Delta_{\Lambda_n}} f \leq \lim_{n\to\infty} e^{t\Delta_{\Omega_n}} f = e^{t\Delta_\Omega^d} f \quad (0 \leq f \in L^2(\Lambda),\, t \geq 0). \tag{3.9}$$

Second step : Perturbation by V^- and U^-

Let $V_n^- = \inf\{V^-,n\}$, $U_n^- = \inf\{U^-,n\}$. Since $V \leq U$ one has $U^- \leq V^-$ and $U_n^- \leq V_n^-$ $(n \in \mathbb{N})$. It follows from the Trotter product formula and (3.9) that

$$e^{t(\Delta_\Lambda^d + U_n^-)} f \leq e^{t(\Delta_\Omega^d + V_n^-)} f \tag{3.10}$$

for all $0 \leq f \in L^2(\Lambda)$, $t \geq 0$. If we show that

$$e^{t(\Delta_\Lambda^d + U^-)} f = \lim_{n\to\infty} e^{t(\Delta_\Lambda^d + U_n^-)} f \tag{3.11}$$

for all $f \in L^2(\Lambda)$, and hence the analogous formula for $\Delta_\Omega^d + V^-$ as well, we can conclude from (3.10) that

$$e^{t(\Delta_\Lambda^d + U^-)} f \leq e^{t(\Delta_\Omega^d + V^-)} f \tag{3.12}$$

for $0 \leq f \in L^2(\Lambda)$, $t \geq 0$.

In order to show (3.11), recall that $\int U_n^- u^2 \leq \int U^- u^2 \leq \alpha \int |\nabla u|^2 + \beta \int |u|^2$ for $u \in H_0^1(\Lambda)$, where $0 < \alpha < 1$. Let $c = \beta + 1$.

Then

$\int |\nabla u|^2 - \int U_n^- u^2 + c \int |u|^2 \geq (1-\alpha) \int |\nabla u|^2 + \int |u|^2 \geq (1-\alpha)\|u\|_{H_0^1(\Lambda)}$.

Denote by b_n the form associated with $-\Delta - U_n^- + cI$ and by b the form associated with $-\Delta - U^- + cI$. Then $b_n \geq b_{n+1}$ and $\lim_{n\to\infty} b_n(u,u) = b(u,u)$ for $u \in D(b) = D(b_n) = H_0^1(\Lambda)$.

Now it follows from [21, Theorem S16, p.373]

$$e^{t(\Delta_\Lambda^d + U^- - c)} f = \lim_{n\to\infty} e^{t(\Delta_\Lambda^d + U_n^- - c)} f, \ f \in L^2(\Lambda).$$

This implies (3.11) and the proof of (3.12) is finished.

Third step : Perturbation by V^+ and U^+.

Let $U_n^+ = \inf\{U^+,n\}$, $V_n^+ = \inf\{V^+,n\}$.

Then $V_n^+ \leq U_n^+$ and thus it follows from (3.12) that

$$e^{t(\Delta_\Lambda^d + U^- - U_n^+)} f \leq e^{t(\Delta_\Omega^d + V^- - V_n^+)} f \tag{3.13}$$

for $0 \leq f \in L^2(\Lambda)$, $t \geq 0$, $n \in \mathbb{N}$. It follows from [21, Theorem S14] that

$$\lim_{n\to\infty} e^{t(\Delta_\Lambda^d + U^- - U_n^+)} f = e^{-t\tilde{H}} f \ \ (f \in L^2(\Lambda)),$$

and

$$\lim_{n\to\infty} e^{t(\Delta_\Omega^d + V^- - V_n^+)} f = e^{-tH} f \ \ (f \in L^2(\Omega)).$$

Hence passing to the limit in (3.13) yields the claim (3.5). □

Proposition 3.4. Assume that Λ is connected. Then $(e^{-t\tilde{H}})_{t\geq 0}$ is an irreducible semigroup on $L^2(\Lambda)$.

Proof. It follows from [12, Theorem 3.3.5] that $e^{t\Delta_\Lambda^d} f \gg 0$ whenever $t > 0$, $0 < f \in L^2(\Lambda)$.

Now we argue as in [5, Proposition 1.3] : we can assume that $U^- = 0$ (since $e^{t(\Delta_\Lambda^d - U^+)} \leq e^{t(\Delta_\Lambda^d - U)}$). It follows from [21, Theorem S16, p.373] that

$$e^{t\Delta_\Lambda^d} f = \lim_{n\to\infty} e^{t(\Delta_\Lambda^d - U^+ + U_n^+)} f \ (f \in L^2(\Lambda)). \tag{3.14}$$

Let $0 < f \in L^2(\Lambda)$. Let $t > 0$. We show that $e^{t(\Delta_\Lambda^d - U^+)} f \gg 0$.
If not, there exists $M \subset \Lambda$ with positive Lebesgue measure and $e^{t(\Delta_\Lambda^d - U^+)} f(x) = 0$ *a.e.* on M.

Since $e^{t(\Delta_\Lambda^d - U^+ + U_n^+)} f \leq e^{tn} e^{t(\Delta_\Lambda^d - U^+)} f$ $(n \in \mathbb{N})$, it follows from (3.14) that $(e^{t\Delta_\Lambda^d} f)(x) = 0$ *a.e.* on M, a contradiction. □

Proof of Theorem 3.1. Since Ω and Λ are connected, both semigroups e^{-tH} and $e^{-t\tilde{H}}$ are irreducible. Assume that $\lambda(H) = \lambda(\tilde{H})$. Replacing U by $U - \lambda(\tilde{H})$ and V by $V - \lambda(H)$ we can assume that $\lambda(H) = \lambda(\tilde{H}) = 0$. It follows from Theorem 2.1 that $(\Omega \setminus \Lambda)$ has zero Lebesgue measure and $H = \tilde{H}$. Denote by b, $\tilde{b}$ the forms associated with H and $\tilde{H}$, respectively. Then $b = \tilde{b}$. In particular :

$$\begin{aligned}\left\{u \in H_0^1(\Omega) : \int_\Omega V^+|u|^2 dx < \infty\right\} &= D(b)\\ &= D(\tilde{b}) = \left\{u \in H_0^1(\Lambda) : \int_\Lambda U^+|u|^2 dx < \infty\right\}.\end{aligned}$$

Assume that $\text{cap}(\Omega \setminus \Lambda) > 0$. Then it follows from Proposition 1.4 that there exists $\varphi \in C_c^\infty(\Omega) \setminus H_0^1(\Lambda)$. Thus $\varphi \in D(b) \setminus D(\tilde{b})$, a contradiction. Thus $\text{cap}(\Omega \setminus \Lambda) = 0$. It follows from Corollary 1.5 that $H_0^1(\Omega) = H_0^1(\Lambda)$.

Since $b = \tilde{b}$ one has

$$\int |\nabla u|^2 dx + \int V u^2 dx = \int |\nabla u|^2 dx + \int U u^2 dx \text{ for all } u \in C_c^\infty(\Lambda)$$

and so

$$\int_\Lambda (V - U) u^2 dx = 0 \text{ for all } u \in C_c^\infty(\Lambda).$$

Since $V - U \geq 0$ *a.e.*, this implies that $V - U = 0$ *a.e.* This completes the proof of the direct implication in Theorem 3.1. The other is obvious. □

References

[1] W. Arendt : Kato's inequality and spectral decomposition for positive C_0-groups. Manuscripta Math. **40** (1982) 277–298.

[2] W. Arendt, C.J.K. Batty : Domination and ergodicity for positive semigroups. Proc. Amer. Math. Soc. **114** (1992) 743–747.

[3] W. Arendt, C.J.K. Batty : Absorption semigroups and Dirichlet boundary conditions. Math. Ann. **295** (1993) 427–448.

[4] W. Arendt, C.J.K. Batty : L'holomorphie du semi-groupe engendré par le Laplacien-Dirichlet sur $L^1(\Omega)$. C. R. Acad. Sci. Paris **315**, Série 1 (1992) 31–35.

[5] W. Arendt, C.J.K. Batty : The spectral function and principal eigenvalues for Schrödinger operators. Preprint.

[6] C.J.K. Batty, D.W. Robinson : Positive one-parameter semigroups on ordered Banach spaces. Acta Appl. Math. **1** (1984) 221–296.

[7] J.M. Borwein, D.T. Yost : Absolute norms on vector lattices. Proc. Edinburgh Math. Soc. **27** (1984) 215–222.

[8] N. Bouleau, F. Hirsch : Dirichlet Forms and Analysis on Wiener Space. W. de Gruyter, Berlin – New-York 1991.

[9] H. Brézis : Analyse fonctionnelle. Masson, Paris 1983.

[10] D. Daners : Pricipal eigenvalues for some periodic-parabolic operators on $\mathbb{R}^N$ and related topics. Preprint 1994.

[11] R. Dautray, J.L. Lions : Analyse mathématique et calcul numérique. Masson 1987, Vol 3.

[12] E.B. Davies : Heat kernels and spectral theory. Cambridge Univ. Press, Cambridge 1989.

[13] J. Deny : Les potentiels d'énergie fine. Acta. Math. **82** (1950) 107–183.

[14] M. Fukushima : Dirichlet forms and Markov processes. North Holland, 1980.

[15] F. Gesztesy, Z. Zhao : Domain perturbation, Brownian motion, capacities, and ground states of Dirichlet Schrödinger operators. Math. Z. **215** (1994) 143–150.

[16] L.I. Hedberg : Spectral synthesis in Sobolev spaces and uniqueness of solutions of the Dirichlet problem. Acta. Math. **147** (1981).

[17] Z.M. Ma, M. Röckner : Introduction to the theory of (non-symmetric) Dirichlet forms. Springer, 1991.

[18] R. Nagel (ed.) : One-parameter Semigroups of Positive Operators. Springer LN **1184**, Berlin 1986.

[19] E.M. Ouhabaz : Propriétés d'ordre et de contractivité des semi-groupes avec applications aux opérateurs elliptiques. Thèse. Besançon 1992.

[20] E.M. Ouhabaz : Invariance of closed convex sets and domination criteria for semigroups. Preprint.

[21] R. Reed, B. Simon : Methods of Modern Mathematical Physics I. Functional Analysis, revised and enlarged ed., Academic Press, New-York 1980.

[22] H.H. Schaefer : Banach lattices and positive operators. Springer, Berlin 1974.

[23] B. Simon : Schrödinger semigroups. Bull. Amer. Soc. **7** (1982) 447–526.

[24] P. Stollmann : Closed ideals in Dirichlet spaces. Preprint (1992).

[25] P. Stollmann, J. Voigt : Perturbation of Dirichlet forms by measures. Preprint (1992).

W. Arendt, S. Monniaux
URA CNRS 741, Équipe de Mathématiques,
16 route de Gray, F-25030 BESANÇON Cedex

Operator Theory:
Advances and Applications, Vol. 78

Geometric Transition for a Class of Hyperbolic Operators with Double Characteristics

E. Bernardi, A. Bove

1 Introduction

The Cauchy problem in the C^∞ category for hyperbolic operators with double characteristics has been studied by many authors under different geometrical perspectives. These results (see e.g. [11], [1] and the references therein) put in evidence the role of both the lower order terms and the symplectic geometry associated with the principal symbol at the double characteristic points.

The purpose of our work is the study of a case in which the set of double points Σ_2 can be written as the union of two nonempty subsets $\Sigma_{2,e}$, $\Sigma_{2,ne}$ i.e. the set of effectively hyperbolic double points and its Σ_2 – complementary respectively. We say that in this case a "transition" occurs, or, roughly speaking, that a real eigenvaluc of the fundamental matrix defined on $\Sigma_{2,e}$ vanishes on $\Sigma_{2,ne}$.

Trying to provide a unifying approach to the first of the open problems mentioned above, R. B. Melrose, [11], in 1984 formulated the following conjecture:

Conjecture. If the Cauchy problem for P (having at most double characteristics) is well posed in Ω_0 then $s(\rho)/\lambda(\rho)$ is uniformly bounded in $\Sigma_{2,e} \cap T^*V \setminus \{0\}$, for some neighborhood V of $0 \in \mathbb{R}^{n+1}$, where

$$s(\rho) = |\text{Im } P^s_{m-1}(\rho)| + \inf\{|\text{Re } P^s_{m-1}(\rho) - s|; |s| \leq \text{Tr}^+ F_{P_m}(\rho)\}.$$

This conjecture, as it stands, has not been proved yet, however it gives a hint of what one should reasonably look for, when trying to formulate Levi conditions on the subprincipal symbol P^s_{m-1}.

The aim of this paper is actually to understand a degenerate effectively hyperbolic case: more precisely we will show that the uniform boundedness of the ratio $s(\rho)/\lambda(\rho)$ may also serve as a sufficient condition for the well posedness of the Cauchy problem, provided certain symplectically invariant geometric conditions are satisfied. Our Assumption (H 4) is therefore a slightly restricted reformulation of Melrose Conjecture.

Even under the usual assumption that the principal symbol P_m vanishes exactly of order two ôn Σ_2, one readily sees that the nature of the set $\Sigma_{2,ne}$, the order of vanishing of λ on $\Sigma_{2,ne}$ and the geometry of the transition from a symplectic point of view can be of a very wild type and may produce nontrivial situations. Therefore our first step to cast some light in this rather entangled matter has been to assume from the very beginning that $\Sigma_{2,ne}$ (henceforth denoted by Σ_2'), as well as Σ_2, be both smooth submanifolds of $T^*\Omega \setminus \{0\}$. In this paper we state a number of results which serve as a symplectic classification of the transition cases that may possibly occur. Assumption (H 2) detects the geometrical framework where our analysis goes on. More precisely (H 2) takes a picture of $F_{P_m}(\rho)$ when ρ is inside or outside Σ_2', avoiding for instance Jordan blocks of size 4 in the canonical form of $F_{P_m}(\rho)$, $\rho \in \Sigma_2'$ and precising that $0 \leq \lambda(\rho) \leq \mathrm{dist}^2_{\Sigma_2}(\rho, \Sigma_2')$, where dist denotes any geodesic distance of $\rho \in \Sigma_2$ from Σ_2'; but this may not yet be enough to prove an energy estimate which eventually leads to existence and uniqueness.

Indeed let us recall that in a non degenerate effectively hyperbolic case (see e.g. Lemma 1.2.1 in [13]), denoting by $\Gamma_{P_m}(\rho)$ the hyperbolicity cone of the localized hyperbolic quadratic form $\sigma(X, F_{P_m}(\rho)X)$, "$P_m$ is effectively hyperbolic at ρ" if and only if "$\Gamma_{P_m}(\rho) \cap \mathrm{ran} F_{P_m}(\rho) \cap [(0, e_0)]^\sigma \neq \emptyset$", where we define the dual with respect to σ, $V^\sigma = \{z \in T_\rho T^*\Omega | \sigma(z, v) = 0, \forall v \in V\}$, for a vector subspace V of $T_\rho T^*\Omega$. Therefore for an effectively hyperbolic operator it is always possible to find a time function f, i.e. a C^∞ function vanishing on the double set of P_m, whose Hamilton vector field $H_f(\rho)$, $\rho \in \Sigma_2$, belongs to $\Gamma_{P_m}(\rho)$ (and for which in addition $\partial f/\partial \xi_0(\rho) = 0$ holds). We may then take advantage of this time function, as in [11] and [13], and build a suitable weight function to be employed in the energy estimates.

On the other hand one quickly realizes that Assumption (H 2) is too static to handle the complexity of the dynamics involved in a degenerate effectively hyperbolic case. The existence of a "generalized" time function adapted to a "good" factorization of P_m is what we need here in order to take care of these difficulties. Roughly speaking we need a "real" time function in $\Sigma_2 \setminus \Sigma_2'$, whose existence is of course guaranteed by (H 2), such that $H_f(\rho)$, when $\rho \in \Sigma_2 \longrightarrow \bar{\rho} \in \Sigma_2'$ has a smooth exit from $\Gamma_{P_m}(\rho)$ to $\ker F_{P_m}(\rho)$. The precise type of requests we are asking on the "time" function f is written down explicitly in Assumption (H 3). We would also like to point out that under the geometrical prescription of Hypothesis (H 2) it is always possible to exhibit a good factorization of the principal symbol of P, according to Definition 2.1 in a way reminiscent of the well–known Ivrii factorization (see e.g. Ivrii [4]). The couple (H 2), (H 3) selects therefore a class of differential operators with double characteristics, whose fundamental matrix admits a non negative real eigenvalue $\lambda(\rho)$, vanishing of order at least two on Σ_2', for which the uniform boundedness of the ratio $s(\rho)/\lambda(\rho)$ in Melrose conjecture proves to be the sufficient condition ensuring the correctness of the associated Cauchy problem.

2 Assumptions and the Result

Let $P(x,D)$ be a differential operator of order m,

$$P(x,D) = P_m(x,D) + P_{m-1}(x,D) + \dots \tag{1}$$

where $P_j(x,D)$ denotes the homogeneous part of order j of P, $j = 0,\dots,m$. We assume that P is a differential operator with C^∞ coefficients in the open set $\Omega \subset \mathbb{R}^{n+1}$, $0 \in \Omega$. In this Section we list our assumptions on the operator P.

The first hypothesis is concerned with the intrinsic properties of the double characteristics of the principal symbol of P.

(H1) (i) The principal symbol $P_m(x,\xi)$ of P is hyperbolic with respect to the level surfaces of x_0.

(ii) The characteristic roots of $\xi_0 \longrightarrow P_m(x,\xi_0,\xi')$ have multiplicity at most two; denoting by $\Sigma_2 = \{(x,\xi) \in T^*\Omega \setminus \{0\} | P_m(x,\xi) = 0, dP_m(x,\xi) = 0, \partial^2 P_m(x,\xi)/\partial\xi_0^2 < 0\}$ the set of the double characteristics of $P_m(x,\xi)$, we assume that Σ_2 is a C^∞ manifold such that the canonical 1–form ω does not vanish on $T\Sigma_2$.

(iii) $\forall\rho \in \Sigma_2$ denote by $V_{i\lambda}(\rho)$ the eigenspace corresponding to the purely imaginary eigenvalue $i\lambda$ of the fundamental matrix $F_{P_m}(\rho)$ of P_m at ρ. Then,

$$\dim \bigoplus_{\substack{i\lambda\neq 0 \\ i\lambda\in \mathrm{sp}\,(F_{P_m}(\rho))}} V_{i\lambda}(\rho) = \text{const.}$$

(iv) $P_m(x,\xi)$ vanishes exactly of order two on Σ_2.

The second assumption deals with the degeneracy of the real eigenvalues of $F_{P_m}(\rho)$ when ρ is close to some critical submanifold Σ_2' of Σ_2.

(H2) There exists a C^∞ submanifold of Σ_2, Σ_2', such that:

(i) $F_{P_m}(\rho)$ has a strictly positive eigenvalue, $\lambda(\rho)$ if and only if $\rho \in \Sigma_2 \setminus \Sigma_2'$.

(ii) Let

$$r(\rho) = \text{rank } F_{P_m}^2(\rho),$$

when $\rho \in \Sigma_2 \setminus \Sigma_2'$, and

$$r'(\rho) = \text{rank } F_{P_m}^2(\rho),$$

when $\rho \in \Sigma_2'$. Then $r(\rho) \equiv r$, $\forall\rho \in \Sigma_2 \setminus \Sigma_2'$, $r'(\rho) \equiv r'$, $\forall\rho \in \Sigma_2'$ and moreover $r = r' + 2$.

(iii) The function λ defined in (i) is a smooth function defined on Σ_2. In particular this implies that $\lambda(\rho) \lesssim \mathrm{d}^2_{\Sigma_2}(\rho, \Sigma_2')$ where $\mathrm{d}_{\Sigma_2}(\cdot, \Sigma_2')$ denotes any geodesic distance in Σ_2.

Assumption (H 2) has essentially a "static" nature, i.e. everything is looked at on a fixed point $\rho \in \Sigma_2$. We also need a more "dynamical" type of assumption, which will be stated in (H 3). To this end let us define what we mean by *factorization* of the principal symbol P_m. Without loss of generality we may assume in the following that $m = 2$.

Definition 2.1. We say that $P_2(x,\xi)$ is factorized (in the sense of Ivrii [4]) if we can find $\ell(x,\xi')$, $m(x,\xi')$ vanishing on Σ_2, homogeneous of degree one, $q(x,\xi') \geq 0$ vanishing on Σ_2 and homogenous of degree two, such that, denoting by $\Lambda(x,\xi) = \xi_0 - \ell(x,\xi')$, $M(x,\xi) = \xi_0 - m(x,\xi')$, we have :

(i) $H_\Lambda(\rho) \in \ker F_{P_2(\rho)} \cap \operatorname{ran} F_{P_2(\rho)}$, $\forall \rho \in \Sigma_2'$.

(ii) $P_2(x,\xi) = -\Lambda(x,\xi)M(x,\xi) + q(x,\xi')$.

(iii) $\{\Lambda, M\}(\rho) = 0$, $\forall \rho \in \Sigma_2$.

(iv) $\forall \rho \in \Sigma_2 \setminus \Sigma_2'$ there exist U_ρ conic neighborhood of ρ, $C_\rho > 0$ such that:

$$|\{\Lambda, q\}| \leq C_\rho \sqrt{q}$$

in U_ρ and $C_\rho \leq c\lambda(\rho)$ for a suitable positive constant c.

(v) $H_\Lambda(\rho) - H_M(\rho) \in \ker F_{P_2(\rho)} \cap \operatorname{ran} F_{P_2(\rho)}$, $\forall \rho \in \Sigma_2$.

(H3) Assume that $P_2(x,\xi)$ is factorized according to Definition 2.1. Then there exists a homogeneous symbol of degree zero $f(x,\xi')$ vanishing on Σ_2 such that:

(i) $\{\Lambda, f\}(\rho) \sim \lambda(\rho)$, $\forall \rho \in \Sigma_2$.

(ii) $0 \leq \{\Lambda, f\}(x,\xi') \leq c_1\{M, f\}(x,\xi')$, for a suitable positive constant c_1.

(iii) There exists $\kappa \in]0, 1[$ such that

$$\{q, f\}^2(x,\xi') \leq 4\kappa\{\Lambda, f\}(x,\xi')\{M, f\}(x,\xi')q(x,\xi'),$$

for any $(x,\xi') \in U_{\bar\rho}$, $\bar\rho \in \Sigma_2'$, $U_{\bar\rho}$ conical neighborhood of $\bar\rho$.

The last hypothesis is concerned with the lower order terms of the operator P.

(H4) There exist two positive constants C, ε such that, $\forall \rho \in \Sigma_2$

$$\operatorname{dist}\left(P_1^s(\rho), \left(\left[-\operatorname{Tr}^+ F_{P_m}(\rho) + \varepsilon, \operatorname{Tr}^+ F_{P_m}(\rho) - \varepsilon\right] \times \{0\}\right) \cup \{(0,0)\}\right) \leq C\lambda(\rho)$$

where dist denotes any distance in $\mathbb{C}$.

We are now in a position to state the main theorem:

Theorem 2.1. Let P be a second order differential operator verifying Assumptions (H 1) — (H 4). Then the Cauchy problem for P in Ω_t, $t \in \mathbb{R}$ is well posed in C^∞.

3 Comments

In this Section we gather a number of results characterizing our geometrical situation as well as the different possible types of degeneracy which may occur. Due to Assumption (H 1) (iv), denoting by $\bar{\rho}$ a point in Σ_2' and $V_{\bar{\rho}}$ a conical neighborhood of $\bar{\rho}$ we can find real valued symbols homogeneous of degree one, $\varphi_j(x,\xi')$, $j = 1, \ldots, k$, defined in $V_{\bar{\rho}}$, such that possibly after a conjugation with a Fourier integral operator leaving the time axis invariant P_2 becomes

$$P_2(x,\xi) = -\xi_0^2 + \sum_{j=1}^{k} \varphi_j^2(x,\xi'), \quad (x,\xi) \in V_{\bar{\rho}}. \tag{2}$$

The following Proposition is essentially well known, however we include it for the sake of completeness and for future reference.

Proposition 3.1. Let P_2 be as in 3.1 above and $\rho \in \Sigma_2$. Put:

$$v_\rho = (\{\xi_0, \varphi_1\}(\rho), \cdots, \{\xi_0, \varphi_k\}(\rho))$$

$$M_\rho = \begin{bmatrix} 0 & \{\varphi_1, \varphi_2\}(\rho) & \cdots & \{\varphi_1, \varphi_k\}(\rho) \\ \{\varphi_2, \varphi_1\}(\rho) & 0 & \cdots & \{\varphi_2, \varphi_k\}(\rho) \\ \vdots & \vdots & \ddots & \vdots \\ \{\varphi_k, \varphi_1\}(\rho) & \{\varphi_k, \varphi_2\}(\rho) & \cdots & 0 \end{bmatrix},$$

and denote by $\pi_\rho : \mathbb{R}^k \longrightarrow \ker M_\rho$ the projection operator onto $\ker M_\rho$. Then either

a) $\pi_\rho v_\rho \neq 0$; in this case P_2 is effectively hyperbolic at $\rho \in \Sigma_2$.

or

b) $\pi_\rho v_\rho = 0$; in this case the following situations occur:

(i) $|M_\rho^{-1} v_\rho| > 1$; then $F_{P_2}(\rho)$ has two real non zero eigenvalues.

(ii) $|M_\rho^{-1} v_\rho| = 1$; then $\ker F_{P_2}^2(\rho) \cap \operatorname{ran} F_{P_2}^2(\rho) \neq \{0\}$, i.e. $F_{P_2}(\rho)$ has a Jordan block of size 4 in its canonical form.

(iii) $|M_\rho^{-1} v_\rho| < 1$; then $\ker F_{P_2}^2(\rho) \cap \operatorname{ran} F_{P_2}^2(\rho) = \{0\}$, i.e. $F_{P_2}(\rho)$ is non effectively hyperbolic.

Here $M_\rho^{-1} v_\rho$ means the unique vector belonging to $(\ker M_\rho)^\perp$ which is mapped to v_ρ by M_ρ.

Proposition 3.2. Assume that (H 1), (H 2) hold. Then $\forall \rho \in \Sigma_2'$,

$$\ker F_{P_2}^2 \cap \operatorname{ran} F_{P_2}^2 = \{0\}.$$

Moreover if $\rho \in \Sigma_2 \setminus \Sigma_2'$, $\pi_\rho v_\rho \neq 0$.

Proposition 3.3. Let $\bar{\rho} \in \Sigma_2'$ and $U_{\bar{\rho}}$ a conical neighborhood of $\bar{\rho}$; it is possible to find a vector $\gamma(x,\xi') = (\gamma_1(x,\xi'), \dots, \gamma_{k-\nu}(x,\xi'))$, where $\gamma_j(x,\xi')$ are real symbols homogeneous of degree zero smoothly dependent on x_0, such that

(i) If

$$\Lambda(x,\xi) = \xi_0 + \langle \gamma(x,\xi'), \varphi''(x,\xi') \rangle,$$
$$M(x,\xi) = \xi_0 - \langle \gamma(x,\xi'), \varphi''(x,\xi') \rangle,$$
$$Q(x,\xi') = |\varphi'(x,\xi')|^2 + \langle (\mathrm{Id} - \gamma \otimes \gamma)\, \varphi''(x,\xi'), \varphi''(x,\xi') \rangle,$$

then

$$P_2(x,\xi) = -\Lambda M + Q;$$

(ii) $|\gamma(\rho)| < 1$, for every $\rho \in \Sigma_2 \cap U_{\bar{\rho}}$;

(iii) $\{\Lambda, Q\} \le C\left(Q + |\partial_0 \varphi'| \sqrt{Q}\right)$;

(iv) $\{\Lambda, M\}(\rho) = 0$, for every $\rho \in \Sigma_2 \cap U_{\bar{\rho}}$.

Remark 3.1. If the manifold Σ_2' is cylindrical with respect to x_0, then $\Lambda = M = \xi_0$ yields a factorization in the sense of Definition 2.1, and condition (iii) of (H 3) becomes:

$$\{\xi_0, f\}^2(x,\xi) P_2(x,\xi) + C\{P_2, f\}^2(x,\xi) \ge 0,$$

for a suitable positive constant C, in a neighborhood of a point $\bar{\rho} \in \Sigma_2'$.

We now exhibit a number of examples and counterexamples to our Assumptions (H 1)—(H 4). Let us start first with some models which do not satisfy at least one of the above mentioned assumptions.

A. Let $P_2(x,\xi) = -\xi_0^2 + \left(x_0 x_2^2 - x_1\right)^2 \xi_n^2 + r\xi_1^2$, near $(0, e_n)$, $r > 0$. Using the notations of Proposition 3.1 we have

$$v = \left(x_2^2, 0\right) \quad \text{and} \quad M = \begin{bmatrix} 0 & \sqrt{r} \\ -\sqrt{r} & 0 \end{bmatrix},$$

hence $\pi v = 0$ and $M^{-1} v = \left(0, x_2^2/\sqrt{r}\right)$ which can coincide with one of the cases of item b) in Proposition 3.1 according to $x_2^2 < \sqrt{r}$, $x_2^2 = \sqrt{r}$ or $x_2^2 > \sqrt{r}$. Therefore Assumption (H 2) is *not* satisfied.

B. Let $P_2(x,\xi) = -\xi_0^2 + \left(x_0\left(x_2^2 - x_3^2\right) - x_1\right)^2 \xi_n^2 + \left(2b x_2 x_3 x_0 - x_4\right)^2 \xi_n^2$, near $(0, e_n)$, $b \in \mathbb{R}$. It is easy to show that Hypotheses (H 1), (H 2) are verified and that $\lambda \sim x_2^2 + x_3^2$. Moreover $\Lambda(x,\xi) = M(x,\xi) = \xi_0$ yields a factorization in the sense of Definition 2.1. On the other hand (i) and (ii) of (H 3) are *not* satisfied, since there is no choice of C^∞ functions α, β such that

$$\partial_0\left[\alpha\left(x_0(x_2^2 - x_3^2) - x_1\right) + \beta(2b x_0 x_2 x_3 - x_4)\right] \ge 0.$$

C. Let $P_2(x,\xi) = -\xi_0^2 + \left(x_0x_2^{2\ell} - x_1\right)^2 \xi_n^2 + k_1\xi_n + ik_2x_2^\ell\xi_n$, near $(0, e_n)$, where $k_1 > 0$, $k_2 \in \mathbb{R}$, $\ell \in \mathbb{N}$. This model satisfies all our assumptions (H 1)—(H 4).

D. Let $P_2(x,\xi) = -\xi_0^2 + \left(x_0x_2^{2\ell} - x_1\right)^2 \xi_n^2 + k(x)\xi_n$, near $(0, e_n)$, where $k(x)$ is a smooth function. After a symplectic dilation in the variables (x_1, ξ_1) we reduce P_2 to the following symbol

$$\tilde{P}_2(x,\xi) = -\xi_0^2 + x_0^2x_2^{4\ell}\xi_n^2 + \tilde{k}(x)\xi_n.$$

Recalling the celebrated necessary conditions of Ivrii – Petkov ([7], [6]) we see that in order that the local Cauchy problem for $\tilde{P}_2$ (and hence for P_2) be well posed we must necessarily have $\tilde{k}(x) = x_2^{2\ell}\tilde{\tilde{k}}(x)$. Hence if k_1 in Example C is zero, we conclude that Assumption (H 4) cannot be improved.

E. In order to give an idea of the microlocal energy estimate we exhibit an elementary estimate for the operator in Example D. We will not enter into the details of the symbolic calculus and proceed only at a formal level. Denote by $f^2(x) = \left(x_0x_2^{2\ell} - x_1\right)^2 + \xi_n^{-1}$; this is the formal time function used in the estimate. The operator P_2 satisfies all our assumptions and f verifies the conditions of Hypothesis (H 3). Denote by $\|v\|$ the L^2–norm of a C_0^∞ function v in the open set Ω. Therefore we compute

$$\begin{aligned}
&2i\mathrm{Im}\,\langle f^{-N}P_2u, -f^{-N}D_0u\rangle \\
&= 2i\mathrm{Im}\,\langle f^{-N}\left(-D_0^2 + f^2D_n + D_n + ikx_2^{2\ell}D_n\right)u, -f^{-N}D_0u\rangle \\
&= D_0\left[\|f^{-N}D_0u\|^2 + \|f^{-N+1}D_nu\|^2 + \|f^{-N}D_n^{1/2}u\|^2\right] \\
&-2iN\mathrm{Re}\left[\|f^{-N-1/2}x_2^\ell D_0u\|^2 + \frac{N-1}{N}\|f^{-N-1/2}x_2^\ell D_nu\|^2\right. \\
&\left. + \|f^{-N-1/2}x_2^\ell D_n^{1/2}u\|^2\right] \\
&-2i\mathrm{Im}\,\langle f^{-N+1/2}ikx_2^\ell D_nu, f^{-N-1/2}x_2^\ell D_0u\rangle \\
&= D_0E_1 - 2iNE_2 + i\mathcal{R}
\end{aligned}$$

It is easy to see that $|\mathcal{R}|$ can be estimated by NE_2, provided N is sufficiently large and that E_1 and E_2 define positive energies. We also note how the positivity of the real eigenvalue of F_{P_2}, $\lambda = x_2^{2\ell}\xi_n$, plays a crucial role in establishing the estimate.

Theorem 2.1 is a consequence of the following Proposition:

Proposition 3.4. The following (local) energy estimate holds:

$$\int_{-\infty}^{0} e^{-2\tau x_0}\|Pu\|_{N+s}^2 dx_0$$
$$\geq C\tau^2 \int_{-\infty}^{0} e^{-2\tau x_0}\left[\|u\|_{s+\frac{1}{2}}^2 + \|D_0 u\|_s^2\right] dx_0$$
$$-C'_\ell \tau \int_{-\infty}^{0} e^{-2\tau x_0}\left[\|u\|_{-\ell}^2 + \|D_0 u\|_{-\ell}^2\right] dx_0,$$

for every $\forall s \in \mathbb{R}$, $\forall \ell \in \mathbb{R}$.

References

[1] E. Bernardi, A. Bove and C. Parenti, Geometric Results for a Class of Hyperbolic Operators with Double Characteristics J. Func. Anal., **116** (1993), 62–82.

[2] L. Hörmander, The Cauchy Problem for differential equation with double characteristics. J. Analyse Math., **32** (1977), 118–196.

[3] L. Hörmander, The Analysis of Linear Partial Differential Operators. Vols. I–IV, Springer Verlag, Berlin, 1985.

[4] V. Ja. Ivrii, The well-posedness of the Cauchy Problem for non-strictly hyperbolic operators, III, The Energy Integral. Trans. Moscow Math. Soc., **34** (1978), 149–168.

[5] V. Ja. Ivrii, Wave Fronts of Solutions of Certain Hyperbolic Pseudodifferential Equations. Trans. Moscow Math. Soc., **39** (1979), 87–119.

[6] V. Ja. Ivrii, Linear Hyperbolic Equations. in Encyclopædia of Mathematical Sciences: Partial Differential Equations, vol. IV **33** (1993), 149–235 (translated from Itogi Nauki i Tekniki, 33, VINITI, Moscow, 1988).

[7] V. Ja. Ivrii and V. M. Petkov, Necessary Conditions for the Well–Posedness of the Cauchy Problem for Non–strictly Hyperbolic Equations, Russian Math. Surveys, **29** (1974), 1–70.

[8] K. Kajitani, S. Wakabayashi, Propagation of Singularities for Several Classes of Pseudodifferential Operators. Bull. Sc. Math. 2e série, **115** (1991), 397–449.

[9] K. Kajitani, T. Nishitani, S. Wakabayashi, The Cauchy Problem for Hyperbolic Operators of Strong Type. Preprint, 1992, Tsukuba University.

[10] B. Lascar, R. Lascar, N. Lerner, Propagation os Singularities for Non Real Pseudo-differential Operators. Preprint, 1993.

[11] R. B. Melrose, The Cauchy Problem and Propagation of Singularities. Seminar on Nonlinear Partial Differential Equations, S.S. Chern Ed., 185–201, Springer-Verlag, M.S.R.I. Publications vol. 2, 1984.

[12] R. B. Melrose, The Cauchy Problem for Effectively Hyperbolic Operators. Hokkaido Math. J., **12** (1983), 371–391.

[13] T. Nishitani, The Effectively Hyperbolic Cauchy Problem. Lecture Notes in Math. , **1505** (1991), 71–167.

[14] T. Nishitani, Local Energy Integrals for Effectively Hyperbolic Operators, I, II. J. Math. Kyoto Univ. **24** (1984), 623–658, 659–666.

Enrico Bernardi and Antonio Bove
Dipartimento di Matematica, Università di Bologna
Piazza di Porta S. Donato 5, 40127 Bologna, Italia.

Operator Theory:
Advances and Applications, Vol. 78

Multi-Quasi-Elliptic Operators in $\mathbb{R}^n$

P. Boggiatto, E. Buzano, L. Rodino

1. Introduction.

Multi-quasi-elliptic polynomials $p(\zeta)$, $\zeta \in \mathbb{R}^d$,are important examples in the class of the hypoelliptic polynomials of Hörmander [12]; they contain as particular cases the elliptic and quasi-elliptic polynomials, as well as any product of polynomials of this type. Their detailed study was carried out in the years 1960–70 by Cattabriga [7], Pini [17], Volevic-Gindikin [20], Friberg [8] and others.

In the next Section 2 we recall their definition, which in turn is based on the notion of complete polyhedron $\mathcal{P} \subset \mathbb{R}^d$. Let us give here the following basic example of multi-quasi-elliptic polynomial in $\mathbb{R}^2$:

$$p(\zeta_1,\zeta_2) = \zeta_1^{2r_1} + \zeta_1^{2s_1}\zeta_2^{2s_2} + \zeta_2^{2r_2}. \tag{1.1}$$

In (1.1) we require

$$0 < s_1 < r_1, \quad 0 < s_2 < r_2 \quad \text{and} \quad r_1 s_2 + r_2 s_1 > r_1 r_2. \tag{1.2}$$

Actually, if $r_1 \leq s_1$ or $r_2 \leq s_2$ the polynomial $p(\zeta_1,\zeta_2)$ would be non-hypoelliptic, whereas when $r_1 s_2 + r_2 s_1 \leq r_1 r_2$ we are in the usual quasi-elliptic case. The complete polyhedron $\mathcal{P}$ related to such $p(\zeta_1,\zeta_2)$ will be the convex hull in $\mathbb{R}^2$ of the vectors $(0,0)$, $(2r_1,0)$, $(0,2r_2)$, $(2s_1,2s_2)$.

We may associate as usual to a hypoelliptic polynomial $p(\zeta)$, $\zeta \in \mathbb{R}^d$, the linear partial differential operator with constant coefficients $p(D)$ in $\mathbb{R}^d$; for the example (1.1) we obtain $p(D_1,D_2) = D_1^{2r_1} + D_1^{2s_1}D_2^{2s_2} + D_2^{2r_2}$, with $D_j = -i\partial_j$. By the very definition of hypoelliptic polynomial, the operator $p(D)$ is also hypoelliptic, in the sense that the distribution-solutions u of $p(D)u = v$ are C^∞ where the datum v is C^∞. We propose here a different construction for the operator corrisponding to $p(\zeta)$, suggested by mathematical questions in quantum mechanics; see Shubin [19], Helffer [10]. Precisely, limiting our attention to multi-quasi-elliptic polynomials in even dimension $d = 2n$, $n \geq 1$, we split $\zeta \in \mathbb{R}^{2n}$ as $\zeta = (x,\xi)$, with $x \in \mathbb{R}^n$ and $\xi \in \mathbb{R}^n$, and give only to ξ the role of dual variables. In this way we associate to $p(\zeta) = p(x,\xi)$ an operator $p(x,D)$ in $\mathbb{R}^n$ with polynomial coefficients, which we regard as a map $p(x,D) : \mathcal{S}(\mathbb{R}^n) \to \mathcal{S}(\mathbb{R}^n)$, $\mathcal{S}'(\mathbb{R}^n) \to \mathcal{S}'(\mathbb{R}^n)$. For example, from the polynomial (1.1), with $\zeta_1 = x$, $\zeta_2 = \xi$, we now obtain the ordinary differential operator on the real axis:

$$P = x^{2r_1} + x^{2s_1}D^{2s_2} + D^{2r_2}.$$

Multi-quasi-elliptic operators $p(x,D)$ of this type have been studied by Boggiatto [5],[6] in the frame of a suitable pseudodifferential and Fourier integral calculus. As one may expect, such $p(x,D)$ turn out to be globally hypoelliptic in $\mathbb{R}^n$; in particular they have finite index as maps from $\mathcal{S}'(\mathbb{R}^n)$ into $\mathcal{S}'(\mathbb{R}^n)$. The results of [5] will be shortly reviewed in Section 2. The class of pseudodifferential operators defined here is included in the classes of Beals [1], Hörmander [12]; with respect to the general calculus, there are however some peculiar properties, due to the particular structure of the weight function.

In Section 3 we first report the results of [6] concerning Fourier integral operators and then, following the ideas of Shubin [19], Helffer [10], we introduce to the main theme of the present paper, namely the asymptotic behaviour of eigenvalues for self-adjoint multi-quasi-elliptic operators $p(x,D)$. Consider for example the Schrödinger operator in $\mathbb{R}^2$

$$-\Delta + x_1^{2r_1} + x_1^{2s_1}x_2^{2s_2} + x_2^{2r_2} \tag{1.3}$$

where Δ is the Laplacian with respect to x_1, x_2, and r_j, s_j satisfy condition (1.2). For $N(\lambda)$, the number of eigenvalues of $p(x,D)$ not exceeding λ, we propose the asymptotic formula

$$N(\lambda) = V(\lambda)\big(1 + O(\lambda^{-\rho})\big) \tag{1.4}$$

with

$$V(\lambda) = (2\pi)^{-n}\int_{p(\zeta)\le\lambda} d\zeta \tag{1.5}$$

and a value of $\rho > 0$ larger then that given by Shubin [19] for general hypoelliptic operators in $\mathbb{R}^n$ (see Theorem 3.3 below).

In Section 4 we present an explicit computation of the Weyl term $V(\lambda)$ for the model (1.3). More precisely we obtain for $\lambda \to +\infty$:

$$V(\lambda) = \frac{s_2}{\pi r_1(1+2s_2)}B\left(\frac{s_2-s_1}{2r_1s_2}, 2+\frac{1}{2s_2}\right)\lambda^{1+\frac{1}{2r_1}+\frac{1}{2s_2}-\frac{s_1}{2r_1s_2}}(1+o(1)) \tag{1.6}$$

if $s_1 < s_2$, and

$$V(\lambda) = \frac{1}{\pi(1+2s)}\left(\frac{s(r_1+r_2)}{r_1r_2} - 1\right)\lambda^{1+\frac{1}{2s}}\log\lambda(1+o(1)) \tag{1.7}$$

if $s = s_1 = s_2$. Here $B(r,s)$ is the Euler Beta function. Of course, if $s_1 > s_2$ one should exchange s_1 and s_2 in formula (1.6).

2. Complete polyhedrons and pseudodifferential operators.

Let $\{t^1,..,t^k\}$ be k vectors in $\mathbb{R}^d$ $(d > 1)$. The convex hull of the set $\{t^1,..,t^k\}$ is said *convex polyhedron* generated by $\{t^1,..,t^k\}$ and the k vectors are said *vertices* if they are convex-linearly independent. Let P_d be the family of the convex polyhedrons $\mathcal{P}$ such that:

$$- \quad \mathcal{P} \subset \mathbb{R}^d_+ = \{t \in \mathbb{R}^d : t_j \geq 0 \quad \text{for} \quad j = 1,...,d\}; \tag{2.1}$$

$$- \quad \mathcal{P} \quad \text{has dimension } d = \dim \mathbb{R}^d; \tag{2.2}$$

$-$ if t^i $(i = 0,...,N(\mathcal{P}))$are the vertices of $\mathcal{P}$ then:

$$Q(t^i) = \{r \in \mathbb{R}^d : 0 \leq r \leq t^i\} \subset \mathcal{P} \tag{2.3}$$

where $r \leq t^i$ means that $r_j \leq t^i_j$ $(j = 1,...,d)$; we set also $t^0 = 0$.

For every $\mathcal{P} \in P_d$ there exists a non empty finite set $A(\mathcal{P}) \subset \mathbb{R}^d_+ \backslash \{0\}$ such that:

$$\mathcal{P} = \bigcap_{a \in A(\mathcal{P})} \{t \in \mathbb{R}^d_+ :< a,t >\leq 1\}$$

(where $< a,t >= \sum_{j=1}^d a_j t_j$).

Let

$$F_a(\mathcal{P}) = \{t \in \mathcal{P} :< a,t >= 1\}, \qquad F(\mathcal{P}) = \bigcup_{a \in A(\mathcal{P})} F_a(\mathcal{P}).$$

A polyhedron $\mathcal{P} \in P_d$ is said to be *complete* if for every $r, t \in \mathbb{R}^d$ with $t \in \mathcal{P}$ and $0 \leq r < t$ we have $r \in \mathcal{P} - F(\mathcal{P})$. That means that the polyhedron has no faces parallel to the coordinate hyperplanes, i.e. $a_j > 0$ $(j = 1,...,d)$ for every $a \in A(\mathcal{P})$.

For $p(\zeta) = \sum a_\alpha \zeta^\alpha$; $\zeta \in \mathbb{R}^d$; $a_\alpha \in \mathbb{C}$; $\alpha \in \mathbb{N}^d_0$ ($\mathbb{N}_0 = \{0,1,2,...\}$), the convex hull of the set $\{0\} \cup \{\alpha \in \mathbb{N}^d_0 : a_\alpha \neq 0\}$ is said *characteristic polyhedron* of $p(\zeta)$. Since we are interested essentially in characteristic polyhedrons of polynomials, we shall consider from now on only polyhedrons with integer vertices also when we won't state it explicitly. A polynomial $p(\zeta)$ is said *multi-quasi-elliptic* if its characteristic polyhedron $\mathcal{P}$ belongs to P_d, is complete and if there exist two constants $C, R > 0$ such that

$$\sum_{\gamma \in V(\mathcal{P})} |\zeta^\gamma| \leq C|p(\zeta)|, \qquad |\zeta| \geq R, \tag{2.4}$$

where $V(\mathcal{P})$ is the set of the vertices of $\mathcal{P}$. In the preceding estimate we may replace $\sum_{\gamma \in V(\mathcal{P})} |\zeta^\gamma|$ with the *weight function*:

$$w_{\mathcal{P}}(\zeta) = \sqrt{\sum_{\gamma \in \mathcal{P}} \zeta^{2\gamma}}. \tag{2.5}$$

Multi-quasi-elliptic polyhedrons $p(\zeta)$ satisfy the algebraic condition of hypoellipticity in Hörmander [12]; in fact there exist positive constants ρ,C,R such that for all α:

$$|D^\alpha p(\zeta)| \leq C|p(\zeta)||\zeta|^{-\rho|\alpha|}, \qquad |\zeta| \geq R. \tag{2.6}$$

More precisely, we can estimate for $|\zeta| \geq R$

$$|D^\alpha p(\zeta)| \leq C w_{\mathcal{P}}^{1-|\alpha|/\mu}(\zeta) \tag{2.7}$$

or equivalently

$$|D^\alpha p(\zeta)| \leq C|p(\zeta)| w_{\mathcal{P}}^{-|\alpha|/\mu}(\zeta) \tag{2.8}$$

where μ is the *formal order* of $p(\zeta)$, given by

$$\mu = \max_{1\leq j\leq d} \max_{a\in A(\mathcal{P})} a_j^{-1}. \tag{2.9}$$

Example 2.1. If $p(\zeta)$ is elliptic of order m, then

$$\mathcal{P} = \big\{t \in \mathbb{R}^d_+ : \sum_{j=1}^d t_j \leq m\big\}.$$

Therefore $A(\mathcal{P}) = \big\{(1/m,...,1/m)\big\}$ and $\mu = m$.

Example 2.2. The polynomial $p(\zeta)$ is said quasi-elliptic when

$$\mathcal{P} = \big\{t \in \mathbb{R}^d_+ : \sum_{j=1}^d m_j^{-1} t_j \leq 1\big\}$$

for suitable positive integers $m_1,...,m_d$. We have then $A(\mathcal{P}) = \big\{(1/m_1,...,1/m_d)\big\}$ and $\mu = \max_{1\leq j\leq d} m_j$.

Example 2.3. The polynomial (1.1) in the Introduction gives an example of multi-quasi-elliptic polynomial which is not quasi-elliptic. The formal order is given by $\mu = \max\{\mu_1,\mu_2\}$ with $\mu_1 = \frac{2r_1 s_2}{r_1 - s_1}$, $\mu_2 = \frac{2r_2 s_1}{r_2 - s_2}$. Note that in this case μ is larger than the true order in (1.1), given by $2(s_1 + s_2)$. We can write

$$\mathcal{P} = \big\{(t_1,t_2) \in \mathbb{R}^2_+ : \frac{1}{2r_1} t_1 + \frac{1}{\mu_1} t_2 \leq 1, \frac{1}{\mu_2} t_1 + \frac{1}{2r_2} t_2 \leq 1\big\}$$

and therefore $A(\mathcal{P})$ consists of the two vectors $(\frac{1}{2r_1}, \frac{1}{\mu_1}), (\frac{1}{\mu_2}, \frac{1}{2r_2})$.

Definition 2.4. Let $\mathcal{P}$ be a complete polyhedron in $\mathbb{R}^d$, and let $w_{\mathcal{P}}(\zeta)$ and μ be defined as in (2.5), (2.9) respectively. For $m \in \mathbb{R}$, $\rho \in]0,1/\mu]$ we denote with $\Lambda^m_{\rho,\mathcal{P}}(\mathbb{R}^d)$ the set of the functions $a(\zeta) \in C^\infty(\mathbb{R}^d)$ that satisfy the following estimates

$$|\partial^\alpha_\zeta a(\zeta)| \leq C_\alpha \quad w_{\mathcal{P}}^{m-\rho|\alpha|}(\zeta), \qquad \zeta \in \mathbb{R}^d$$

for any multi-index α and for suitable constants $C_\alpha > 0$. Observe that $w_{\mathcal{P}} \in \Lambda^1_{1/\mu,\mathcal{P}}(\mathbb{R}^d)$.

In the case of even dimension $d = 2n$, $n \in \mathbb{N}$, $\zeta = (x,\xi)$, $x \in \mathbb{R}^n$, $\xi \in \mathbb{R}^n$, we associate to each symbol $a \in \Lambda^m_{\rho,\mathcal{P}}(\mathbb{R}^{2n})$ the pseudodifferential operator:

$$Au(x) = (2\pi)^{-n} \int_{\mathbb{R}^n} e^{ix\xi} a(x,\xi)\hat{u}(\xi)d\xi, \qquad u \in \mathcal{S}(\mathbb{R}^n). \tag{2.10}$$

We say that $a(x,\xi)$ is the symbol of the operator A (writing $a = \sigma(A)$, $A = Op(a)$) and we indicate by $\mathrm{L}^m_{\rho,\mathcal{P}}(\mathbb{R}^n)$ the class of operators with symbol in $\Lambda^m_{\rho,\mathcal{P}}(\mathbb{R}^{2n})$.

Definition 2.5. A symbol $a(\zeta) \in \Lambda^m_{\rho,\mathcal{P}}(\mathbb{R}^d)$ is said to be multi-quasi-elliptic if there exist positive constants C,R such that

$$w^m_{\mathcal{P}}(\zeta) \leq C|a(\zeta)|, \qquad |\zeta| \geq R.$$

We shall also then write $a \in H\Lambda^m_{\rho,\mathcal{P}}(\mathbb{R}^d)$. If $d = 2n$, $n \in \mathbb{N}$, then $H\mathrm{L}^m_{\rho,\mathcal{P}}(\mathbb{R}^n)$ denotes the correspondent class of pseudodifferential operators.

A multi-quasi-elliptic polynomial $p(\zeta)$ is in $H\Lambda^1_{1/\mu,\mathcal{P}}(\mathbb{R}^d)$ in view of (2.4), (2.7); we also have obviously $w_{\mathcal{P}} \in H\Lambda^1_{1/\mu,\mathcal{P}}(\mathbb{R}^d)$. If $a \in \Lambda^m_{\rho,\mathcal{P}}(\mathbb{R}^{2n})$, then the corresponding pseudodifferential operator A defines a continuous map $A : \mathcal{S}(\mathbb{R}^n) \to \mathcal{S}(\mathbb{R}^n)$, which extends to a continuous map $A : \mathcal{S}'(\mathbb{R}^n) \to \mathcal{S}'(\mathbb{R}^n)$. Furthermore, if $a \in \Lambda^0_{\rho,\mathcal{P}}(\mathbb{R}^{2n})$, then A extends to a continuous map $A : L^2(\mathbb{R}^n) \to L^2(\mathbb{R}^n)$; if $a \in \Lambda^{-m}_{\rho,\mathcal{P}}(\mathbb{R}^n)$, $m > 0$, then $A : L^2(\mathbb{R}^n) \to L^2(\mathbb{R}^n)$ is a compact operator. The standard rules of the symbolic calculus are valid in $\mathrm{L}^m_{\rho,\mathcal{P}}(\mathbb{R}^n)$; in particular we have asymptotic formulas for symbols of adjoints and products as usual.

Let us be more precise concerning hypoellipticity of the multi-quasi-elliptic operators, and define associated Sobolev spaces according to Beals [1], Bony-Chemin [4].

Definition 2.6. A continuous map $R : \mathcal{S}'(\mathbb{R}^n) \to \mathcal{S}(\mathbb{R}^n)$ is said *regularizing*. Operators $R \in \bigcap_{m\in\mathbb{R}} \mathrm{L}^m_{\rho,\mathcal{P}}(\mathbb{R}^n)$ are regularizing.

Proposition 2.7. (existence of the parametrix) Let $A \in H\mathrm{L}^m_{\rho,\mathcal{P}}(\mathbb{R}^n)$; then there exist an operator $B \in H\mathrm{L}^{-m}_{\rho,\mathcal{P}}(\mathbb{R}^n)$ and two regularizing operators R_1 and R_2 such that:

$$AB = I + R_1, \qquad BA = I + R_2.$$

B is said *parametrix* of A.

Proposition 2.8. (regularity of multi-quasi-elliptic operators) Let $A \in H\mathrm{L}^m_{\rho,\mathcal{P}}(\mathbb{R}^n)$. If $Au \in \mathcal{S}(\mathbb{R}^n)$ for some $u \in \mathcal{S}'(\mathbb{R}^n)$ then necessarily $u \in \mathcal{S}(\mathbb{R}^n)$.

Definition 2.9. Let $\mathcal{P} \in P_{2n}$ be a fixed complete polyhedron and $a_m \in H\Lambda^m_{\rho,\mathcal{P}}(\mathbb{R}^{2n})$ be a multi-quasi-elliptic symbol; then we set:

$$H^m_{\mathcal{P}}(\mathbb{R}^n) = A_m^{-1}(L^2(\mathbb{R}^n)), \qquad \text{where} \qquad A_m = Op(a_m).$$

We remark that $H^m_{\mathcal{P}}(\mathbb{R}^n)$ depends neither on ρ nor on the particular symbol $a_m \in H\Lambda^m_{\rho,\mathcal{P}}(\mathbb{R}^{2n})$ chosen, but only on $m \in \mathbb{R}$ and $\mathcal{P} \in P_{2n}$.

Proposition 2.10. $H^m_{\mathcal{P}}(\mathbb{R}^n)$ has a Hilbert space structure given by the inner product:

$$(u,v)_{H^m_{\mathcal{P}}(\mathbb{R}^n)} = (A_m u, A_m v)_{L^2} + (Ru, Rv)_{L^2}$$

where A_m is a multi-quasi-elliptic operator defining the space $H^m_{\mathcal{P}}(\mathbb{R}^n)$ according to Definition 2.9 and R is a regularizing operator, $R = I - \widetilde{A}_m A_m$, with $\widetilde{A}_m$ parametrix of A_m according to Definition 2.6 and Proposition 2.7 above. We have the following continuous immersions:

$$\mathrm{id} : \mathcal{S}(\mathbb{R}^n) \to H^m_{\mathcal{P}}(\mathbb{R}^n), \qquad \mathrm{id} : H^m_{\mathcal{P}}(\mathbb{R}^n) \to \mathcal{S}'(\mathbb{R}^n)$$

for any $m \in \mathbb{R}$, and the compact immersions:

$$\mathrm{id} : H^t_{\mathcal{P}}(\mathbb{R}^n) \to H^s_{\mathcal{P}}(\mathbb{R}^n)$$

if $t > s$.

Proposition 2.11. If $A \in \mathrm{L}^m_{\rho,\mathcal{P}}(\mathbb{R}^n)$, then it defines a continuous operator:

$$A : H^s_{\mathcal{P}}(\mathbb{R}^n) \to H^{s-m}_{\mathcal{P}}(\mathbb{R}^n).$$

Proposition 2.12. Let $A \in H\mathrm{L}^m_{\rho,\mathcal{P}}(\mathbb{R}^n)$; then it defines a Fredholm operator

$$A : H^s_{\mathcal{P}}(\mathbb{R}^n) \to H^{s-m}_{\mathcal{P}}(\mathbb{R}^n).$$

We recall from [5] that an equivalent definition of the spaces $H_{\mathcal{P}}(\mathbb{R}^n) = H^1_{\mathcal{P}}(\mathbb{R}^n)$ is:

$$H_{\mathcal{P}}(\mathbb{R}^n) = \{u \in \mathcal{S}'(\mathbb{R}^n) : x^\alpha D^\beta u \in L^2(\mathbb{R}^n) \text{ for } (\alpha,\beta) \in \mathcal{P}\}$$

and the inner product

$$< u,v >_{\mathcal{P}} = \sum_{(\alpha,\beta)\in\mathcal{P}} (x^\alpha D^\beta u, x^\alpha D^\beta v)_{L^2}$$

gives $H_{\mathcal{P}}(\mathbb{R}^n)$ an equivalent Hilbert space structure.

Taking advantage from the properties of the weight function $w_{\mathcal{P}}(\zeta)$, and using in particular the fact that $w^2_{\mathcal{P}}(\zeta)$ is a polynomial, one can go further in the study of the classes $\mathrm{L}^m_{\rho,\mathcal{P}}(\mathbb{R}^n)$, $H\mathrm{L}^m_{\rho,\mathcal{P}}(\mathbb{R}^n)$. Namely, in [5] a formula is given for the index of the multi-quasi-elliptic operators in dimension $n = 1$. Such computation, as well as its possible extension to multi-quasi-elliptic systems in arbitrary dimension n, are valuable, since no general index formula is available from the Weyl calculus of Hörmander [12].

Other interesting problems, whose study we leave to the future, concern an abstract characterization of $\mathrm{L}^m_{\rho,\mathcal{P}}(\mathbb{R}^n)$ in the spirit of Beals [2], and applications to the spectral invariance of Gramsch-Ueberberg-Wagner [9]. Though Bony-Chemin [4] have already discussed these questions for general Weyl operators, we may expect in our case a more direct formulation of the results in term of the polyhedron $\mathcal{P}$.

3. Fourier integral operators and spectral asymptotics.

Let us consider Fourier integral operators in $\mathbb{R}^n$ of the form

$$Fu(x) = (2\pi)^{-n} \int_{\mathbb{R}^n} e^{i\phi(x,\xi)} a(x,\xi)\hat{u}(\xi)d\xi; \tag{3.1}$$

the phase and the amplitude functions $\phi(\zeta)$, $a(\zeta)$, $\zeta = (x,\xi) \in \mathbb{R}^{2n}$, are as in the assumptions of the following two theorems, which summarize the results of [6].

We write $f \sim g$ if $f(\zeta)$ and $g(\zeta)$ are two positive functions satisfying for suitable positive constants C_1, C_2:

$$C_1 f(\zeta) \le g(\zeta) \le C_2 f(\zeta), \qquad \zeta \in \mathbb{R}^{2n}.$$

Theorem 3.1. Let F be a Fourier integral operator as in (3.1) with amplitude $a(\zeta) \in \Lambda^m_{\rho,\mathcal{P}}(\mathbb{R}^{2n})$ ($m \in \mathbb{R}$, $\rho \in]0,1/\mu]$, $\zeta = (x,\xi) \in \mathbb{R}^{2n}$) and phase $\phi(\zeta)$ satisfying:

- ϕ is a real valued function such that: $\partial^\gamma_\zeta \phi \in \Lambda^0_{\rho,\mathcal{P}}(\mathbb{R}^{2n})$ for $|\gamma| = 2$;
- $w_{\mathcal{P}}(\nabla_\xi \phi(x,\xi),\xi) \sim w_{\mathcal{P}}(x,\xi)$;
- $w_{\mathcal{P}}(x,\nabla_x \phi(x,\xi),) \sim w_{\mathcal{P}}(x,\xi)$.

Then F defines a continuous map from $\mathcal{S}(\mathbb{R}^n)$ to $\mathcal{S}(\mathbb{R}^n)$, which extends to a map $F : \mathcal{S}'(\mathbb{R}^n) \to \mathcal{S}'(\mathbb{R}^n)$. Moreover, let P be a pseudodifferential operator with symbol $p(\zeta) \in \Lambda^t_{\rho,\mathcal{P}}(\mathbb{R}^{2n})$. Then $P \circ A$ is a Fourier integral operator of phase $\phi(\zeta)$ and ampitude $h(\zeta) \in \Lambda^{m+t}_{\rho,\mathcal{P}}(\mathbb{R}^{2n})$ defined, modulo regularizing terms, by the following asymptotic expansion:

$$h(x,\xi) \sim \sum_{\alpha \in \mathbb{N}^n_0} c_\alpha(x,\xi)$$

where

$$c_\alpha(x,\xi) = \left(\frac{1}{\alpha!}\right)\partial_\xi^\alpha p(x,\nabla_x\phi(x,\xi),)\,D_z^\alpha\left[e^{i\psi(x,z,\xi)}a(z,\xi)\right]_{z=x}$$

with $\psi(x,z,\xi) = \phi(z,\xi) - \phi(x,\xi) - < \nabla_x\phi(x,\xi), z - x >$.

Theorem 3.2. Let $a(x,\xi)$ and $\phi(x,\xi)$ satisfy the same hypotheses as in Theorem 3.1 and suppose further that:

$$\inf_{(x,\xi)\in\mathbb{R}^{2n}} |\det[\partial_{x_j}\partial_{\xi_k}\phi(x,\xi)]| \geq \delta_0$$

for some positive constant δ_0.

Then the Fourier integral operator F defines a continuous linear map:

$$F : H_\mathcal{P}^s(\mathbb{R}^n) \to H_\mathcal{P}^{s-m}(\mathbb{R}^n)$$

for any $s \in \mathbb{R}$.

Results analogous to those in Theorem 3.1 have been proved by other authors; see for instance Helffer [10], Helffer-Robert [11] for the case when $w_\mathcal{P}(\zeta) \sim |\zeta|$, Parenti-Segala [16], Rodino [18], Liess-Rodino [14] for non homogeneous settings in the local frame, Mohamed [15] for more general amplitudes under restrictive assumptions on the phase function. We do not know whether Theorems 3.1, 3.2 can be extended to the general case, when $w_\mathcal{P}(\zeta)$ is replaced by an arbitrary weight function of Beals [1], Hörmander [12], globally defined in $\mathbb{R}^n$. In fact, the peculiar property that $w_\mathcal{P}^2(\zeta)$ is a polynomial was used in the proof in [6].

Here we present an application of Theorem 3.1 to the study of the asymptotic behaviour of the eigenvalues of the multi-quasi-elliptic operators. Precisely, consider $p(\zeta) \in H\Lambda^m_{1/\mu,\mathcal{P}}(\mathbb{R}^{2n})$, where we assume $m \geq 2/\mu$, with μ formal order of $\mathcal{P}$ defined as in (2.9). We suppose the corresponding operator $P = Op(p)$ is formally self-adjoint in $\mathbb{R}^n$, i.e.

$$\int (Pu)\bar{v}dx = \int u\overline{Pv}dx \qquad \text{for all} \quad u,v \in \mathcal{S}(\mathbb{R}^n).$$

We assume also for the symbol $p(\zeta)$:

$$p(\zeta) = p_0(\zeta) + p_1(\zeta), \tag{3.2}$$

where $p_1(\zeta) \in \Lambda^{m-2/\mu}_{1/\mu,\mathcal{P}}(\mathbb{R}^{2n})$ and for some $C > 0$

$$w_\mathcal{P}^m(\zeta) \leq Cp(\zeta), \qquad \zeta \in \mathbb{R}^{2n}.$$

The unbounded operator P, from $L^2(\mathbb{R}^n)$ to $L^2(\mathbb{R}^n)$ with domain $H^m_\mathcal{P}(\mathbb{R}^n)$, is then self-adjoint, semi-bounded from below. Its spectrum in $L^2(\mathbb{R}^n)$ is discrete, with eigenvalues $\lambda_j \to +\infty$. Taking into account possible multiplicity of the eigenvalues, we write $N(\lambda)$ for the number of them not exceeding $\lambda > 0$, i.e.

$$N(\lambda) = \sum_{\lambda_j < \lambda} 1.$$

Theorem 3.3. Under the preceding assumptions, we have

$$N(\lambda) = (2\pi)^{-n}\left[1 + O(\lambda^{-\frac{2}{\mu m}})\right] \int_{p_0(\zeta)\leq\lambda} d\zeta, \qquad \lambda \to +\infty. \tag{3.3}$$

The formula (3.3) with $\frac{2}{\mu m}$ replaced by $\frac{1}{\mu m}$ follows from the results of Shubin [19] (see in particular Theorem 30.1), in view of Definition 2.5. To obtain the stronger remainder in our Theorem 3.3 we may argue as in Helffer [10], Helffer-Robert [11]. Precisely, we consider the power $Q = P^{\frac{2}{\mu m}}$, which is in $HL^{2/\mu}_{1/\mu,\mathcal{P}}(\mathbb{R}^n)$; i.e. Q has formal order 2, the formal order of P being μm . We then use the standard remark that the fundamental solution $u(x,y,t)$ of the operator $D_t - Q$ is the Schwartz' kernel of the operator e^{itQ} and it is connected with $N(\lambda)$ by the formula

$$\int u(x,x,t)\,dx = \int e^{it\lambda}\, d_\lambda N(\lambda). \tag{3.4}$$

For $Q \in HL^{2/\mu}_{1/\mu,\mathcal{P}}(\mathbb{R}^n)$, this fundamental solution can be expressed in the form (3.1), for small values of the paramater t. The conclusion is then obtained by applying Theorem 3.1 to compute explicitly phase and amplitude functions, and using (3.4). We omit here the details.

On the other hand, we mention that Theorem 3.3 can be regarded as consequence of a result of Mohamed [15] for general Weyl operators, if our symbol $p_0(\zeta)$ is verified to satisfy the technical assumption (1.4) of [15]. Moreover, according to Ivrii [13], optimal remainder for Schrödinger operators of the form (1.3) can be deduced from his forthcoming book on spectral asymptotics. In the next Section 4 we shall address to explicit computations of the Weyl term for model operators; we believe such results are new with respect to the existing literature.

4. Computation of the Weyl term.

We end with an explicit computation of the Weyl term $V(\lambda)$ for the Schrödinger operator (1.3) with symbol

$$p(x,\xi) = \xi_1^2 + \xi_2^2 + x_1^{2r_1} + x_1^{2s_1}x_2^{2s_2} + x_2^{2r_2}$$

where, according to (1.2):

$$0 < s_1 < r_1, 0 < s_2 < r_2 \quad \text{and} \quad r_1s_2 + r_2s_1 > r_1r_2.$$

Our result has been stated in (1.6) and (1.7) of Section 1. Here we show how to compute (1.6); we refer to [3] for the derivation of (1.7) together with a generalization to multi-quasi-elliptic potentials in $\mathbb{R}^2$ and better remainder estimates. From (1.5) we have:

$$V(\lambda) = \frac{1}{4\pi^2}\int_{p(x,\xi)\leq\lambda} d\xi dx.$$

Let

$$\begin{cases} \xi_1 = \lambda^{1/2}\eta_1 \\ \xi_2 = \lambda^{1/2}\eta_2 \\ x_1 = \lambda^{\frac{1}{2r_1}} \operatorname{sgn}(y_1)|y_1|^{\frac{1}{r_1}} \\ x_2 = \lambda^{\frac{1}{2s_2}-\frac{s_1}{2r_1s_2}} \operatorname{sgn}(y_1y_2)|y_1|^{-\frac{s_1}{r_1s_2}}|y_2|^{\frac{1}{s_2}}, \end{cases}$$

we obtain

$$V(\lambda) = \frac{1}{4\pi^2 r_1 s_2}\lambda^{1+\frac{1}{2r_1}+\frac{1}{2s_2}-\frac{s_1}{2r_1s_2}}\int_{\mathbb{R}^4}\chi_\lambda(\eta,y)|y_1|^{\frac{s_2-s_1}{r_1s_2}-1}|y_2|^{\frac{1}{s_2}-1}d\eta dy,$$

where $\chi_\lambda(\eta,y)$ is the characteristic function of the set

$$\{(\eta,y)\in\mathbb{R}^4 : |\eta|^2+|y|^2+\lambda^{\frac{r_2}{s_2}-\frac{r_2s_1}{r_1s_2}-1}|y_1|^{-2\frac{r_2s_1}{r_1s_2}}|y_2|^{2\frac{r_2}{s_2}}\leq 1\}$$

Thanks to the dominate convergence theorem, we have:

$$\begin{aligned}\lim_{\lambda\to+\infty}&\int_{\mathbb{R}^4}\chi_\lambda(\eta,y)|y_1|^{\frac{s_2-s_1}{r_1s_2}-1}|y_2|^{\frac{1}{s_2}-1}d\eta dy\\ &=\int_{|\eta|^2+|y|^2\leq 1}|y_1|^{\frac{s_2-s_1}{r_1s_2}-1}|y_2|^{\frac{1}{s_2}-1}d\eta dy\end{aligned}$$

On the other hand an elementary computation yields

$$\begin{aligned}\int_{|\eta|^2+|y|^2\leq 1}&|y_1|^{\frac{s_2-s_1}{r_1s_2}-1}|y_2|^{\frac{1}{s_2}-1}d\eta dy\\ &=\pi\int_{|y|^2\leq 1}|y_1|^{\frac{s_2-s_1}{r_1s_2}-1}|y_2|^{\frac{1}{s_2}-1}(1-|y|^2)dy\\ &=4\pi\int_0^1\int_0^{(1-y_1^2)^{1/2}}y_1^{\frac{s_2-s_1}{r_1s_2}-1}y_2^{\frac{1}{s_2}-1}(1-y_1^2-y_2^2)dy_1dy_2\\ &=\frac{8\pi s_2^2}{1+2s_2}\int_0^1 y_1^{\frac{s_2-s_1}{r_1s_2}-1}(1-y_1^2)^{\frac{1}{2s_2}+1}dy_1\\ &=\frac{4\pi s_2^2}{1+2s_2}B\left(\frac{s_2-s_1}{2r_1s_2},2+\frac{1}{2s_2}\right).\end{aligned}$$

This completes the proof of (1.6).

References

[1] Beals R., A general calculus of pseudodifferential operators, Duke Math. J., **42** (1975), 1–42.

[2] Beals R., Characterization of pseudodiffernetial operators and applications, Duke Math., **44** (1977), 45–57.

[3] Boggiatto P., Buzano E., in preparation.

[4] Bony J.M., Chemin J.Y., Espaces fonctionnels associés au calcul de Weyl-Hörmander, Bull. Soc. Math. France, **122** (1994), 77–118.

[5] Boggiatto P., Spazi di Sobolev associati ad un poliedro ed operatori pseudodifferenziali multi-quasi-ellittici in $\mathbb{R}^n$, Boll. U.M.I. **7–B** (1993), 511–548.

[6] Boggiatto P., Sobolev spaces associated to a polyhedron and Fourier integral operators in $\mathbb{R}^n$, Submitted to Ann. Mat. Pura Appl.

[7] Cattabriga L., Moltiplicatori di Fourier e teoremi di immersione per certi spazi funzionali, Ann. Scuola Norm. Sup. Pisa Cl. Sc., **25** (1970), 147–194.

[8] Friberg J., Multi-quasi-elliptic polynomials, Ann. Scuola Norm. Sup. Pisa Cl. Sc., **21** (1967), 239–260.

[9] Gramsch B., Ueberberg J., Wagner K., Spectral invariance and submultiplicativity for Fréchet algebras with applications to pseudo-differential operators and Φ^*-quantization, Operator Theory: Advances and Appl., **57** (1992), 71–98.

[10] Helffer B., Théorie Spectrale pour des Opérateurs Globalement Elliptiques, Soc. Math. de France, Astérisque, 1984.

[11] Helffer B., Robert D., Comportement asymptotique précisé du spectre d'opérateures globalement elliptiques dans $\mathbb{R}^n$ Séminaire Goulaouic-Meyer-Schwartz, (1980), exposé no. 11.

[12] Hörmander L., The analysis of linear partial differential operators, Vols. II and III, Springer-Verlag, Berlin, 1983.

[13] Ivrii, Preprints Ecole Polytechnique, 1990–92.

[14] Liess O., Rodino L., Fourier integral operators and inhomogeneous Gevrey classes, Ann. Mat. Pura e Appl., **110** (1988), 167–262.

[15] Mohamed A., Théorie spectrale pour des opérateurs pseudodifférentiels, Thèse à Nantes (1983).

[16] Parenti C., Segala F., Propagation and reflection of singularities for a class of evolution equations, Comm. Part. Diff. Eq., **6** (1981), 741–782.

[17] Pini B., Osservazioni sulla ipoellitticità, Boll. U.M.I., 3 (1963), 420–432.

[18] Rodino L., Microlocal analysis for spatially inhomogeneous pseudodifferential operators, Ann. Scuola Norm. Sup. Pisa, Cl. Sc., serie IV **9** (1982), 221–253.

[19] Shubin M.A., Pseudodifferential Operators and Spectral Theory, Springer-Verlag, Berlin, 1987.

[20] Volevic L. R., Gindikin S. G., On a class of hypoelliptic polynomials, Mat. Sbornik **75** (1968), 369–383.

P. Boggiatto, E. Buzano, L. Rodino
Dipartimento di Matematica
Universita' di Torino, Via Carlo Alberto 10, 10123 Torino, Italy

Operator Theory:
Advances and Applications, Vol. 78

Real Analogue of the Bergman Kernel

L. Boutet de Monvel

1 The Bergman Kernel of a Complex Domain

In this short report I wish to recall some of my work on the singularity of the Bergman kernel, and some recent developments by my thesis student A. Attioui concerning a real analogue.

Let Ω be an open set in $\mathbb{C}^n$. The Bergman kernel $B_\Omega = B$ is the Schwartz kernel of the orthogonal projector $\mathfrak{B}$ of $L^2(\Omega)$ onto its subspace of square integrable holomorphic functions. More generally B_Ω is defined for any complex analytic space as the Schwartz kernel of the orthogonal projector from $L^2(\Omega)^{n,0}$, the space of square integrable $(n,0)$ differential forms (which has an intrinsic Hilbert norm: $||\omega||^2 = 2^{-n}\int |\overline{\omega}\wedge\omega|$) on its subspace of holomorphic forms; intrinsically B is a differential form of type (n,n) on $\Omega \times \overline{\Omega}$.

We now suppose that Ω has an analytic strictly pseudo-convex boundary, i.e. it can be defined by an inequality $u < 0$, where $u(z)$ is a real analytic function of z in a neighborhood of the closure of Ω, $du \neq 0$ on the boundary $\partial\Omega$, and the Levi matrix $(\partial^2 u/\partial z_j \partial \overline{z}_k)$ is hermitian $\gg 0$. Then it was proved by Sjöstrand and myself [6], and Kashiwara [9] in the analytic case, that the Bergman kernel $B(x,\overline{y})$ has a holonomic form:

i) it is analytic in $\Omega\times\overline{\Omega}$ (holomorphic with respect to x and antiholomorphic with respect to y), and extends analytically across all points of the boundary, except those of the diagonal of $\partial\Omega$.

ii) near the diagonal of $\partial\Omega$ (for x,y close to each other and close to $\partial\Omega$) B is of the form

$$B(x,\overline{y}) = a(x,\overline{y})/u(x,\overline{y})^{n+1} + b(x,\overline{y})\, Log\, u(x,\overline{y}) \tag{1.1}$$

where a and b are analytic (holomorphic with respect to x and antiholomorphic with respect to y), and $u(x,\overline{y})$ is the holomorphic extension of the real analytic function u (this is well defined if x and y are close to each other; $u = u(z,\overline{z})$).

The method of proof in [6] shows that the singularity of B is "local", more precisely that the Taylor coefficients of a and b which contribute to the singularity are polynomials of those of u (i.e. of the jet of $\partial\Omega$) (and of the inverse of the Monge-Ampère determinant

$$\det\begin{bmatrix} u & \partial u/\partial z \\ \partial u/\partial\overline{z} & \partial^2 u/\partial z\partial\overline{z} \end{bmatrix} \tag{1.2}$$

whose non-vanishing follows from the strict pseudoconvexity).

The differential invariants of real pseudo-convex hypersurfaces in a complex space were analyzed by E. Cartan [7] (in the case $n = 2$), Chern and Moser [8] (cf. also the article of M. Kuranishi [10] for a complete account). Since the Bergman kernel is biholomorphically invariant, it is natural to try to express the Taylor coefficients of its singularity in terms of these differential invariants. It is possible to give such a description, using the fact that the Bergman kernel is given by the calculus of Fourier integral operators and the asymptotic formula of the stationary phase. This leads to a complete algorithm, but rather heavy to handle.

Another method to describe the coefficients of B in terms of invariants was described in [4] and we recall it here. The basic example is that of the sphere, or of the paraboloid (which is equivalent to the standard unit sphere via a homographic transformation) of equation $u_0 = 0$ with

$$u_0 = t + \bar{t} + z\bar{z}. \tag{1.3}$$

The corresponding Bergman kernel is

$$B_0 = c(u_0)^{-n-1} \text{ (intrinsically } B_0 = c(\partial\bar{\partial}\, Log\, u_0)^n, \text{ with } c \text{ a constant).} \tag{1.4}$$

It is convenient to fit expressions such as B or B_0 in a space of microfunctions of the form:

$$\sum b_k(t,z,\bar{z})U_k \text{ with } U_k = u_0^k \text{ if } k < 0, \text{ resp. } u_0^k\, Log\, u_0 \text{ if } k \geq 0 \tag{1.5}$$

on which act microdifferential operators

$$P(t,z,\partial_t,\partial_z) = \sum p_j(t,z,\partial_z)\partial_t^j \text{ defined near } t = z = \zeta = 0, \tau > 0. \tag{1.6}$$

For distribution theory in the usual sense, there should be only finitely many terms containing U_0^k or ∂_z^j with $k < 0$ or $j \geq 0$. Here this is not enough and we need expansions of infinite order. To give these a meaning, we endow t resp. z with the weight 2, resp. 1 (and $\partial_t, \partial_z, \tau, \zeta$, etc, with the corresponding weights $w(\partial_t) = w(\tau) = -2$, $w(\partial_z) = w(\zeta) = -1$, etc). Our expansions will converge formally in the sense of these weights, and this will allow to compute formally the coefficients of the singularity of B.

Let now Ω be strictly pseudoconvex as above and choose a point of the boundary $\partial\Omega$. It is always possible to choose local coordinates $t = z_1$, $z = (z_2, \ldots, z_n)$ so that near this point a function u defining the boundary has the form

$$u = t + \bar{t} + z \cdot \bar{z} + \varrho(t,z,\bar{z}) = u_0 + \varrho \tag{1.7}$$

where ϱ is independent of $\bar{t}$, and vanishes of order 4 at the origin, resp. 6 if $n = 2$ (e.g. we may choose "normal coordinates" i.e. such that $\varrho = O(z^2\bar{z}^2)$, and $\Delta\varrho$ (resp. $\Delta^2\varrho, \Delta^3\varrho$) vanishes of order 3 (resp. 2, 1) for $z = \bar{z} = 0$, with $\Delta = \sum \partial^2/\partial_{z_j}\partial_{\bar{z}_j}$; this is not quite the same normalization as that of Chern and Moser, and here ϱ

will usually not be real because the condition on t and $\bar{t}$ is not symmetric; but it is formally equivalent and is well adapted for the computation of coefficients of B).

The recipe for the coefficients of B is the following. Let us set

$$a(t,z,\tau,\zeta) = \varrho(t,z,\zeta/\tau)\tau \tag{1.8}$$

and let

$$\mathbf{A} \approx A(t,z,\partial_t,\partial_z) \tag{1.9}$$

be the microdifferential operator with total symbol

$$A(t,z,\tau,\zeta) = e^a = 1 + a + a^2/2 + \dots \tag{1.10}$$

(all these are well defined in the sense of weights since $w(a) \geq 2$, resp. 4 if $n = 2$). Then the Bergman kernel can be expanded in terms of weights:

$$B \approx \sum b_{mnpq} t^m z^p \bar{z}^q U_n \quad (n \in \mathbb{Z}, m,p,q \geq 0) \tag{1.11}$$

with U_n as in (1.5) above, and we have

$$B = \text{cste}\, {}^t\mathbf{A}^{-1}(B_0) \approx \text{cste}\, {}^t\mathbf{A}^{-1}(u_0^{-3}). \tag{1.12}$$

The reason for this is the following: the Taylor formula $Log\, u = \sum \frac{1}{k!}\varrho^k \partial_t^k u_0$ obviously implies

$$Log\, u \approx \mathbf{A}\, Log\, u_0. \tag{1.13}$$

Now the analysis in [9], [3] shows that the Bergman projector behaves formally (up to a constant factor) as the inverse of the operator with kernel $Log\, u$ (or $Y(u)$, Y the Heaviside function). The relation $Op\, Log\, u = Op\, Log\, u_0 \circ \mathbf{A}$ then gives $(Op\, Log\, u)^{-1} = \mathbf{A}^{-1}(Op\, Log\, u_0)^{-1}$, hence (11).

In [3] [4] I used this to get an infinitesimal relation between the Bergman kernel end the solution of the Monge-Ampère equation, and to prove that in dimension 2 the absence of a logarithmic term in B can only happen if the boundary is equivalent to a sphere near the given point (this answers positively to a conjecture of I. Ramadanoff [11], and generalizes and localizes the result of Boichu and Cœuré [2] for Reinhardt domains; it is in fact sufficient that the logarithmic coefficient (b in the formula above) vanishes of order 2 on the boundary near a point x$\in\delta\Omega$). However, in spite of the apparent simplicity of formula (12), the coefficients of B remain long to compute by this method and hard to understand in dimension $\geq$3.

2 The real analogue

Let X and Y be two real manifolds of the same dimension n, and $\Sigma \subset X \times Y$ a smooth hypersurface with equation $u = 0$. We define a microfunction k:

$$k = Log(u \pm i0) \text{ (resp. } Y(u), Y \text{ the Heaviside function)} \tag{2.1}$$

(as microfunctions these are the same up to a microlocally constant factor; $Log(u \pm i0)$ is supported by one half of the conormal bundle of Σ). We let K be the operator with symbol k. Since k is a function K acts from forms on Y to functions on X. It is obviously a Fourier integral operator whose Lagrangian is the conormal bundle $N = T^*_\Sigma(X \times Y)$ of Σ. Its symbol is elliptic, so K is (microlocally) invertible if its canonical relation is. This means that the projection $N \to T^*X$ (or T^*Y) is locally a diffeomorphism, i.e. has a nonzero jacobian. It is elementary to verify that this is equivalent to the nonvanishing of the Monge-Ampère determinant:

$$\det \begin{bmatrix} u & \partial u/\partial x \\ \partial u/\partial y & \partial^2 u/\partial x \partial y \end{bmatrix} \tag{2.2}$$

The real analogue of the Bergman is the microlocal inverse B of K. It should be obvious from the description above that its singularity is given formally by the same formulas as for the Bergman kernel.

The basic model here (analogue of the sphere or paraboloid in the complex case) is the hypersurface with equation

$$u_0 = x_1 + y_1 + \sum_2^n x_i y_i. \tag{2.3}$$

Modulo a change of coordinates this is the same as the incidence relation ($\Sigma = \{x \in y\}$) in the product $P \times P^*$ of two dual projective spaces, and in that case K is related to the Gel'fand or Radon transform.

In the real analogue the admissible changes of coordinates are product diffeomorphisms $(x,y) \to (\varphi(x),\psi(y))$ (the real analogue of "biholomorphic" maps $(z,\overline{z}) \to (F(z),\overline{F}(\overline{z}))$). The group of the model (incidence relation on $P \times P^*$) is the group $PGL(n+1)$, acting by $(x,y) \to (Ux,{}^tU^{-1}y)$ (this is another real form of the group $PU(n,1)$ of automorphisms of the complex ball). We may still pick "normal" coordinates, i.e. such that

$$u = u_0 + \varrho,\ \varrho = \sum \varrho_{pq}(t) x^p y^q,\ (|p| \geq 2, |q| \geq 2),\ \varrho \text{ independent of } \overline{t}, \tag{2.4}$$

and $\Delta\varrho$ (resp. $\Delta^2\varrho, \Delta^3\varrho$) vanishes of order 3 (resp. 2, 1) for $x = y = 0$, with $\Delta = \sum \partial^2/\partial_{x_j}\partial_{y_j}$ (normal coordinates are unique up to a transformation of $PGL(n+1)$.

However there are some slight differences when we go further in the analysis. In [4] I showed that, when $n = 2$, the product $\varrho_{24}\varrho_{42}$ is a linear combination of the first coefficients of the logarithmic term of B. In the complex case the coefficients ϱ_{24} and ϱ_{42} are always complex conjugate so they both vanish if their product vanishes. In the real case this is of course no longer true, and the real analogue of Ramadanoff's conjecture, i.e. that Σ is locally equivalent to the incidence relation if B has no logarithmic term, is harder to prove. It was proved by Attioui using more terms of the expansion of B in the case $n = 2$.

In dimension 3 A. Attioui gives the following beautifully simple counterexample: let us consider the function

$$u = x_1 + y_1 + x_2 y_2 + x_3 y_3 + x_2^2 y_3^2, \qquad \varrho = x_2^2 y_3^2 \tag{2.5}$$

If we compute the symbols and operators $a, A, \mathbf{A}$ as above we get

$$a(x,\xi) = x_2^2\xi_3^2/\xi_1$$

and since the operators $x_2, \partial_3, \partial_1$ commute

$$\mathbf{A} = A(x,\partial_x) = \exp(x_2^2\partial_3^2\partial_1^{-1})$$

since $x_2\partial_3\partial_1$ commute we also have ${}^t(x_2^2\partial_3^2\partial_1^{-1}) = -x_2^2\partial_3^2\partial_1^{-1}$, so

$${}^t\mathbf{A}^{-1} = (\exp{}^t(x_2^2\partial_3^2\partial_1^{-1}))^{-1} = (\exp(-x_2^2\partial_3^2\partial_1^{-1}))^{-1} = \exp(x_2^2\partial_3^2\partial_1^{-1}) = \mathbf{A}$$

and finally

$$\text{cste}\,\mathbf{A}(u_0 - 3) = \text{cste}\,u^{-3}$$

So in this case B has no logarithmic term, although Σ is not equivalent to the incidence relation (real anlogue of complex sphere) (this is obvious since u is already in normal form, and the model incidence relation has only one normal form).

References

[1] A. Attioui, Version réelle de la conjecture de Ramadanoff et équation de Monge-Ampère, *C.R.Acad.Sci.Paris* **317**, série I, 1983, 283–287, et thèse Paris 6, 1994.

[2] D. Boichu, G. Cœuré, Sur le noyau de Bergman des domaines de Reinhardt, *Invent. Math.* **72** (1983), 131–152

[3] L. Boutet de Monvel, Complément sur le noyau de Bergman, *Séminaire E.D.P. Ecole Polytechnique*, 1985–86, exposé no 20.

[4] L. Boutet de Monvel, Le noyau de Bergman en dimension 2 (suite), *Séminaire E.D.P. Ecole Polytechnique*, 1987–88, exposé no 22.

[5] L. Boutet de Monvel, The Bergman kernel: computation and invariants, Osaka, décembre 1990, *Lecture Notes in Pure and Appl. Math.* **143**, Marcel Dekker, 1992, 13–29.

[6] L. Boutet de Monvel, J. Sjöstrand, Sur la singularité des noyaux de Bergman et de Szegö, *Astérisque* **34–35** (1976), 123–164.

[7] E. Cartan, Sur la géométrie pseudo-conforme des hypersurfaces de deux variables complexes I, *Ann. Math. Pures Appl.* (4), **11** (1932), 17–90, et II, *Ann. Sc. Norm. Sup. Pisa* **2**, 1, (1932), 333–354.

[8] S.S. Chern, J. Moser, Real hypersurfaces in complex manifolds, *Acta Math.* **133** (1974), 219–271.

[9] M. Kashiwara, Analyse microlocale du noyau de Bergman, *Séminaire Goulaouic-Schwartz* 1976–77, exposé no 8, Ecole Polytechnique.

[10] M. Kuranishi, Cartan connections and CR structures with non-degenerate Levi-form, *Astérisque*, hors série, 1985 (Elie Cartan et et les Mathématiques d'aujourd'hui), 273–288.

[11] I. Ramadanoff, A characterization of the balls in $\mathbb{C}^n$ by means of the Bergman kernel, *C.R. Acad. Bulgare des Sciences* **34**, no 7 (1981).

Louis Boutet de Monvel
Institut de Mathématiques (UMR 9994 du CNRS), Université de Paris 6, Tour 46-0, 5-ième étage, boîte 172, 4 place Jussieu, F-75252, Paris Cedex 05

Operator Theory:
Advances and Applications, Vol. 78

Attractors of Non-Autonomous Evolution Equations with Translation-Compact Symbols[1]

V. V. Chepyzhov, M. I. Vishik

0. Introduction.

We consider Cauchy problems of the type:

$$\partial_t u = A(u,t), u|_{t=\tau} = u_\tau, t \geq \tau, \tau \in \mathbb{R}. \tag{1}$$

Here for any $t \in \mathbb{R}$, operators $A(.,t) : E_1 \mapsto E_0$ forms a family of nonlinear mappings, $E_1 \subseteq E_0$, E_1 and E_0 are Banach spaces. Let E be a Banach space such that $E_1 \subseteq E \subseteq E_0$. Let us assume that for any $\tau \in \mathbb{R}$ and for arbitrary $u_\tau \in E$ there exists a unique solution $u(t), t \geq \tau$, of the problem (1) such that $u(t) \in E$ for any $t \geq \tau$. Any solution $u(t)$ can be represented in the form:

$$u(t) = U_\sigma(t,\tau)u_\tau = u(t), t \geq \tau, (\sigma = \sigma(t)).$$

The family of operators $\{U_\sigma(t,\tau)\} = \{U_\sigma(t,\tau), t \geq \tau, \tau \in \mathbb{R}\}$: $U_\sigma(t,\tau) : E \mapsto E, t \geq \tau, \tau \in \mathbb{R}$, is said to be the process corresponding to the problem (1). The process $\{U_\sigma(t,\tau)\}$ acts in the space E.

We use the notion of a uniform attractor to describe the limit behavior of a process $\{U_\sigma(t,\tau)\}$ when $t-\tau$ tends to infinity. An attractor $\mathcal{A}$ of a process $\{U(t,\tau)\}$ is a minimal closed attracting set of the process. The attracting property of $\mathcal{A}$ means that for any bounded set $B \subseteq E$ $\text{dist}_E(U(t,\tau)B, \mathcal{A}) \to 0$ $(t-\tau \to +\infty)$. The property of minimality is the natural generalization of the invariance property in the definition of a semigroup's attractor.

The problem (1) depends on a functional parameter $\sigma(t), t \in \mathbb{R}$. Values of the function $\sigma(t)$ belong to some Banach space $\mathcal{E}$ or metric space $\mathcal{M}$. The function $\sigma(t)$ is called the time symbol of the problem (1). One says also that $\sigma = \sigma(t)$ is the symbol of the corresponding process $\{U_\sigma(t,\tau)\}$. Functional classes of $\sigma(t)$ will be described in the next section. In applications to evolution partial differential equations, a time symbol $\sigma(t)$ consists of all time-dependent terms of an equation under consideration: coefficients, right-hand sides, nonlinear interaction functions etc.

In works [2,3,4] we have studied non-autonomous nonlinear partial differential equations with almost periodic symbols. As compared with these works, in the present paper we shall extend significantly classes of symbols under consideration. We shall investigate equations with translation-compact symbols. (Definitions see in sec.1.)

[1]Research supported in part by grant N MR5000 from International Scientific Foundation and by Russian Foundation of Fundamental Researches.

To construct an attractor of the problem (1) we consider a whole family of equations (1) with symbols $\sigma(t)$ belonging to some translation-invariant set Σ, i.e. the set Σ, together with any element $\sigma(t)$, contains also the element $T(h)\sigma(t) = \sigma(t+h)$ for any $h \in \mathbb{R}$. The essential role is played by the topology in the enveloping space of Σ. To use a known construction of an attractor, the set Σ must be compact in this enveloping topological space. In the sequel we shall restrict our consideration to the translation-compact functions in the (uniform or mean) convergence topology on any bounded segment form the time axis $\mathbb{R}$. (See sec. 1).

In the present work we study the questions concerning the existence and the structure of uniform attractors for the non-autonomous 2D-Navier-Stokes system and for the reaction-diffusion system with translation-compact terms.

1. Translation-compact functions.

We shall introduce two classes of translation-compact functions. The first class is based on the mean convergence topology on any bounded segment of the time axis $\mathbb{R}$, and the second one is based on uniform convergence topology on any bounded segment of $\mathbb{R}$.

Let $p \geq 1$. Consider the space $L_p^{\mathrm{loc}}(\mathbb{R}, \mathcal{E})$, where $\mathcal{E}$ is a Banach space. By the definition, the space $L_p^{\mathrm{loc}}(\mathbb{R}, \mathcal{E})$ consists of all functions $\sigma(t), t \in \mathbb{R}$, with values in $\mathcal{E}$, p-power integrable in Bochner sense on any bounded segment, i.e. for any $[t_1, t_2] \subset \mathbb{R}$

$$\int_{t_1}^{t_2} \|\sigma(s)\|_{\mathcal{E}}^p ds < +\infty.$$

The space $L_p^{\mathrm{loc}}(\mathbb{R}, \mathcal{E})$ is supplemented by the p-power mean convergence topology on any bounded segment of $\mathbb{R}$, i.e. $\sigma_n \to \sigma$ $(n \to +\infty)$ in $L_p^{\mathrm{loc}}(\mathbb{R}, \mathcal{E})$, if for any $[t_1, t_2] \subset \mathbb{R}$

$$\int_{t_1}^{t_2} \|\sigma_n(s) - \sigma(s)\|_{\mathcal{E}}^p ds \to 0 \qquad (n \to +\infty).$$

The space $L_p^{\mathrm{loc}}(\mathbb{R}, \mathcal{E})$ with the above topology is a linear countably normable topological space. Evidently, this space is metrizable, for example, by use of Frechét metric, and the corresponding metric space is complete. Let us formulate the compactness criterion in the space $L_p^{\mathrm{loc}}(\mathbb{R}, \mathcal{E})$.

Theorem 1.1. A set Σ is precompact in $L_p^{\mathrm{loc}}(\mathbb{R}, \mathcal{E})$ if and only if

i) for any $[t_1, t_2] \subset \mathbb{R}$ a set $\{\int_{t_1}^{t_2} \psi(s)ds | \psi \in \Sigma\} \subset\subset \mathcal{E}$ is precompact in $\mathcal{E}$;

ii) for any $[t_1, t_2] \subset \mathbb{R}$ there exists a positive function $\alpha(t_1, t_2, s), \alpha(t_1, t_2, s) \to 0$ $(s \to 0+)$, such that for any $\psi \in \Sigma$ one has:

$$\int_{t_1}^{t_2} \|\psi(s) - \psi(s+l)\|_{\mathcal{E}}^p ds \leq \alpha(t_1, t_2, |l|).$$

Now we introduce the concept of translation-compact function in $L_p^{\mathrm{loc}}(\mathbb{R},\mathcal{E})$.

Definition 1.1. A function $\sigma(t)\in L_p^{\mathrm{loc}}(\mathbb{R},\mathcal{E})$ is said to be translation-compact (tr.c.) in $L_p^{\mathrm{loc}}(\mathbb{R},\mathcal{E})$ if the set $\{\sigma(t+h)|h\in\Sigma\}$ is precompact in $L_p^{\mathrm{loc}}(\mathbb{R},\mathcal{E})$. The closure of this set in $L_p^{\mathrm{loc}}(\mathbb{R},\mathcal{E})$ is designated by $\mathcal{H}(\sigma)$:

$$\mathcal{H}(\sigma)=\overline{\{\sigma(t+h)|h\in\Sigma\}}.$$

$\mathcal{H}(\sigma)$ is called the hull of the function σ in $L_p^{\mathrm{loc}}(\mathbb{R},\mathcal{E})$.

First of all notice that any tr.c. function σ in $L_p^{\mathrm{loc}}(\mathbb{R},\mathcal{E})$ is mean bounded on any segment having the finite length, i.e. the value

$$\eta(h)=\sup_{t\in\mathbb{R}}\int_t^{t+h}\|\sigma(s)\|_{\mathcal{E}}^p ds<+\infty \tag{1.1}$$

is finite for any $h\geq 0$. From Theorem 1.1 it follows the translation-compactness criterion.

Theorem 1.2. A function $\sigma(t)$ is tr.c. in $L_p^{\mathrm{loc}}(\mathbb{R},\mathcal{E})$ if an only if

i) for any $h\geq 0$ the set $\left\{\int_t^{t+h}\sigma(s)ds|t\in\mathbb{R}\right\}\subset\subset\mathcal{E}$ is precompact in $\mathcal{E}$,

ii) there exists a function $\alpha(s),\alpha(s)\to 0$ $(s\to 0+)$, such that for any $t\in\mathbb{R}$:

$$\int_t^{t+1}\|\sigma(s)-\sigma(s+l)\|_{\mathcal{E}}^p ds\leq\alpha(|l|).$$

Let us formulate some important properties of tr.c. functions and their hulls.

Proposition 1.3. Let $\sigma_0(t)$ be a tr.c. function in $L_p^{\mathrm{loc}}(\mathbb{R},\mathcal{E})$. Then

i) for any $\sigma\in\mathcal{H}(\sigma_0)$ the function σ is tr.c. in $L_p^{\mathrm{loc}}(\mathbb{R},\mathcal{E})$, $\mathcal{H}(\sigma)\subseteq\mathcal{H}(\sigma_0)$ (notice, inclusion can be strict);

ii) translation group $\{T(h),h\in\mathbb{R}\},(T(h)\sigma(t)=\sigma(t+h))$ is continuous on $\mathcal{H}(\sigma_0)$ in the topology of $L_p^{\mathrm{loc}}(\mathbb{R},\mathcal{E})$;

iii) $T(h)\mathcal{H}(\sigma_0)=\mathcal{H}(\sigma_0)$ for any $h\in\mathbb{R}$;

iv) the function $\eta_{\sigma_0}(h)$ (replace σ by σ_0 in (1.1)) is greater than $\eta_\sigma(h)$ for any $\sigma\in\mathcal{H}(\sigma_0)$: $\eta_\sigma(h)\leq\eta_{\sigma_0}(h)\ \forall\, h\in\mathbb{R}$.

For the specific spaces $\mathcal{E}$ one can formulate simple and convenient sufficient conditions of translation-compactness in $L_p^{\mathrm{loc}}(\mathbb{R},\mathcal{E})$. Let, for example, $\mathcal{E}=L_2(\Omega)$, where $\Omega \subset\subset \mathbb{R}^n$, let also $Q = \Omega \times [0,1]$ and $H_s(Q)$ is a Sobolev space of order $s > 0$. Assume that $\sigma(x,t) \in L_2^{\text{loc}}(\Omega \times \mathbb{R}) = L_2^{\text{loc}}(\mathbb{R}, L_2(\Omega))$. Now if

$$\|\sigma(x, t+h)\|_{H_s(Q)} \leq M < +\infty \; \forall\, h \in \mathbb{R}, \tag{1.2}$$

where M does not depend on $h \in \mathbb{R}$, then σ is tr.c. function in $L_2^{loc}(\mathbb{R}, L_2(\Omega))$.

The second class of tr.c. functions is based on the space $C(\mathbb{R}, \mathcal{M})$ of continuous functions $\sigma(t), t \in \mathbb{R}$, with values in a complete metric space $\mathcal{M}$. In the space $C(\mathbb{R}, \mathcal{M})$ we introduce the uniform convergence topology on any bounded segment of the time axis. In this topology a sequence of functions $\{\sigma_n(t)\}_{n \in \mathbb{N}}$ converges to a function $\sigma(t)$ if for any $[t_1, t_2] \subset \mathbb{R}$

$$\max_{t \in [t_1, t_2]} \rho_{\mathcal{M}}(\sigma_n(t), \sigma(t)) \to 0 \qquad (n \to +\infty),$$

where $\rho_{\mathcal{M}}(.,.)$ means the metric on $\mathcal{M}$. It is clear, that the topology under the consideration is metrizable and the corresponding metric space $C(\mathbb{R}, \mathcal{M})$ is complete.

Definition 1.2. A function $\sigma(t) \in C(\mathbb{R}, \mathcal{M})$ is said to be translation-compact (tr.c.) in $C(\mathbb{R}, \mathcal{M})$ if the hull $\mathcal{H}(\sigma) = \overline{\{\sigma(t+h) | h \in \Sigma\}}$ of σ is compact in $C(\mathbb{R}, \mathcal{M})$. Here overlying means the closure in $C(\mathbb{R}, \mathcal{M})$.

Arzelá-Ascoli theorem implies the following translation-compactness criterion in $C(\mathbb{R}, \mathcal{M})$.

Theorem 1.4. A function $\sigma(t)$ is tr.c. in $C(\mathbb{R}, \mathcal{M})$ if an only if

i) the set $\{\sigma(t) | t \in \mathbb{R}\} \subset\subset \mathcal{M}$ is precompact in $\mathcal{M}$;

ii) $\sigma(t)$ is uniformly continuous on $\mathbb{R}$, i.e. there exists a function $\alpha(s), \alpha(s) \to 0$ $(s \to 0+)$, such that $\rho_{\mathcal{M}}(\sigma(t_1), \sigma(t_2)) \leq \alpha(|t_1 - t_2|)$ for any $t_1, t_2 \in \mathbb{R}$.

Remark. Proposition 1.3 is valid also for tr.c. functions $\sigma_0(t)$ in $C(\mathbb{R}, \mathcal{M})$ (excluding the property (iv) when $\mathcal{M}$ is not a Banach space).

Finally notice, that any almost periodic function $\sigma(t) \in C_b(\mathbb{R}, \mathcal{M})$ is tr.c. in $C(\mathbb{R}, \mathcal{M})$. Recall, that the definition of almost periodic function is analogous of that of given in Definition 1.2. One has to use more strong uniform convergence topology in $\mathbb{R}$ of the space $C_b(\mathbb{R}, \mathcal{M})$. The weaker topology of the space $C(\mathbb{R}, \mathcal{M})$ (or even $L_p^{\text{loc}}(\mathbb{R}, \mathcal{E})$) makes possible to extend significantly the class of almost periodic functions.

2. Uniform attractors of families of processes.

We recall some definitions concerning attractors of families of processes and formulate the main existence and structure theorem. A two-parametric family of mappings $\{U(t,\tau)\} = \{U(t,\tau), t \geq \tau, \tau \in \mathbb{R}\}, U(t,\tau) : E \mapsto E$, acting in a Banach space E is said to be a process if the following conditions are valid:

$$U(t,\tau) = U(t,s)U(s,\tau), \quad U(\tau,\tau) = I \quad \forall \tau \in \mathbb{R},\ \tau \leq s \leq t.$$

Let we are given a family of processes $\{U_\sigma(t,\tau)\}$, depending on a functional parameter σ. Parameters σ belong to a complete metric space Σ, $\sigma \in \Sigma$. A parameter σ is called a symbol of a process $\{U_\sigma(t,\tau)\}$, and Σ is a symbol space.

Definition 2.1. A set P_Σ is said to be uniformly (w.r.t. $\sigma \in \Sigma$) attracting for the family of processes $\{U_\sigma(t,\tau)\}, \sigma \in \Sigma$, if for any $\tau \in \mathbb{R}$ and arbitrary bounded in E set B:

$$\sup_{\sigma \in \Sigma} \operatorname{dist}_E(U_\sigma(t,\tau)B, P_\Sigma) \to 0 \qquad (t \to +\infty).$$

As usually, $\operatorname{dist}_E(x,Y) = \sup_{x\in X} \operatorname{dist}_E(X,Y) = \sup_{x\in X} \inf_{y\in Y} \|x-y\|_E$.

Definition 2.2. A family of processes $\{U_\sigma(t,\tau)\}, \sigma \in \Sigma$, possessing a compact uniformly attracting set is called uniformly (w.r.t. $\sigma \in \Sigma$) asymptotically compact.

Definition 2.3. A set $\mathcal{A}_\Sigma$ is said to be the uniform attractor of the family of processes $\{U_\sigma(t,\tau)\}, \sigma \in \Sigma$, if it is minimal closed uniformly (w.r.t. $\sigma \in \Sigma$) attracting set of the family of processes. Minimality means, that any closed uniformly attracting set contains $\mathcal{A}_\Sigma$.

One can easily show that any uniformly (w.r.t. $\sigma \in \Sigma$) asymptotically compact family of processes $\{U_\sigma(t,\tau)\}, \sigma \in \Sigma$, possesses a uniform (w.r.t. $\sigma \in \Sigma$) attractor $\mathcal{A}_\Sigma$. (See, for example [2]).

Now we shall investigate the structure of a uniform attractor more closely under some additional conditions. We assume that some strictly invariant semigroup $\{T(h), h \geq 0\}$ acts on Σ: $T(h) : \Sigma \mapsto \Sigma, T(h)\Sigma = \Sigma,\ \forall h \geq 0$. Let also the following translation identity be valid for the family of processes $\{U_\sigma(t,\tau)\}, \sigma \in \Sigma$, and for the semigroup $\{T(h)\}$:

$$U_\sigma(t+h,\tau+h) = U_{T(h)\sigma}(t,\tau),\ \forall \sigma \in \Sigma, t \geq \tau;\ t,\tau \in \mathbb{R}, h \geq 0. \tag{2.1}$$

Consider the following family of mappings $\{S(t), t \geq 0\}$ acting in the extended phase space $E \times \Sigma$ by the rule:

$$S(t)(u,\sigma) = (U_\sigma(t,0)u, T(t)\sigma), t \geq 0, (u,\sigma) \in E \times \Sigma. \tag{2.2}$$

One can easily check using (2.1) that $\{S(t)\}$ is a semigroup on $E \times \Sigma$: $S(t) : E \times \Sigma \mapsto E \times \Sigma, t \geq 0, S(t_1)S(t_2) = S(t_1+t_2), S(0) = I, \forall t_1, t_2 \geq 0$ (see [2]).

A family of operators $\{U_\sigma(t,\tau)\}, \sigma \in \Sigma$, is said to be $(E \times \Sigma, E)$- continuous, if for any fixed t and τ the mapping $(u,\sigma) \mapsto U_\sigma(t,\tau)u$ is continuous from $E \times \Sigma$ into E.

We shall describe the structure of uniform attractors in terms of kernel sections of processes.

Definition 2.4. A curve $\{u(s), s \in \mathbb{R}\}, u(s) \in E \ \forall s \in \mathbb{R}$; is said to be a complete trajectory of the process $\{U(t,\tau)\}$, if

$$U(t,\tau)u(\tau) = u(t) \ \forall t \geq \tau; t, \tau \in \mathbb{R}.$$

Definition 2.5. The kernel $\mathcal{K}$ of the process $\{U(t,\tau)\}$ consists of all bounded complete trajectories of the process $\{U(t,\tau)\}$:

$$\mathcal{K} = \{u(\cdot) : u(t), t \in \mathbb{R}, \text{ is a complete trajectory of}$$

$$\text{the process } \{U(t,\tau)\} \text{ and } \|u(t)\|_E \leq C_u \ \forall t \in \mathbb{R}\}.$$

The set $\mathcal{K}(s) = \{u(s) : u(\cdot) \in \mathcal{K}\}$ is called the kernel section at time $t = s, s \in \mathbb{R}$. In the sequel Π_1 and Π_2 denotes the following projectors:

$\Pi_1 : E \times \Sigma \mapsto E, \Pi_2 : E \times \Sigma \mapsto \Sigma, \Pi_1(u,\sigma) = u, \Pi_2(u,\sigma) = \sigma$. Let us formulate the main theorem on the attractors of families of processes.

Theorem 2.1. Let a family of processes $\{U_\sigma(t,\tau)\}, \sigma \in \Sigma$, acting in the space E be uniformly (w.r.t. $\sigma \in \Sigma$) asymptotically compact and $(E \times \Sigma, E)$-continuous. Also let Σ be a compact metric space, $\{T(h)\}$ be a continuous invariant $(T(h)\Sigma = \Sigma, \ \forall h \geq 0)$ semigroup on Σ satisfying the translation identity (2.1). Then the semigroup $\{S(t)\}$ corresponding to the family of processes $\{U_\sigma(t,\tau)\}, \sigma \in \Sigma$, and acting on $E \times \Sigma$ by the formula (2.2) possesses the compact in $E \times \Sigma$ attractor $\mathcal{A} : S(t)\mathcal{A} = \mathcal{A} \ \forall t \geq 0$. Moreover,

i) $\Pi_1 \mathcal{A} = \mathcal{A}_\Sigma$ is the uniform (w.r.t. $\sigma \in \Sigma$) attractor of the family of processes $\{U_\sigma(t,\tau)\}, \sigma \in \Sigma$;

ii) $\Pi_2 \mathcal{A} = \Sigma$;

iii) $\mathcal{A}_\Sigma = \bigcup_{\sigma \in \Sigma} \mathcal{K}_\sigma(0)$. Here $\mathcal{K}_\sigma$ is the kernel of the process $\{U_\sigma(t,\tau)\}$ with the symbol $\sigma \in \Sigma$.

Note that the section $\mathcal{K}_\sigma(0)$ in iii) can be replaced by any $\mathcal{K}_\sigma(s)$, where s is an arbitrary real number, $s \in \mathbb{R}$.

Theorem 1 was proved in [2]. It follows from the general theorem on the semigroup's attractors (see, for examples, [1,9,13]) being applied to the semigroup (2.2). Works [2,3,4] dealt with non-autonomous dynamical systems with almost periodic symbols and cascade systems, as it is called there. Let us give some references of important investigations for non-autonomous evolution equations and proceeesses: [12], [6], [7], [9], [10]. Now we consider more general classes of symbols.

3. Examples of equations with tr.c. symbols, arising in mathematical physics.

Let us return to the problem (1). Assume, that the symbol $\sigma(t)$ of the equation (1) belongs to the symbol space $\Sigma = \mathcal{H}(\sigma_0)$, where $\sigma_0(t)$ is some fixed tr.c. function (in $L_p^{\rm loc}(\mathbb{R},\mathcal{E})$ or $C(\mathbb{R},\mathcal{M})$). If the problem (1) is correct for any symbol $\sigma \in \mathcal{H}(\sigma_0)$ then to the problem (1) there corresponds a family of processes $\{U_\sigma(t,\tau)\}, \sigma \in \mathcal{H}(\sigma_0)$, acting in the corresponding space E. To construct the uniform attractor of such family of processes, we use the following theorem.

Theorem 3.1. Let the family of processes $\{U_\sigma(t,\tau)\}, \sigma \in \mathcal{H}(\sigma_0)$, corresponding to the problem (1) with tr.c. symbol $\sigma_0(t)$ (in $L_p^{\rm loc}(\mathbb{R},\mathcal{E})$ or $C(\mathbb{R},\mathcal{M})$) be uniformly (w.r.t. $\sigma \in \mathcal{H}(\sigma_0)$) asymptotically compact and $(E\times\mathcal{H}(\sigma_0), E)$-continuous. Then the uniform attractor of this family of processes coincides with $\mathcal{A}_{\mathcal{H}(\sigma_0)} = \bigcup_{\sigma\in\mathcal{H}(\sigma_0)} \mathcal{K}_\sigma(s)$. Here s is fixed and any, $\mathcal{K}_\sigma$ is the kernel of the process $\{U_\sigma(t,\tau)\}$ with the symbol $\sigma \in \mathcal{H}(\sigma_0)$. As this take place, the kernel $\mathcal{K}_\sigma$ is not empty for any $\sigma \in \mathcal{H}(\sigma_0)$ and a kernel section $\mathcal{K}_\sigma(s)$ is compact in E for any $t \in \mathbb{R}$.

Theorem 3.1 follows from Theorem 2.1 and Proposition 1.3. Evidently, the translation identity (2.1) takes place for the family of processes corresponding to the problem (1) and for the translation group $\{T(h), h \in \mathbb{R}\}$, $(T(h)\sigma(t) = \sigma(t+h))$.

As we shall see in Example 3.2, sometimes it is natural to consider mixed symbols consisting of components from different classes. For example, let $\sigma_0(t) = (\sigma_0^{(1)}(t), \sigma_0^{(2)}(t))$, where $\sigma_0^{(1)}(t) \in C(\mathbb{R},\mathcal{M}))$, $\sigma_0^{(2)}(t) \in L_p^{\rm loc}(\mathbb{R},\mathcal{E})$ and let functions $\sigma_0^{(1)}(t)$, $\sigma_0^{(2)}(t)$ be tr.c. in their spaces. In this case, we consider the following symbol space: $\Sigma = \mathcal{H}(\sigma_0^{(1)})\times\mathcal{H}(\sigma_0^{(2)})$. The Theorem 3.1 is valid with $\mathcal{H}(\sigma_0)$ being replaced by $\mathcal{H}(\sigma_0^{(1)}) \times \mathcal{H}(\sigma_0^{(2)})$.

Example 3.1. 2D Navier-Stokes system in a bounded domain with tr.c. external force. Excluding the pressure, the system can be written in the form:

$$\partial_t u + Lu + B(u,u) = g(x,t), \quad x = (x_1,x_2) \in \Omega \subset\subset \mathbb{R}^2, t \in \mathbb{R}. \qquad (3.1)$$

Here

$$L = -\nu\Pi\Delta, \quad B(u,u) = \Pi\sum_{i=1}^{2} u_i\partial_{x_i}u, \quad \varphi = \Pi\varphi_0, \quad u|_{\partial\Omega} = 0,$$

$u = (u^1,u^2)$, $g = (g^1,g^2)$. (see [11,14,1]). By H and H_1 we denote, respectively, the closure of the set $\mathcal{V}_0 = \{v : v \in (C_0^\infty(\Omega))^2, (\nabla, v) = 0\}$ in the norms $\|\|$ and $\|\|_1$ of the space $(L_2(\Omega))^2$ and $(H_1(\Omega))^2$, by Π the orthogonal projector on H onto $(L_2(\Omega))^2$ and its different extensions. The system (3.1) is non-autonomous and the external force $g(\cdot,t)$ is considered as the symbol of this system. We assume that $g(\cdot,t) \in \mathcal{H}(\varphi)$, where $\varphi(x,t)$ is a tr.c. function in $L_2^{\rm loc}(\mathbb{R},H)$. The translation-compactness criterion is given by the Theorem 1.1. The sufficient condition (1.2) is convenient in applications. Another sufficient condition follows from Dubinskij result (see [8]). The function φ is tr.c. in $L_2^{\mathrm{loc}}(\mathbb{R}, H)$ if $\varphi \in L_2^{\mathrm{loc}}(\mathbb{R}, H_1)$, $\partial_t\varphi \in L_2^{\mathrm{loc}}(\mathbb{R}, H_{-1})$, and

$$\int_t^{t+1} \|\varphi(s)\|^2 ds \le C_0, \quad \int_t^{t+1} |\varphi(s)|_{-1}^2 ds \le C_1 \; \forall t \in \mathbb{R}.$$

We supplement the system (3.1) by the initial conditions:

$$u|_{t=\tau} = u_\tau, \quad u_\tau \in H_1, \quad (E = H_1). \tag{3.2}$$

The problem (3.1), (3.2) is uniquely solvable and it generates the family of processes $\{U_g(t,\tau)\}, g \in \mathcal{H}(\varphi)$, acting in H_1. The proof of this assertion is similar to known results for the autonomous system (when $g(x,t) \equiv g(x)$) (see [1,14]).

Theorem 3.2. If the function $\varphi(x,t)$ is tr.c. in $L_2^{\mathrm{loc}}(\mathbb{R}, H)$ then the family of processes $\{U_g(t,\tau)\}, g \in \mathcal{H}(\varphi)$ is uniformly (w.r.t. $g \in \mathcal{H}(\varphi)$) asymptotically compact and $(H_1 \times \mathcal{H}(\varphi), H_1)$-continuous.

The proof of Theorem 3.2 is rather long, so the complete proof we shall present in other publication (see [5]). Here we consider only some elements of it. Let us show the existence of bounded in H uniformly (w.r.t. $g \in \mathcal{H}(\varphi)$) absorbing set. Similarly to (1.1), we introduce the function

$$\eta(T) = \sup_{t\in\mathbb{R}} \int_t^{t+T} |\varphi(s)|^2 ds < +\infty \; \forall T \ge 0.$$

Proposition 1.3 implies that for any $g \in \mathcal{H}(\varphi)$ and arbitrary $t \in \mathbb{R}$

$$\int_t^{t+T} |g(s)|^2 ds \le \eta(T). \tag{3.3}$$

As it usually is, (3.1) implies the inequalities (see [3]):

$$\frac{d}{dt}|u|^2 + \nu\|u\|^2 \le \frac{1}{\nu}|g(t)|_{-1}^2,$$

or

$$\frac{d}{dt}|u|^2 + \lambda_1\nu|u|^2 \le \frac{1}{\nu\lambda_1}|g(t)|^2, \tag{3.4}$$

where λ_1 is the first eigenvalue of the operator L.

Lemma 3.3. Let $z(t)$ be positive absolutely continuous function and $a(t) \in L_1^{\mathrm{loc}}(\mathbb{R})$, $a(t) \ge 0, \alpha > 0$. Let also

$$\int_t^{t+T} a(s) ds \le b(T) \; \forall t \in \mathbb{R}, T \ge 0, \tag{3.5}$$

where $b(T)$ is nondecreasing function and

$$\frac{d}{dt}z(t) \le -\alpha z(t) + a(t) \tag{3.6}$$

almost everywhere in $\mathbb{R}$. Then for every t and τ such that $t-\tau \geq 1$ one has:

$$z(t)e^{\alpha t} - z(\tau)e^{\alpha\tau} \leq b(2)\frac{e^{\alpha}}{e^{\alpha}-1}\left(e^{\alpha t} - e^{\alpha\tau}\right).$$

Proof. It follows from (3.6) that

$$\frac{d}{dt}\left(z(t)e^{\alpha t}\right) \leq a(t)e^{\alpha t},$$

Put $t_i = \tau + (t-\tau)i/n, i = 0, 1, \ldots, n$. Then

$$z(t)e^{\alpha t} - z(\tau)e^{\alpha\tau} \leq \int_{\tau}^{t} a(s)e^{\alpha s}ds = \sum_{i=1}^{n-1}\int_{t_i}^{t_{i+1}} a(s)e^{\alpha s}ds \leq$$

$$b\left(\frac{t-\tau}{n}\right)\sum_{i=1}^{n-1} e^{\alpha t_i} = b\left(\frac{t-\tau}{n}\right)e^{\alpha\tau}\sum_{i=1}^{n-1} e^{\alpha(t-\tau)n/i} =$$

$$b\left(\frac{t-\tau}{n}\right)\frac{e^{\alpha(t-\tau)/n}}{e^{\alpha(t-\tau)/n}-1}\left(e^{\alpha t} - e^{\alpha\tau}\right). \tag{3.7}$$

Now if $t-\tau \geq 1$, then $1 \leq (t-\tau)/n \leq 2$ for some n. Therefore, $b\left(\frac{t-\tau}{n}\right) \leq b(2)$ (see (3.5)). Function $e^x/(e^x-1)$ decrease when x increase. Hence, $\frac{e^{\alpha(t-\tau)/n}}{e^{\alpha(t-\tau)/n}-1} \leq e^{\alpha}/(e^{\alpha}-1)$. Finally, (3.7) yields

$$z(t)e^{\alpha t} - z(\tau)e^{\alpha\tau} \leq b(2)\frac{e^{\alpha}}{e^{\alpha}-1}\left(e^{\alpha t} - e^{\alpha\tau}\right).\Box$$

Applying Lemma 3.3 to inequality (3.4), we obtain:

$$|u(t)|^2 \leq |u(\tau)|^2 e^{-a(t-\tau)} + \alpha^{-1}b(2)e^{\alpha}/(e^{\alpha}-1)\left(1-e^{-a(t-\tau)}\right),$$

where $\alpha = \nu\lambda_1$. Consequently, the set $B_0 = \{|u| \leq R\}$, where $R > R_0, R_0 = \alpha^{-1}b(2)e^{\alpha}/(e^{\alpha}-1)$, is uniformly absorbing for the family of processes $\{U_g(t,\tau)\}$, $g \in \mathcal{H}(\varphi)$, corresponding to the problem (3.1), (3.2) and acting in H. The set B_o is bounded in H. Closer analysis permits to construct from the set B_0 another uniformly absorbing set B_1 compact in H_1. This analysis uses mostly the smoothing property of the Navier-Stokes system.

The $(H_1 \times \mathcal{H}(\varphi), H_1)$-continuity of the family of processes we shall prove in [5]. Notice that weaker property of $(H \times \mathcal{H}(\varphi), H)$-continuity have been proved in [2] in the case when the external force φ is almost periodic function.

According to Theorem 3.2, the family of processes $\{U_g(t,\tau)\}, g \in \mathcal{H}(\varphi)$, satisfies all the conditions of Theorem 3.1. It follows that the set

$$\mathcal{A}_{\mathcal{H}(\varphi)} = \bigcup_{g\in\mathcal{H}(\varphi)} \mathcal{K}_g(0)$$

is the uniform attractor of this family. Here $\mathcal{K}_g$ is the kernel of the process $\{U_g(t,\tau)\}$ with a symbol g. Notice that the kernel $\mathcal{K}_g$ consists of all bounded (in H_1) solutions $u(t), t \in \mathbb{R}$, of the equation (3.1).

Example 3.2. Non-autonomous reaction-diffusion system with tr.c.terms. The following system is considered:

$$\partial_t u = a\Delta u - f(u,t) + \varphi(x,t), \quad u|_{\partial\Omega} = 0, \quad t \in \mathbb{R}, \tag{3.8}$$

$$u|_{t=\tau} = u_\tau, u_\tau \in (L_2(\Omega))^N = H, t \geq \tau. \tag{3.9}$$

Here, as usually, $x \in \Omega \subset\subset \mathbb{R}^n$, $u \in \mathbb{R}^N, u = u(x,t), u = (u^1, \ldots, u^N), \varphi = (\varphi^1, \ldots, \varphi^N)$, $f = (f^1, \ldots, f^N).a = \{a_{ij}\}_{i=1,\ldots,N}^{j=1,\ldots,N}$ is a real $N \times N-$matrix with a positive symmetric part $a+a^* \geq \beta^2 I, \beta > 0$. It is assumed that $\varphi(\cdot,t) \in L_2^{\text{loc}}(\mathbb{R}, H)$. Also let $f, f'_{vj} \in C(\mathbb{R}^N \times \mathbb{R}, \mathbb{R}^N)$, $(j = 1,.., N)$, and let the following conditions hold:

$$\gamma_2|v|^p - C_2 \leq (f(v,t),v) \leq \gamma_1|v|^p + C_1, \gamma_i > 0, p \geq 2, \tag{3.10}$$

$$(f'_v(v,t)z,z) \geq -C_3(z,z) \; \forall z \in \mathbb{R}^N, \tag{3.11}$$

$$|f'_v(v,t)| \leq C_4\left(|v|^{p-2} + 1\right) \tag{3.12}$$

for any $v \in \mathbb{R}^N, t \in \mathbb{R}$.

Problem (3.8), (3.9) has (for all u_τ) a unique solution $u(t)$ belonging to $C_b([\tau,+\infty), H) \bigcap L_2^{\text{loc}}((\tau, +\infty), (H_1^0(\Omega))^N) \bigcap L_p^{\text{loc}}((\tau, +\infty), (L_p(\Omega))^N)$ (see [1,2]). Therefore, there exists the process $\{U_{\sigma_0}(t,\tau)\}$, acting in the space H and corresponding to problem (3.8), (3.9): $U_{\sigma_0}(t,\tau) : H \mapsto H, t \geq \tau, \tau \in \mathbb{R}$. System (3.8) has the time symbol $\sigma_0(t) = (f(v,t), \varphi(x,t))$.

Suppose that the function $\varphi(x,t)$ is tr.c. in $L_2^{\text{loc}}(\mathbb{R}, H)$, and $f(v,t)$ is tr.c. function in the space $C(\mathbb{R}, \mathcal{M}_1)$, where $\mathcal{M}_1$ is the following metric space. The space $\mathcal{M}_1$ consists of functions $\psi(v) = (\psi^1(v), \ldots, \psi^N(v)), v = (v^1, \ldots, v^N) \in \mathbb{R}^N, \psi \in C^1(\mathbb{R}^N, \mathbb{R}^N)$, with the finite norm

$$\|\psi\|_{\mathcal{M}_1} = \sup_{v \in \mathbb{R}^N}\left(\frac{|\psi(v)|}{1+|v|^{p-1}} + \frac{|\psi'_v(v)|}{1+|v|^{p-2}}\right)$$

Evidently, $\mathcal{M}_1$ is a Banach space. By Theorem 1.4. the function f is tr.c. in $C(\mathbb{R}, \mathcal{M}_1)$ if and only if:

$$\{f(.,t)|t \in \mathbb{R}\} \subset\subset \mathcal{M}_1 \text{ is a precompact set in } \mathcal{M}_1,$$

$$|f(v,t_1) - f(v,t_2)| \leq \alpha_1(|t_1 - t_2|)(1 + |v|^{p-1}), \tag{3.13}$$

$$|f'_v(v,t_1) - f'_v(v,t_2)| \leq \alpha_2(|t_1 - t_2|)(1 + |v|^{p-2}),$$

$$\text{where } \alpha_1(s), \alpha_2(s) \to 0 \quad (s \to 0+).$$

Consider the symbol space $\Sigma = \mathcal{H}(f) \times \mathcal{H}(\varphi)$, where hulls are taken in the corresponding spaces. Let $\sigma = (q,g) \in \Sigma$. Evidently, $g(x,t) \in L_2^{\text{loc}}(\mathbb{R}, H)$, and $q(v,t)$ satisfies conditions (3.10), (3.11), (3.12) with one and the same constants. The system (3.8) where f and φ are replaced by q and g has the symbol $\sigma = (q, g)$. Hence this system generate a process $\{U_\sigma(t, \tau)\}$ acting in H_1. Thus, to the problem (3.8), (3.9) there corresponds the family of processes $\{U_\sigma(t, \tau)\}, \sigma \in \mathcal{H}(f) \times \mathcal{H}(\varphi)$.

Theorem 3.4. The family of processes $\{U_\sigma(t, \tau)\}, \sigma \in \mathcal{H}(f) \times \mathcal{H}(\varphi)$ is uniformly (w.r.t. $\sigma \in \mathcal{H}(f) \times \mathcal{H}(\varphi)$) asymptotically compact and $(H \times \mathcal{H}(f) \times \mathcal{H}(\varphi), H)$-continuous.

We omit the proof of Theorem 3.4. The proof uses Lemma 3.3 and the general reasoning scheme is analogous to that of given in [2]. In [2] we have considered a particular case, when φ and f are almost periodic (in time) function with values, respectively, in H and $\mathcal{M}_1$.

Now we conclude, that Theorem 3.1 is applicable to the family of processes corresponding to the problem (3.8), (3.9). As an example of function $f(v, t)$ satisfying the conditions (3.10)-(3.13), consider the function

$$f(v, t) = \sum_{i=1}^{m} \zeta_i(t) f_i(v), \tag{3.14}$$

where all $f_i(v)$ satisfies (3.10)-(3.12) and functions $\zeta_i(t) \in C_b(\mathbb{R}, \mathbb{R}_+), i = 1, \ldots, m,$ are tr.c. in $C(\mathbb{R}, \mathbb{R}_+)$,i.e., by theorem 1.3, they are bounded and uniformly continuous on $\mathbb{R}$. The example (3.14) can be generalized by replacing finite sum into infinite one. Another examples of functions f, we shall present in the other paper (see [5]). Also we shall prove sufficient conditions for f to be tr.c. in the space $C(\mathbb{R}, \mathcal{M}_1)$.

References

[1] A.V. Babin and V.B. Vishik, Attractors of Evolution Equations. Amsterdam, London, New-York, Tokyo, North-Holland, 1992.

[2] V.V. Chepyzhov and M.I. Vishik, Attractors of non-autonomous dynamical systems and their dimension, J. Math. Pures Appl., 73, N3, 1994, pp. 279–333.

[3] V.V. Chepyzhov and M.I. Vishik, Non-autonomous evolution equations and their attractor, Russ. J. Math. Physics, Vol. 1, No. 2, 1993, pp. 165–190.

[4] V.V. Chepyzhov and M.I. Vishik, Nonautonomous evolutionary equations with almost periodic symbols, Rendiconti del Seminario Matematici e Fisico di Milano, Vol. LXII, 1992, 185–213.

[5] V.V. Chepyzhov and M.I. Vishik, Non-autonomous partial differential equations with translation-compact symbols, to appear.

[6] C.M. Dafermos, Semi-flows associated with compact and almost uniform processes, Math. Systems Theory, Vol. 8, 1974, pp. 142–149.

[7] C.M. Dafermos, Almost periodic processes and almost periodic solutions of evolutional equations, Proceedings of a University of Florida, International Symposium, New York Academic Press, 1977, pp. 43–57.

[8] Yu.A. Dubinsky, Weak convergence in nonlinear elliptic and parabolic equations, Mat.Sbornik, 67(4), 1965, pp. 609–642.

[9] J.K. Hale, Asymptotic Behavior of Dissipative Systems, Mathematical Surveys and Monographs, Vol. 25, A.M.S., Providence, RI, 1987.

[10] A. Haraux, Systèmes dinamiques dissipatifs et applications. Paris, Milan, Barcelona, Rome, Masson, 1991.

[11] O.A. Ladyzhenskaya, Mathematical problems in the dynamics of a viscous incompressible liquid, Moscow, Nauka, 1970.

[12] G.R. Sell, Non-autonomous differential equations and topological dynamics I, II, Amer. Math. Soc., Vol. 127, 1967, pp. 241–262, pp. 263–283.

[13] R. Temam, Infinite-dimensional dynamical systems in mechanics and physics, Springer-Verlag, 1988.

[14] R. Temam, On the theory and numerical analysis of the Navier-Stokes equations, Springer-Verlag, 1974.

V. V. Chepyzhov and M. I. Vishik
Institute for Problem of Information Transmission,
Russian Academy of Sciences, Ermolovoy 19,
Moscow 101447, GSP-4, Russia
E-mail: chep@ippi.msk.ac.su

Operator Theory:
Advances and Applications, Vol. 78

Localization for 2-dimensional Random Schrödinger Operators with Magnetic Fields

J.M. Combes[1], P.D. Hislop[2]

Abstract

We sketch the proof of the existence of localized states at the edges of the bands for the two-dimensional Landau Hamiltonian with a random potential, of arbitrary disorder, provided that the magnetic field is sufficiently large. The corresponding eigenfunctions decay exponentially with the magnetic field and distance. We also prove that the integrated density of states is Lipschitz continuous away from the Landau energies. The proof relies on a Wegner estimate for the finite-area magnetic Hamiltonians with random potentials and exponential decay estimates for the finite-area Green's functions. The proof of the decay estimates for the Green's functions uses fundamental results from two-dimensional bond percolation theory.

This is an extended version of a talk given by P. D. Hislop at the Conference "Partial Differential Equations", Holzhau, Germany, July 1994.

1. Introduction: the Model and Results.

The existence of localized states for a two-dimensional non-interacting gas of electrons in a constant, transverse, magnetic field plays a role in the theory of the integer quantum Hall effect. We refer to the book of Prange and Girvin [4] for a detailed discussion and many references. We also refer to the article of Kunz [1] for a calculation of the quantum Hall conductivity for models of the type considered here and to the article of Bellissard [2] for a review of the approach using noncommutative geometry. In this article, we outline our proof [6] of the existence of localized states for the corresponding single-electron model. The general lines of the proof follow [5] but there are some interesting new aspects. First, we use a more general spectral averaging theorem in the proof of Wegner's estimate and to obtain a version of Kotani's trick. This method, based on the use of differential inequalities, was developed by us in [16]. Secondly, percolation theory is used to obtain the existence, with good probability, of an effective magnetic barrier which acts as a classically forbidden region in the sense that we obtain the exponential decay of the finite-area Green's function across that region. We consider a one-particle Hamiltonian which describes an electron in two-dimensions (x_1, x_2) subject to a

[1]Supported in part by CNRS.
[2]Supported in part by NSF grants INT 90-15895 and DMS 93-07438.

constant magnetic field of strength $B > 0$ in the perpendicular x_3-direction, and a random potential V_ω. The Hamiltonian H_ω has the form

$$H_\omega = (p - A)^2 + V_\omega, \tag{6}$$

on the Hilbert space $L^2(\mathbb{R}^2)$, where $p \equiv -i\nabla$, and the vector potential A is

$$A = \frac{B}{2}(x_2, -x_1), \tag{7}$$

so the magnetic field $B = \nabla \times A$ is in the x_3-direction. The random potential V_ω is Anderson-like having the form

$$V_\omega(x) = \sum_{i \in \mathbb{Z}^d} \lambda_i(\omega) u(x - i). \tag{8}$$

We make the following assumptions on the single-site potential u and the coupling constants $\{\lambda_i(\omega)\}$.

(V1) $u \geq 0$, $u \in C^2$, supp $u \subset B(0, \frac{1}{\sqrt{2}})$, and $\exists C_0 > 0$ and $r_0 > 0$ s.t. $u|B(0, r_0) > C_0$.

(V2) $\{\lambda_i(\omega)\}$ is an independent, identically distributed family of random variables with common distribution $g \in C^2([-M, M])$ for some $0 < M < \infty$, s.t. $\int \lambda g(\lambda) d\lambda = 0$ and $g(\lambda) > 0$ Lebesgue a.e. $\lambda \neq 0$.

We denote by $H_A \equiv (p - A)^2$, the Landau Hamiltonian. As is well-known, the spectrum of H_A consists of an increasing sequence $\{E_n(B)\}$ of eigenvalues, each of infinite multiplicity, given by

$$E_n(B) = (2n + 1)B, \; n = 0,1,2, \ldots \tag{9}$$

Note that $D(H_\omega) = D(H_A) \; \forall \; \omega \in \Omega$. We will call $E_n(B)$ the nth Landau level and denote by P_n the projection onto the corresponding subspace. The orthogonal projection is denoted by $Q_n \equiv 1 - P_n$. Let $M_0 \equiv \sup_{x,\omega} |V_\omega(x)| < \infty$. Then, $\sigma(H_\omega) \subset \bigcup_{n \geq 0} \sigma_n$, where $\sigma_n \equiv [E_n(B) - M_0, \, E_n(B) + M_0]$, which we call the nth Landau band. Because the model is ergodic with respect to the group of magnetic translations, standard results (cf. [12]) can be used to show that $\sigma(H_\omega)$ is a deterministic subset of $\bigcup_{n \geq 0} \sigma_n$.

We can now state our main result.

Theorem 1.1. Let H_ω be the family given in (6) with vector potential A satisfying (7), $B > 0$, and the random potential V_ω as in (8) and satisfying (V1)-(V2). Let

$$I_n(B) \equiv \left[E_n(B) + \mathcal{O}(B^{-1}), \, E_{n+1}(B) - \mathcal{O}(B^{-1}) \right].$$

There exists $B_0 \gg 0$ such that for $B > B_0$ and all $n = -1,0,1,2,\dots$,

$$\sigma(H_\omega) \cap I_n(B)$$

is pure point almost surely and the corresponding eigenfunctions decay exponentially. The integrated density of states is Lipschitz continuous away from $\sigma(H_A)$.

We note that the above theorem holds at arbitrary disorder. For large disorder, the techniques of [5] apply directly to show that, without the percolation estimates, $\sigma(H_\omega)$ is almost surely pure point in each Landau band.

As is clear from the Wegner estimate, Theorem 2.1, our method fails to give information about the integrated density of states at the Landau energies. However, we can improve the result if we make a stronger hypothesis in (V1) on supp u.

Corollary 1.1. If, in addition to the hypothesis of Theorem 1.1, we have $u \geq C_0 \chi_{\Lambda_1(0)}$, $C_0 > 0$, then the integrated density of states is Lipschitz continuous.

If the hypothesis of Corollary 1.1 does not hold, then a large portion of configuration space is unaffected by the potential. It is not, therefore, surprising that there is a discontinuity in the integrated density of states at the Landau energies as there is for the Landau Hamiltonian. A phenomenon of this type has been observed by Brezin et. al. [3] for a Poisson distribution of impurities at low energy. So we do not expect that the IDS is Lipschitz continuous at the Landau energies without a condition of the support of u which implies that the zero set of V_ω is in some sense "small".

We mention the work of W.M. Wang [13] who proved an asymptotic expansion in the semi-classical limit of large B for the density of states away from the Landau levels. We have also recently learned of related results of Wang [14] and of Pulé [20] on the existence of localized states.

We conclude this section with some simple observations on the Landau projections P_n. The projection P_n on the nth Landau level of H_A has a kernel given by

$$P_n(x,y) = Be^{-\frac{iB}{2} x \wedge y} p_n \left(B^{\frac{1}{2}}(x-y) \right), \tag{10}$$

where $p_n(x)$ is of the form

$$p_n(x) = \left\{ \text{n}^{\text{th}} \text{ degree polynomial in } x \right\} e^{-\frac{|x|^2}{2}}, \tag{11}$$

and independent of B. We will make repeated use of the following elementary lemma, the proof of which follows by direct calculation using the kernel (5)–(6).

Lemma 1.1. Let χ_1,χ_2 be functions of disjoint, compact support with $|\chi_i| \leq 1$ and let $\delta \equiv$ dist (supp χ_1, supp χ_2) > 0. Then,

(1) $\|\chi_1 P_n \chi_1\|_1 \leq C_n B |\text{supp } \chi_1|^2$;

(2) $||\chi_1 P_n \chi_1||_{HS} \leq C_n B|\text{supp } \chi_1|$;

(3) $||\chi_1 P_n \chi_2||_{HS} \leq C_n B|p_n\left(B^{\frac{1}{2}}\delta\right)|^{\frac{1}{2}} \{|\text{supp } \chi_1||\text{supp } \chi_2|\}^{\frac{1}{2}}$;

(4) $||\chi_1 P_n||_{HS} \leq C_n B$,

where C_n varies from line to line and depends only on χ_i and n, and HS denotes the Hilbert Schmidt norm. Note that $|p_n\left(B^{\frac{1}{2}}\delta\right)|^{\frac{1}{2}} \leq C_0 e^{-\epsilon B}$, for any $\epsilon > 0$ and B large enough.

2. Wegner Estimate.

We define local Hamiltonians as relatively compact perturbations of the Landau Hamiltonian $H_A = (p - A)^2$, as defined in section 1. Let $\Lambda \subset \mathbb{R}^2$ denote an open connected region in $\mathbb{R}^2$. We let $\Lambda_\ell(x)$ denote a square of side ℓ centered at $x \in \mathbb{R}^2$,

$$\Lambda_\ell(x) \equiv \{y \in \mathbb{R}^2 |\ |x_i - y_i| < \ell,\ i = 1{,}2\}.$$

Given $\Lambda \subset \mathbb{R}^2$, the local potential V_Λ is defined as follows. Freeze all $\lambda_j(\omega) \in \mathbb{Z}^2 \cap (\mathbb{R}^2 \backslash \Lambda)$ and consider $\widetilde{V}$ so obtained. This potential depends on the external, fixed coupling constants and on all $\lambda_i(\omega) \in \mathbb{Z}^2 \cap \Lambda$. We define $V_\Lambda \equiv \widetilde{V}|\Lambda$ and define $H_\Lambda \equiv H_A + V_\Lambda$ on $L^2(\mathbb{R}^2)$. These Hamiltonians are not independent of the external configurations but we will prove estimates uniform in the external random variables. We will use the conditional probability law

$$\overline{P}(A \cap B) \leq \overline{P}(A)\overline{P}(B), \tag{12}$$

where $\overline{P}$ is the probability conditioned on the external variables and A & B are any two events in Λ. Note that $\sigma_{\text{ess}}(H_\Lambda) = \sigma_{\text{ess}}(H_A)$, since V_Λ is relatively compact.

We prove the following form of Wegner's Theorem which, under the general condition of (V1), is not valid at the Landau energies $E_n(B)$.

Theorem 2.1. $\exists B_0 > 0$ and a constant $C_W > 0$ such that for all $B > B_0$ and for any $E \notin \sigma(H_A)$,

$$\mathbf{P}\{\text{dist } (\sigma(H_\Lambda), E) < \delta\} \leq C_W[\text{dist } (\sigma(H_A), E)]^{-2} ||g||_\infty \delta B|\Lambda|.$$

This theorem will follow from the properties of the spectral projectors for H_A as presented in Lemma 1.1 and a spectral averaging theorem. Since H_Λ depends analytically on the coupling constants λ_i, we need only a simple version which we state without proof (cf. [5], [15], [16]).

Lemma 2.1. Let $H_\lambda \equiv H_0 + \lambda u$, $\lambda \in \mathbb{R}$, be a self-adjoint family on $D(H_0)$ with $C_0 D^2 \leq u \leq M < \infty$. Let $E_\lambda(\cdot)$ be the spectral family for H_λ. For any $h \geq 0$, supp h compact, and $h \in L^\infty(\mathbb{R})$, and for any $L \subset \mathbb{R}$ measurable, we have

$$|| \int_{\mathbb{R}} h(\lambda) D E_\lambda(L) D d\lambda || \leq C_0^{-1} |L|\ ||h||_\infty .$$

For simplicity, we will work with the case $n = 0$ below, although the calculation is uniform in n. As can be easily checked, the calculations depend only on the difference between the energy E that we are considering and the nearest Landau energy $E_n(B)$ and hence is independent of n. To begin the proof of Theorem 2.1, we need a simple estimate. Let Δ be an interval in the first Landau band. Let E_Δ be the spectral projector for H_Λ associated with Δ.

Lemma 2.2. $||E_\Delta Q_0 E_\Delta|| \leq d_\Delta^{-\frac{1}{2}} \left(1 - (2d_\Delta)^{-1}|\Delta|\right)^{-\frac{1}{2}} M^{\frac{1}{2}}$, where $d_\Delta \equiv \text{dist } (\sigma(H_A)\backslash\{B\}, \Delta) = \mathcal{O}(\mathcal{B})$.

We refer to [6] for the simple proof. Note that since $d_\Delta = \mathcal{O}(B)$, we obtain,

$$||E_\Delta Q_0 E_\Delta|| = \mathcal{O}\left(B^{-\frac{1}{2}}\right). \tag{13}$$

Sketch of the Proof of Theorem 2.1. We can assume without less of generality that the closest point in $\sigma(H_A)$ to E is $E_0(B) = B$. Let $\Delta \subset B_0\backslash\{E_0(B)\}$ be a connected interval containing E and let E_Δ be the spectral projection for H_Λ and Δ. Recall from Chebyshev's inequality that

$$\mathbf{P}_\Lambda\{\text{dist } (\sigma(H_\Lambda), E) < \eta\} \leq \mathbb{E}_\Lambda(TrE_\Delta), \tag{14}$$

where $\mathbf{P}_\Lambda$ and $\mathbb{E}_\Lambda$ denote the probability and expectation with respect to the variables in $\Lambda \cap \mathbb{Z}^2$ and Tr denotes the trace on $L^2(\mathbb{R}^2)$. We first note that

$$TrE_\Delta \leq 2Tr(P_0 E_\Delta P_0). \tag{15}$$

This follows from the identity

$$TrE_\Delta = TrE_\Delta P_0 E_\Delta + TrE_\Delta Q_0 E_\Delta,$$

the bound

$$TrE_\Delta Q_0 E_\Delta \leq ||E_\Delta Q_0 E_\Delta||(TrE_\Delta),$$

and Lemma 2.2. Let us now suppose $\inf \Delta > B$ for definiteness. From (15), and positivity we obtain

$$\begin{aligned} TrE_\Delta P_0 E_\Delta &\leq Tr(E_\Delta(H_\Lambda - B)P_0(H_\Lambda - B)E_\Delta) \cdot \text{dist } (\Delta, B)^{-2} \\ &\leq Tr(P_0 V_\Lambda E_\Delta V_\Lambda P_0) \cdot \text{dist } (\Delta, B)^{-2}. \end{aligned} \tag{16}$$

Writing $V_\Lambda = \sum_i \lambda_i u_i$ for short, the trace in (16) is

$$\sum_{i,j} \lambda_i \lambda_j Tr(P_0 u_i E_\Delta u_j P_0), \tag{17}$$

where $i,j \in \Lambda \cap \mathbb{Z}^2$. Defining $A^{ij} \equiv u_i^{\frac{1}{2}} A u_j^{\frac{1}{2}}$ for any $A \in B(\mathcal{H})$, we have from (17),

$$\sum_{i,j} \lambda_i \lambda_j Tr\left(P_0^{ij} E_\Delta^{ij}\right). \tag{18}$$

We must estimate

$$\begin{aligned}\mathbb{E}_\Lambda\left(\sum_{i,j}\lambda_i\lambda_j Tr\left(P_0^{ij}E_\Delta^{ij}\right)\right) &\leq \sum_{i,j}\mathbb{E}_\Lambda\left(|\lambda_i\lambda_j|\ |Tr\left(P_0^{ij}E_\Delta^{ij}\right)|\right) \\ &\leq \tfrac{1}{2}M^2\sum_{i,j}||P_0^{ij}||_1\mathbb{E}_\Lambda\left\{||E_\Delta^{ii}||+||E_\Delta^{jj}||\right\}.\end{aligned} \tag{19}$$

Since $E_\Delta^{ii} \equiv u_i^{\frac{1}{2}}E_\Delta u_i^{\frac{1}{2}} \geq 0$ and self-adjoint, we have

$$\begin{aligned}\mathbb{E}_\Lambda\left(||E_\Delta^{ii}||\right) &\leq \sup_{\psi,\ ||\psi||=1}\left\{\mathbb{E}_\Lambda\left(\langle\psi,\ E_\Delta^{ii}\psi\rangle\right)\right\} \\ &\leq C_0^{-1}||g||_\infty|\Delta|,\end{aligned}$$

by Lemma 2.1. Consequently, (19) is bounded above by

$$\frac{1}{2}C_0^{-1}M^2||g||_\infty|\Delta|\sum_{i,j}||P_0^{ij}||_1. \tag{20}$$

To evaluate the trace norm, we divide the sum in (15) into three pieces: a) $i = j$, b) $i \neq j$ and $supp\ u_i \cap supp\ u_j \neq 0$, and c) $supp\ u_i \cap supp\ u_j = 0$. The sum in each case is estimated using the properties of the projection as given in Lemma 1.1. We finally obtain for all B large enough,

$$\mathbb{E}(TrE_\Delta) \leq C_W||g||_\infty|\Delta|B|\Lambda|,$$

where C_W depends on M^2, C_0^{-1}, and $|\text{supp } u|^2$. □.

The estimate of Theorem 2.1 suffices to prove the Lipschitz continuity of the integrated density of states away from the Landau levels, as stated in Theorem 1.1. We refer to [6] for the proof of Corollary 1.1 which shows how the additional hypothesis on supp u allows the improvement.

Percolation Theory and Decay Estimates.

In his study of the density of states for Landau Hamiltonians with a white noise potential, Wegner [11] introduced a one Landau band approximation. That is, the calculations are done in the subspace which is the range of the projection P_0. The effective one-band Hamiltonian obtained from (1) is $H_{eff} = B + P_0VP_0$. The coupling to the other levels is given by P_0VQ_0 and its adjoint. This is usually considered to be small in the case of bounded disorder and large magnetic field (see the estimates after (40)), so we will simply write V. Then, for an electron with energy E, the effective Hamiltonian is $E = B + V(x)$ in the first Landau band. The classical electron motion is constrained to these curves. If these equipotential lines remain in a bounded region of space, we expect that the wavefunction remains localized. Since the potential is random, we can ask what the probability is that these lines percolate to infinity. If they do not have a large probability to percolate to infinity, we would expect that the spectrum of this single-band approximation to H_ω is almost surely pure point, near E. The interband interaction should not change this picture significantly. We will prove that the Green's function decays exponentially in x and B through regions where $|V(x) + B - E| > a > 0$. This decay is sufficient to establish the a. s. pure point nature of $\sigma(H_\omega)$ near E.

We now sketch the technical estimates required to justify the one-Landau band approximation. We consider for simplicity the first Landau band $\sigma_0 \equiv [B - M_0, B + M_0]$, but all other bands can be analysed using the same techniques.

3.1. Percolation Estimates.

We first summarize the results of two-dimensional bond percolation theory and show that in annular regions between boxes of side ℓ and $\ell/3$ there exist closed, connected ribbons where the condition $|V(x)+B-E| > a > 0$ is satisfied, provided $E \neq B$, with a probability which converges exponentially to 1 as ℓ tends to infinity. Obviously, the existence of such a ribbon is equivalent to the impossibility for equipotential lines at energy E to percolate from the center of the box to its boundary.

Recall that $V_\omega(x) = \sum_{i\in\mathbb{Z}^2} \lambda_i(\omega)u(x-i)$, where the single-site potential $u \geq 0$ and has support inside a ball of radius $r_u < \frac{1}{\sqrt{2}}$. Consider a new square lattice $\Gamma \equiv e^{i\pi/4}\sqrt{2}\mathbb{Z}^2$. The midpoint of each bond of Γ is a site of $\mathbb{Z}^2$. We will denote by b_j the bond of Γ having $j \in \mathbb{Z}^2$ as it's midpoint. For definiteness, we assume $E \in (B, B + M_0)$. The other energy interval can be treated similarly.

Definition 3.1. The bond b_j of Γ is *occupied* if $\lambda_j(\omega) < \frac{E-B}{2}$. The probability $\mathbf{P}\left\{\lambda_j(\omega) < \frac{E-B}{2}\right\} \equiv p$ is the probability that b_j is occupied (p is independent of j by the *iid* assumption).

Let us assume that the bond b_j is occupied and consider the small region about b_j defined by,

$$\mathcal{R}_j \equiv \left\{x|\text{dist }(x, b_j) < \frac{1}{\sqrt{2}} - r_u \equiv r_1\right\}. \tag{21}$$

Obviously, $\mathcal{R}_j$ does not intersect the support of the other single-site potentials centered on $\mathbb{Z}^2\backslash\{j\}$ so that $V(x) = \lambda_j(\omega)u(x-j)\ \forall\, x \in \mathcal{R}_j$. Then, if b_j is occupied, one has $V(x) < \frac{E-B}{2}\ \forall\, x \in \mathcal{R}_j$ (recall that $\frac{E-B}{2} > 0$). We now assume that there is a closed circuit of occupied bonds $\mathcal{C} = \bigcup_{j\in\gamma} b_j$, $\gamma \subset \mathbb{Z}^2$ (i.e. a connected union occupied bonds). We call $\mathcal{R} \equiv \bigcup_{j\in\gamma} \mathcal{R}_j$ the closed *ribbon* associated with $\mathcal{C}$. For all $x \in \mathcal{R}$, we have $V(x) < \frac{E-B}{2}$. If we take $a \equiv \frac{E-B}{2}$, then

$$V(x) + B - E < -a\ \forall\, x \in \mathcal{R}. \tag{22}$$

The existence of a closed ribbon $\mathcal{R}$ so that V satisfies condition (22) is a consequence of the existence of a closed circuit $\mathcal{C}$ in Γ of occupied bonds. In order to estimate the probability that $\mathcal{C}$ exists, we use some standard results of percolation theory (see, e.g. [17] and [18]). On the lattice Γ defined above, it follows from Definition 3.1 that the probability that any bond is occupied is given by

$$p = \int_{-M}^{a} g(\lambda) d\lambda.$$

Under our assumptions on the density g, if $a > 0$ then $p > p_c = \frac{1}{2}$, and we are above the critical percolation threshold $p_c = \frac{1}{2}$. Note that when $E = B$, $a = 0$ so $p = \frac{1}{2} = p_c$, the critical probability.

Proposition 3.1. Assume (V1) and (V2). Let $r_u \equiv \frac{1}{2}$diam (supp u), $\ell > \sqrt{2}$, $E \in \sigma_0 \backslash \{B\}$, and $a > 0$. Then there exists constants $m > 0$, depending on p, and C_0 such that $\exists$ a ribbon $\mathcal{R}$ satisfying

$$\text{diam}\mathcal{R} \geq 2\left(\frac{1}{\sqrt{2}} - r_u\right); \tag{23}$$

$$\begin{aligned} &\text{dist } (\mathcal{R}, \partial r_{3\ell}), \text{ dist } (\mathcal{R}, \partial r_\ell) \geq \tfrac{1}{\sqrt{2}} + r_u; \\ &\mathcal{R} \subset a_\ell, \end{aligned} \tag{24}$$

and s.t.

$$V(x) + B - E > -a, \forall\, x \in \mathcal{R}, \tag{25}$$

with a probability larger than

$$1 - 12\, C_0\, e^{-m\ell}, \tag{26}$$

where

$$p \equiv \int_{-M}^{a} g(\lambda) d\lambda. \tag{27}$$

3.2. Decay Estimates.

The effective one Landau band Hamiltonian $B+V$ localizes electrons at energies E where the equipotential lines $E = V(x) + B$ don't percolate to infinity. The effect of the interband interaction is to induce some tunneling through the "Classically Forbidden" ribbon $\mathcal{R}$ of Proposition 3.1. As a consequence, instead of localization in the compact subsets of R^2 bounded by $\mathcal{R}$, one expects exponential decay of the Green's function in x and B across such ribbons $\mathcal{R}$. Such an estimate is the starting point of the inductive, multi-scale analysis described in section 4. By the geometric resolvent equation, we show there that it suffices to consider the following ideal situation, where for some $a > 0$,

$$V(x) + B - E < -a, \quad \forall\, x \in R^2, \tag{28}$$

or, alternately,

$$V(x) + B - E > a, \quad \forall\, x \in R^2. \tag{29}$$

A condition such as (23) with $E > B$ is satisfied, with a probability given in Proposition 3.1, by a smoothing of the potential $V_{\mathcal{R}}$ defined as

$$V_{\mathcal{R}}(x) = \begin{cases} V(x) & x \in \mathcal{R} \\ 0 & x \in R^2 \setminus \mathcal{R}. \end{cases} \tag{30}$$

Here we obtain decay estimates on

$$H = H_A + V$$

with V having compact support with non-empty interior and satisfying (23) or (24). The strategy is to exploit a type of translation-analyticity of H and employ a Combes–Thomas type argument.

Let $\mathcal{O}$ be an open, bounded, connected set in R^2 with smooth boundary and define $\rho(x) = \text{dist}\,(x, \mathcal{O})$. By a suitable smoothing, we can assume that $\rho \in C^2$ and that $||\nabla\rho||_\infty < C_0$ and $||\Delta\rho||_\infty < C_1$, for constants $C_0,\ C_1 > 0$ depending only on $\mathcal{O}$. We consider one-parameter families of operators defined for $\alpha \in R$ as

$$H_A(\alpha) \equiv e^{i\alpha\rho} H_A e^{-i\alpha\rho}; \tag{31}$$

$$H(\alpha) \equiv H_A(\alpha) + V; \tag{32}$$

$$P(\alpha) \equiv e^{i\alpha\rho} P e^{-i\alpha\rho}, \quad \text{etc.}, \tag{33}$$

and similarly for the local Hamiltonian $H_\Lambda(\alpha) \equiv H_A(\alpha) + V_\Lambda$. For $\alpha \in R$, these families are unitarity equivalent with the $\alpha = 0$ operators.

Lemma 3.1. The family $H(\alpha)$ *(and similarly for* $H_\Lambda(\alpha)$*)*, $\alpha \in R$, has an analytic continuation into the strip

$$S \equiv \{\alpha \in \mathbb{C} \mid |\text{Im}\,\alpha| < \eta_\rho B^{1/2}\}, \tag{34}$$

as a type A analytic family with domain $D(H)$. The positive constant η_ρ depends only on the distance function ρ. Furthermore, in this strip S, one has $P(\alpha)^2 = P(\alpha)$ and for some constant C_1 independent of α,

$$||P(\alpha)|| < C_1 \tag{35}$$

and

$$||Q(\alpha)(H_A(\alpha) - z)^{-1}|| < C_1 B^{-1}, \quad \textit{if}\ \text{dist}\,(z, B) \le B. \tag{36}$$

Sketch of the Proof. The analyticity of the family $H_A(\alpha)$ can be seen by analyzing the formula, valid for $\alpha \in R$,

$$H_A(\alpha) = H_A + \alpha^2|\nabla\rho|^2 + i\alpha\Delta\rho - 2\alpha\nabla\rho \cdot (p - A). \tag{37}$$

Using standard resolvent formulas, we can estimate the resolvent of $H_A(\alpha)$ by

$$||(H_A(\alpha) - z)^{-1}|| < C_0 B^{-1}, \tag{38}$$

for some constant C_0 uniform in $\alpha \in S$ and $z \in C(B) \equiv \{z| \, |z - B| = B\}$, circle of radius B centered at B. It follows that the family $P(\alpha)$, $\alpha \in R$, has an analytic continuation in S given by the contour integral

$$P(\alpha) = \frac{-1}{2\pi i} \int_{C(B)} (H_A(\alpha) - z)^{-1} \, dz. \tag{39}$$

The boundedness of $P(\alpha)$ and the decay of $Q(\alpha)$ follow from this representation and (33). □

We now come to the main estimate of this section.

Theorem 3.1. Assume that (V, E, B) satisfy (23) or (24) for some $a > 0$ and $E \in \sigma_0 \setminus \{B\}$. Furthermore, assume that supp$V$ is compact with non-empty interior. There exists constants $C_2 \leq \eta_\rho$, C_3, and B_1, depending only on $M_0 \equiv ||V||_\infty$, $||\nabla\rho||_\infty, and$ $||\nabla V||_\infty$, such that if we define $\gamma \equiv C_2 \min\{B^{1/2}, aB\}$, and u is a solution of

$$(H_A + V - z)u = v, \quad z \equiv E + i\epsilon, \quad \epsilon > 0, \quad E > 0, \tag{40}$$

for some $v \in D(e^{\gamma\rho})$, then for $B > B_1$, $\forall \alpha \in \mathbb{C}$, $|\operatorname{Im}\alpha| < \gamma$, we have

$$u \in D(e^{i\rho\alpha}), \tag{41}$$

$$||e^{\alpha\rho} P u|| \leq C_3 a^{-1} ||e^{\alpha\rho} v||, \tag{42}$$

and

$$||e^{\alpha\rho} Q u|| \leq C_3 B^{-1} ||e^{\alpha\rho} v||. \tag{43}$$

Sketch of the Proof. We obtain *a priori* estimates on the solution u of (35) using the family of complex transformations introduced in above. Projecting the equation (35) for u along $P(\alpha)$ gives

$$(B + V - z)(Pu)(\alpha) = (Pv)(\alpha) + ([QVP - PVQ]u)(\alpha). \tag{44}$$

Taking the scalar product of (39) with $(Pu)(\alpha)$ results in the inequality,

$$\begin{aligned} a||(Pu)(\alpha)||^2 \; &\leq ||(Pu)(\alpha)|| \; ||(Pv)(\alpha)|| \\ &+\{||(P^*Q)(\alpha)|| \; ||(QVP)(\alpha)||\}||(Pu)(\alpha)||^2 \\ &+||(PVQ)(\alpha)|| \; ||(Pu)(\alpha)|| \; ||(Qu)(\alpha)||. \end{aligned} \tag{45}$$

We need two technical estimates proved in [6]: for B large enough,

$$||(QVP)(\alpha)|| \leq C_4 B^{-1/2},$$

and

$$||(P^*Q)(\alpha)|| \leq C_5 \; |Im\alpha| \; B^{-1/2}.$$

With these estimates, we obtain from (40),

$$\begin{aligned}(a - C_6\gamma B^{-1})||(Pu)(\alpha)||^2 \;\; &\leq ||(Pu)(\alpha)|| \; ||(Pv)(\alpha)|| \\ &+C_7B^{-1/2}||(Pu)(\alpha)|| \; ||(Qu)(\alpha)||,\end{aligned} \tag{46}$$

where the constants C_6 and C_7 depend only on $||V||_\infty$, $||\nabla V||_\infty$, and $||\nabla\rho||_\infty$. To estimate $||(Qu)(\alpha)||$, it follows from the resolvent equation and the equation

$$u(\alpha) = (H(\alpha) - z)^{-1}v(\alpha) \equiv ((H-z)^{-1}v)(\alpha),$$

that

$$\begin{aligned}||(Qu)(\alpha)|| \;\; &\leq ||(Q(H_A - z)^{-1}V)(\alpha)|| + ||\{Q(H_A - z)^{-1}QV(Q+P)u\}(\alpha)|| \\ &\leq C_1B^{-1}||V(\alpha)|| + C_1B^{-1}M||(Qu)(\alpha)|| \\ &+C_1B^{-1}M||(QVPu)(\alpha)||,\end{aligned} \tag{47}$$

with $M \equiv ||V||_\infty < \infty$. Using the estimate on QVP given above and taking $B > 2MC_1$, we obtain,

$$||(Qu)(\alpha)|| \leq 2C_1B^{-1}||v(\alpha)|| + C_7B^{-3/2}||(Pu)(\alpha)||, \tag{48}$$

where $C_7 \equiv 2MC_1C_2$. These estimates suffice to prove the result. □

This theorem on the decay of projections yields the following corollary on the decay of the Green's function which is our main result.

Corollary 3.1. Let $\mathcal{O}$ be an open, connected, bounded subset of R^2 with smooth boundary and suppose $\mathcal{E} \subset R^2 \setminus \mathcal{O}$. Let $E \in \sigma_0 \setminus \{B\}$ and assume that (B, E, V) satisfy (23) or (24) for some $a > 0$. Let χ_X, $X = \mathcal{O}$ and $\mathcal{E}$, be bounded functions with support in X and s.t. $||\chi_X||_\infty \leq 1$. Then,

$$\sup_{\epsilon \neq 0} ||\chi_\mathcal{E}(H_A + V - E - i\epsilon)^{-1}\chi_\mathcal{O}|| \leq C \max\{a^{-1}, B^{-1}\}e^{-\gamma d}, \tag{49}$$

where C and γ are as in Theorem 3.1 and $d \equiv \text{dist}\,(\mathcal{O}, \mathcal{E})$.

4. Green's Function Estimates.

We will show below that Corollary 3.1 implies hypothesis [H1](γ_0, ℓ_0) of [5]. This hypothesis, along with Wegner's estimate, Theorem 2.1, are the main starting points of the multi-scale analysis described in [5]. The goal of this analysis is to verify the main assumption (A2) of Theorem 3.2 of [5], which gives sufficient conditions for pure point spectrum in an interval and exponential decay of eigenfunctions. A version of Kotani's trick, necessary to control the singular continuous spectrum for the models studied here, follows from Lemma 2.1.

Let us recall that we first reduce the problem to one involving a one-parameter family of operators. For the family H_ω in (1.1), we consider variations $\omega' \in \Omega$ for which only λ_0 changes. For ω fixed and $\lambda \equiv \lambda_0(\omega') - \lambda_0(\omega)$, we have

$$H_{\omega'} = H_\omega + \lambda u \equiv H_0 + \lambda u = H_\lambda, \tag{50}$$

The main condition (A2) of [5] on the family H_λ is as follows. Let $I \subset R$ be an interval of energies on which we hope to prove a. s. localization. We then require that $\exists I_0 \subset I$ with $|I_0| = |I|$ s.t. $\forall\ E \in I_0$,

$$\sup_{\epsilon \neq 0} \|R_0(E + i\epsilon)u^{\frac{1}{2}}\| < \infty. \tag{51}$$

Multi-scale analysis is used to verify this condition for a.e. ω (recall $H_0 = H_\omega(A)$) and a.e. energy E in appropriate subsets of each Landau band. The starting point for this analysis is Wegner's estimate and an estimate on the decay of the finite-area Green's function with suitable probability. This estimate, called condition [H1] (γ_0, ℓ_0) in [5], is the following

[H1] (γ_0, ℓ_0) For some $\gamma_0 > 0$, $\ell_0 > 1$, $\exists \xi > 4$ s.t.

$$\mathbf{P}\left\{\sup_{\epsilon > 0} \|W(\chi_{\ell,\delta})R_{\Lambda_{\ell_0}}(E + i\epsilon)\chi_{\ell_0/3}\| \leq e^{-\gamma_0 \ell_0}\right\} \geq 1 - \ell_0^{-\xi}.$$

The main theorem, which we recall in the present context, concerning condition (51), is the following.

Theorem 4.1. (Theorem 2.3 of [5]). Let $\gamma_0 > 0$. $\exists$ a minimum length scale $\ell^\star \equiv \ell^\star(\gamma_0, C_W)$, s.t. : if [H1]$(\gamma_0, \ell_0)$ holds at energy E for $\ell_0 > \ell^\star$, then for $\mathbf{P}$ - a.e. ω $\exists$ a finite constant $d_\omega > 0$ s.t.

$$\sup_{\epsilon \neq 0} \|(H_\omega - E - i\epsilon)^{-1} u^{\frac{1}{2}}\| < d_\omega \delta(u),$$

where $\delta(u)$ depends only on u.

We prove below that [H1](γ_0, ℓ_0) holds at each energy in $[B - M_0, B - \mathcal{O}(B^{-1})] \cup I_0(B) \cap \sigma_0$ with a suitable γ_0 (see Proposition 4.1) and for all ℓ_0 large enough provided B is large. By Theorem 3.2 of [5], this theorem, and the compactness result proved in [6], imply that H_λ in (50) has pure point spectrum in this set for a.e. λ. By the probabilistic arguments of [5], we conclude that H_ω has only pure point spectrum in this set for $\mathbf{P}$ - a.e. ω.

The second main theorem of [5] which we recall here allows us to prove exponential decay of the eigenfunctions.

Theorem 4.2. (Theorem 2.4 of [5]). Let χ_x be the characteristic functions of a unit cube centered at $x \in \mathbb{R}^2$. Under the assumptions of Theorem 4.1, for **P**-a.e. ω $\exists$ a finite constant $d_\omega > 0$ s.t. for all x, $||x||$ large enough,

$$\sup_{\epsilon>0} ||\chi_x(H_\omega - E - i\epsilon)^{-1} u^{\frac{1}{2}}|| \leq d_\omega e^{-\gamma_1 ||x||},$$

where $\gamma_1 \equiv (1/6\sqrt{2})\gamma_0$, γ_0 as in Theorem 4.1.

Let us remark that for our problem, $\gamma_0 \sim B^\sigma$, for some $\sigma > 0$ so there is exponential decay in the B-field also. We now turn to the ideas of the proof of $[\mathrm{H1}](\gamma_0, \ell_0)$.

To begin, we introduce some geometry. In this section, we work with subregions of the lattice $\Gamma \equiv e^{i\pi/4}\sqrt{2}\mathbb{Z}^2$, introduced in section 3, rather than in $\mathbb{Z}^2$. Recall that there is a 1:1 correspondence between bonds $b_j \in \Gamma$ and vertices $j \in \mathbb{Z}^2$. We arbitrarily choose a vertex of Γ as the origin and define boxes $\Lambda_\ell \subset \Gamma$ relative to this point,

$$\Lambda_\ell \equiv \left\{x \in \mathbb{R}^2 | |x_i| \leq \ell/2 \;\; for \;\; i = 1, 2\right\}.$$

For convenience, we fix points so the bond b_0 has one of its ends at $0 \in \Gamma$. For any $\delta > 0$, consider $\Lambda_{\ell,\delta} \equiv \{x \in \Lambda_\ell | \mathrm{dist}\,(x, \partial\Lambda_\ell) < \delta\}$. Let $\chi_{\ell,\delta}$ be the C^2-function which satisfies $\chi_{\ell,\delta} \geq 0$, $|\nabla\chi_{\ell,\delta}| \subset \Lambda_\ell \backslash \Lambda_{\ell,\delta}$ and $\chi_{\ell,\delta}|\Lambda_{\ell,\delta} = 1$. Let $W(\chi) \equiv [\chi, H_A]$, for any $\chi \in C^2$. Let $V_\Lambda \equiv V|\Lambda$, $\Lambda \subset \mathbb{R}^2$ and $H_\Lambda \equiv H_A + V_\Lambda$, as in section 3.

The main result of this section is to prove that for all $E \in [B - M_0, B - \mathcal{O}(B^{-1})] \cup I_0(B) \cap \sigma_0$ and for all ℓ_0 sufficiently large, the condition $[\mathrm{H1}](\gamma_0, \ell_0)$ holds. The proof of this fact relies on the Wegner estimate, Theorem 2.1, and the existence of a ribbon implying the decay estimates of Corollary 3.1.

Proposition 4.1. Let χ_2 be any function, $||\chi_2||_\infty \leq 1$, supported on $\Lambda_\ell \cap \mathrm{Ext}\mathcal{R}$, where $\mathrm{Ext}\mathcal{R} \equiv \left\{x \in \mathbb{R}^2 | \lambda x \notin \mathcal{R} \;\forall\; \lambda \geq 1\right\}$, so that, in particular, supp $\chi_2 \cap \mathcal{R} = \emptyset$. For any $E \in \sigma_0 \backslash \{B\}$, $\delta > 0$ and $a > 0$, we have

$$\begin{aligned} \sup_{\epsilon\neq 0} ||\chi_2 R_{\Lambda_\ell}(E + i\epsilon)\chi_{\ell/3}|| \leq \;& Ce^{-\gamma d} \max\left\{a^{-1}, B^{-1}\right\} \cdot \max\left\{\delta^{-1}, \right. \\ & \left. (2M + |E|)\delta^{-2}\right\}, \end{aligned} \tag{52}$$

where C and γ are as in Theorem 3.1 and $d \equiv r_1 - 3\epsilon$ $\;(r_1 \equiv \mathrm{diam}\mathcal{R})$, and for any $\epsilon > 0$, with a probability larger than

$$1 - \left\{Ce^{-M\ell} + C_W[\mathrm{dist}\;(E,B)]^{-2} ||g||_\infty \delta B \ell^2\right\}. \tag{53}$$

In particular, for $\chi_{\ell,\delta}$ defined above and $E \in \sigma_0$ with $a = \frac{E-B}{2} = \mathcal{O}\left(B^{-1+\sigma}\right)$, any $\sigma > 0$, we have that for any $\ell_0 > \sqrt{2}$ and large enough, and any $\xi > 4$ $\;\exists B(\ell_0) > 0$ s.t. $\forall\; B > B(\ell_0)$, [H1] (γ_0, ℓ_0) holds for some $\gamma_0 > \frac{1}{8}d > 0$.

Sketch of the Proof. We will outline the proof of (47). By Corollary 3.1, $\exists$ B_0 s.t. $B > B_0$ implies $\exists$ a ribbon $\mathcal{R} \subset \Lambda_\ell \backslash \Lambda_{\ell/3}$ (with a probability given by (26)) satisfying

$$\mathrm{dist}\;(\mathcal{R}, \partial\Lambda_\ell), \;\; \text{and} \;\; \mathrm{dist}(\mathcal{R}, \partial\Lambda_{\ell/3}) > \frac{1}{\sqrt{2}} + r_u > 0, \tag{54}$$

and

$$r_1 \equiv \text{diam}\mathcal{R} > 2\left(\frac{1}{\sqrt{2}} - r_u\right), \tag{55}$$

and such that

$$V(x) + B - E > -a \quad \forall\, x \in \mathcal{R},\, a = \frac{E-B}{2}. \tag{56}$$

(We assume $E > B$; similar arguments hold for $E < B$). We will work with smoothed characteristic functions and omit the details in the construction of such functions.

Let $\chi_{\mathcal{R}} \geq 0$ be chosen such that $\chi_{\mathcal{R}} = 1$ on $\mathcal{R}$, except in a small neighborhood of the boundary, which we call $\mathcal{R}_\epsilon$, and such that supp $|\nabla \chi_{\mathcal{R}}| \subset \mathcal{R}_\epsilon$. For any $\epsilon > 0$, $3\epsilon \ll r_1$, we define $\mathcal{C}_\epsilon$ to be a small region near the center of the ribbon $\mathcal{R}$ such that dist $(\mathcal{R}_\epsilon, \mathcal{C}_\epsilon) \geq r_1 - 3\epsilon$. Let χ_1 satisfy $\chi_1 | \Lambda_{\ell/3} = 1$ and supp $|\nabla \chi_1| \subset \mathcal{C}_\epsilon$. By simple commutation, we have (with χ_2 as in the proposition),

$$\begin{aligned} \chi_2 R_{\Lambda_\ell}(E + i\epsilon)\chi_{\ell/3} &= \chi_2 R_{\Lambda_\ell} \chi_1 \chi_{\ell/3} \\ &= \chi_2 R_{\Lambda_\ell} W(\chi_1) R_{\Lambda_\ell} \chi_{\ell/3} \\ &= \chi_2 R_{\Lambda_\ell} \chi_{\mathcal{R}} W(\chi_1) R_{\Lambda_\ell} \chi_{\ell/3}. \end{aligned} \tag{57}$$

Next, denote by $R_{\mathcal{R}}$ the resolvent of $H_{\mathcal{R}} \equiv H_A + V_{\mathcal{R}}$, where $V_{\mathcal{R}}$ is defined in (25). The geometric resolvent equation relating R_{Λ_ℓ} and $R_{\mathcal{R}}$ is

$$R_{\Lambda_\ell} \chi_{\mathcal{R}} = \chi_{\mathcal{R}} R_{\mathcal{R}} + R_{\Lambda_\ell} W(\chi_{\mathcal{R}}) R_{\mathcal{R}}. \tag{58}$$

Substituting (58) into (57) and noting that $\chi_2 \chi_{\mathcal{R}} = 0$, we obtain

$$\chi_2 R_{\Lambda_\ell} \chi_{\ell/3} = \chi_2 R_{\Lambda_\ell} W(\chi_{\mathcal{R}}) R_{\mathcal{R}} W(\chi_1) R_{\Lambda_\ell} \chi_{\ell/3}. \tag{59}$$

Note that from (55) and the choice of $\chi_{\mathcal{R}}$ and χ_1, we obtain that

$$\text{dist (supp } W(\chi_{\mathcal{R}}), \text{ supp } W(\chi_1)) \geq r_1 - 3\epsilon. \tag{60}$$

The main idea in estimating (54) is to apply Wegner's estimate, Theorem 2.1, to control the two R_{Λ_ℓ} factors in (59), and the decay estimate, Corollary 3.1, to control the factor $R_{\mathcal{R}}$, which is possible due to the localization of $W(\chi_{\mathcal{R}})$ and $W(\chi_1)$ and (60).

To estimate the $R_{\mathcal{R}}(E + i\epsilon)$ contribution to (54), we use Corollary 3.1 with $\mathcal{O} \equiv \mathcal{C}_\epsilon$ and $\mathcal{E} = \mathcal{R}_\epsilon$. Let χ_X, $X = \mathcal{O}$ and $\mathcal{E}$, be a characteristic function on these sets. Then $W(\chi_{\mathcal{R}})\chi_{\mathcal{E}} = W(\chi_{\mathcal{R}})$ and $\chi_{\mathcal{O}} W(\chi_1) = W(\chi_1)$. Inserting these localization functions into (59), we obtain from Corollary 3.1,

$$||\chi_{\mathcal{E}} R_{\mathcal{R}}(E + i\epsilon)\chi_{\mathcal{O}}|| \leq C \max\left\{a^{-1}, B^{-1}\right\} e^{-\gamma d}, \tag{61}$$

with probability larger than

$$1 - Ce^{-m\ell}. \tag{62}$$

The factor d satisfies

$$d \geq r_1 - 3\epsilon, \tag{63}$$

where r_1 is given in (50). Next, we turn to

$$W(\chi_1) R_\ell(E + i\epsilon) \chi_{\ell/3}, \tag{64}$$

and

$$\chi_2 R_\ell(E + i\epsilon) W(\chi_{\mathcal{R}}), \tag{65}$$

where we write R_ℓ for R_{Λ_ℓ} for short. Ignoring the fact that W contains a gradient, we will bound R_ℓ by Wegner's estimate. From Theorem 2.1, we have for any $\delta > 0$,

$$||R_\ell(E + i\epsilon)|| < \delta^{-1}, \tag{66}$$

with probability larger than

$$1 - C_W [\text{dist } (E, B)]^{-2} ||g||_\infty \delta B \ell^2. \tag{67}$$

Using the estimate $P(A \cap B) \geq P(A) + P(B) - 1$, and the above estimates, we find

$$\begin{aligned} ||\chi_2 R_\ell(E + i\epsilon) \chi_{\ell/3}|| \leq\ & 2C \max\left\{a^{-1}, B^{-1}\right\} \cdot \max\left\{\delta^{-1},\right. \\ & \left.(2M + |E|)\delta^{-2}\right\} \cdot e^{-\gamma d}, \end{aligned} \tag{68}$$

with probability at least

$$1 - \left\{C e^{-m\ell} + C_W [\text{dist } (E, B)]^{-2} ||g||_\infty \delta B \ell^2\right\}. \tag{69}$$

This finishes the sketch of the proof of the first part of the proposition. □

References

[1] H. Kunz, The quantum Hall effect for electrons in a random potential, Commun. Math. Phys. **112**, 121–145 (1987).

[2] J. Bellissard, Ordinary quantum Hall effect and non-commutative cohomology, in Localization in Disordered Systems, W. Weller and P. Zieche, eds., Leipzig: Teubner 1988.

[3] E. Brezin, D. J. Gross, C. Itzykson, Nuclear Phys. B **235**, 24 (1984).

[4] R. E. Prange and S. M. Girvin (eds.), The Quantum Hall Effect, Graduate Texts in Contemporary Physics, New York: Springer-Verlag (1987).

[5] J. M. Combes and P. D. Hislop, Localization for some continuous, random Hamiltonians in d-dimensions, J. Funct. Anal. **124**, 149–180 (1994).

[6] J. M. Combes and P. D. Hislop, Landau Hamiltonians with random potentials: localization and the density of states, submitted to Commun. Math. Phys.

[7] J. Howland, Perturbation theory of dense point spectra, J. Funct. Anal. **74** (1987), 52–80.

[8] B. Simon and T. Wolff, Singular continuous spectrum under rank one perturbation and localization for random Hamiltonians. Comm. Pure Appl. Math. **39**, (1986), 75–90.

[9] H. von Dreifus and A. Klein, A new proof of localization in the Anderson tight binding model. Commun. Math. Phys. **124**, (1989), 285–299.

[10] T. Spencer, Localization for random and quasi-periodic potentials , J. Statist. Phys. **51** , 1009–1019 (1988).

[11] F. Wegner, Z. Physik B **51**, 279 (1983).

[12] R. Carmona and J. Lacroix, Spectral theory of random Schrödinger operators, Birkhäuser, Boston, 1990.

[13] W. M. Wang, Développement asymptotique de la densité d'états pour l'opérateur de Schrödinger aléatoire avec champ magnetique, Séminaire des Equations aux Dérivées Partielles 1992–1993, Ecole Polytechnique, Exposé XVIII (1993).

[14] W. M. Wang, private communication, July 1994.

[15] S. Kotani and B. Simon, Localization in general one-dimensional systems. II, Commun. Math. Phys. **112**, (1987), 103–120.

[16] J. M. Combes, P. D. Hislop and E. Mourre, Spectral Averaging, Perturbation of Singular Spectrum, and Localization, preprint 1994, submitted to Trans. Amer. Math. Soc.

[17] G. Grimmett, Percolation, New York: Springer-verlag, 1989.

[18] J. L. Chayes and L. Chayes, Percolation and Random Media, in Critical Phenomena, Random Systems, and Gauge Theories, Proc. les Houches Summer School, Session XLIII, 1000–1142 (1984).

[19] T. Kato, Perturbation theory for linear operators, Second Edition, Berlin: Springer-Verlag, 1980.

[20] J. Pulé, communication at the IAMP satellite conference on Disordered Systems, Paris, July 1994.

J. M. Combes
Département de Mathématiques, Université de Toulon et du Var
83130 La Garde, France.

P. D. Hislop
Mathematics Department, University of Kentucky
Lexington KY 40506-0027, USA.

Operator Theory:
Advances and Applications, Vol. 78

Some Problems of Calculus of Variations in Infinite Dimensions[1]

Yu. L. Daletskiĭ, V. R. Steblovskaya

Introduction.

In [9,10] we try to generalize some basic results of classical calculus of variations to the infinite-dimensional case. We consider an extremal problem for the functional of the form

$$F(u,v) = \int_X \Phi\big(u(x),v(x),x\big)\,\mu(dx) \tag{0.1}$$

where $X = H_- \supset H_0 \supset H_+$ is an rigged Hilbert space with Hilbert-Schmidt imbeddings; μ is a finite non-negative Borel measure on X; maps $u : X \to Y_1$ and $v : X \to Y_2$ (where Y_j $(j = 1,2)$ are Hilbert spaces) belong to Hilbert spaces B_1 and B_2 respectively;

$$B_k = \mathcal{L}_2(X,Y_k,\mu) =$$
$$= \left\{ y : X \to Y_k;\, \|y\|^2_{B_k} = \int_X \big(y(x),y(x)\big)_Y\,\mu(dx) < \infty \right\} \quad (k = 1,2);$$

$v(x) = A(u)(x)$, where $A : B_1 \to B_2$ is a differentiable map.

We suppose the measure μ to be smooth one, i. e. it allows integration by parts (see section 1). Trying to extend classical field theory to the infinite dimensional case one necessarily comes to considering the variational problems related to smooth measures. On the other hand some problems of stochastic optimal control lead to the same considerations.

For the functional (0.1) the analogues of the Euler-Lagrange equation, the Noether theorem, and the canonical Hamilton system are obtained.

In this work we briefly recall the necessary information related to the Euler-Lagrange equation and to the canonical Hamilton system and then concentrate on examples.

We consider the examples in which A is a differential operator, in particular, $A(u)(t,y) = \left(\frac{d}{dt},\nabla_y\right) u(t,y)$ and more special $A(u)(x) = \langle z(x),\nabla\rangle u(x)$, where $\langle z(x),\nabla\rangle$ is a differentiation along the vector field $z(x)$. In the first example with the special choice of Φ the Euler-Lagrange equation turns into the hyperbolic partial differential equation associated with the Dirichlet operator. In the second one we solve the infinite dimensional Hamilton system by the method of characteristics.

[1]The research described in this publication was made possible in part by Grant No U44000 from the International Science Foundation.

1. Some facts from the theory of smooth measures.

Gaussian measures play a very important role in analysis, but this class is not invariant under smooth non-linear transformations. That's why a larger class of smooth (differentiable) measures possessing the mentioned property and containing the class of Gaussian ones was introduced in [2,11]. Later the theory of smooth measures was developed by many mathematicians (for references see [5]).

Let's introduce some definitions and facts we use in the sequel. Consider an rigged Hilbert space $X = H_- \supset H_0 \supset H_+$ with Hilbert-Schmidt imbedding operators. By $\mathcal{M}(X)$ we denote the Banach space of real-valued bounded measures on X. By $\mathcal{M}_0(X) \subset \mathcal{M}(X)$ denote a class of measures such that for each direction $h \in H_+$ the *integration by parts formula*

$$\int_X \langle h,\nabla\rangle\phi(x)\,\mu(dx) = -\int_X \phi(x)\rho_\mu(h;x)\,\mu(dx) \qquad (\phi \in C_1(H_-,\mathbb{R}^1)) \tag{1.1}$$

holds (we suppose that the integral in the righthand side of (1.1) exists). Here the function $\rho_\mu(h;x)$ is called the *logarithmic derivative of μ along h* (see [3]), $\langle h,\nabla\rangle\phi(x) = \langle h,\phi'(x)\rangle$ is the derivative of ϕ along h, brackets $\langle\,,\rangle$ mean pairing between H_+ and H_-.

It follows from (1.1) that $\rho_\mu(h;x)$ depends on h linearly (a.e.). Suppose that there exists a μ-measurable locally bounded map $\lambda : H_- \to H_-$ such that $\rho_\mu(h;x) = \langle\lambda(x),h\rangle$. The function $\lambda(x)$ is called the *logarithmic gradient of μ*.

For example, a canonical Gaussian measure with characteristic functional $\chi(\theta) = \exp(-\frac{1}{2}\|\theta\|^2)$ belongs to $\mathcal{M}_0(X)$ with $\lambda(x) = -x$.

Then pass to the concept of a *logarithmic derivative of measure along a vector field.*

Let H_1, H_2 be Hilbert spaces, G be a set from H_1. Denote by $C_k(G,H_1,H_2)$ a function space such that the following conditions hold:

(i) any function $f \in C_k(G,H_1,H_2)$ takes G to H_2;

(ii) f is differentiable along any direction $h \in H_1$ up to order k inclusively;

(iii) f is bounded on G together with its derivatives $f^{(j)}(x)$ $(j = 1,\dots,k)$, where $f^{(j)}(x) \stackrel{\text{def}}{=} (f^{(j-1)}(x))'$ is a j-linear map from H_1 to H_2.

It's shown in [3] that for $\mu \in \mathcal{M}_0(X)$ and $z \in C_1(H_-,H_-,H_+)$ the *integration by parts formula*

$$\int_X \langle z(x),\nabla\rangle\phi(x)\,\mu(dx) = -\int_X \phi(x)\rho_\mu(z;x)\,\mu(dx) \tag{1.2}$$

holds. Here

$$\rho_\mu(z;x) = \langle\lambda(x),z(x)\rangle + \operatorname{div} z(x) = \langle\lambda(x) + \nabla, z(x)\rangle \cdot 1$$

is a *logarithmic derivative of μ along z.*

It's possible to extend the concept of logarithmic derivative of measure to a larger class of vector fields. First we introduce some notations.

By $\mathcal{M}_n^m(H_-,H_0) \subset \mathcal{M}_0(H_-)$ we denote the class of positive measures with logarithmic gradient $\lambda(x)$ being differentiable along H_0 up to order n inclusively and

(i) $\lambda'(x)$ is bounded in H_0;

(ii) $\lambda^{(k)}(x)$ $(k > 1)$ is a k-linear Hilbert-Schmidt operator on H_0;

(iii) $\int_X \|\lambda'(x)\|^m \,\mu(dx) < \infty$;

(iv) $\int_X \sigma_2^m\big(\lambda^{(k)}(x)\big)\,\mu(dx) < \infty$ $(k = 2,\dots,n)$, where $\sigma_2\big(\lambda^{(k)}(x)\big)$ is the Hilbert-Schmidt norm of the k-linear map $\lambda^{(k)}(x)$.

By $W_n^m(H_-,H_0,\mu)$ denote the Banach space of functions $z : H_- \to H_0$ differentiable along H_0, the completion of $C_n(H_-,H_-,H_+)$ with respect to the norm

$$\sigma_{n,m}(z) = \sum_{j=0}^{n}\left\{\int_X \sigma_2^m\big(z^{(j)}(x)\big)\,\mu(dx)\right\}^{\frac{1}{m}}$$

It's proved in [3] that if $\mu \in \mathcal{M}_1^2(H_-,H_0)$ and $z \in C_1(H_-,H_0,H_+)$, then $\rho_\mu(z;x)$ exists and it is square integrable. In addition the following relations hold:

$$\int_X |\rho_\mu(z;x)|^2\,\mu(dx) = \int \left\{\mathrm{Tr} z'(x)^2 - \langle \lambda'(x)\cdot z(x), z(x)\rangle\right\}\,\mu(dx) \tag{1.3}$$

$$\left\{\int_X |\rho_\mu(z;x)|^2\,\mu(dx)\right\}^{\frac{1}{2}} \le \mathrm{const}\cdot\sigma_{1,4}(z) \tag{1.4}$$

Note that the righthand side of (1.3) makes sense also for lager class of vector fields, in particular for $z \in W_1^4(H_-,H_0,\mu)$. It follows from (1.4) that the linear differential operator $\rho_\mu : z \mapsto \langle\lambda(x),z(x)\rangle + \mathrm{div} z(x)$ is a continuous map from $C_1(H_-,H_0,H_+)$ to $\mathcal{L}_2(H_-,\mu)$. Hence it's possible to extend it to $W_1^4(H_-,H_0,\mu)$. We denote this extension by the same symbol $\rho_\mu(z;x)$.

In this case the differential operator $-\langle\lambda(x) + \nabla, z(x)\rangle$ is the adjoint of the differential operator $\langle z(x),\nabla\rangle$ in the space $\mathcal{L}_2(X,\mu)$ of μ-square integrable functions on X.

Consider again $\mu \in \mathcal{M}_1^2(H_-,H_0)$. Let a vector field $z \in W_1^4(H_-,H_0,\mu)$ passes an integral flow G_t:

$$\frac{d(G_t x)}{dt} = z(G_t x); \quad G_0 x = x.$$

Consider the first order differential operator associated with z:

$$A = \langle z(x),\nabla\rangle$$

and the following Cauchy problem

$$\frac{\partial u(t,x)}{\partial t} + Au(t,x) = 0 \quad u(0,x) = \phi(x) \in C_1(H_-,\mathbb{R}^1)) \tag{1.5}$$

Denote by $\mathcal{U}(t)$ an evolution family, giving the solution of (1.5)

$$\mathcal{U}(t) : \phi(x) \mapsto u(t,x).$$

It's easy to show that $\mathcal{U}(t)$ realizes the shift of the initial function along the trajectories of z:

$$u(t,x) = \phi(G_t x).$$

In addition a measure μ_t "shifted" along the trajectories of z arises:

$$\mu_t = \mu \circ G_t^{(-1)}.$$

It's shown in [8] that under these conditions on μ and z measure μ_t is absolutely continuous with respect to μ: $\mu_t \prec \mu$, and the density

$$\rho(x) = \frac{\mu_t(dx)}{\mu(dx)}$$

is given by the following formula

$$\rho(x) = \exp\left\{\int_0^t \rho_\mu(z,G_{t-s}x)\,ds\right\}, \tag{1.6}$$

where ρ_μ is the logarithmic derivative of μ along z.

2. An infinite-dimensional analogues for some results of classical calculus of variations.

Instead of (0.1) we consider here a functional of more general form.

Let B_j $(j = 1,2)$ be a Banach space, $A : B_1 \to B_2$ be a differentiable map, and $F : B_1 \times B_2 \to \mathbb{R}^1$ be a smooth real-valued function.

In this section we consider the extremal problem for a functional

$$f(u) = F\big(u,A(u)\big). \tag{2.1}$$

We obtain the main relations for (2.1). Then in section 3 we reduce them to the concrete examples.

All computations in section 2 look very simple, however considering the variational problem posed in such a way seems to be interesting even for the finite dimensional case.

The infinite dimensions specific arises when considering the integral type functionals. The essential role in this case belongs to the measure that (we recall) is supposed to be smooth. Therefore the infinite dimensional analogues of the classical relations (the Euler-Lagrange equation, the canonical Hamilton system) contain the additional terms related to the logarithmic derivative of the measure (see section 3).

2.1. The Euler-Lagrange Equation.

The necessary condition for an extremum for (2.1) is

$$\frac{d}{d\epsilon}\bigg|_{\epsilon=0} f(u+\epsilon h) = 0 \tag{2.2}$$

for any $h \in B_1$. Computing this derivative, we get

$$\frac{d}{d\epsilon}\bigg|_{\epsilon=0} f(u+\epsilon h) = \langle F_1'(u,A(u)),h\rangle_{B_1} + \langle F_2'(u,A(u)),A'(u)\cdot h\rangle_{B_2}$$

$$= \langle F_1'(u,A(u)) + A'(u)^* \cdot F_2'(u,A(u)),h\rangle_{B_1} = 0 \tag{2.3}$$

where $F_k'(u,A(u))$ ($k=1,2$) is the derivative of F with respect to the first or second argument respectively; brackets $\langle\cdot,\cdot\rangle_{B_k}$ ($k=1,2$) denote pairing between B_k and a dual space B_k^* (in particular, $F_1'(u,A(u)) \in B_1^*$, $F_2'(u,A(u)) \in B_2^*$); $A'(u)$ is the derivative of the operator A ($A'(u) \in L(B_1,B_2)$); $A'(u)^*$ is the corresponding dual operator ($A'(u)^* \in L(B_2^*,B_1^*)$) for $A'(u)$; we always assume that $A'(u)^*$ exists.

Since h is an arbitrary element of B_1, (2.3) readily implies the following relation

$$F_1'(u,A(u)) + A'(u)^* \cdot F_2'(u,A(u)) = 0 \tag{2.4}$$

We call (2.4) an *analogue of the classical Euler-Lagrange equation* (see [9,10]).

2.2. Canonical Hamilton's system.

Here we introduce an analogue of the classical Legendre transforms and obtain the analogue of the canonical Hamilton's system of equations.

Consider the functional (2.1) again. An element

$$p = F_2'(u,A(u)) \in B_2^* \tag{2.5}$$

is said to be a *generalized impulse*. Denote $A(u)$ by v and assume that it's possible to solve (2.5) for v:

$$v = V(u,p).$$

Then introduce an *analogue of the hamiltonian*

$$H(u,p) = \langle p,v\rangle_{B_2} - F(u,v)\bigg|_{v=V(u,p)} \tag{2.6}$$

Differentiating (2.6) we have

$$dH(u,p) = \langle dp,v\rangle_{B_2} + \langle p,dv\rangle_{B_2} - \langle F_1'(u,v),du\rangle_{B_1} - \langle F_2'(u,v),dv\rangle_{B_2}\Big|_{v=V(u,p)}$$

By (2.5) and (2.4), we get the *analogue of the classical Hamilton's system of equations*

$$A(u) = H_2'(u,p)$$

$$A'(u)^* \cdot p = H_1'(u,p) \tag{2.7}$$

Like in the classical theory, these equations are equivalent to the *Euler-Lagrange equation* (2.4).

Suppose $B_1 = B_2 = B$ and consider a pair of functions $\Phi,\Psi \in C_1(B\times B,\mathbb{R}^1)$. An operation

$$\{\Phi,\Psi\} = \big\langle \Phi_1'(u,p), \Psi_2'(u,p)\big\rangle_B - \big\langle \Psi_1'(u,p), \Phi_2'(u,p)\big\rangle_B \tag{2.8}$$

is said to be an *analogue of Poisson's brackets.* We say that $\Phi(u,p)$ and $\Psi(u,p)$ *commute*, if

$$\{\Phi,\Psi\} = 0.$$

Along with (2.7), let us consider the associated system of equations

$$\frac{du}{dt} = A(u)$$

$$\frac{dp}{dt} = -A'(u)^* \cdot p \tag{2.9}$$

This system is Hamilton's one with the hamiltonian

$$K(u,p) = \langle p,A(u)\rangle_B \tag{2.10}$$

Assume there exists an integral flow T_t of (2.9):

$$T_t\ (u,p) \mapsto \big(\phi(t,u),\psi(t,u)\cdot p\big) \tag{2.11}$$

where

$$\frac{d\phi(t,u)}{dt} = A(u); \quad \frac{d\psi(t,u)\cdot p}{dt} = -A'(u)^*\cdot p$$

Denote the map related to T_t and defined on $C_1(B\times B,\mathbb{R}^1)$ by

$$\widetilde{T}_t\ \Phi(u,p) \mapsto \Phi\big(\phi(t,u),\psi(t,u)\cdot p\big) \tag{2.12}$$

Theorem 2.1. Let (u,p) be a solution of a Hamilton's system (2.7) and let $\Phi(u,p)$ be a function from $C_1(B\times B,\mathbb{R}^1)$ commuting with the hamiltonian $H(u,p)$ (2.6). Then the pair (u,p) is a stationary point of the map $\widetilde{T}_t\cdot\Phi(u,p)$ at the point $t=0$.

Proof. Let (u,p) be a solution of (2.7). Compute the derivative

$$\frac{d}{dt}\bigg|_{t=0} \widetilde{T}_t\Phi(u,p) = \left\langle \Phi_1'(u,p), \frac{d\phi(t,u)}{dt}\bigg|_{t=0} \right\rangle_B + \left\langle \Phi_2'(u,p), \frac{d\psi(t,u)\cdot p}{dt}\bigg|_{t=0} \right\rangle_B$$

Using the associated system (2.9), we obtain

$$\frac{d}{dt}\bigg|_{t=0} \widetilde{T}_t\Phi(u,p) = \langle \Phi_1'(u,p), A(u)\rangle_B - \langle \Phi_2'(u,p), A'(u)^* \cdot p\rangle_B$$

Further, by (2.7)

$$\frac{d}{dt}\bigg|_{t=0} \widetilde{T}_t\Phi(u,p) = \langle \Phi_1'(u,p), H_2'(u,p)\rangle_B - \langle \Phi_2'(u,p), H_1'(u,p)\rangle_B = \{\Phi, H\} \quad (2.13)$$

As $\{\Phi, H\} = 0$, (2.13) proves the theorem.

3. Examples.

3.1 Consider a space $X = [0;\tau] \times Y$, where $Y = H_- \supset H_0 \supset H_+$ is a rigged Hilbert space. Let $\mu = m \times \nu$ be a measure on X, where m is a Lebesgue measure on $\mathbb{R}^1$, $\nu \in \mathcal{M}_1^2(H_-,H_0)$.

Suppose $u(x) = u(t,y)$ is a smooth real-valued function on X, i. e.

(i) $u(\cdot,y) \in C_2(\mathbb{R}^1)$ for any $y \in Y$;

(ii) $u(t,\cdot) \in C_2(H_-,H_0,\mathbb{R}^1)$ for any $t \in [0;\tau]$.

Without loss of generality we assume that $u(0,y) = u(\tau,y) = 0$. Furthermore, suppose

$$v(x) = (v^1(x),v^2(x)) = A(u)(x) = \left(\frac{d}{dt}; \nabla_y\right) \cdot u(x) = (\dot{u}(x),u'(x)).$$

Thus, we consider the functional of the form

$$\int_0^\tau \int_Y \Phi(u(t,y),\dot{u}(t,y),u'(t,y),t,y)\, \nu(dy)dt \quad (3.1)$$

In this case

$$Y_1 = \mathbb{R}^1, Y_2 = \mathbb{R}^1 \times H_0, B_1 = \mathcal{L}_2(X,\mu),$$

$$B_2 = \mathcal{L}_2(X,\mathbb{R}^1 \times H_0,\mu),\ A'(u) = A = \left(\frac{d}{dt}; \nabla_y\right)$$

Φ is a real-valued function differentable with respect to all its arguments. Using the integration by parts, it's easy to compute the adjoint operator

$$A'(u)^* = A^* = \left(-\frac{d}{dt}; -(\lambda(x) + \nabla)\right) \quad (3.2)$$

where $\lambda : H_- \to H_-$ is a logarithmic gradient of ν. The domain of A^* is the subspace $W \subset B_2$, consisting of the functions

$$w(t,y) = \big(w^1(t,y), w^2(t,y)\big)$$

such that

(i) $\forall y \in Y$ $w^1(\cdot,y)$ is a differentiable function;

(ii) $\forall t \in [0;\tau]$ $w^2(t,\cdot) : H_- \to H_0$ is a vector field of class $W_1^2(H_-,H_0,\nu)$.

Suppose the vector field $\Phi_2' \in W$. Then the extremals of (3.1) satisfy the analogue of the Euler-Lagrange equation which in this situation has the following form

$$\Phi_1'\big(u(x),\dot u(x),u'(x),x\big) - \frac{d}{dt}\big(\Phi_2'\big)^1\big(u(x),\dot u(x),u'(x),x\big) - \\ \langle \lambda(x) + \nabla, \big(\Phi_2'\big)^2\big(u(x),\dot u(x),u'(x),x\big)\rangle \cdot 1 = 0 \tag{3.3}$$

Consider now a more concrete example of the functional (3.1). First introduce some notations. By $C_2^b(H_-,\mathbb{R}^1)$ denote the subspace of $C_2(H_-,\mathbb{R}^1)$ with the norm

$$\|u\|_{C_2^b} = \sup_{y\in H_-}\{|u(y)| + \|u'(y)\| + \|u''(y)\|\}.$$

Denote by $C_k^{\text{pol}}(H_-,\mathbb{R}^1)$ the subset of $C_k(H_-,\mathbb{R}^1)$ consisting of all "polynomially bounded" mappings (see [1]), i. e. which satisfy

$$\|u^{(l)}(y)\| \le C(1+\|y\|)^p \quad (y \in H_-)$$

in the corresponding operator norms of the derivatives $u^{(l)}$ $(l = 0,\dots,k)$.

Let ν be a Borel measure of class $\mathcal{M}_0(H_-)$ with the following properties:

(i)

$$\int_{H_-} \|y\|^p \nu(dy) < \infty \quad (\forall p \in \mathbb{N}) \tag{3.4}$$

(ii)

$$\int_{H_-} \|\lambda(y)\|^2 \|y\|^p\, \nu(dy) < \infty \quad (\forall p \in \mathbb{N}) \tag{3.5}$$

where $\lambda : H_- \to H_-$ is the logarithmic gradient of ν.

Remark that under the condition (3.4) it's sufficient for (3.5) that

$$\exists \delta > 0 : \int_{H_-} \|\lambda(y)\|^{2+\delta}\nu(dy) < \infty. \tag{3.6}$$

Consider an extremal problem for the functional (3.1) with

$$\Phi\big(u(t,y),\dot u(t,y),u'(t,y),t,y\big) = \dot u(t,y)^2 + (u'(t,y),u'(t,y)) \tag{3.7}$$

$$\int_0^T \int_Y \Big(\dot{u}(t,y)^2 + (u'(t,y),u'(t,y))\Big)\, \nu(dy)dt \to \text{extr} \tag{3.8}$$

$$u(0,y) = u_0(y), \qquad \dot{u}(0,y) = u_1(y) \tag{3.9}$$

The Euler-Lagrange equation (3.3) turns into

$$\ddot{u}(t,y) = -\langle \lambda(y) + \nabla, u'(t,y)\rangle \cdot \nVdash$$

or

$$\ddot{u}(t,y) = -\mathrm{Tr}u''(t,y) - \langle \lambda(y), u'(t,y)\rangle \tag{3.10}$$

There is a Dirichlet operator H_ν with the domain $\mathcal{D}(H_\nu) = C_2^{\mathrm{pol}}(H_-,\mathbb{R}^1)$ associated with the measure ν in the righthand side of (3.8):

$$(H_\nu u)(t,y) = -\mathrm{Tr}\, u''(t,y) - \langle \lambda(y), u'(t,y)\rangle .$$

The following result was proved in [1]

Theorem 3.1. Let measure $\nu \in \mathcal{M}_0(H_-)$ satisfy the conditions (3.4) and (3.5). Suppose there exists a sequence of maps $\ell_n : H_- \to H_-$ $(n \in \mathbb{N})$, such that

(i) $\ell_n \in C_2^{\mathrm{pol}}(H_-,H_-)$;

(ii) $\int_{H_-} \|\lambda(y) - \ell_n(y)\|^2\, \nu(dy) \to 0 \quad (n \to \infty)$;

(iii) $\forall n \in \mathbb{N}\ \exists C_n \geq 0 : \forall y \in H_-$

$$(\ell_n(y),y)_- \leq C_n(1 + \|y\|_-^2);$$

(iv) $\exists C \geq 0 : \forall h \in H_-\ \forall y \in H_-\ \forall n \in \mathbb{N}$

$$(\ell_n'(y)h,h)_- \leq C\|h\|_-^2 .$$

Then $C_2^b(H_-,\mathbb{R}^1)$ is a domain of essential self-adjointness for H_ν.

Corollary. Under the conditions of Theorem 3.1 there exists a unique extremal for the functional (3.8).

Proof. Under the condition of essential self-adjointness of H_ν for any $u_0 \in \mathcal{D}(H_\nu)$, $u_1 \in \mathcal{D}(H_\nu^{\frac{1}{2}})$ there exists a unique solution of the Euler-Lagrange equation (3.10) (see [4]).

3.2. Let $X = H_- \supset H_0 \supset H_+$, $\mu \in \mathcal{M}_1^2(H_-,H_0)$, and $u(t,\cdot) \in C_2(H_-,H_0,\mathbb{R}^1)$. Further, suppose

$$v(x) = A(u)(x) = \langle z(x), \nabla\rangle u(x)$$

where $z : H_- \to H_0$ is a vector field of class $W_1^2(H_-,H_0,\mu)$.

So, we consider the variational problem for the functional

$$f(u) = \int_X \Phi\big(u(x), \langle z(x),\nabla\rangle u(x),x\big)\ \mu(dx). \tag{3.11}$$

In this case $B_1 = B_2 = \mathcal{L}_2(X,\mu)$,

$$A'(u)^* = A^* = -\langle \lambda(x) + \nabla, z(x)\rangle \tag{3.12}$$

where λ is the logarithmic gradient of μ .

The extremals of (3.11) satisfy the Euler-Lagrange equation

$$\Phi_1'\big(u(x), \langle z(x),\nabla\rangle u(x),x\big) - \langle\lambda(x)+\nabla, z(x)\rangle\cdot\Phi_2'\big(u(x), \langle z(x),\nabla\rangle u(x),x\big) = 0 \tag{3.13}$$

Let us complete the second order differential equation (3.13) by "initial conditions"

$$u(x)\big|_\Gamma = u_0(x), \tag{3.14}$$

where Γ is some smooth submanifold of X.

Now, introduce the generalized impulse

$$p(x) = \Phi_2'\big(u(x),v(x),x\big),$$

where

$$v(x) = \langle z(x),\nabla\rangle u(x)$$

and the hamiltonian

$$\begin{aligned} H(u,p) &= \int_X h\big(u(x),p(x),x\big)\ \mu(dx) \\ &= \int_X \{p(x)\cdot v(x) - \Phi\big(u(x),v(x),x\big)\}\Big|_{v(x)=V(u,p)(x)} \mu(dx) \end{aligned}$$

Thus, we transform (3.13) into the following Hamilton's system

$$\langle z(x),\nabla\rangle u(x) = h_2'\big(u(x),p(x),x\big)$$

$$\langle\lambda(x) + \nabla,z(x)\rangle\cdot p(x) = -h_1'\big(u(x),p(x),x\big) \tag{3.15}$$

It's possible to solve (3.15) analogously to the classical method of characteristics:

$$\frac{dx}{dt} = z(x)$$

$$\frac{du}{dt} = h_2'\big(u(x),p(x),x\big)$$

$$\frac{dp}{dt} = -h_1'\big(u(x),p(x),x\big) - p(x)\cdot\langle\lambda(x) + \nabla, z(x)\rangle\cdot 1 \tag{3.16}$$

Let us complete (3.16) by one more "initial condition"

$$p(x)\big|_{\Gamma} = p_0(x) \tag{3.17}$$

The integral surface of (3.16) is formed by the characteristic curves that cross Γ.

In this case the associated Hamilton's system (2.9) has very simple form:

$$\frac{du}{dt} = \langle z(x), \nabla\rangle u$$

$$\frac{dp}{dt} = \langle \lambda(x) + \nabla, z(x)\rangle \cdot p \tag{3.18}$$

It's possible to compute the integral flow of (3.18) by the method of characteristics:

$$u(x,t) = u_0(G_t x)$$

$$p(x,t) = p_0(G_t x) \cdot \exp\left\{ \int_0^t \rho_\mu(z, G_{t-s}x)\, ds \right\} \tag{3.19}$$

where G_t is the integral flow of z:

$$\frac{dG_t x}{dt} = z(G_t x); \qquad G_0 x = x.$$

Introduce an *impulse measure*

$$\mathcal{P}_t(\Delta) = \int_\Delta p(x,t)\, \mu(dx).$$

It follows from (3.19) and (1.6) that

$$\mathcal{P}_t = \mathcal{P}_0 \circ G_{-t}.$$

Thus, the integral flow of the associated Hamilton's system (3.18) realizes the shifts of the coordinate u and the impulse measure $\mathcal{P}$ along the trajectories of the vector field z:

$$u \mapsto u \circ G_t; \qquad \mathcal{P} \mapsto \mathcal{P} \circ G_{-t}.$$

By Theorem 1.1, if the pair (u,p) is the solution of (3.15), then

$$\frac{d}{dt}\bigg|_{t=0} \psi\Big(x, u \circ G_t, \frac{\mathcal{P}(G_{-t}dx)}{\mathcal{P}(dx)}\Big) = 0,$$

where ψ is a function commuting with h (in particular, $\psi = h$).

References

[1] S. Albeverio, Yu. Kondratjev, M. Röckner, Dirichlet operators via stochastic analysis, BiBoS Preprint N 571/4/93.

[2] V. I. Averbukh, O. G. Smolyanov, S. V. Fomin, Extended functions and differential equations in linear spaces, Proc. Moscow Math. Soc. **24**, 1971, 1133–1174, (Translated from Russi.an)

[3] Ya. I. Belopolskaya and Yu. L. Dalecky, Stochastic equations and differential geometry, Kluwer Academic Publishers, 1989.

[4] Yu. M. Berezansky and Yu. G. Kondratjev, Spektral methods in infinite-dimensional analysis, Naukova dumka, Kiev, 1988.

[5] V. I. Bogachev, O. G. Smolyanov, Analytic properties of infinite-dimensional distributions, Uspekhi Mat. Nauk, **45**,no. 3, 1990, 3–83, Russian.

[6] I. M. Gelfand and S. V. Fomin, Calculus of variations, Richard A. Silverman, Prentice-Hall, 1963.

[7] Yu. L. Dalecky and S. V. Fomin, Measures and differential equations in infinite-dimensional spases, Kluwer Academic Publishers, 1991.

[8] Yu. L. Dalecky and V. R. Steblovskaya, Smooth measures, Yu. L. Dalecky and S. V. Fomin Measures and differential equations in infinite-dimensional spaces, Kluwer Academic Publishers, 1991, 283–308.

[9] Yu. L. Dalecky and V. R. Steblovskaya, Smooth measures: absolute continuity, stochastic integrals, variational problems, Proc. of VI USSR-Japan Symposium on Probability Theory (A. N. Shiryaev, V. S. Korolyuk, S. Watanabe, N. Fukushima, eds.), World Scientific Publishing Co., 1992, pp. 52–69.

[10] Yu. L. Dalecky and V. R. Steblovskaya, On infinite-dimensional variational problems, (to appear in Stochastic Analysis and Applications in 1995). See also SFB 237. Preprint N 166, Institut für Mathematik Ruhr-Universität-Bochum, 1992.

[11] A. V. Skorohod, Integration of Hilbert space, Springer-Verlag, 1974.

Operator Theory:
Advances and Applications, Vol. 78

Finite Capacities in Spectral Theory

M. Demuth, F. Gesztesy, J. van Casteren, Z. Zhao

Abstract

The notion of capacity comprises a natural link between the spectrum of selfadjoint Feller operators in $L^2(\Sigma), \Sigma \subseteq \mathbb{R}^d$, with Dirichlet boundary conditions on $\partial\Sigma$ and geometric properties of the region Σ. Here we describe two complementary results. On changing the boundary of Σ, the lowest eigenvalue (ground state) turns out to be shifted if and only if the capacity of the difference set is positive. On the other hand the absolutely continuous spectra of Feller operators with Dirichlet conditions are not affected by changing the perturbations arbitrarily on sets of finite capacities, because the corresponding scattering systems turn out to be complete.

1. The notion of capacity.

Because there are different notions of capacity in the literature we first of all fix the definition used in the present article. Let K_0 be a free Feller operator, i.e., the generator of a Feller semigroup with a density p_0 such that

$$\left(e^{-tK_0} f\right)(x) = \int_{\mathbb{R}^d} p_0(t,x,y)\ f(y)\ dy. \tag{1}$$

For p_0 we assume the basic assumptions of stochastic spectral analysis (see [4]), i.e., p_0 is continuous in t, x, y, satisfies the Chapmann-Kolmogorov equation, has total mass one, is symmetric in x and y, the semigroup e^{-tK_0} is continuous at $t = 0$ and satisfies the Feller property. Hence K_0 is the L^2-generator of a Markov process

$$\left\{\mathbb{R}_+, (\Omega, \mathfrak{F}, P_x), X(t), (\mathbb{R}^d, \mathfrak{B}^d)\right\}. \tag{2}$$

K_0 is positive and self-adjoint.

For any Borel set Γ in $\mathbb{R}^d$ its capacity is given by

$$\operatorname{cap}(\Gamma) := \inf\Big\{(K_0^{1/2} f, K_0^{1/2} f) + (f,f), f \in \operatorname{dom} K_0^{1/2}, f \geq 1\!\!1_O,$$

$$O \text{ open and } \Gamma \subset O\Big\}. \tag{3}$$

This infimum is realized by the 1-equilibrium potential

$$v_\Gamma(x) := E_x\left\{e^{-T_\Gamma}, T_\Gamma < \infty\right\}, \tag{4}$$

where T_Γ is the first hitting time of Γ, i.e.,

$$T_\Gamma := \inf\{s > 0, X(s) \in \Gamma\}. \tag{5}$$

In particular we have

$$\int_{\mathbb{R}^d} v_\Gamma(x)\, dx = \text{cap}(\Gamma). \tag{6}$$

(see also Fukushima [5], Lemma 2.2.6). The capacity gives a classification of Borel sets which is more appropriate than the Lebesgue measure to the systems described by K_0. This can be illustrated in connection with the Laplace operator. If H_0 is the self-adjoint realization of $-\Delta$ in $L^2(\mathbb{R}^d)$, and if $K_0 = H_0$, the capacity of order zero of a Borel set Γ is given by

$$\text{cap}_{(0)}(\Gamma) = \inf\left\{(\nabla f, \nabla f), f \in W^{1,2}(\mathbb{R}^d), f \geq 1\!\!\mathrm{K}_O \text{a.e.}, O \text{ open }, \Gamma \subset O\right\}. \tag{7}$$

By definition only the behaviour of the admissible functions outside Γ contributes to the zero order capacity.

For instance if Γ is a ball $B \subset \mathbb{R}^d$ of radius r we have

$$\text{cap}_{(0)}(B) = c\, r^{d-2}.$$

On the other hand

$$\inf\left\{(f, f), f \geq 1\!\!\mathrm{K}_B, f \in L^2\right\} = \text{meas}(B) = c\, r^d.$$

It turns out that for small r the capacity of order zero plays the dominant role.

2. Dirichlet Feller and Schrödinger operators

Let Σ be an open connected region in $\mathbb{R}^d$. Let K_0 be a free Feller operator. Let V be a Kato-Feller-potential, that is, a real valued Borel measureable function, satisfying

$$\lim_{\tau \to 0} \sup_x \int_0^\tau ds \int_{\mathbb{R}^d} dy\; p_0(s, x, y)\; [1\!\!\mathrm{K}_B V_+(y) + V_-(y)] = 0. \tag{8}$$

where B is any compact set in $\mathbb{R}^d$. For such V we can define the Feller operator $K = K_0 \dot{+} V$, which is self-adjoint in $L^2(\mathbb{R}^d)$.

By definition the operator $K = K_0 \dot{+} V$ is the generator of the Feynman-Kac semigroup $\{\exp(-tK)\}_{t \geq 0}$ represented by

$$\left(e^{-tK} f\right)(x) = E_x\left\{e^{-\int_0^t V(X(s))ds} f(X(t))\right\}.$$

This kind of representation can be used to introduce the Dirichlet Feller operator. Define

$$(U(t)f)(x) := E_x\left\{ e^{-\int_0^t V(X(s))ds} f(X(t)), t < T_{\mathbb{R}^d \setminus \Sigma} \right\}; \tag{9}$$

then $U(t)$ restricted to $L^2(\Sigma)$ forms a strongly continuous symmetric semigroup on $L^2(\Sigma)$. We denote its generator by K_Σ and call it Dirichlet Feller operator. The authors believe that under suitable conditions K_Σ coincides with the self-adjoint semibounded operator associated with the closure of the form

$$\left\langle K_0^{1/2} f, K_0^{1/2} g \right\rangle + \int dx\, V(x) f(x) \overline{g(x)}, \tag{10}$$

where f, g belong to $\operatorname{dom} K_0^{1/2}$ and where these functions vanish on $\mathbb{R}^d \setminus \Sigma$. In the Schrödinger case, e.g. $K_0 = H_0$, K_Σ coincides with the associated operators of the Dirichlet Schrödinger form on $W_0^{1,2}(\Sigma) \cap \operatorname{dom} V_+^{1/2}$.

3. General remarks on capacity and spectrum.

The aim of the present article is to demonstrate that sets of finite and positive capacity play a crucial role for spectral properties of K_Σ. With $\sigma(K_\Sigma), \sigma_{\text{ess}}(K_\Sigma)$, $\sigma_{\text{ac}}(K_\Sigma)$ we denote the spectrum, the essential spectrum, and the absolutely continuous spectrum of K_Σ, respectively. Let $\lambda(\Sigma)$ be the bottom of the spectrum, i.e., $\lambda(\Sigma) := \inf \sigma(K_\Sigma)$. Let Σ_{small} and Σ_{large} be two open connected regions, such that $\Sigma_{\text{small}} \subset \Sigma_{\text{large}}$. For $f \geq 0$ it follows

$$\left(e^{-t(K_0)_{\Sigma_{\text{large}}}} f\right)(x) \geq \left(e^{-t(K_0)_{\Sigma_{\text{small}}}} f\right)(x),$$

$x \in \mathbb{R}^d$. Hence

$$\left\| e^{-t(K_0)_{\Sigma_{\text{large}}}} f \right\|_2 \geq \left\| e^{-t(K_0)_{\Sigma_{\text{small}}}} f \right\|_2.$$

Because of

$$\left\| e^{-t(K_0)_\Sigma} \right\| = e^{-t\lambda_\Sigma}$$

for any possible Σ, it follows

$$\lambda(\Sigma_{\text{large}}) \leq \lambda(\Sigma_{\text{small}}), \tag{11}$$

i.e., the bottom of the spectrum can only increase if we shrink Σ. On the other hand the domains of $(K_0)^{1/2}_{\Sigma_{\text{small}}}$ and $(K_0)^{1/2}_{\Sigma_{\text{large}}}$ are the same if we assume that $\operatorname{cap}(\Sigma_{\text{large}} \setminus \Sigma_{\text{small}}) = 0$. Hence this condition implies $\lambda(\Sigma_{\text{large}}) = \lambda(\Sigma_{\text{small}})$.

One interesting question is wether the converse is true, i.e., does $\operatorname{cap}(\Sigma_{\text{large}} \setminus \Sigma_{\text{small}}) > 0$ imply a positive difference between the infima of the spectra. In general, this can not be true. For many free Feller operators K_0 it is known that

$\sigma_{\mathrm{ac}}(K_0) = \sigma_{\mathrm{ac}}((K_0)_\Sigma)$ and $\sigma_{\mathrm{ac}}(K_0) = \sigma(K_0)$ as well as $\sigma_{\mathrm{ac}}((K_0)_\Sigma) = \sigma((K_0)_\Sigma)$ if for instance $\mathbb{R}^d \setminus \Sigma$ is bounded. Thus in general the question has a negative answer. Nevertheless, the question can be answered affirmatively (see section 4) if $\lambda(\Sigma_{\mathrm{small}})$ is an eigenvalue of $(K_0)_{\Sigma_{\mathrm{small}}}$.

The other problem studied in this article is the dependence of $\sigma_{\mathrm{ac}}(K_\Sigma)$ on the region Σ. The two operators K_0 and K_Σ are selfadjoint operators in the Hilbert spaces $L^2(\mathbb{R}^d)$ and $L^2(\Sigma)$, respectively.

Let $J : L^2(\mathbb{R}^d) \to L^2(\Sigma)$ be the restriction operator given by $Jf := f/_\Sigma$, such that $J^*J = \mathbb{1}_\Sigma$, the projection operator in $L^2(\mathbb{R}^d)$ onto Σ, and $JJ^* = \mathrm{Id}_{L^2(\Sigma)}$, the identity operator in $L^2(\Sigma)$.

The scattering system $\{K_\Sigma, J, K_0\}$ is called complete if the wave operators

$$\Omega_\pm(K_\Sigma, J, K_0) := s\text{-}\lim_{t\to\pm\infty} e^{itK_\Sigma} J e^{-itK_0} P_{\mathrm{ac}}(K_0), \tag{12}$$

$$\Omega_\pm(K_0, J^*, K_\Sigma) := s\text{-}\lim_{t\to\pm\infty} e^{itK_0} J^* e^{-itK_\Sigma} P_{\mathrm{ac}}(K_\Sigma) \tag{13}$$

exist and if

$$s\text{-}\lim_{t\to\pm\infty} \mathbb{1}_{\mathbb{R}^d\setminus\Sigma}\, e^{-itK_0} P_{\mathrm{ac}}(K_0) = 0. \tag{14}$$

Here $P_{\mathrm{ac}}(A)$ denotes the projection operator onto the absolutely continuous subspace of the operator A.

The complement of Σ is called the obstacle region. As mentioned above it is known for many free Feller operators K_0 that the scattering theory is not affected by perturbations on bounded obstacle regions. Hence the next question is whether we can allow sets of finite capacities as an obstacle. A positive answer to this question is given in section 5.

4. Capacities and eigenvalues.

In this section we shall extend the main theorem in [6] on ground states of Dirichlet Schrödinger operators (see Theorem 1 in [6]) to the present setting of Feller generators. Let $\{\mathbb{R}_+, (\Omega, \mathfrak{F}, P_x), X(t), (\mathbb{R}^d, \mathfrak{B}^d)\}$ be the Markov process (see (2)) corresponding to the density p_0 given by (1). We assume that $p_0(t,x,y)$ is strictly positive for all $t > 0$, $x, y \in \Sigma$, where Σ is an open connected region and that $\{X_t\}_{t\geq 0}$ has continuous sample paths. Define the Dirichlet Feller operator $K_\Sigma := (K_0 \dot{+} V)_\Sigma$ as in section 2 and let τ_Σ be the first exit time of $\{X_t\}_{t\geq 0}$ from Σ.

Theorem 1. Assume Σ_{small} and Σ_{large} are two open connected regions in $\mathbb{R}^d$ satisfying $\Sigma_{\mathrm{small}} \subset \Sigma_{\mathrm{large}}$. Let $K_{\Sigma_{\mathrm{small}}}, K_{\Sigma_{\mathrm{large}}}$ be the corresponding Dirichlet Feller operators and suppose $\lambda(\Sigma_{\mathrm{small}}) = \inf \sigma(K_{\Sigma_{\mathrm{small}}})$ to be an eigenvalue of $K_{\Sigma_{\mathrm{small}}}$. Then

$$\lambda(\Sigma_{\mathrm{large}}) < \lambda(\Sigma_{\mathrm{small}})$$

if and only if

$$\mathrm{cap}(\Sigma_{\mathrm{large}} \setminus \Sigma_{\mathrm{small}}) > 0. \tag{15}$$

Remark. Note that there are no spectral assumptions on $K_{\Sigma_{\text{large}}}$. There are several possibilities to satisfy the assumption that $\lambda(\Sigma_{\text{small}})$ is an eigenvalue. The case where K_0 is the Laplacian and the negative potential part is large enough is discussed, e.g., by Arendt, Batty [1]. If $V \equiv 0$ and $K_0 = -\Delta$ the region Σ_{small} cannot be to large. Of course, $\lambda(\Sigma_{\text{small}})$ is an eigenvalue if Σ_{small} is compact. However, also unbounded Σ_{small} are allowed. Necessary and sufficient conditions for $e^{t\Delta_\Sigma}$ (Δ_Σ - the Dirichlet Laplacian defined as the self-adjoint operator in $L^2(\Sigma)$ given by the Laplacian with Dirichlet boundary condition on $\partial\Sigma$) to be a trace class operator are given by Davies [2]. In a recent preprint, McGillivray [7] gave quantitative upper and lower bounds for $\lambda(\Sigma_{\text{small}}) - \lambda(\Sigma_{\text{large}})$ in terms of the capacity of order zero for the difference set.

Proof. One can follow the proof of Theorem 1 in [6] step by step. Hence we restrict ourselves to a brief sketch here. First we assume that $\text{cap}(\Sigma_{\text{large}} \setminus \Sigma_{\text{small}}) = 0$. That means $\Sigma_{\text{large}} \setminus \Sigma_{\text{small}}$ is a polar set for the Markov process (2). This is equivalent to

$$P_x\{\tau_{\Sigma_{\text{small}}} < \tau_{\Sigma_{\text{large}}}\} = 1$$

for $x \in \Sigma_{\text{small}}$. Therefore

$$\left(e^{-tK_{\Sigma_{\text{large}}}} f\right)(x) = \left(e^{-tK_{\Sigma_{\text{small}}}} f\right)(x)$$

for $f \in L^2(\Sigma_{\text{large}}), x \in \Sigma_{\text{small}}, t \geq 0$, implying the equality of the infinitesimal generators.

To show the converse we assume $\text{cap}(\Sigma_{\text{large}} \setminus \Sigma_{\text{small}}) > 0$. Next we extend the normalized eigenfunction φ of $K_{\Sigma_{\text{small}}}$ associated with $\lambda(\Sigma_{\text{small}})$ to $\widetilde{\varphi}$ in Σ_{large} by defining $\widetilde{\varphi}/_{\Sigma_{\text{large}} \setminus \Sigma_{\text{small}}} = 0$. Then one obtains $\|\widetilde{\varphi}\| = \|\varphi\| = 1$.

If $\text{meas}(\Sigma_{\text{large}}) > \text{meas}(\Sigma_{\text{small}})$, the strict positivity of $p_0(t, x, y)$ for $t > 0$, $x, y \in \Sigma_{\text{large}}$ together with the Feynman-Kac formula then yields

$$\left(e^{-tK_{\Sigma_{\text{large}}}} \widetilde{\varphi}\right)(x) > 0,\ t > 0, x \in \Sigma_{\text{large}} \setminus \Sigma_{\text{small}}.$$

Hence one infers

$$\begin{aligned} \left\| e^{-tK_{\Sigma_{\text{large}}}} \right\|^2 &\geq \int_{\Sigma_{\text{large}}} dx \left| \left(e^{-tK_{\Sigma_{\text{large}}}} \widetilde{\varphi}\right)(x) \right|^2 \\ &> \int_{\Sigma_{\text{small}}} dx \left| \left(e^{-tK_{\Sigma_{\text{small}}}} \varphi\right)(x) \right|^2 \qquad (16) \\ &= \left\| e^{-tK_{\text{small}}} \right\|^2. \end{aligned}$$

If $\text{meas}(\Sigma_{\text{large}}) = \text{meas}(\Sigma_{\text{small}})$ one can use $\text{cap}(\Sigma_{\text{large}} \setminus \Sigma_{\text{small}}) > 0$ and properties of capacity to show the existence of $t > 0$ such that

$$P_x\{t < \tau_{\Sigma_{\text{large}}}\} > P_x\{t < \tau_{\Sigma_{\text{small}}}\},\ x \in \Sigma_{\text{small}}.$$

Since $\operatorname{meas}(\Sigma_{\text{large}} \setminus \Sigma_{\text{small}}) = 0$, $X_t \in \Sigma_{\text{small}} \quad P_x$ - a.s., the Feynman-Kac formula implies

$$\left(e^{-tK_{\Sigma_{\text{large}}}}\widetilde{\varphi}\right)(x) > \left(e^{-tK_{\Sigma_{\text{small}}}}\varphi\right)(x),\ x \in \Sigma_{\text{small}}$$

which also yields (16).

Thus in both cases we have the strict inequality

$$\left\|e^{-tK_{\Sigma_{\text{large}}}}\right\| > \left\|e^{-tK_{\Sigma_{\text{small}}}}\right\|.$$

Because of $\|e^{-tK_\Sigma}\| = e^{-t\lambda(\Sigma)}$, one infers

$$\lambda(\Sigma_{\text{large}}) < \lambda(\Sigma_{\text{small}}).$$

q.e.d.

As demonstrated in Example 7 in [6], Theorem 1 does not extend to higher (degenerate) eigenvalues.

5. Capacities and absolutely continuous spectra.

Here we study perturbed Feller operators constructed from the basic assumptions for the density function p_0. That means K_0 is the L^2-generator of the strong Markov process $\{\mathbb{R}_+, (\Omega, \mathfrak{F}, P_x), X(t), (\mathbb{R}^d, \mathfrak{B}^d)\}$ where Ω denotes the set of cadlag functions starting in x. V is a Kato-Feller-potential such that $K_0 \dot{+} V$ is well defined in $L^2(\mathbb{R}^d)$. Finally, $K_\Sigma = (K_0 \dot{+} V)_\Sigma$ is the Dirichlet Feller operator, i.e., the generator of

$$E_x\left\{e^{-\int\limits_0^t V(X(s))ds} f(X(t)), t < T_{\mathbb{R}^d \setminus \Sigma}\right\}$$

restricted to $L^2(\Sigma)$. Set $Jf := f|\Sigma$.

Theorem 2. (Demuth, van Casteren [4])

a) The two-space scattering system $\{(K_0)_\Sigma, J, K_0\}$ is complete if

$$\operatorname{cap}(\mathbb{R}^d \setminus \Sigma) < \infty.$$

b) The two-space scattering system $\{(K_0 \dot{+} V)_\Sigma, J, K_0 \dot{+} V\}$ is complete if the $(\lambda + V)$-harmonic function

$$h_{\lambda+V}(x) := E_x\left\{e^{-\int\limits_0^{T_\Gamma} (\lambda+V(X(s)))ds}, T_\Gamma < \infty\right\}$$

is in $L^1(\mathbb{R}^d)$.

c) For $V \in L^1(\mathbb{R}^d)$ the scattering system $\{K_0 \dot{+} V, K_0\}$ is complete.

Corollary. Suppose that the conditions in a), b) and c) are satisfied. Then we obtain the invariance of the absolutely continuous spectra, i.e.,

$$\sigma_{\mathrm{ac}}(K_0) = \sigma_{\mathrm{ac}}(K_0 \dot{+} V) = \sigma_{\mathrm{ac}}((K_0 \dot{+} V)_\Sigma) = \sigma_{\mathrm{ac}}((K_0)_\Sigma). \tag{17}$$

Remark. Sets of finite capacity can be unbounded. This can be seen in the Laplace situation. Let $K_0 = H_0 = -\Delta$. Assume that $\mathbb{R}^d \setminus \Sigma$ is the union of balls B_m of radius r_m. In this case

$$\mathrm{cap}(\mathbb{R}^d \setminus \Sigma) \leq c \sum_m r_m^{d-2}.$$

For $d > 4$ there are unbounded obstacles $\mathbb{R}^d \setminus \Sigma$ for which $\sum_m r_m = \infty$ but which have finite capacities because $\sum_m r_m^2 < \infty$.

That means the behaviour of potentials on sets of finite capacity is irrelevant for the scattering theory. Theorem 2 proves a non-local freedom in mathematical scattering theory.

Proof. The details of the proof are given in [4]. Here we provide a brief sketch. The proof is based on a generalized trace class criterion for two space scattering systems and on a trace class criterion for products of integral operators given by Demuth, Stollmann, Stolz, van Casteren [3], abbreviated as CDS^2-trace class criterion. Let K_1, K_2 be two self-adjoint operators in different Hilbert spaces $\mathfrak{H}_1$ and $\mathfrak{H}_2$, respectively. Let T be a bounded identification operator mapping $\mathfrak{H}_1$ into $\mathfrak{H}_2$. The system $\{K_2, T, K_1\}$ forms a complete scattering system if

$$e^{-K_2}\Big(e^{-K_2}T - Te^{-K_1}\Big)e^{-K_1} \in \mathfrak{B}_1(\mathfrak{H}_1, \mathfrak{H}_2), \tag{18}$$

$$e^{-K_2}T - Te^{-K_1} \in \mathfrak{B}_\infty(\mathfrak{H}_1, \mathfrak{H}_2), \tag{19}$$

$$(T^*T - \mathrm{Id}_{\mathfrak{H}_1})e^{-K_1} \in \mathfrak{B}_\infty(\mathfrak{H}_1, \mathfrak{H}_1), \tag{20}$$

$$(TT^* - \mathrm{Id}_{\mathfrak{H}_2})e^{-K_2} \in \mathfrak{B}_\infty(\mathfrak{H}_2, \mathfrak{H}_2). \tag{21}$$

Here $\mathfrak{B}_1$ is the set of trace class operators and $\mathfrak{B}_\infty$ is the set of compact operators. The most stringent condition is the trace class assumption in (18). In our present situation we know in addition that all the semigroups are integral operators. Hence we can use the CDS^2-trace class criterion [3].

Let A, B be two integral operators with measurable kernels $A(\cdot,\cdot)$ and $B(\cdot,\cdot)$. Their product is a trace class operator if

$$\int_{\mathbb{R}^d} dx \; \|A(\cdot, x)\|_{L^2} \; \|B(x, \cdot)\|_{L^2} = M < \infty. \tag{22}$$

In that case we have the estimate

$$\|AB\|_{\mathrm{tr}} \leq M.$$

Here $\|.\|_{\mathrm{tr}}$ denotes the trace norm. (Note that M is always smaller than the product of the Hilbert-Schmidt norms of A and B.) For the proof of (18) we can use the decomposition

$$\begin{aligned} A &= e^{-\frac{1}{2}K_2}, \\ B &= e^{-\frac{1}{2}K_2} D e^{-K_1} \end{aligned}$$

with

$$D = e^{-K_2}T - Te^{-K_1}.$$

By our basic assumptions of stochastic spectral analysis we have always

$$\sup_{x,y} \left(e^{-\frac{1}{2}K_2}\right)(x,y) = c < \infty.$$

Hence $e^{-K_2}De^{-K_1}$ is a trace class operator if

$$\int dx \int dy \, |D(x,y)| < \infty, \tag{23}$$

where $D(.,.)$ denotes the integral kernel of D.

Part a) of Theorem 2 now follows because

$$\int dy \, |D(x,y)| = E_x\{T_{\mathbb{R}^d \setminus \Sigma} < t\}$$

and

$$\begin{aligned} \int dy \, E_x\{T_{\mathbb{R}^d\setminus\Sigma} < t\} &\leq e^t \int dx \, E_x\left\{e^{-T_{\mathbb{R}^d\setminus\Sigma}}, T_{\mathbb{R}^d\setminus\Sigma} < \infty\right\} \\ &\leq e^t \int dx \, v_{\mathbb{R}^d\setminus\Sigma}(x) \\ &\leq c \cdot e^t \operatorname{cap}(\mathbb{R}^d \setminus \Sigma), \end{aligned}$$

see [4].

The function $h_{\lambda+V}(x)$ is the natural generalization of $v_{\mathbb{R}^d\setminus\Sigma}(x)$ if the Kato-Feller-potential V is not equal to zero. That means the arguments for part b) of Theorem 2 are identical.

For part c) we have a single space scattering system. The conditions in (20) and (21) are irrelevant, and T is the identity operator in $L^2(\mathbb{R}^d)$. The main condition is

$$e^{-(K_0 \dot{+} V)}\left(e^{-(K_0 \dot{+} V)} - e^{-K_0}\right)e^{-K_0} \in \mathfrak{B}_1.$$

Using again the CDS²-trace class criterion this is seen to be satisfied for Kato-Feller-potential in $L^1(\mathbb{R}^d)$. q.e.d.

References

[1] W. Arendt, Ch. Batty: Exponential stability of Schrödinger semigroups. Diff. Int. Equations **6** (1993), 1009–1024.

[2] E. B. Davies: Trace properties of the Dirichlet Laplacian. Math. Z. **188** (1985), 245–251.

[3] M. Demuth, P. Stollmann, G. Stolz, J. van Casteren: Trace norm estimates for products of integral operators and diffusion semigroups. Preprint, Max-Planck-Institut, Bonn, MPI 93-89 (1993).

[4] M. Demuth, J. van Casteren: Completeness of scattering systems with obstacles of finite capacity. Preprint, TU Clausthal, Mathematische Berichte 94/7 (1994).

[5] M. Fukushima, Y. Oshima, M. Takeda: Dirichlet forms and symmetric Markov processes, de Gruyter, Berlin-New York, 1994.

[6] F. Gesztesy, Z. Zhao: Domain perturbations, Brownian motion, capacities, and ground states of Dirichlet Schrödinger operators. Math. Z. **215** (1994), 143–150.

[7] I. McGillivray: A capacity estimates for the ground-state shift. Preprint TU Clausthal, 1995.

Remark. In article by W. Arendt and S. Monniaux in this proceedings (p. 9–19) a similar result is obtained by an analytic proof.

M. Demuth
Inst. of Mathematics, TU Clausthal
Erzstr. 1, D-38678 Clausthal-Zellerfeld, Germany

F. Gesztesy
Dept. of Mathematics, Univ. of Missouri
Columbia, MO 65211, USA

J. van Casteren
Dept. of Math. and Comp. Sciences, Univ. of Antwerp (UIA)
Universitetsplein 1, 2610 Antwerp-Wilrijk, Belgium

Z. Zhao
Dept. of Mathematics, Univ. of Missouri
Columbia, MO 65211, USA.

Operator Theory:
Advances and Applications, Vol. 78

Classical N-Body Scattering

J. Dereziński

Abstract

We compare what is known on the scattering theory of classical and quantum N-body systems.

In this talk we would like to review results on the large-time behavior of trajectories of N-body systems. We will compare them with what is known for quantum N-body systems. We will try to find as many analogies as possible.

In the literature on classical N-body systems, a considerable amount of work has been devoted to phenomena which are due to the singularity of the potential $\frac{1}{|x|}$. We will not be concerned with this class of problems and we will always assume that potentials are bounded.

The results on the quantum N-body case are taken from [1, 2]. The results on the classical N-body case are taken from [3], [7] and [4]. All of these results will be described in detail in [4].

1. 2-body case.

Let us start with the 2-body case. We will assume, both in quantum and classical cases, that the potential is a function on $X := \mathbb{R}^n$ satisfying

$$|\partial^\alpha V(x)| \leq C\langle x\rangle^{-\mu-|\alpha|}, \ \ |\alpha| = 0,1,2, \ \ \mu > 0. \tag{70}$$

In the quantum case the Hamiltonian is a self-adjoint operator on $L^2(X)$ given by

$$H := -\frac{1}{2}\Delta + V(x).$$

In the classical case the Hamiltonian is a function on phase space $X \times X'$

$$H(x,\xi) := \frac{1}{2}\xi^2 + V(x),$$

It generates the dynamics

$$\phi(t)(y,\eta) = (x(t,y,\eta),\xi(t,y,\eta)).$$

We have the following basic results (see eg. [4] and references therein).

Theorem 1.1. (Quantum case)

1) There exists (in the so-called strong C_∞ sense-see [4]) the following limit:

$$\lim_{t\to\infty} \mathrm{e}^{itH}\frac{x}{t}\mathrm{e}^{-itH} =: P^+$$

It satisfies

$$[P^+,H]=0,$$
$$\mathrm{E}_{\{0\}}(P^+) = E^{\mathrm{pp}}(H).$$

2) For an appropriate function $S(t,\xi)$ there exists

$$s-\lim_{t\to\infty} \mathrm{e}^{itH}\mathrm{e}^{-iS(t,D)} =: \Omega^+.$$

Besides,

$$\operatorname{Ran}\Omega^+ = \operatorname{Ran}\mathrm{E}^{\mathrm{c}}(H) = \mathrm{E}_{X\backslash\{0\}}(P^+).$$

(In the case $\mu > 1$, which is the short-range case, we can take $S(t,\xi) = \frac{1}{2}t\xi^2$).

The classical 2-body scattering theory can be formulated as follows (see [14], [9] and [4]).

Theorem 1.2. (Classical case)

1) There exists

$$\lim_{t\to\infty}\frac{x(t,y,\eta)}{t} =: \xi^+(y,\eta).$$

2) Set

$$\tilde{\phi}(t)(x,\xi) := (x+\nabla_\xi S(t,\xi),\xi).$$

Then on $X\times(X'\backslash\{0\})$ there exists the classical wave transformation

$$\lim_{t\to\infty}\phi(-t)\tilde{\phi}(t) =: \mathcal{F}^+.$$

Besides,

$$\operatorname{Ran}\mathcal{F}^+ = (\xi^+)^{-1}(X\backslash\{0\}).$$

Thus we have classified $(\xi^+)^{-1}(X\backslash\{0\})$ — the union of all "scattering trajectories".

Let us define

$$\mathcal{B} := \{\text{ union of bounded trajectories }\},$$

$$\mathcal{B}^+ := \{\text{ union of trajectories bounded for } t>0\}.$$

Then the following fact is true (see eg. [4]).

Propositon 1.3.

$$H(\mathcal{B}) = H(\mathcal{B}^+) \tag{71}$$

The above set will be called the set of trapping energies and will be denoted by σ.

We also define

$$\sigma_{\text{reg}} := \{\lambda \ : \ V^{-1}(]-\infty,\lambda[) \text{ has a non-empty bounded connected component }\}.$$

Obviously,

$$\sigma_{\text{reg}} \subset \sigma.$$

2. N-body systems.

We will use the formalism of generalized N-body systems.

X will denote a Euclidean space. It will have the meaning of the configuration space of an N-body system. $\{X_a : \ a \in \mathcal{A}\}$ is a certain finite family of subspaces of X closed wrt intersection. We will assume that $X_{a_{\min}} := X$ belongs to this family. We will write $a_1 \leq a_2$ iff $X_{a_1} \supset X_{a_2}$ and $b = a_1 \vee a_2$ iff $X_b = X_{a_1} \cap X_{a_2}$. We will write $a_{\max} := \bigcup_{a \in \mathcal{A}} a$.

The orthogonal complement of X_a in X is denoted X^a.

There will be also special symbols for the sets

$$Z_a := X_a \backslash \bigcup_{b \not\leq a} X_b \tag{72}$$

Note that X is the disjoint union of $\{Z_a : \ a \in \mathcal{A}\}$.

The phase space of an N-body system is $X \times X'$. An element of this space will be usually denoted (x,ξ) or (y,η).

D, D_a and D^a denote the operators $\frac{1}{i}\nabla$, $\frac{1}{i}\nabla_a$ and $\frac{1}{i}\nabla^a$ respectively. Likewise, Δ, Δ_a and Δ^a denote the Laplacians corresponding to the variables x, x_a and x^a respectively.

We assume that for every $a \in \mathcal{A}$ we are given a real function (called a potential) $X^a \ni x^a \mapsto v^a(x^a)$. We will assume that

$$|\partial_{x^a}^\alpha v^a(x^a)| \leq C\langle x^a\rangle^{-\mu-|\alpha|}, |\alpha| = 0,1,2.$$

We set

$$V(x) := \sum_{a \in \mathcal{A}} v^a(x^a)$$

and

$$V^a(x) := \sum_{b \subset a} v^b(x^b).$$

We define H and H_a to be the self-adjoint operators on $L^2(X)$ such that

$$H := -\frac{1}{2}\Delta + V(x)$$

and

$$H_a := -\frac{1}{2}\Delta + V^a(x).$$

Clearly, $H = H_{a_{\max}}$.

We may identify $L^2(X)$ with $L^2(X_a) \otimes L^2(X^a)$. Then we can write:

$$H_a = -\frac{1}{2}\Delta_a \otimes 1 + 1 \otimes H^a,$$

where

$$H^a := -\frac{1}{2}\Delta^a + V^a(x^a)$$

is a self-adjoint operator on $L^2(X^a)$.

In the classical case we define

$$H(x,\xi) := \frac{1}{2}\xi^2 + V(x)$$

and

$$H_a(x,\xi) := \frac{1}{2}\xi^2 + V^a(x).$$

Clearly, $H(x,\xi) = H_{a_{\max}}(x,\xi)$.

We may identify $X \times X'$ with $X_a \times X'_a \times X^a \times (X^a)'$. Then we can write:

$$H_a(x,\xi) = \frac{1}{2}\xi_a^2 + H^a(x^a,\xi^a)$$

where

$$H^a(x^a,\xi^a) := \frac{1}{2}(\xi^a)^2 + V^a(x^a)$$

is a Hamiltonian on $X^a \times (X^a)'$

If C is a self-adjoint operator and I is a Borel subset of $\mathbb{R}$ then $\mathrm{E}_I(C)$ will denote the spectral projection of C onto I. $\mathrm{E}^{\mathrm{pp}}(C)$ and $\mathrm{E}^{\mathrm{cont}}(C)$ denote the spectral projection of C onto its pure point and continous spectrum respectively. $\sigma(C)$, $\sigma^{\mathrm{pp}}(C)$ and $\sigma^{\mathrm{cont}}(C)$ denote the spectrum, the pure point spectrum and the continuous spectum of C respectively. $\mathrm{Ran}C$ denotes the range of C.

If $C = (C_1,\dots,C_n)$ is an n-tuple of commuting self-adjoint operators and Θ is a Borel subset of $\mathbb{R}^n$ then $E_\Theta(C)$ denotes the spectral projection of C onto Θ. $\sigma(C) \subset \mathbb{R}^n$ denotes the joint spectrum of $(C_1,\dots,C_n)$.

Now we are going to present a number of theorems on the scattering theory of N-body systems. Most of the time they will appear in pairs: the statement on quantum systems and then a property which we think should be regarded as its classical analog.

Localization of thresholds/trapping energies.

The following result follows from the Mourre estimate (see [11] and [12]).

Theorem 2.4. (Quantum case) The bound states of an N-body system accumulate at thresholds:

$$\overline{\sigma^{\rm pp}(H)}\backslash\sigma^{\rm pp}(H) \subset \cup_{b<a_{\max}}\sigma^{\rm pp}(H^b).$$

To state its classical analog let us denote by σ^a the set of trapping energies of $H^a(x^a,\xi^a)$. We will write σ instead of $\sigma^{\max}$. The following theorem has been proven in [4] using the so-called classical Mourre estimate of [15] and [6].

Theorem 2.5. (Classical case) The trapping energies of an N-body system accumulate at trapping energies of its subsystems:

$$\overline{\sigma}\backslash\sigma \subset \cup_{b<a_{\max}}\sigma^b.$$

Existence of asymptotic velocity

The following result comes from [1] and [2], its proof is based on the ideas of [8].

Theorem 2.6. (Quantum case) There exists (in the so-called strong C_∞-sense) the following limit:

$$\lim_{t\to\infty} {\rm e}^{itH}\frac{x}{t}{\rm e}^{-itH} =: P^+.$$

The vector of self-adjoint operators P^+ is called the (quantum) asymptotic velocity.

The following result comes from [3], its proof is a modification of the ideas used in the proof of the analogous quantum result.

Theorem 2.7. (Classical case) There exists the following limit:

$$\lim_{t\to\infty}\frac{x(t,y,\eta)}{t} =: \xi^+(y,\eta).$$

The function $\xi^+(y,\eta)$ is called the (classical) asymptotic velocity.

Joint spectrum / joint image of the energy and the asymptotic energy

The following result is taken from [1] and [2]. Its proof is based on the ideas of [8].

Theorem 2.8. (Quantum case) The joint spectrum of the asymptotic velocity and the energy can be determined if we know all the bound states and thresholds:

$$\sigma(P^+,H) = \bigcup_{a\in\mathcal{A}}\{(\xi_a,\lambda+\frac{1}{2}\xi_a^2)\ :\ \xi_a\in X_a,\ \lambda\in\cup_{b\leq a}\sigma^{\rm pp}(H^b)\}.$$

The classical analog of the above theorem that we know is not as satisfactory. It comes from [4]. In order to state it we introduce

$$\sigma^a_{\rm reg} := \{\lambda\ :\ (V^a)^{-1}(]-\infty,\lambda[) \text{ has a non-empty bounded connected component }\}.$$

Obviously,

$$\sigma^a_{\rm reg} \subset \sigma^a.$$

Theorem 2.9. (Classical case) The closure of the joint image of the asymptotic velocity and the energy has the following properties:

$$\begin{aligned}&\bigcup_{a\in\mathcal{A}}\{(\xi_a,\lambda+\tfrac{1}{2}\xi_a^2)\ :\ \xi_a\in X_a,\ \lambda\in\cup_{b\leq a}\sigma^b_{\rm reg}\}\\ &\subset \overline{(\xi^+,H)(X\times X')}\\ &\subset \bigcup_{a\in\mathcal{A}}\{(\xi_a,\lambda+\tfrac{1}{2}\xi_a^2)\ :\ \xi_a\in X_a,\ \lambda\in\cup_{b\leq a}\sigma^b\}.\end{aligned}$$

Asymptotic completeness in the free region

In the free region, that is the set $Z_{a_{\rm min}}$ one can analyze both quantum and classical scattering rather completely. It has essentially the same properties as the 2-body scattering.

Theorem 2.10. (Quantum case) For an appropriate function $S(t,\xi)$ there exists

$$\text{s}-\lim_{t\to\infty} {\rm e}^{itH}{\rm e}^{-iS(t,D)} =: \Omega^+_{a_{\rm min}}.$$

and we have

$$\text{Ran}\,\text{E}_{Z_{a_{\rm min}}}(P^+) = \text{Ran}\,\Omega^+_{a_{\rm min}}.$$

Theorem 2.11. (Classical case) Let

$$\tilde\phi(t)(x,\xi) := (x+\nabla_\xi S(t,\xi),\xi).$$

Then on $X\times Z_{a_{\rm min}}$ there exists

$$\lim_{t\to\infty}\phi(-t)\tilde\phi(t) =: \mathcal{F}^+_{a_{\rm min}}.$$

and we have

$$(\xi^+)^{-1}(Z_{a_{\rm min}}) = \text{Ran}\,\mathcal{F}^+_{a_{\rm min}}.$$

Asymptotic completeness / existence of asymptotic external position

The following theorem comes from [2]. The milestones of the history of the proof of the asymptotic completeness are the proof of Enss for 3-body systems including long-range systems for $\mu > \sqrt{3}-1$ (see [5] and references therein), the proof of the N-body case for short-range systems with an arbitrary N [13], and a later simpler proof of the same statement due to Graf (see [8], also [1]).

Note that this is the first time when we make an additional restriction on the value of μ, so far all the results were true for $\mu > 0$.

Theorem 2.12. (Quantum case) Let $\mu > \sqrt{3} - 1$. Choose appropriate functions $S_a(t,\xi_a)$. Then for any $a \in \mathcal{A}$ there exists

$$\mathrm{s-}\lim_{t\to\infty} \mathrm{e}^{itH}\mathrm{e}^{-iS_a(t,D_a)}\mathrm{E}^{\mathrm{pp}}(H^a) =: \Omega_a^+.$$

We have

$$\Omega_a^{+*} = \mathrm{s-}\lim_{t\to\infty} \mathrm{e}^{iS_a(t,D_a)}\mathrm{e}^{-itH}\mathrm{E}_{Z_a}(P^+)$$

and

$$L^2(X) = \sum_{a\in\mathcal{A}}^{\oplus} \Omega_a^+.$$

The above theorem gives a complete classification of states in the Hilbert space. Unfortunately, it seems that a good classical analog of the asymptotic completeness does not exists. Naively we might try to fully classify classical trajectories of an N-body system. The functions $H(x,\xi)$ and $\xi^+(x,\xi)$ give a partial classification. The following theorem, which assumes essentially the same decay of potentials as the previous one, defines another asymptotic quantity: the asymptoic external position, which allows a finer classification. This theorem comes from [3].

Theorem 2.13. (Classical case) Let $\mu > \sqrt{3} - 1$. Let $\xi^+(y,\eta) \in Z_a$. Then there exists

$$\lim_{t\to\infty} (\nabla_{\xi_a} S_a(t,\xi_a(t,y,\eta)) - x_a(t,y,\eta)) =: x_a^+(y,\eta).$$

Asymptotics of internal motion

In the quantum case, at least for $\mu > \sqrt{3} - 1$ we understand rather well the asymtotics of the internal motion: it becomes asymptotic to the evolution of bound states of internal Hamiltonians. In the classical case in general we have little control on the asymptotics of the internal motion. Nevertheless, there are two cases where the internal motion is asymptotic to a trajectory of the internal Hamiltonian.

Without stating the precise form of these results, let us describe these cases.

First of all we can classify all the trajectories if potentials decay superexponentially, more precisely if they satisfy

$$|\nabla v^a(x^a)| \leq C, \quad \forall\, \theta \text{ there exists } C_\theta \text{ such that } |\nabla v^a(x^a)| \leq C_\theta \mathrm{e}^{\theta|x^a|}.$$

The large time asymptotics of N-body trajectories for such potentials was first studied in [3]. An analogous result for potentials of compact support comes from [10].

Another case is that of 3-body systems with radial potentials with $\mu > 1$. The spherical symmetry of subsystems guarantees that the internal angular momentum has a limit as $t \to \infty$. Besides, the internal Hamiltonian is completely integrable, so we can use the action-angle coordinates. This easily allows one to prove that on a large part of the phase space the internal coordinates of a trajectory are asymptotic to a trajectory of the internal Hamiltonian. This result comes from [7] and [4].

References

[1] J. Dereziński: Algebraic approach to the N-body long range scattering, Rev. in Math. Phys., Vol 3 n°1 (1991) p 1–62.

[2] J. Dereziński: Asymptotic Completeness for N-particle long-range Quantum Systems, Ann. Math. 138, 427–476 (1993).

[3] J. Dereziński: Large time behavior of classical N-body systems, Comm. Math. Phys. 148, 503–520 (1992).

[4] J. Dereziński, C.Gerard: Asymptotic Completeness of N-body Systems, in preparation.

[5] V. Enss: Completeness of two-body and three-body quantum scattering, Journees Equations aux Derivees Partielles, Saint-Jean-de-Monts 1989, 1–31.

[6] C. Gérard: Semiclassical resolvent estimates for two and three-body Schrödinger operators, Comm. Part. Diff. Eq., 15 (1990) 1161–1178.

[7] C. Gérard: Asymptotic behavior of scattering trajectories for classical 3-body problems, preprint, Ecole Polytechnique 1993.

[8] G.M. Graf: Asymptotic completeness for N-body short range quantum systems : A new proof, Comm. in Math. Phys. Vol 132 (1990) p 73–101.

[9] I. Herbst: Classical Scattering with Long Range Forces, Comm. in Math. Phys. Vol 35 (1974) p 193–214.

[10] W. Hunziker: The S-Matrix in Classical Mechanics, Comm. in Math. Phys. Vol 8 (1968) p 282–299.

[11] E. Mourre: Absence of singular continuous spectrum for certain selfadjoint operators, Comm. in Math. Phys., 78 (1981) 519–567.

[12] P. Perry, I. Sigal, B. Simon: Spectral analysis of N-body Schrödinger operators, Ann. Math. Vol 114 (1981), p 519–567.

[13] I. Sigal and A. Soffer: The N-particle problem: asymptotic completeness for short-range systems, Annals of Math. 125, 35–108 (1987).

[14] B. Simon: Wave Operators for Classical Particle Scattering, Comm. in Math. Phys. Vol 23 (1971) p 37–48.

[15] Wang: Semiclassical resolvent estimates for N-body Schrödinger operators, J. of Funct. Anal. Vol 96 n°2, (1991) p 466–483.

Jan Dereziński
Division of Mathematical Methods in Physics
Warsaw University
Hoża 74, 00-682 Warszawa, Poland

Operator Theory:
Advances and Applications, Vol. 78

Quantum Fermi Accelerators with Pure-Point Quasi-Spectrum

P. Duclos, P. Šťovíček

Abstract

We consider a quantized version of the Fermi accelerator as well as a simplified model. We show that if the time dependent external force is periodic, sufficiently regular, with small enough fluctuations and a "non resonant" frequency then the corresponding Floquet operators is pure point.

1 Problems, methods and results

1.1 What are the Fermi accelerators?

What one usually calls Fermi Accelerator (FA) is a model describing a one dimensional classical particle moving freely in a box which has one of its wall oscillating in time. This model was precisely formulated by Ulam [3] and takes its origin in an attempt by Fermi [2] to understand the formation of cosmic rays. Latter a simplification of this model was proposed by Lichtenberg and Lieberman [LL] which goes naturally under the name Simplified Fermi Accelerator (SFA).

The quantum counterpart of the FA, (QFA) was considered e.g. by Šeba [Se]; it can be formulated as follows: let $t \to a(t)$, $a(t) > 0$ be the T-periodic function describing the moving boundary and $\psi(t,.) \in L^2(0,a(t))$ the wave function of the quantum particle at time t. ψ is the solution of the following Schrödinger equation:

$$(-i\partial_t - \partial_x^2)\psi = 0, \quad \psi(t,0) = \psi(t,a(t)) = 0 \ \ \forall t \in \mathbb{R}, \quad \psi(0,.) = \psi_0. \tag{73}$$

Since $\psi(t,.)$ belongs to a time-dependent Hilbert space the above problem requires some care to be meaningfull. This question is discussed in [Ka1] where other interesting boundary conditions beside the Dirichlet one chosen above, are considered. Karner [Ka2] has proposed the following quantization of the SFA: here $\psi(t,.)$ is an element of $L^2(0,1)$,

$$(-i\partial_t + H(t))\psi = 0, \quad H(t) := H_0 + g(t)\delta, \quad H_0 = -\partial_x^2 + \text{NBC}, \quad \psi(0,.) = \psi_0. \tag{74}$$

NBC stands for Neumann boundary condition and the formal potential $g(t)\delta(x)$ means that $H(t)$ is $-\partial_x^2$ equipped with the time dependent T-periodic boundary condition: $(\partial_x\psi)(t,0) = g(t)\psi(t,0)$.

For the forthcoming discussion we adopt the loose notation $H(t) = H_0 + V(t)$ for the time dependent Hamiltonian of both systems; H_0 stands for the kinetic energy operator which is roughly $-\partial_x^2$ and $V(t)$ is the time dependent T-periodic potential. Precise definitions of these objects will be made in Section 2.

1.2 Dynamical stability

What are these models good for? Apart for the physical phenomena they are supposed to model, they are good laboratories to explore the questions of dynamical stability and chaos both in classical and quantum mechanics. The system is said dynamically instable if it is able to pump energy from the driven force by an amount which grows indifinitely in time. More precisely one wants to investigate the behaviour of the classical and quantum kinetic energies, $p^2(t)$ and $(H_0\psi_t,\psi_t)$ respectively, as the time t goes to infinity; $p(t)$ denotes the classical momenta at time t of the trajectory with a given initial condition (q_0,p_0) and ψ_t is the wave function at time t of the quantum system with a given initial state ψ_0. We shall not consider here the chaotic aspects of these models. The question of dynamical stability of the classical FA and SFA accelerators are fairly well understood at least numerically and for not too big fluctuations of the external force (see [LL]). To discuss the quantum case we shall use the following tool.

1.3 Quasi-spectrum

It turns out that we are not able to answer the question concerning the behaviour of $(H_0\psi_t,\psi_t)$. Instead we shall consider the following quantity $(F(H_0 > e)\psi_t,\psi_t)$ which is the probability that the system explores the spectral subspaces of (the kinetic energy opertor) H_0 corresponding to (energy) values bigger than e; $F(A > a)$ denotes the spectral projector of the selfadjoint operator A associated to the interval (a,∞). An extension of the RAGE theorem to periodically time dependent quantum systems due to Enss and Veselić [EV] (see also [YKi]) asserts the following: let $U(T) := U(T,0)$ be the propagator over one period associated to the quantum system then

$$U(T) \text{ is pure point} \quad \Longrightarrow \quad \lim_{e\to\infty}\sup_{t\in\mathbb{R}} \|F(H_0 > e)\psi_t\| = 0. \tag{75}$$

We recall that a selfadjoint operator is said to be pure point if its spectral measure is purely atomic.

Often in the literature one can see that dynamical stability is associated to pure-pointness of $U(T)$; this is true if stability means boundeness in time of $(F(H_0 > e)\psi_t,\psi_t)$ but this is not yet proven if we understand it as the boundedness of $(H_0\psi_t,\psi_t)$. This last statement may even be wrong (see [16] and the discussion of this question in [3]). So using the Howland-Yajima trick (see [Ho1, 13]) saying that $1\otimes U(T)$ is unitarily equivalent to $\exp(-iTK)$ where K is the selfadjoint operator with symbol $-i\partial_t + H(t)$+PBC (periodic boundary condition) in time we get an equivalent criterium by replacing in (75) $U(T)$ by K. Spectral values of K are usually called the quasi-energies of the system, and they form the quasi-spectrum. $U(t)$ and K are named Floquet operator and Floquet Hamiltonian respectively.

1.4 Purely singular quasi-spectrum

This denomination means that the Floquet Hamiltonian K has no absolutely continuons spectrum or put differently: K has only point and/or singular continuous spectrum. The question of finding sufficient conditions under which this property occurs has received much attention since the seminal paper of Howland [Ho2] (see [7, 17]). They all deal with K of the form $-i\partial_t + H_0 + V(t)$+PBC on $L^2(0,T)\otimes\mathcal{H}$, where $\mathcal{H}$ is an abstract Hilbert space and use a crucial condition on the spectrum of H_0 of the type:

$$\exists\alpha > 0, \quad \inf\{(E^0_{n+1} - E^0_n)n^{-\alpha},\, n \in \mathbb{N}\} > 0 \tag{76}$$

where $\sigma(H_0) = \{E^0_1 < E^0_2 < \ldots\}$. The proof is based on the "big divisors method", (see Subsection 1.6). We use the same method for proving pure pointness but under a slighly different form better suited to our aim.

Except in [Ho3], the case of unbounded V is not considered in the literature. To face this difficulty for the QSFA we also use the big divisors method (see Subsection 2.2).

1.5 Stability of pure point quasi-spectrum

On $L^2(0,2\pi)\otimes\mathcal{H}$, let $K_0 := -i\omega\partial_t + H_0$+PBC in time, $\mathcal{H}$ being an abstract Hilbert space. We have made a scaling in time so that, the period is now 2π and the circular frequency ω appears an explicit parameter that we take once and for all in an open bounded interval Ω such that $\inf\Omega > 0$. Obviously K_0 is essentially selfadjoint on $\mathcal{H}^1_p\otimes D(H_0)$; $\mathcal{H}^1_p$ denotes the subspace of functions of $\mathcal{H}^1(0,2\pi)$, the usual Sobolev space, which have the same value at 0 and 2π. The same is true for $K := K_\omega := K_0 + V$, if V is bounded. We assume that H_0 has only discrete spectrum with simple multiplicity. This implies in particular that K_0 is pure point with spectrum being the closure of $\omega\mathbb{Z}+\sigma(H_0)$. By considering K in the eigenbasis of K_0 it becomes an operator on $L^2(\mathbb{Z}\times\mathbb{N})$; the image of K_0 is denoted D_0 whereas for K and V we keep the same letters:

$$K = D_0 + V, \quad D_0(i,j) = (\omega i_1 + E_{i_2})\delta_{i,j}.$$

The following assumption on V plays an important role: let

$$\forall\alpha > 0, \quad \tau_\star(\alpha) := \begin{cases} 6 + \frac{4}{\alpha} & \text{if } \alpha \le 1, \\ 6 + \frac{8}{1+\alpha} & \text{if } \alpha > 1, \end{cases} \tag{77}$$

then

$$\exists\tau > \tau_\star(\alpha), \quad \sup_{d\in\mathbb{Z}^2}(1 + |d|^\tau)\sup_{i-j=d}|V(i,j)| =: C_V < \infty. \tag{78}$$

Theorem 1.1. Let K be as above, then there exist two constants $\gamma_\star > 0$, $C_\star > 0$ such that the inequalities $0 < \gamma < \gamma_\star$ and $0 \le C_V \le C_\star\gamma^2$ imply that one can find a subset $\Omega' \subset \Omega$ fulfilling: $|\Omega\setminus\Omega'| < \gamma$ and for every ω in Ω', K_ω is pure point.

This theorem is proven in [4]. The proof is based on an idea borrowed from the KAM theorem. Bellisard [14] was the first to derive such a theorem in the case of the pulsed rotor where the perturbation was chosen analytic in time and space; this means that he considered the case $\alpha = 1$ for the gap condition (76) and that the decay of the matrix elements in (78) was of exponential type. Latter on, Combescure [Co] proposed an extension of this theorem to non analytic perturbations; her method was inspired by the corresponding extension of the KAM theorem to non analytic perturbations due to Moser. She has treated the case $\alpha = 0$ (H_0 being a one dimensional harmonic oscillator) and was therefore obliged to consider rather special type of perturbations.

The above theorem is however not able to cope with our two models since both perturbations being not at all regular in the space variable (see (81), (74)) the corresponding matrices V have no decay of the type required by (78) in the second index. Even worse the perturbation is not bounded for the QSFA. This is why we have to use the following tool to find a way out of these difficulties.

1.6 The big divisors method

We brielfly sketch how one can use the adiabatic method together with the growing gap condition of H_0, (76), to improve the decay of $V(i,j)$ in the variable $i_2 - j_2$ at the cost of loosing differentiabilty in time that is decay in the variable $i_1 - j_1$. All the detailed proofs of this subsection can be found in [4].

Here we make a partial Fourier analysis of K, we keep the time variable but we identify $\mathcal{H}$ with $L^2(\mathbb{N})$ using the eigenbasis of H_0. If V is bounded and strongly C^r time differentiable, $r \geq 1$, then one can show (see [Ho2] and section 2.2) that there exits a unitary mapping J which diagonalize $H_0 + V$ for each time: $H_0 + V = JH^dJ^\star$, and J is also strongly C^r differentiable; H^d stands for the diagonal matrix with entries being the eigenvalues of $H_0 + V$. Then we perform the following manipulations:

$$\begin{aligned} K &= -i\partial_t + H_0 + V = -i\partial_t + JH^dJ^\star \\ &= J(-i\partial_t + H^d - iJ^\star J')J^\star = J(-i\partial_t + H_0 + \tilde{V})J^\star =: J\tilde{K}J^\star \end{aligned}$$

where $\tilde{V} := H^d - H_0 - iJ^\star J'$. The first part of this new perturbation, $H^d - H_0$ is diagonal, and from regular perturbation theory one can see that it is bounded and strongly C^r differentiable. The second one is constructed with J which has the following property:

$$JH^dJ^\star = H_0 + V \quad \Longrightarrow \quad [H_0,J] = J(H^d - H_0) - VJ. \tag{79}$$

So from the boundedness of V we deduce that the commutator of J with H_0 is also bounded. This last property combined with the gap property of H_0 means something on the decay of J:

$$|(E_n^0 - E_m^0)J_{n,m}| \leq \|[H_0,J]\|.$$

This shows that the matrix elements of J decays like const$|n-m|^{-(1+\alpha)}$. This is the *big divisors* phenomenon. Of course the same property is true for $J^\star$ and by derivating with respect to the time the basic equality on the right of (79) we find that it also holds true for all the derivatives of J, up to order r. So the second part of $\tilde{V}$ which is a product of two such matrices will also have the same decay if we prove that this property is conserved by multiplication. This motivates the following definitions:

$$\mathcal{A}_0 := \mathcal{B}(L^2(\mathbb{N})), \quad X \in \mathcal{A}_k :\Leftrightarrow X \text{ and } \mathrm{ad}^k_{H_0} X \in \mathcal{A}_0.$$

$\mathcal{B}(\mathcal{H})$ deontes the algebra of bounded operators on $\mathcal{H}$ and $\mathrm{ad}_{H_0}X := [H_0,X]$. It is not difficult to prove the

Lemma 1.2. For every $k \in \mathbb{Z}_+$, $\mathcal{A}_k$ is a $\star$-subalgebra of $\mathcal{B}(L^2(\mathbb{N}))$ and $\mathcal{A}_k \subset \mathcal{A}_{k+1}$. It is now easy to see that $\tilde{V}$ and in fact all its derivatives up to order $r-1$ belong to $\mathcal{A}_{k+1}$ for every time t if V is in $\mathcal{A}_k$, $k \geq 0$.

We introduce the following convenient

Definition. A family of bounded operators $X(t)$ depending periodically on t (with a given period) belongs to $C^r(\mathcal{A}_k)$, $k \in \mathbb{Z}_+$, if and only if $X(t)$ is strongly C^r and $\mathrm{ad}^k_{H_0}(d/dt)^s X(t)$ is bounded and stongly continuous for $s = 0,1,\ldots,r$.

We summarize the above discussion in the

Theorem 1.3. *If V belongs to $C^{r+1}(\mathcal{A}_k)$, $r \geq 0$, then $\tilde{V}$ is in $C^r(\mathcal{A}_{k+1})$.*

1.7 The main theorem and applications

Combining the results of the two previous sections we arrive at our main

Theorem 1.4. Let H_0 be a selfadjoint operator acting on a separable Hilbert space which has a simple discrete spectrum $\{E^0_1 < E^0_2 < \ldots\}$ which obeys the gap condition (76) for a certain $\alpha > 0$. Let $t \to V(t)$ be a 2π-periodic strongly continuous family of selfadjoint bounded operators in $\mathcal{H}$. Let furthermore $K_{\omega,\beta} := -i\partial_t + H_0 + \beta V(\omega t)$ be the corresponding Floquet Hamiltonian acting in $L^2((0,T),\mathcal{H})$, $T := 2\pi/\omega$, with periodic boundary condition in t and with the frequency ω lying in an interval $\Omega = (a,b)$, $0 < a < b < \infty$.

If $V(t)$ is $N(\alpha)$ strongly differentiable with:

$$N(\alpha) := [\tau_\star(\alpha)] + [\tau_\star(\alpha)/(1+\alpha)] + 2 \tag{80}$$

($\tau_\star(\alpha)$ is given by (77) and $[x]$ means the integer part of x) then there exist two constants $\gamma_\star > 0$ and $\beta_\star > 0$ such that the inequalities

$$0 < \gamma < \gamma_\star, \quad 0 \leq \beta \leq \beta_\star\gamma^2,$$

imply that one can find $\Omega' := \Omega'(\gamma,\beta) \subset \Omega$ with the properties

$$|\Omega \setminus \Omega'| < \gamma \quad \textit{and} \quad \forall\omega \in \Omega',\ K_{\omega,\beta} \textit{ is pure point}.$$

In section 2 we transform our two QFA accelerators into equivalent forms suited to this theorem. Using these forms it is then straightforward to derive the

Theorem 1.5. Let $a(t) = a_0 + \beta a_1(\omega t)$ and $g(t) = g_0 + \beta g_1(\omega t)$ be the functions characterising the QFA and QSFA respectively. We assume: (i) ω belongs to Ω, a compact interval or $\mathbb{R}$ which does not contain 0, (ii) a_0 is a strictly positive constant and (iii) a_1 and g_1 are both 2π-periodic functions.

If a_1 is C^{19} then there exist two constants $\gamma^\star > 0$ and $\beta^\star > 0$ such that the inequalities,

$$0 < \gamma < \gamma^\star, \quad 0 < \beta < \beta^\star \gamma^2,$$

imply that one can find $\Omega' := \Omega(\gamma,\beta) \subset \Omega$ with the properties: $|\Omega \setminus \Omega'| < \gamma$ and for each ω in Ω' the Floquet operators of the QFA is pure point. The same conclusion holds true for the SQFA if g_1 is C^{18}.

We finish this section by the following

Remarks. (a) In the preparation of the QSFA we have considered only the case $g_0 = 0$. There is however no difficulty to extend the analysis of section 2.2 to non vanishing g_0.

(b) By working a little bit more on lemma 2.3 of section 2.2 one can see that not only the regularised potential $\tilde{V}$ is bounded but it also exhibits some decay property: $|E_n - E_m|(J^{(s)}\varphi_n^0,\varphi_m^0) = \mathcal{O}(1)$, uniformly in n and m, for all its derivatives $J^{(s)}$, $s = 0,1,\ldots,r$. This could be used to show that the conclusion of theorem 1.5 holds as well if g_1 is only C^{17}. Similarly, a_1 being C^{18} is sufficient for the QFA since its regularized potential $\tilde{V}$ (see (81)) has already some decay: $[H_0^r, \tilde{V}]$ is bounded for every $r < 1/2$.

2 Preparation of the two models

2.1 The QFA

Here we want first, to give a definition of the Floquet Hamiltonian K associated to the QFA problem (73) and second, to find an equivalent form adapted to our method for showing that K is pure point. K acts in $L^2(M)$ with $M := \{(t,x), 0 < t < T, 0 < x < a(t)\}$ and is defined by $K := -i\partial_t - \partial_x^2$ with a periodic boundary condition in time and Dirichlet boundary conditions (DBC) in space. We shall require here that $a \in C^2(\mathbb{R})$. The method below is inspired from [Se].

Let $(0,T) \times (0,1) \ni (t,x) \to s(t,x) := (t,a(t)x) \in M$; the unitary implementation of this C^2-diffeormorphism is: $L^2(M) \ni \psi \to S\psi := |s'|^{\frac{1}{2}}\psi \circ s \in L^2((0,T) \times (0,1))$. The image $K_1 := SKS^{-1}$ of K under S is

$$K_1 = -i\partial_t + a^{-2}(H_0 - a\dot{a}A), \quad H_0 := -\partial_x^2 + \text{DBC}, \quad A := \frac{1}{2i}(x\partial_x + \partial_x x).$$

The dot means time derivative. With such a form one can show in particular (see [RS2, pp 285]) the existence of the propagator over one period $U_1(T)$ and using the

Howland-Yajima trick (see the proof of the lemma below) one can associate canonically to $U_1(t)$ a selfadjoint realisation of K_1. But this form is still not convenient for our purpose. Let $\varphi(t,x) := -\frac{1}{4}a\dot{a}x^2$ then:

$$K_2 := e^{i\varphi}K_1 e^{-i\varphi} = -i\partial_t + a^{-2}(H_0 + V), \quad V := \frac{1}{4}a^3\ddot{a}x^2.$$

V is now bounded but we have to get rid of the time dependent factor a^{-2}. We use the following time rescaling: $(\mathcal{T}\psi)(t,x) = \dot{h}^{\frac{1}{2}}\psi(h(t),x)$, with h being the inverse function of $f : t \to \int_0^t a^{-2}(s)ds$. Then

$$K_3 := \mathcal{T}K_2\mathcal{T}^{-1} = \dot{h}^{-\frac{1}{2}}\tilde{K}_3\dot{h}^{-\frac{1}{2}}, \quad \tilde{K}_3 := -i\partial_t + H_0 + \tilde{V}, \quad \tilde{V}(t,x) := V(h(t),x), \tag{81}$$

K_3 and $\tilde{K}_3$ acting on $L^2((0,\tilde{T}) \times (0,1))$ with $\tilde{T} := f(T)$. It is well known that $\tilde{K}_3$ is essentially selfadjoint on $\mathcal{H}_p^1 \otimes D(H_0)$; this vector space being invariant under the multiplication by $\dot{h}^{\frac{1}{2}}$ and its inverse we deduce that this also the case for K_3. So composing successively by $\mathcal{T}^{-1}$, $\exp(-i\varphi)$ and S^{-1} we get an essentially selfadjoint realisation of K. Moreover one also has the

Lemma 2.1. K is pure-point if and only if $\tilde{K}_3$ is pure-point.

Proof. Consider the two evolution equations:

$$K_2U_2(t) = 0,\ U_2(0) = 1 \quad \text{and} \quad \tilde{K}_3\tilde{U}_3(t) = 0,\ \tilde{U}_3(0) = 1.$$

Obviously $U_2 \circ h = \tilde{U}_3$ so that $U_2(T) = \tilde{U}_3(\tilde{T})$. Due to the Howland-Yajima trick (see [Ho1, 13]) one knows that $1 \otimes U_2(T)$ is unitarily equivalent to $\exp(-iTK_2)$ and the same is true for $1 \otimes \tilde{U}_3(\tilde{T})$ and $\exp(-i\tilde{T}\tilde{K}_3)$. This together with K_2 unitarily equivalent to K shows the lemma.

2.2 The QSFA

It is convenient to consider $H := H_g := H_0 + g\delta$ as an entire analytic family of type B [9, ch.VII §4] in the variable g. Then this is very elementary to verify that H is purely discrete for all g in $\mathbb{C}$; each eigenvalue $E_n := E_n^g$ is simple, analytic in a neighbourhood of the real axis and one has

$$\forall n \geq 0,\ E_n^0 = n^2\pi^2 \text{ and } \quad E_n = E_0 + 2g + \mathcal{O}(n^{-2}g^2), \text{ as } n \to \infty. \tag{82}$$

Since their associated eigenprojections $P_n := P_n^g$ are also analytic we may define analytic families of eigenvectors of H by

$$\varphi_n := \varphi_n^g := \frac{P_n\varphi_n^0}{\|P_n\varphi_n^0\|}, \quad \varphi_0^0 := 1, \quad \varphi_n^0(x) := \sqrt{2}\cos(n\pi x)\ \forall n \geq 1$$

provided we verify that $P_n\varphi_0 \neq 0$. Assume that $P_n\varphi_n^0 = 0$ and let ϕ_n be any vector in the range of P_n; since one has $(E_n - E_n^0)(\phi_n,\varphi_n^0) = g\phi_n(0)\varphi_n^0(0)$ this would mean that $\phi_n(0)$ and therefore $\phi_n'(0)$ vanish. So $\phi_n = 0$ by the unique continuation principle which leads to the wrong statement: $P_n = 0$.

Armed with these eigenvectors we define the following linear map on $\mathcal{H}$:

$$\forall n \geq 0, \quad J\varphi_n^0 := \varphi_n,$$

for which we have the

Lemma 2.2. J is an analytic family of bounded operators which is unitary on the real axis.

Proof. The unitarity of J on the real axis is obvious. To verify the boundedness of J we note that for every fixed g one has using lemma 2.4 below: $\forall n,m$,

$$(E_n - E_m)(J^\star J\varphi_n^0,\varphi_m^0) = (g - \bar{g})\varphi_n(0)\overline{\varphi_m(0)} = \mathcal{O}(1), \text{ and } |(J^\star J\varphi_n^0,\varphi_n^0)| \leq 1$$

uniformly in n and m. These estimates show with the help of the Schur-Holmgren criterium and the fact that $|E_n - E_m|^{-1} = \mathcal{O}(|n-m|^{-2})$ that $J^\star J$ is bounded.

Since $g \to (J\varphi_n^0,\varphi_m^0) = (\varphi_n,\varphi_m^0)$ is clearly analytic on $\mathbb{C}$ for all fixed couple of eigenvectors of H_0 this proves the statement concerning the analyticity.

We are now able to find a more convenient setting for our QSFA.

Lemma 2.3. If g is in $C^r(\mathbb{R},\mathbb{R})$, $r \geq 1$, and T-periodic, there exists a selfadjoint realisation of the QSFA Floquet Hamiltonian which is unitarily equivalent to $\tilde{K} := -i\partial_t + H_0 + \tilde{V}$ with $\tilde{V}$ being bounded and C^{r-1} norme differentiable in time and T-periodic.

Proof. Of course with g being such a function of time we get from the previous lemma that J is now a unitary family r time continuously norm differentiable and T-periodic; this implies that $-iJ^\star J'$ is $r-1$ time continuously norm differentiable and T-periodic. The eigenvalues of H are themselves T-periodic functions of class C^r. Let $H^d := J^\star HJ$, then $H^d - H_0$ is diagonal in the eigenbasis of H_0 with eigenvalues $E_n - E_n^0$. This shows that $H^d - H_0$ is bounded (see (82)), C^r norme differentiable in time and T-periodic. So as in section 1.6 we have at least formally $K = J\tilde{K}J^\star$ with $\tilde{V} := H^d - H_0 - iJ^\star J'$. Since $\tilde{K}$ is essentially selfadjoint on $\mathcal{H}_p^1 \otimes D(H_0)$ this gives a way to define K.

We end up this section by some technical results.

Lemma 2.4. For all φ_n as defined above one has as n tends to infinity: $\varphi_n(0) = \mathcal{O}(1)$ uniformly with respect to g in a given compact.

Sketch of proof. It is convenient to view the operator H as $H_0 + g\tau^\star\tau$, where $\tau : \mathcal{H} \to \mathbb{C}$ is the trace operator $\tau u := u(0)$. We shall use the resolvent formula, $R(z) = R_0(z) - gR(z)\tau^\star\tau R_0(z)$, where $R(z) := R_g(z) := (H-z)^{-1}$ and the following estimates: let $\Gamma_n := \{z \in \mathbb{C}, |\sqrt{z} - \sqrt{E_n^0}| = \frac{\pi}{6}\}$, then $\tau R(z)\tau^\star = \mathcal{O}(n^{-1})$, as well as $\tau R(z)$, as n tends to infinity, uniformly in $z \in \Gamma_n$ and g in a given compact. To get these estimates we remark that one can compute explicitely:

$$\tau R(z)\tau^\star = \frac{\tau R_0(z)\tau^\star}{1 + g\tau R_0(z)\tau^\star} = (g - \sqrt{z}\tan\sqrt{z})^{-1}.$$

Clearly $\tau R_0(z)\tau^\star = \mathcal{O}(n^{-1})$ and the same is true for $\tau R_(z)\tau^\star$. Then using

$$\|\tau R(z)\|^2 = \tau R(z)R(\overline{z})\tau^\star = \frac{\operatorname{Im}\tau R(z)\tau^\star}{\operatorname{Im} z}$$

we find by straightforward calculus the second statement. This allows to get:

$$\begin{aligned}\|P_n - P_n^0\| &= \frac{|g|}{2\pi}\int_{\Gamma_n} \|R(z)\tau^\star\tau R_0(z)dz\| \\ &\leq |g|\frac{|\Gamma_n|}{2\pi} \sup_{z\in\Gamma_n} \|\tau R(z)\|\, \|\tau R_0(z)\| = \mathcal{O}(n^{-1})\end{aligned}$$

where we have used that $|\Gamma_n|$ grows like n. This shows in particular that $\|P_n\varphi_n^0\| = 1 + \mathcal{O}(n^{-1})$.

Finally with

$$\varphi_n(0)\,\|P_n\varphi_n^0\| = \tau P_n\varphi_n^0 = \varphi_n^0(0) - g\frac{i}{2\pi}\int_{\Gamma_n} \tau R(z)\tau^\star\tau R_0(z)dz$$

and all the above results one may conclude easily.

References

[B] Bellissard J.: Stability and instability in quantum mechanics, in "Trends and Developments in the Eighties", Albeverio and Blanchard Eds., Word Scientific, Singapore, 1985, pp. 1–106

[C] Combes J.-M.: Connection between quantum dynamics and spectral properties of time-evolution operators, in "Differential Equations and Applications to Mathematical Physics", W. F. Ames et al Eds., Mathematics in Sciences and Engineering **192**, Academic Press, 1993, pp. 59–68

[Co] Combescure M.: The quantum stability problem for time-periodic perturbations of the harmonic oscillator, Ann. Inst. Henri Poincaré **47** (1987) 62–82; Erratum: Ann. Inst. Henri Poincaré **47** (1987) 451–454

[DS] Duclos P., Šťovíček P.: Floquet Hamiltonians with pure point spectrum, to appear

[EV] Enss V., Veselić K.: Bound states and propagating states for time-dependent Hamiltonians, Ann. Inst. Henri Poincaré **39** (1983) 159–191

[F] Fermi S.: On the Origin of the Cosmic Radiations, Phys. Rev. **75**(8), (1949), 1169–1174

[G] Guarneri I.: Spectral Properties of Quantum Diffusion on Discrete Lattice, Europh. Lett. **10**(2), (1989), 95–100

[Ho1] Howland J. S.: Scattering theory for Hamiltonians periodic in time, Indiana J. Math. **28** (1979) 471–494

[Ho2] Howland J. S.: Floquet operators with singular spectrum I,II, Ann. Inst. Henri Poincaré **49** (1989) 309–323; Ann. Inst. Henri Poincaré **49** (1989) 325–334

[Ho3] Howland J. S.: Stability of quantum oscillators, J. Math. Phys. **25** (1992) 5177–5181

[J] Joye A.: Absence of absolutely continuous spectrum of Floquet operators, J. Stat. Phys. **75** (1994) 929–952

[K] Kato T.: Perturbation theory of linear operators, Springer-Verlag, New York, 1966

[Ka1] Karner B.: On the quantum Fermi accelerator and its relevance to "quantum chaos", Lett. Math. Phys. **17** (1989) 329–339

[Ka2] Karner G.: The simplified Fermi accelerator in classical and quantum mechanics, to appear in J. Stat. Phys.

[YKi] Yajima K., Kitada H.: Bound states and scattering states for time periodic Hamiltonians, Ann. Ins. Henri Poincaré **39**(2) (1983) 145–157

[Ko] Kotani S.: Ljapunov indices determine absolutely continuous spectra of stationary random one-dimensional Schrödinger operators, in "Stochastic Analysis", K. Ito Ed., North Holland, Amsterdam, 1984, pp. 225–248

[LL] Lichtenberg A.J., Lieberman M.A.: Regular and Stochastic Motion, Applied Mathematical Science **38**, Springer-Verlag 1983

[N] Nenciu G.: Floquet operators without absolutely continuous spectrum, Ann. Inst. Henri Poincaré **59** (1993) 91–97

[RS2] Reed M., Simom B.: Modern Methods of Mathematical Physics II. Fourier Analysis and Selfadjointness. Academic Press 1975

[Se] Seba P.: Quantum chaos in Fermi-accelerator model, Phys. Rev. A **41** (1990) 2306–2310

[U] Ulam S.: On some statistical properties of dynamical systems, Proc. 4th Berkeley Symp. Math. Stat. Prob. **3**, 315, Univ. California Press (1961)

[Y] Yajima K.: Scattering Theory for Schrödinger Equations with Potential Periodic in Time, J. Math. Soc. Japan **29** (1977) 729–743

P. Duclos
Centre de Physique Théorique, CNRS
13288 Marseille–Luminy and PHYMAT, Université de Toulon et du Var,
BP 132, 83957 La Garde Cedex, France

P. Šťovíček
Department of Mathematics and Doppler Institute
Faculty of Nuclear Science, CTU, Trojanova 13
120 00 Prague, Czech Republic

Operator Theory:
Advances and Applications, Vol. 78

On Moments of Negative Eigenvalues of an Elliptic Operator

Yu.V. Egorov, V.A. Kondratiev

Let Ω be a domain in $\mathbb{R}^n$, let

$$L = \sum_{|\alpha|\leq m, |\beta|\leq m} D^\alpha a_{\alpha\beta}(x) D^\beta$$

be an elliptic symmetric positive operator of order $2m$ with measurable coefficients, considered with zero Dirichlet boundary conditions, such that

$$\int_\Omega \sum_{|\alpha|\leq m, |\beta|\leq m} a_{\alpha\beta}(x) D^\alpha u \overline{D^\beta u} dx \geq a_0 \int_\Omega \sum_{|\alpha|=m} |D^\alpha u(x)|^2 dx,$$

where $a_0 = const > 0$ for all functions $u \in C_0^\infty(\Omega)$. Let $V(x)$ be a real-valued function defined in Ω.

We consider the problem: to estimate in terms of V the sum

$$S_\gamma = \sum_{\lambda_j < 0} |\lambda_j|^\gamma,$$

where λ_j are negative eigenvalues of the operator $H = L - V$. We will assume for the operator H the max-min principle is valid.

Such sums are very important for Mathematical Physics, they were considered (for the Schrödinger operator) in the works [1]–[3] of Ruelle, Lieb and Thirring and others. In particular, Lieb and Thirring have obtained the estimate

$$S_\gamma \leq C \int V(x)^{\gamma+n/2} dx$$

in the case $m = 1, \gamma + n/2 > 1$. Note also Lieb's paper [4], where sharp values of the constant C were indicated.

We obtain here more general estimates. Our proofs are rather elementary, they are valid for any dimension n. We use the methods of our works [5]-[6].

Our attention to this problem was attired by Professor Heinz Siedentop in relation to the Lieb's question on the possibility to estimate the sum $S_{1/2}$ in terms of the integral $\int V(x)dx$ in the case $m = 1/2$. For the operator

$$Ly = -y'' + V(x)y, \quad -\infty < x < \infty$$

with $V(x) \geq 0$ the estimate $S_\gamma \leq C \int V(x)^{\gamma+1/2} dx$ is true for $\gamma > 1/2$. If $\gamma = 1/2$, we can prove that $|\lambda_j| = o(j^{-1/2})$ for $\lambda_j \to 0-$, if $\int V(x)dx < \infty$. But the convergence of the series $\sum_{\lambda_j<0}|\lambda_j|^{1/2}$ has not been proved. Theorem 1 in below says that $S_{1/2}$ cannot be estimated by the integrals $\int V_+(x)^{1+r/2}|x-x_0|^r dx$ for any $r>0$.

The principal result here is the following theorem.

Theorem 1. Let $r\geq 0$, $\gamma>0$, $\gamma+(n+r)/2m>1+r/n$. Then

$$S_\gamma \leq C_{m,n,s,\gamma}\int_\Omega V_+(x)^{(n+r)/2m+\gamma}|x-x_0|^r dx,$$

where $V_+(x)=max(V(x),0), x_0\in\mathbb{R}^n$. If $1+r/n\geq\gamma+(n+r)/2m>1$ and $\int_\Omega V_+(x)^{(n+r)/2m+\gamma}|x-x_0|^r dx<\infty$, then $S_\gamma<\infty$. However, in this case for any $\varepsilon>0, N>0$ there is a potential $V(x)$ such that $\int_\Omega V_+(x)^{(n+r)/2m+\gamma}|x-x_0|^r dx<\varepsilon$, and $S_\gamma>N$.

Remark that with $\gamma=0$ Theorem 1 gives our result of [5] for the counting function. Our proof is based on the following Lemma, having its own interest.

Lemma 1. Let $\alpha>0$, N_α be the number of eigenvalues which are smaller than $-\alpha$. Then

$$N_\alpha\leq C_{m,n,s}\alpha^{-q+n/2m+r/2m}\int_\Omega V_\alpha(x)^q|x-x_0|^r dx+K_{m,n},$$

where $V_\alpha(x)=V(x)$, if $V(x)>\alpha$ and $V_\alpha(x)=0$ if $V(x)\leq\alpha$; $q>1$, $0\leq r<2mq-n$ and $K_{m,n}$ is the number of different derivatives D^β of all orders $0\leq\beta\leq m-n/2$. In particular, $K_{m,n}=0$, if $2m\leq n$.

Proof of Lemma. Remark that N_α coincides with the codimension of the subspace of functions $u\in C_0^\infty(\Omega)$, for which the estimate

$$\int_\Omega V(x)u(x)^2dx\leq(Lu,u)+\alpha\int u(x)^2dx$$

is valid. It is clear that this estimate is equivalent to the inequality

$$\int_\Omega V_\alpha(x)u(x)^2dx\leq(Lu,u)+\alpha\int u(x)^2dx. \tag{1}$$

Let at first $n>2m$. For an unit cube the estimate

$$\left(\int_{Q_0}\frac{u(x)^{2p}}{|x-x_0|^s}dx\right)^{1/p}\leq c_0\int_{Q_0}(\sum_{|\beta|=m}|D^\beta u|^2+|u|^2)dx,\ u\in C^\infty(Q_0) \tag{2}$$

is valid, if

$$0\leq s<n,\ 1\leq p<(n-s)/(n-2m). \tag{3}$$

Indeed, by the Hölder inequality we have

$$\int_{Q_0}\frac{u(x)^{2p}}{|x-x_0|^s}dx\leq\left(\int_{Q_0}u(x)^{2pp_1}dx\right)^{1/p_1}\left(\int_{Q_0}|x-x_0|^{-sq_1}dx\right)^{1/q_1},$$

where $p_1^{-1} + q_1^{-1} = 1$. The latter integral is finite if $sq_1 < n$. On the other hand, the Sobolev estimate

$$\left(\int_{Q_0} u(x)^{2pp_1} dx \right)^{1/pp_1} \leq c_0 \int_{Q_0} \left(\sum_{|\beta|=m} |D^\beta u|^2 + |u|^2 \right) dx$$

is true for $2pp_1 \leq 2n/(n-2m)$. Therefore, such p_1 and q_1 do exist if the condition (3) is satisfied.

Therefore, the estimate for a cube Q' with an edge b

$$\left(\int_{Q'} u(x)^{2p} |x - x_0|^{-s} dx \right)^{1/p} \tag{4}$$

$$\leq c_0 b^{2m-n/q-s/p} \int_{Q'} \left(\sum_{|\beta|=m} |D^\beta u(x)|^2 + b^{-2m} |u(x)|^2 \right) dx,$$

where $p^{-1} + q^{-1} = 1$, holds under the condition (3).

On the other hand, consider the class of smooth functions in Q_0, satisfying the conditions

$$\int_{Q_0} D^\beta u(x) dx = 0, 0 \leq |\beta| \leq m - 1.$$

Repeating the above estimations, one can see that the estimate

$$\left(\int_{Q_0} \frac{u(x)^{2p}}{|x - x_0|^s} dx \right)^{1/p} \leq c_0 \int_{Q_0} \sum_{|\beta|=m} |D^\beta u|^2 dx,$$

is true for functions of this class under the condition (3). Therefore, the estimate

$$\left(\int_{Q'} \frac{u(x)^{2p}}{|x - x_0|^s} dx \right)^{1/p} \leq c_0 b^{2m-n/q-s/p} \int_{Q'} \sum_{|\beta|=m} |D^\beta u(x)|^2 dx, \tag{5}$$

where $p^{-1} + q^{-1} = 1$, is valid for functions of this class in a cube Q' with its edge equal to b under the condition (3).

Let $\varepsilon > 0$ be a positive number, which will be indicated later. By Rosenblum's lemma (see [6]), there is a covering of the support K of V (which we can assume to be compact) by a finite number N of cubes with their edges parallel to the coordinate axes, such that each point of K is covered by at most c_n cubes,

$$\int_{Q_j} V_\alpha(x)^q |x - x_0|^{s(q-1)} dx \leq \varepsilon |\alpha|^{q-n/2m-s(q-1)/2m} I, \tag{6}$$

where

$$I = \alpha^{-q+n/2m+s(q-1)/2m} \int_\Omega V_\alpha(x)^q |x - x_0|^{s(q-1)} dx$$

and $N \leq c_n \varepsilon^{-1}$. Let b_j be the edge of the cube Q_j.

If $2|\alpha|^{-1/2m} > b_j \geq |\alpha|^{-1/2m}$, then $b_j^{-2m} \leq |\alpha|$ and we have

$$\left(\int_{Q_j} u(x)^{2p}|x-x_0|^{-s}dx\right)^{1/p} \tag{7}$$

$$\leq c_0|\alpha|^{-1+n/2mq+s/2mp}\int_{Q_j}\left(\sum_{|\beta|=m}|D^\beta u(x)|^2+\alpha|u(x)|^2\right)dx.$$

If $b_j \geq 2|\alpha|^{-1/2m}$, we can bisect the cube Q_j several times and represent it as a sum of intersecting cubes Q_j' having size b_j' such that

$$2|\alpha|^{-1/2m} > b_j' \geq |\alpha|^{-1/2m}.$$

We can write the above estimates (7) for each cube Q_j' and sum them over all such cubes. It gives (7) for the cube Q_j.

If $b_j \leq |\alpha|^{-1/2m}$ and the condition (3) is fulfilled, then the estimate

$$\left(\int_{Q_j}\frac{u(x)^{2p}}{|x-x_0|^s}dx\right)^{1/p} \leq c_0|\alpha|^{-1+n/2mq+s/2mp}\int_{Q_j}\sum_{|\beta|=m}|D^\beta u(x)|^2dx \tag{8}$$

is true under the supplementary conditions

$$\int_{Q_j} D^\beta u(x)dx = 0,\ 0\leq|\beta|\leq m-1. \tag{9}$$

Now using the estimates (6), (7), (8), we have

$$\int_\Omega V_\alpha(x)|u(x)|^2dx \leq \sum_{j=1}^N\int_{Q_j}V_\alpha(x)|u(x)|^2dx$$

$$\leq \sum_{j=1}^N\left(\int_{Q_j}\frac{u(x)^{2p}}{|x-x_0|^s}dx\right)^{1/p}\left(\int_{Q_j}V_\alpha(x)^q|x-x_0|^{s(q-1)}dx\right)^{1/q}$$

$$\leq c_0(\varepsilon I)^{1/q}\sum_{j=1}^N\int_{Q_j}\left(\sum_{|\beta|=m}|D^\beta u(x)|^2+\alpha|u(x)|^2\right)dx$$

$$\leq (Lu,u)+\alpha\int_\Omega u^2dx,$$

if ε is such that $c_0(\varepsilon I)^{1/q} = \min(a_0,1)$. Thus the estimate (1) holds, if the conditions (9) are fulfilled. Therefore, the number N_α does not exceed the number of the conditions (9), i.e.

$$N_\alpha \leq C\varepsilon^{-1} \leq C'I = C'\alpha^{-q+n/2m+s(q-1)/2m}\int_\Omega V_\alpha(x)^q|x-x_0|^{s(q-1)}dx,$$

where the constants C, C' depend on m,n,s. Putting $r = s(q-1)$, we obtain the result. Remark that

$$r = s(q-1) < n(q-1) \tag{10}$$

and r can be arbitrary number such that $0 \leq r < nq - n$. It is easy to see that the condition (3) is equivalent to the condition

$$0 \leq r < 2mq - n.$$

If $n \leq 2m$, then the estimates (2) or (5) are also true for $0 \leq s < n$. However, if we assume that

$$D^\beta u(x_0) = 0 \text{ for } 0 \leq |\beta| < m - n/2, \tag{11}$$

then applying the Hardy inequality (see, for example [6]), we can obtain the inequalities (4) and (5) for $0 \leq s < n + (2m-n)p$, i.e. if $0 \leq r < 2mq - n$. The proof stated above remains true, but the number N_α must be enlarged by $K_{m,n}$, equal to the number of conditions (11). Remark, that, in particular, $K_{m,n} = 0$, if $2m \leq n$ and $K_{m,n} = 1$, if $2m = n + 1$.

Proof of Theorem 1. Remark that

$$S_\gamma = \sum_{\lambda_j < 0} \int_0^\infty \alpha^\gamma dN_\alpha = -\gamma \int_0^\infty \alpha^{\gamma-1}(N_\alpha - K_{m,n}) d\alpha.$$

If $\gamma - q + n/2m + r/2m > 0$, then applying Lemma, we have

$$S_\gamma \leq C \int_0^\infty \alpha^{\gamma-1-q+n/2m+r/2m} \int_\Omega V_\alpha(x)^q |x - x_0|^r dx d\alpha.$$

Taking into account the condition of Lemma 1, we can state that for any $\gamma > 0$, satisfying the condition

$$\gamma + (n+r)/2m > 1 + r/n,$$

there is a $q > 1$ such that

$$\gamma + (n+r)/2m > q > 1 + r/n$$

and

$$\begin{aligned} S_\gamma &\leq C \int_\Omega |x - x_0|^r dx \int_0^{V_+(x)} \alpha^{\gamma-1-q+n/2m+r/2m} V_\alpha(x)^q d\alpha \\ &\leq C_1 \int_\Omega |x - x_0|^r V_+(x)^{\gamma+n/2m+r/2m} dx. \end{aligned}$$

If $\gamma + (n+r)/2m \leq 1 + r/n$, but $\gamma + (n+r)/2m > 1$, we cannot estimate the finite sum $\sum |\lambda_j|^\gamma$ corresponding to the first $K_{m,n}$ negative eigen-values, but the sum of $|\lambda_j|^\gamma$ for other negative eigen-values is estimated as above.

Let $(n+r)/2m+\gamma < 1+r/n$. Put $V(x) = |x-x_0|^{-n-\kappa}$ if $\delta^2 < |x-x_0| < \delta$, $V(x) = \delta^{-2n-2\kappa}$, if $\delta^2 > |x-x_0|$ and $V(x) = 0$ if $\delta < |x-x_0|$. Here $\kappa > 0$ is such that $(1+\kappa/n)((n+r)/2m+\gamma) < 1+r/n$. Then $\int V(x)dx \to \infty$ as $\delta \to 0$, but

$$\int_{R^n} |x-x_0|^r V_+(x)^{\gamma+n/2m+r/2m} dx < \varepsilon,$$

if δ is sufficiently small. Since the first negative eigen-value λ_1 (with the maximal value of its module) can be defined as

$$\lambda_1 = \inf_u \int \Big(\sum_{|\beta|=m} |D^\beta u(x)|^2 - V(x)u^2 \Big) dx \Big(\int u^2 dx\Big)^{-1},$$

we can write that

$$\lambda_1 \le \int \Big(\sum_{|\beta|=m} |D^\beta u_0(x)|^2 - V(x)u_0^2 \Big) dx \Big(\int u_0^2 dx \Big)^{-1} \le \int u_0'^2 dx - \int V(x) dx,$$

if u_0 is a function from C_0^∞, equal 1 for $|x-x_0| < 1/2$ and such that $\int u_0^2 dx = 1$. Therefore, $\lambda_1 \to -\infty$ and $S_\gamma \to +\infty$ as $\delta \to 0$.

If $(n+r)/2m+\gamma = 1+r/n$, then we can get the same effect, putting $V(x) = |x-x_0|^{-n}/\ln(1/|x-x_0|)$ for $\delta < |x-x_0| < 1/ln(1/\delta)$.

The proof is complete.

Theorem 2. Let Ω be a domain in $\mathbb{R}^n$ such that its intersection with the sphere $|x-x_0| = R$ for $R > 1$ has a diameter not exceeding AR^σ with $\sigma < 1$. Let $r \ge 0$, $\gamma > 0$, $\gamma + (n+r)/2m > 1 + r/n$. Then

$$S_\gamma \le C_{m,n,s,\gamma} \int_\Omega V_+(x)^{(n+r)/2m+\gamma} |x-x_0|^{\sigma r} dx,$$

where $V_+(x) = \max(V(x), 0), x_0 \in \mathbb{R}^n$.

Proof. As in the proof of Theorem 1 we show at first that

$$N_\alpha \le C_{m,n,r} I, \text{ where } I = \alpha^{-q+n/2m+r/2m} \int_\Omega V_\alpha(x)^q |x-x_0|^{\sigma r} dx. \tag{12}$$

After it the proof of Theorem 1 can be repeated.

To prove the inequality (12) we follow the same schema as above. However, we choose the cubes Q_j in such a way that

$$(2A)^r \int_{Q_j} V_\alpha(x)^q |x-x_0|^{\sigma s(q-1)} dx \le \varepsilon \alpha^{q-n/2m-s(q-1)/2m} I.$$

Let x_j be the center of a cube Q_j and b_j its size. We need no conditions (9) for cubes of size $b_j > 2A|x_j - x_0|^\sigma$, since for these cubes the estimate (5) is true, because $u(x) = 0$ on a subset of the cube of measure more than $c_0 b_j^n$, $c_0 > 0$.

Let Q_j one of other cubes. Then

$$\int_{Q_j} V_\alpha(x)|u(x)|^2 dx \leq \left(\int_{Q_j} \frac{u(x)^{2p}}{|x-x_j|^s} dx\right)^{1/p} \left(\int_{Q_j} V_\alpha(x)^q |x-x_j|^{s(q-1)} dx\right)^{1/q}.$$

If $n \leq 2m$, then we assume that

$$u(x_j) = 0, \ D^\beta u(x_j) = 0 \text{ for } |\beta| < m - n/2.$$

If $s < n + p(2m-n)$, then the estimate

$$\left(\int_{Q_j} \frac{u(x)^{2p}}{|x-x_j|^s} dx\right)^{1/p} \leq c_0 b_j^{2m-n/q-s/p} \int_{Q_j} \sum_{|\beta|=m} |D^\beta u(x)|^2 dx$$

is true under the conditions (3) and since

$$\int_{Q_j} V_\alpha(x)^q |x-x_j|^r dx \leq (2A)^r \int_{Q_j} V_\alpha(x)^q |x-x_0|^{\sigma r} dx \leq \varepsilon \alpha^{q-n/2m-r/2m} I,$$

we can repeat the proof of Lemma 1.

Remark. If the point x_0 is outside of the closure of the domain Ω, then the estimates in Theorems 1 and 3 are valid for $\gamma + (n+r)/2m > 1$.

We would like to thank Professor Heinz Siedentop who attired our attention to this problem.

References

[1] V. Ruelle, Large volume limit of the distribution of characteristic exponents in turbulence, Comm. Math. Phys., **87**(1982), 287–302.

[2] E. Lieb, W. Thirring, Inequalities for the moments of the eigenvalues of the Schrödinger Hamiltonian and their relation to Sobolev inequalities. In: Studies in mathematical physics: essays in honor of V. Bargmann; E. Lieb, B. Simon, A. Wightman (eds.), 269–303. Princeton, NJ: Princeton University Press 1976.

[3] E. Lieb, W. Thirring, Bounds for the kinetic energy of fermions which proves the stability of matter, Phys. Rev. Lett., **687**(1975), 35.

[4] E. Lieb, On characteristic exponents in turbulence, Comm. Math. Phys., **92**(1984), 473–480.

[5] Yu. V. Egorov, V.A. Kondratiev, On the negative spectrum of an elliptic operator, Mat. Sb., 181(1990), 147–166. English transl. in Math. USSR-Sb. **69** (1991), 155–177.

[6] Yu. V. Egorov, V.A. Kondratiev, Estimates of the negative spectrum for an elliptic operator, Lectures of XIV School on Operator Theory on Functional Spaces, Novgorod, 1990; English transl. in Amer. Math. Soc., Providence, RI. (1991).

Yu. V. Egorov
Université Paul Sabatier, UFR MIG, 118 Route de Narbonne
31062 Toulouse, France

V. A. Kondratiev
Lomonossov University, Meh-mat Faculty
119 899 Moscow, Russie

Operator Theory:
Advances and Applications, Vol. 78

Magnetic Lieb-Thirring Inequalities and Stochastic Oscillatory Integrals

L. Erdős[1]

Abstract

We study the generalizations of the well-known Lieb-Thirring inequality for the magnetic Schrödinger operator with a nonconstant magnetic field. We use stochastic methods to prove estimates on the moments of the negative eigenvalues.

1 Introduction

The classical Lieb-Thirring inequality [10], [11] estimates the moments of the negative eigenvalues ($E_1 \leq E_2 \leq \ldots \leq 0$) of a general Schrödinger operator $-\Delta + V$, acting on $L^2(\mathbb{R}^d)$, as follows

$$\sum_i |E_i|^\gamma \leq L_{\gamma,d} \int_{\mathbb{R}^d} |V(x)|_-^{\gamma+d/2} \mathrm{d}x \tag{83}$$

($|V(x)|_- := \max(0, -V(x))$ denotes the negative part), where $L_{\gamma,d}$ is a universal finite constant ($\gamma \geq 0$ if $d \geq 3$, $\gamma > 0$ if $d = 2$ and $\gamma > 1/2$ if $d = 1$). In the applications, the $\gamma = 0$ and $\gamma = 1$ cases are especially important; the corresponding eigenvalue moment equals the maximal number of bound *noninteracting* electrons (subject to the Pauli principle) and their ground state energy, respectively.

We remark two essential features of (83). First, the right hand side, apart from the constant, agrees with the corresponding semiclassical formula. Second, there is no extra condition imposed on the potential V (apart from the natural integrability assumption which makes the semiclassical formula finite). Both properties are crucial in the applications.

Our main concern is to generalize this inequality for operators with a magnetic field. In general, the magnetic field is a 2-form, $\mathbf{B}$, most conveniently given by a vector potential $\mathbf{A} : \mathbb{R}^d \to \mathbb{R}^d$, which, under the canonical identification, can be viewed as a 1-form, and $\mathbf{B} = \mathrm{d}\mathbf{A}$.

The Schrödinger operator with a magnetic field describing a *spinless* particle in an external potential V is

$$H_{\text{spinless}} := (\mathbf{p} - \mathbf{A}(x))^2 + V(x), \tag{84}$$

[1]Work supported by the NSF grant PHY90-19433 A02, by the Alfred Sloan Foundation dissertation Fellowship and by the Erwin Schrödinger Institute for Mathematical Physics in Vienna.

where $\mathbf{p} := -i\nabla$, $x \in \mathbb{R}^d$. From the Lieb-Thirring point of view, (84) is less interesting, since, by a trivial application of the diamagnetic inequality, one can easily prove the same bound (83) for the negative eigenvalues of H_{spinless} as well.

In the present work we shall be concerned with the Pauli operator, which is more relevant physically and, at the same time, more challenging mathematically. The three-dimensional Pauli Hamiltonian is

$$H_{\text{Pauli}} := [(\mathbf{p} - \mathbf{A}(x)) \cdot \boldsymbol{\sigma}]^2 + V(x) = (\mathbf{p} - \mathbf{A}(x))^2 - \boldsymbol{\sigma} \cdot \mathbf{B}(x) + V(x) \tag{85}$$

acting on $L^2(\mathbb{R}^3, \mathbf{C}^2)$, the Hilbert space of a spin-1/2 particle. Here $\boldsymbol{\sigma} = (\sigma_1, \sigma_2, \sigma_3)$ is the vector of Pauli matrices, and in three dimensions one can view $\mathbf{B} = \mathrm{d}\mathbf{A} = \mathrm{rot}\mathbf{A}$ as a vector. We shall assume that $\mathbf{A}$, $\mathrm{div}\mathbf{A}$ and $\mathbf{B}$ are in L^2_{loc}.

The following Lieb-Thirring inequality for the sum of the negative eigenvalues of the Pauli operator with a *constant* magnetic field $\mathbf{B}(x) = (0,0,B)$ has been obtained in [12]:

$$\sum_i |E_i| \le C_1 B \cdot \int_{\mathbb{R}^3} |V(x)|_-^{3/2} \mathrm{d}x + C_2 \int_{\mathbb{R}^3} |V(x)|_-^{5/2} \mathrm{d}x \tag{86}$$

(C_1 and C_2 are universal constants).

The main goal is to generalize this inequality for nonhomogeneous magnetic field, maintaining the two basic features mentioned above (comparability with the semiclassical formula and no extra condition on the external potential). We mention that in the related works ([16], [11], [12], etc.) special attention is devoted to the case of the strong magnetic field. We found physically interesting and at the same time mathematically more difficult, therefore challenging, to treat the strong nonhomogeneous magnetic field.

There is a vast literature of various spectral studies in the case of the homogeneous magnetic field, but results, especially quantitative ones, are fairly rare for nonhomogeneous field (see [1], [3], [1], [14], [15], [18]). The technical reason for this (apart from the obvious physical relevance of the constant magnetic field) is twofold. First, the Schrödinger operator with a constant magnetic field (without external potential) is exactly solvable, and, after decomposing the operator according to the Landau levels, one obtains a simplified (lower dimensional) setup, so the additional effect of the external potential becomes easier. Some version of this strategy has almost always been used in the works concerning homogeneous magnetic field.

The second technical difficulty is that perturbations of the magnetic field can be much less controlled than that of the external potential. Naively one would expect that a local change of the magnetic field does not affect too much local quantities observed far away, but the vector potential, appearing in the operator, is a nonlocal quantity (i.e. it undergoes a nonlocal change with a long tail even under local perturbation of the field itself).

Our basic method is stochastic via the Feynman-Kac formula, which is valid under fairly general conditions on the magnetic field.

2 Various Lieb-Thirring inequalities

The most general naive conjecture for the Pauli operator is the following:

NAIVE CONJECTURE. For any $\gamma > 1/2$ there exist two universal constants $C_1(\gamma)$ and $C_2(\gamma)$, such that for the moments of the negative eigenvalues of H_{Pauli} we have

$$\sum_i |E_i|^\gamma \le C_1(\gamma) \int_{\mathbb{R}^3} |\mathbf{B}(x)|\, |V(x)|_-^{\gamma+1/2} \mathrm{d}x + C_2(\gamma) \int_{\mathbb{R}^3} |V(x)|_-^{\gamma+3/2} \mathrm{d}x. \qquad (87)$$

This conjecture is based on a heuristic argument (see [12], [8]); namely, that the contribution from the 'lowest Landau level' should be treated separately (giving rise to the first term in (87)) and the higher 'levels' do not give essentially bigger contribution than the usual estimate for the nonmagnetic Lieb-Thirring inequality (83) (second term in (87)). The Lieb-Thirring inequality for a constant magnetic field was proved along this line [12]. Strictly speaking, the Landau levels do not exist for a nonhomogeneous magnetic field, nevertheless a strong magnetic field has a localization effect which makes reasonable, at least locally, the comparison of the original operator with a Pauli operator with a frozen constant magnetic field. Therefore, one can define the Landau levels locally and one can heuristically extend the constant field argument for a nonconstant field.

It has turned out that the conjecture (87) is wrong for certain magnetic fields (a counterexample is found in [8]). If the magnetic field changes very abruptly in space, then the heuristic argument (locally constant field approximation) is not precise enough; one can present a magnetic field $\mathbf{B}$ and an external potential V such that $\mathit{supp}\,\mathbf{B} \cap \mathit{supp}\,V$ is empty (but $\mathbf{B}$ is huge where $V = 0$), therefore the right hand side of (87) is independent of $\mathbf{B}$, while the left hand side blows up as the magnetic field increases.

We shall present three Lieb-Thirring type theorems.

Theorem 1. For arbitrary, bounded magnetic field and $\gamma \ge 1$, the moments of the eigenvalues of the Pauli operator H_{Pauli} satisfy

$$\sum_i |E_i|^\gamma \le C_3(\gamma) \|\mathbf{B}\|_\infty^{3/2} \int_{\mathbb{R}^3} |V(x)|_-^\gamma \mathrm{d}x + C_4(\gamma) \int_{\mathbb{R}^3} |V(x)|_-^{\gamma+3/2} \mathrm{d}x, \qquad (88)$$

where $C_3(\gamma)$ and $C_4(\gamma)$ are explicit constants.

If we assume that the magnetic field has *constant direction*, then the lowest Landau-level (which is the most important level for Lieb-Thirring type inequalities) can be identified without localization (this fact is closely related to the Aharonov-Casher phenomenon [1], [5], namely that the two-dimensional Pauli operator usually has a ground state with a large degeneracy). This observation helps us to prove the following stronger version of Theorem 1:

Theorem 2. Assume that $\mathbf{B}$ is bounded and it has constant direction, $\mathbf{B}(x) = (0,0,B(x))$. Then, for $\gamma > 1/2$, the moments of the negative eigenvalues of the H_{Pauli} satisfy

$$\sum_i |E_i|^\gamma \leq C_5(\gamma)\|B\|_\infty \int_{\mathbb{R}^3} |V(x)|_-^{\gamma+1/2}\mathrm{d}x + C_6(\gamma)\int_{\mathbb{R}^3} |V(x)|_-^{\gamma+3/2}\mathrm{d}x. \tag{89}$$

(A straightforward Schwarz inequality shows that (89) is stronger than (88)).

Theorems 1 and 2 are robust, they do not exploit the local properties of $\mathbf{B}(x)$. Intuitively, the global boundedness of the magnetic field should not be necessary (actually, we can treat integrable singularities of $\mathbf{B}$ with a simple method, see [8]). But the counterexample warns us that some condition is needed to exclude magnetic fields increasing from 0 to a high value in a short distance. The conditions of the following theorem are of this spirit. For technical reasons, they are much more restrictive than the physical intuition would predict, nevertheless, under these conditions, we do prove the original form (87) of the Lieb-Thirring inequality.

Theorem 3. Assume that the magnetic field with constant direction ($\mathbf{B}(x) = (0,0,B(x))$) has a positive lower bound $0 < B_0 \leq B(x)$ and $B(x)$ is continuously differentiable. Let

$$c(B) := B_0^{-1/2} \cdot \sup_x(|\nabla b(x)| \cdot |b(x)|^{31/6}) \tag{90}$$

be finite, where $b(x) := B(x)/B_0$. Then (87) is valid with constants $C_1(\gamma)$ and $C_2(\gamma)$ which depend on $c(B)$ in an increasing way (especially, they blow up as $B_0 \to 0$).

Remark. The condition on the finiteness of $c(B)$ essentially imposes an upper bound on the gradient of B. Only small gradient is allowed in the regions where $B(x)$ is large. We also remark that this condition is almost homogeneous in the magnetic field, therefore we have a semiclassical statement as well. If we include the Planck constant in the original Pauli Hamiltonian, $[(h\mathbf{p} - \mathbf{A}) \cdot \boldsymbol{\sigma}]^2 + V = h^2[(\mathbf{p} - h^{-1}\mathbf{A}) \cdot \boldsymbol{\sigma}]^2 + V$, then the magnetic field should be rescaled by h^{-1} (so $b(x)$ is unchanged and $c(Bh^{-1}) \to 0$ as $h \to 0$), thus we obtain the following estimate:

$$\sum_i |E_i|^\gamma \leq C_2(\gamma)\cdot h^{-3}\int |V(x)|_-^{\gamma+3/2}\mathrm{d}x + C_1(\gamma)\cdot h^{-2}\int B(x)|V(x)|_-^{\gamma+1/2}\mathrm{d}x. \tag{91}$$

Since $C_2(\gamma)$ is not the semiclassical constant, the second term becomes relevant only for large magnetic field $(B \geq h^{-1})$.

3 Stochastic oscillatory integrals

In the proofs of these theorems, we first use modified Birman-Schwinger arguments (see [12] and [13]) to separate the external potential. The next step and the real novelty is to give precise estimates on the heat kernel (therefore on the resolvent kernel as well) of the free operator (with a nonhomogeneous magnetic field but without external potential). In the corresponding Feynman-Kac formula, the effect of the magnetic field appears as an oscillatory phase factor. To illustrate this phenomenon, consider the simplest case, the two-dimensional Pauli operator on the spin-up subspace; $H_0 := (\mathbf{p} - \mathbf{A})^2 - B$ (this operator naturally arises in the case of the three-dimensional problem with a field with constant direction).

According to the Feynman-Kac formula

$$e^{-tH_0}(x,y) = (4\pi)^{-1}e^{-\frac{(x-y)^2}{4t}}\mathbf{E}_{0,x}^{2t,y}e^{-i\int_0^{2t}\mathbf{A}(W(s))\circ dW(s)+\frac{1}{2}\int_0^{2t}B(W(s))ds}, \tag{92}$$

where $\mathbf{E}_{0,x}^{2t,y}$ stands for the expectation value of the two-dimensional Brownian bridge $W(s)$ ($W(0) = x$, $W(2t) = y$), and the circle ($\circ$) denotes the Stratonovich stochastic integral.

The point is that the two terms in the exponent have roughly the same effect which neutralize each other; the oscillation decreases the expectation value, while the $\int B$ term increases it. A widely used method to give an upper estimate for this type of expressions is to take the absolute value of the exponential (diamagnetic inequality). We use this idea in the proof of Theorems 1 and 2. But in this way one loses the full oscillation effect, which approach is simply too rough in Theorem 3. We have developed a new localization technique in the path space that is powerful enough to take into account almost the whole oscillation (under certain conditions on the magnetic field). The simplest application of this method, independently of the Lieb-Thirring inequalities, is found in [6].

Finally we remark, that in the proof of Theorem 1 we use a generalized Feynman-Kac formula for the full Pauli operator (see [4]). In addition to the Wiener process, this involves a Poisson jump process as well to describe the interaction between the spin-up and spin-down subspaces.

Acknowledgement. The author wishes to express his gratitude to Elliott H. Lieb, Michael Aizenman and Jan Philip Solovej for their continuous help and encouragement.

References

[1] Y. Aharonov, A. Casher, Ground state of spin-1/2 charged particle in a two-dimensional magnetic field. Phys. Rev. **A19** (1979), 2461–2462.

[2] J. Avron, I. Herbst and B. Simon, Schrödinger operators with magnetic fields. I. General interactions. Duke Math. J. **45** (1978), 847–883.

[3] Y. Colin de Verdiére, L'asymptotique de Weyl pour les bouteilles magnétiques. Commun. Math. Phys. **105** (1986), 327–335.

[4] G. F. De Angelis, G. Jona-Lasinio and M. Sirugue, Probabilistic solution of Pauli type equations. J. Phys. A: Math. Gen. **16** (1983), 2433–2444.

[5] L. Erdős, Ground state density of the two-dimensional Pauli operator in the strong magnetic field. Lett. Math. Phys. **29** (1993), 219–240.

[6] L. Erdős, Estimates on stochastic oscillatory integrals and on the heat kernel of the magnetic Schrödinger operator. Accepted for Duke Math. J.

[7] L. Erdős, Magnetic Lieb-Thirring inequalities and stochastic oscillatory integrals. Ph.D. Thesis, Princeton University, 1994.

[8] L. Erdős, Magnetic Lieb-Thirring inequalities. Submitted to Commun. Math. Phys.

[9] E. H. Lieb, The number of bound states of one-body Schrödinger operators and the Weyl problem. pp. 241–251 in Proceedings of Symposia in Pure Mathematics. Vol. 36, 1980.

[10] E. H. Lieb and W. Thirring, Inequalities for the moments of the eigenvalues of the Schrödinger Hamiltonian and their relation to Sobolev inequalities. pp. 269–304 in Studies in Mathematical Physics, Essays in Honor of Valentine Bargmann. Eds. E. H. Lieb, B. Simon and A. S. Wightman, Princeton Univ. Press, Princeton, New Jersey, 1976.

[11] E. H. Lieb, J. P. Solovej and J. Yngvason, Asymptotics of heavy atoms in high magnetic fields: I. Lowest Landau band regions. To appear in Commun. Pure Appl. Math.

[12] E. H. Lieb, J. P. Solovej and J. Yngvason, Asymptotics of heavy atoms in high magnetic fields: II. Semiclassical regions. Commun. Math. Phys. **161** (1994), 77–124.

[13] E. H. Lieb, J. P. Solovej and J. Yngvason in preparation. An announcement of the result was made in Quantum dots to appear in the proceedings of the international conference on partial differential equations and mathematical physics held at the University of Alabama, Birmingham, March 1994.

[14] H. Matsumoto, The short time asymptotics of the traces of the heat kernels for the magnetic Schrödinger operators. J. Math. Soc. Japan. **42** (1990), 677–689.

[15] H. Matsumoto, Classical and non-classical eigenvalue asymptotics for the magnetic Schrödinger operators. J. of Funct. Anal. **95** (1991), 460–482.

[16] A. Sobolev, The quasiclassical asymptotics of local Riesz means for the Schrödinger operator in a strong homogeneous magnetic field. Duke Math J. (to appear).

[17] H. Tamura, Asymptotic distribution of eigenvalues for Schrödinger operators with homogeneous magnetic fields. Osaka J. Math. **25** (1988), 633–647.

Operator Theory:
Advances and Applications, Vol. 78

On the Trace Density in Deformation Quantization

B. V. Fedosov

Let $Q : C^\infty(M)[[h]] \to W_D$ be a quantization map considered in the author's papers [2, 3, 4]. Here M is a symplectic manifold of dimension $2n$, W_D the algebra of flat sections of the Weyl algebras bundle W with respect to an Abelian connection D on W. The coefficient bundle of W is supposed to be Hom(E,E) for some m-dimensional complex vector bundle E. The Abelian connection D depends thus on the symplectic connection ∂^S on M and the connection ∂^E on E.
The construction of deformation quantization in [2, 3, 4] provides a trace functional on the algebra W_D. We introduce a *trace density* as a function $t(x,h) \in C^\infty(M)[[h]]$ such that for any flat section $\widehat{a} = Q(a(x))$ with a scalar compactly supported function $a(x)$ we have

$$\operatorname{tr}\widehat{a} = (2\pi h)^{-n} \int_M a(x)t(x,h)\frac{\omega^n}{n!}.$$

Our main result consists in the following theorem.

Theorem 1. There exists a trace density

$$t(x,h) = m + ht_1(x) + h^2 t_2(x) + \dots$$

for the algebra W_D. The coefficients $t_1(x), t_2(x), \dots$ are polynomials in the curvature tensors R^S_{ijkl} and R^E_{ij} of the connections ∂^S and ∂^E and their covariant derivatives.

We give a procedure which allows in principle to determine successively the coefficients $t_1, t_2, \dots$.
We apply this result to the calculation of the integral

$$\operatorname{tr} 1 = (2\pi h)^{-n} \int_M t(x,h)\frac{\omega^n}{n!} = \operatorname{ind}\{1,0\}$$

for a compact manifold M. This calculation yields a new approach to the proof of the index theorem for deformation quantization similar to the Atiyah–Bott–Patodi proof of the ordinary index theorem for elliptic operators. It is based on the result of Tamarkin [6] concerning topological invariants of connections over symplectic manifolds generalizing the results of Abramov [1] and Gilkey [6] for Riemannian manifolds. Unfortunately we are able to perform this calculation only for the relatively simple case when the symplectic connection has zero curvature while the curvature $R^E_{ij} \neq 0$.

References

[1] A.A. Abramov. On topological invariants of the riemannian spaces obtained by integration of tensor fields. Dokl. Acad. Nauk SSSR, 81, 1951.

[2] B.V. Fedosov. Index theorems, volume 65 of Sovremennye problemy matematiki. Fundamental'nye napravleniya, pages 165–268. VINITI, 1991.

[3] B.V. Fedosov. Proof of the index theorem for deformation quantization. In Advances in Partial Differential Equations. Akademie Verlag, Berlin, 1994. To appear.

[4] B.V. Fedosov. A simple geometrical construction of deformation quantization. Journal of differential geometry, 1994.

[5] P.B. Gilkey. Invariance theory, the heat equation and the Atiyah-Singer index theorem. Math. Lecture Series, 11. Publish of Perish, Inc., Wilington, Del., 1984.

[6] D. Tamarkin. Topological invariants of connections on symplectic manifolds. Funktsional'nyi Analiz i Ego Prilozheniya. To appear.

Boris V. Fedosov
Moscow Institute of Physics and Technology

Operator Theory:
Advances and Applications, Vol. 78

The stationary Phase Method with Remainder Estimate as Dimension of the Space goes to Infinity

Daisuke Fujiwara

1 Introduction

Stationary phase method concerns with oscillatory integrals over R^d of the following form:

$$I(S,a,\nu) = \int_{R^d} e^{-i\nu S(x)} a(x) dx,$$

where the phase function $S(x)$ is a real valued C^∞ function, the amplitude $a(x)$ is of class C^∞, too and ν is a large positive parameter. It is a method to evaluate $I(S,a,\nu)$ asymptotically, as $\nu \to \infty$. In the simplest case that $a(x) \in C_0^\infty(R^d)$ and that $S(x)$ has only one critical point x^*, where $HessS(x^*)$ is non-degenerate, it gives

$$I(S,a,\nu) = \left(\frac{2\pi}{i\nu}\right)^{d/2} [\det\{HessS(x^*)\}]^{-1/2}(e^{-i\nu S(x^*)}a(x^*) + r_d(\nu))$$

and an estimate of the remainder term

$$r_d(\nu) = O(\nu^{-1}).$$

Since the stationary phase method is closely related to the mathematical theory of Feynman path integrals (cf. Feynman [1], Feynman-Hibbs [9], Pauli [8], Fujiwara [3], [2]), we need to know the behaviour of $\nu r_d(\nu)$ as $d \to \infty$. We ask the following question:

Can one control $\nu r_d(\nu)$ *as* $d \to \infty$ *?*

We give a positive answer to this question. Detailed discussions can be found in Fujiwara [6]. Applications are discussed in Fujiwara [2], [4], [7] and [5].

2 Results

Oscillatory integrals we will treat in this note is of the following type:

$$I(\{t_j\},S,a,\nu)(x_L,x_0) = \prod_{j=1}^{L}\left(\frac{\nu}{2\pi i t_j}\right)^{\frac{1}{2}} \int_{R^{L-1}} e^{i\nu S(x_L,\dots,x_0)} a(x_L,\dots,x_0) \prod_{j=1}^{L-1} dx_j,$$

where S is the phase function, a the amplitude, ν a large parameter, and t_j (j=1,2,...,L) are small positive parameters. This type of oscillatory integral appears when one treats Feynman path integrals. Our aim is to give an estimate of the remainder term which is uniform with respect to the parameters $\{t_j\}$.

Our assumption for the phase function is

Assumption 2.1 $S(x_L,\ldots,x_0)$ is real-valued and is of the form

$$S(x_L,\ldots,x_0)=\sum_{j=1}^{L}S_j(t_j,x_j,x_{j-1}), \tag{93}$$

where

$$S_j(t_j,x_j,x_{j-1})=\frac{|\,x_j-x_{j-1}\,|^2}{2t_j}+t_j\omega_j(t_j,x_j,x_{j-1}).$$

For any $m\geq 2$ there exists a positive constant κ_m such that

$$\sup_{x_j,x_{j-1}}|\,\partial_{x_j}^{\alpha}\partial_{x_{j-1}}^{\beta}\omega_j(t_j,x_j,x_{j-1})\,|\leq\kappa_m$$

if $2\leq\alpha+\beta\leq m$.

Remark 2.1 I call the property (93) *locality of the phase.*

We give two examples of phase functions satisfying Assumption 2.1.

Example 2.1 Let $L(\xi,x)=\frac{1}{2}\xi^2-V(x)$, $(\xi,x)\in R^2$, be a Lagrangian with a potential $V(x)$. Assume that the potential $V(x)$ is a real-valued C^∞-function satisfying estimates:

$$\sup_{x}|\,V^{(k)}(x)\,|<\infty \quad \textit{for any } k\geq 2. \tag{94}$$

Then for a small $T>0$, there exists a unique classical orbit $\gamma^{cl}(t)$ such that $\gamma^{cl}(0)=y$, $\gamma^{cl}(T)=x$. Let

$$S^{cl}(T,x,y)=\int_0^T L(\dot{\gamma}^{cl}(t),\gamma^{cl}(t))dt$$

be the classical action. Then $S^{cl}(T,x,y)$ is of the form

$$S^{cl}(T,x,y)=\frac{|\,x-y\,|^2}{2T}+T\phi^{cl}(T,x,y)$$

and for any $m\geq 2$ there exists a constant C_m such that

$$\sup_{x,y}|\,\partial_x^{\alpha}\partial_y^{\beta}\phi^{cl}(T,x,y)\,|\leq C_m$$

if $2\leq\alpha+\beta\leq m$. Therefore, $S(x_L,\ldots,x_0)=\sum_{j=1}^{L}S(t_j,x_j,x_{j-1})$ satisfies Assumption 2.1.

Example 2.2 . Let $L(\xi,x)$ be the same lagrangian. Let $\gamma^{ln}(t)$ be the straight line connecting $(0,y)$ and (T,x) in the time-space, i.e.,

$$\gamma^{ln}(t) = \frac{t}{T}x + \frac{T-t}{T}y.$$

Using $\gamma^{ln}(t)$ in place of classical orbit γ^{cl}, we obtain $S^{ln}(x_L,\ldots,x_0) = \sum_{j=1}^{L} S^{ln}(t_j,x_j,x_{j-1})$. This satisfies Assumption 2.1.

If $\sum_{j=1}^{L} t_j$ is small, the critical point, which we denote by $(x^*_{L-1},\ldots,x^*_1)$, of the function $(x_{L-1},\ldots,x_1) \to S(x_L,x_{L-1},\ldots,x_1,x_0)$ is unique under Assumption 2.1 We abbreviate $S(x_L,x^*_{L-1},\ldots,x^*_1,x_0)$ to $S(\overline{x_L,x_0})$. We can write the Hessian of S at the critical point as $H+W$, where

$$H = \begin{pmatrix} \frac{1}{t_1}+\frac{1}{t_2} & -\frac{1}{t_2} & 0 & 0 & \cdots \\ -\frac{1}{t_2} & \frac{1}{t_2}+\frac{1}{t_3} & -\frac{1}{t_3} & 0 & \cdots \\ 0 & -\frac{1}{t_3} & \frac{1}{t_3}+\frac{1}{t_4} & -\frac{1}{t_4} & \cdots \\ \vdots & \vdots & \vdots & \vdots & \vdots \end{pmatrix}$$

and

$$W = \begin{pmatrix} t_1\partial_{x_1}^2\omega_1 + t_2\partial_{x_1}^2\omega_2 & t_2\partial_{x_1}\partial_{x_2}\omega_2 & 0 & \cdots \\ t_2\partial_{x_1}\partial_{x_2}\omega_2 & t_2\partial_{x_2}^2\omega_2 + t_3\partial_{x_2}^2\omega_3 & t_3\partial_{x_2}\partial_{x_3}\omega_3 & \cdots \\ 0 & t_3\partial_{x_2}\partial_{x_3}\omega_3 & t_3\partial_{x_3}^2\ omega_3 + t_4\partial_{x_3}^2\omega_4 & \cdots \\ \vdots & \vdots & \vdots & \vdots \end{pmatrix}$$

It is clear that

$$\det H = \frac{T_L}{t_1 t_2 \ldots t_L} \neq 0,$$

where $T_L = \sum_{j=1}^{L} t_j$.

Our first result is

Theorem 2.1 Under the Assumption 2.1 there exists a positive constant δ_1 independent of L such that if $T_L = t_1 + \ldots + t_L \leq \delta_1$ then

$$\begin{aligned} &I(\{t_j\},S,1,\nu)(x_L,x_0) \\ &= \left(\frac{\nu i}{2\pi T_L}\right)^{1/2} e^{-i\nu S(\overline{x_L,x_0})} \left[\det(I+WH^{-1})\right]^{-1/2} (1 + r(\nu,x_L,x_0)), \end{aligned}$$

where the remainder term $r(\nu,x_L,x_0)$ satisfies the estimate: For any $K \geq 0$ there exists positive constant C_K such that if $\mid \alpha_0 \mid, \mid \alpha_L \mid \leq K$

$$\mid \partial_{x_0}^{\alpha_0}\partial_{x_L}^{\alpha_L} r(\nu,x_L,x_0) \mid \leq C_K T_L^3 \nu^{-1}.$$

Remark 2.2 δ_1 and C_K are independent of L as far as T_L is bounded. Therefore, we can control $r(\nu,x_L,x_0)$ even when L tends to ∞.

In order to state the result for general integral with amplitude $a(x)$, we require a little more preparations. Let $1 \leq k \leq l \leq L$. Then the critical point of the function $(x_{l-1},\dots,x_{k+1}) \to \sum_{j=k+1}^{l} S_j(t_j,x_j,x_{j-1})$ is unique if $t_{k+1}+\dots+t_l$ is small. Let $(x^*_{l-1},\dots,x^*_{k+1})$ denote the critical point, which is a function of x_l and x_k. We abbreviate $a(x_L,\dots,x_l,x^*_{l-1},\dots,x^*_{k+1},x_k,\dots,x_0)$ to $a(x_L,\dots,x_{l+1},\overline{x_l,x_k},x_{k-1},\dots,x_0)$.

Our assumption concerning the amplitude function is the following:

Assumption 2.2 For any integer $K \geq 0$ there exists a positive constant A_K with the following properties: (i) If $\mid \alpha_j \mid \leq K$ for $j = 0,1,\dots,L$, then

$$\mid \prod_{j=0}^{L} \partial_{x_j}^{\alpha_j} a(x_L,\dots,x_0) \mid \leq A_K.$$

(ii) For any sequence of positive integers $\{j_1,\dots,j_s\}$ satisfing

$$0 = j_0 < j_1 - 1 < j_1 < j_2 - 1 < \dots < j_s - 1 < j_s < L$$

we have

$$\mid \partial_{x0}^{\alpha_0} \partial_{xL}^{\alpha_L} \prod_{k=1}^{s} \partial_{x_{j_k-1}}^{\alpha_{j_k-1}} \partial_{x_{j_k}}^{\alpha_{j_k}} a(\overline{x_L,x_{j_s}},\overline{x_{j_s-1},x_{j_{s-1}}},\dots,\overline{x_{j_1-1},x_{j_0}}) \mid \leq A_K,$$

as far as $\mid \alpha_j \mid \leq K$ for $j = 0, j_1-1, j_1,\dots,j_s-1, j_s, L$.

Now we can state our second result:

Theorem 2.2 Under the Assumption 2.1 and Assumption 2.2 there exists a positive constant δ_1 such that if $0 < T_L \leq \delta_1$

$$I(\{t_j\},S,a,\nu)(x_L,x_0)$$
$$= \left(\frac{\nu i}{2\pi T_L}\right)^{1/2} e^{-i\nu S(\overline{x_L,x_0})} \left[\det(I + WH^{-1})\right]^{-1/2} \left(a(\overline{x_L,x_0}) + r(\nu,x_L,x_0)\right),$$

where $r(\nu,x_L,x_0)$ satisfies the estimate: For any $K \geq 0$ there exists positive constants C_K and $M(K)$ such that if $\mid \alpha_0 \mid, \mid \alpha_L \mid \leq K$ we have

$$\mid \partial_{x_0}^{\alpha_0} \partial_{x_L}^{\alpha_L} r(\nu,x_L,x_0) \mid \leq C_K T_L \nu^{-1} A_{M(K)}.$$

Remark 2.3 δ_1 , C_K and $M(K)$ of Theorem 2.2 are independent of L as far as T_L is bounded by δ_1. Therefore, we can control $r(\nu,x_L,x_0)$ even when L tends to ∞.

We give an example of amplitude functions satisfying Assumption 2.2.

Example 2.3 Let $b_j(x_j,x_{j-1})$, $j = 1,\dots,L$, be functions bounded together with their derivatives of all order, i.e., for any positive integer K there exists C_K such that

$$\sup_x \mid \partial_{x_j}^{\alpha_j} \partial_{x_{j-1}}^{\alpha_{j-1}} b_j(x_j,x_{j-1}) \mid \leq C_K \qquad 0 \leq \alpha_j, \alpha_{j-1} \leq K.$$

Then $a(x_L,\dots,x_0) = \exp\left(\sum_{j=1}^{L} t_j b_j(x_j,x_{j-1})\right)$ satisfies Assumption 2.2 above.

Remark 2.4 Recently, Tsuchida [10] proved similar results for oscillatory integrals including the effect of magnetic field. His phase stems from the lagrangian of the following form:

$$L(\xi,x) = \frac{1}{2}\xi^2 + \xi A(x) - V(x),$$

where $(\xi,x) \in R^2$ and $A(x)$ is the vector potential which satisfy suitable assumptions.

Remark 2.5 We can discuss the case that $t_1 + t_2 + \dots + t_L > \delta_1$ in the following way. Assume, for instance, that $\delta_1 < t_1 + t_2 + \dots + t_L < 2\delta_1$. We choose m so that $t_1 + t_2 + \dots + t_m < \delta_1$ and that $t_{m+1} + t_{m+2} + \dots + t_L < \delta_1$. We first integrate by variables $x_1,x_2,\dots,x_{m-1}$ and $x_{m+1},x_{m+2},\dots,x_L$ and apply stationary phase method of Theorem 2.1 or 2.2 and finally we integrate by the variable x_m and apply usual stationary phase method.

References

[1] R.P. Feynman, Space time approach to non relativistic quantum mechanics, Rev. of Modern Phys.,**20**, 367–387, 1948.

[2] D. Fujiwara, The Feynman path integrals as an improper integral over the Sobolev space. In Proceedings of Journées d'equations aux dérivés partielles, St. Jean de Monts 1990 Société Mathématiques de France, 1990.

[3] D. Fujiwara, Remarks on convergence of the Feynman path integrals, Duke Math. J., **47**, 559–600, 1980,

[4] D. Fujiwara, Some Feynman path integrals as oscillatory integrals over a Sobolev manifolds. In Proc. International conference on Functional Analysis in memory of Professor Kôsaku Yosida, 1992.

[5] D. Fujiwara, Some Feynman path integrals as oscillatory integrals over a Sobolev manifolds, 1992, Preprint.

[6] D. Fujiwara, The stationary phase method with an estimate of the remainder term on a space of large dimension, Nagoya Math. J.,**124**, 61–97, 1991,

[7] D. Fujiwara, Stationary phase method with an estimate of the remainder term over a space of large dimension. In Spectral and Scattering Theory and Related Topics, 1993, Advanced Studies in Pure Mathematics 23.

[8] W. Pauli, Pauli Lectures on Physics, MIT press, Cambridge, Mass. U.S.A., 1973.

[9] R.P. Feynman and A.R. Hibbs, Quantum mechanics and path integrals, Mcgraw-Hill, New York, 1965.

[10] T. Tsuchida, Remarks on Fujiwara's stationary phase method on a space of large dimension with a phase function involving electromagnetic fields, preprint, Department of Mathematics, Kanazawa University, Kanazawa, Ishikawaken, 920-11 Japan, 1993.

Daisuke Fujiwara
Department of Mathematics, Gakushuin University
1-5-1 Mejiro, Toshimaku, Tokyo 171, Japan

Operator Theory:
Advances and Applications, Vol. 78

The Eta Invariant, Equivariant Spin Bordism and Metrics of Positive Scalar Curvature

P. B. Gilkey[1], B. Botvinnik

Abstract

We use the eta invariant and equivariant spin bordism group to study the space of metrics of positive scalar curvature. MSC numbers: 58G12, 58G25, 53A50, 53C25, 55N22.

§1 Introduction.

There are many relationships between algebraic topology, differential geometry, and the theory of elliptic partial differential operators. In this note, we will study one example. We will use the eta invariant (which is an analytic invariant) and equivariant spin bordism (which is a topological invariant) to study when a compact spin manifold M of dimension $m \geq 5$ admits a metric of positive scalar curvature (this is a geometrical question).

If $m \equiv 0 \bmod 8$ or if $m \equiv 4 \bmod 8$, let $\hat{A}(M) \in \mathbb{Z}$ be the index of the spin complex; by the Atiyah-Singer index theorem, $\hat{A}(M)$ is the integral over M of the A-roof genus. Let $\mathcal{H}$ be the space of harmonic spinors; this is the kernel of the Dirac operator. If $m \equiv 1 \bmod 8$, let $\hat{A}(M) \in \mathbb{Z}_2$ be the mod 2 reduction of $\dim(\mathcal{H})$. If $m \equiv 2 \bmod 8$, $\dim(\mathcal{H})$ is always even and we let $\hat{A}(M) \in \mathbb{Z}_2$ be the mod 2 reduction of $\frac{1}{2}\dim(\mathcal{H})$. We set $\hat{A}(M) = 0$ if $m \equiv 3, 5, 6, 7 \bmod 8$. By the Lichnerowicz formula [13], there are no harmonic spinors if the metric g has positive scalar curvature; this is one of the crucial links between differential geometry and elliptic operator theory. Since $\hat{A}(M)$ is independent of the metric, $\hat{A}(M) \neq 0$ implies M does not admit a metric of positive scalar curvature. Rosenberg has defined a more refined invariant $\alpha(M) \in KO(C_r^*(\pi_1 M))$; if $\alpha(M) \neq 0$, M does not admit a metric of positive scalar curvature. Note that $\alpha(M) = \hat{A}(M)$ if M is simply connected or if $\pi_1(M)$ is a cyclic group of odd order. If $\pi_1(M)$ is a spherical space form group, $\alpha(M)$ is just $\hat{A}$ for all possible spin structures on M and takes values in a $\mathbb{Z}_2$ vector space if $m \equiv 1$ or if $m \equiv 2 \bmod 8$.

Conjecture 1.1 (Gromov-Lawson-Rosenberg): Let M be a connected spin manifold of dimension $m \geq 5$. Then M admits a metric of positive scalar curvature if and only if $\alpha(M) = 0$.

Conjecture 1.1 has been established in some cases:

[1]Research partially supported by the NSF (USA) and IHES (France)

Theorem 1.2: Let M be a connected spin manifold of dimension $m \geq 5$ with finite fundamental group G.

(a) **(Stolz [18]).** If M is simply connected, then M admits a metric with positive scalar curvature if and only if $\alpha(M) = 0$.

(b) **(Kwasik and Schultz [12]):** If G is cyclic of odd order, then M admits a metric of positive scalar curvature if and only if $\alpha(M) = 0$.

(c) **(Botvinnik, Gilkey, and Stolz [5]):** If G is a spherical space form group, then M admits a metric of positive scalar curvature if and only if $\alpha(M) = 0$.

Remark: A group G is a spherical space form group if and only if the p Sylow subgroups for odd primes are cyclic and the 2 Sylow subgroup is either cyclic or quaternion; these are the groups with periodic cohomology.

Remark: Conjecture 1.1 has also been established if $\pi_1(M)$ is one of a short list of infinite groups, including free groups, free Abelian groups, and fundamental groups of orientable surfaces [16].

Remark: The situation is quite different if M is not spin. For example, if M is not spin but is simply connected, Gromov and Lawson [11] have shown that M always admits a metric of positive scalar curvature. We shall not consider this case further and shall assume henceforth that M is spin.

Remark: The metric in Theorem 1.2 (b,c) is highly non-unique in general. If $m \equiv 3$ mod 4, the results of [3] show there are an infinite number of non concordant metrics of positive scalar curvature on M.

We now recall the concept of equivariant bordism from algebraic topology. Let BG be the classifying space of a finite group G and let $\sigma : M \to BG$ define a G structure on M. If M is connected, σ is equivalent to giving a representation of the fundamental group of M into G. The canonical structure is given by taking $G = \pi_1(M)$ and letting the representation be the identity map.

The group $\Omega_m^{spin}(BG)$ classifies spin manifolds with a G structure up to bordism. We consider triples (M,s,σ) where M is a compact smooth manifold without boundary with a spin structure s and a G structure σ. We introduce the equivalence relation $(M_1,s_1,\sigma_1) \simeq (M_2,s_2,\sigma_2)$ on the set of such triples if there exists a compact Riemannian manifold N with boundary $M_1 \dot{\sqcup} -M_2$ such that the spin structures s_i and G structures σ_i extend from the M_i to N. The set of equivalence classes is the equivariant bordism group $\Omega_m^{spin}(BG)$ which we shall be studying. Disjoint union makes $\Omega_m^{spin}(BG)$ into an Abelian group. We denote an element of $\Omega_m^{spin}(BG)$ by $[(M,s,\sigma)]$ or just $[M]$ when (s,σ) are understood. Let $\Omega_m^{spin} = \Omega_m^{spin}(\{1\})$ be defined by the trivial group so there is no G structure involved; Cartesian product makes Ω_*^{spin} into a graded ring and $\Omega_*^{spin}(BG)$ into a Ω_*^{spin} module. Let the reduced bordism group $\tilde\Omega_m^{spin}(BG)$ be the kernel of the forgetful functor which goes from $\Omega_m^{spin}(BG)$ to Ω_m^{spin}; there is a natural splitting

$$\Omega_m^{spin}(BG) = \tilde\Omega_m^{spin}(BG) \oplus \Omega_m^{spin}. \tag{1.1}$$

The following Theorem provides the crucial link between differential geometry and algebraic topology. It reduces the question of constructing a metric of positive scalar curvature to a question in equivariant spin bordism. We refer to [11, 17] for the proof of (a) and to [14, 16] for the application of (a) to bordism which is contained in (b).

Theorem 1.3:

(a) **(Gromov-Lawson, Schoen-Yau)** Let M be a compact manifold, which is not necessarily connected, which admits a Riemannian metric of positive scalar curvature. Then any manifold which can be obtained from M by performing surgeries in codimension at least 3 also admits a metric with positive scalar curvature.

(b) **(Miyazaki-Rosenberg)** Let M be a connected spin manifold of dimension $m \geq 5$ with finite fundamental group. Let σ be an isomorphism from $\pi_1(M)$ to G and let $[M,s,\sigma] = [M_1,s_1,\sigma_1]$ in $\Omega_m^{spin}(BG)$. If M_1 admits a metric of positive scalar curvature, then so does M.

Remark: It is the restriction that the codimension of the surgery be at least 3 in (a) that leads to the requirement that m be at least 5 in (b).

In §2, we discuss the eta invariant; this is an analytic invariant which is crucial to the understanding of these bordism groups if m is odd. Fix a metric g on M, and let A_ρ be the Dirac operator with coefficients in a representation ρ of G. Let

$$\eta(g,M)(\rho) := \eta(A_\rho) = \tfrac{1}{2}\{\mathrm{Tr}_{L^2}(A_\rho^{-s}) + \dim\ker(A_\rho)\}|_{s=0} \tag{1.2}$$

be the associated eta invariant. Since the eta invariant is additive with respect to the direct sum of representations, we may extend $\eta(g,M)(\rho)$ to the group representation ring $R(G)$. We will show that if we restrict ρ to the augmentation ideal $R_0(G)$ of virtual representations of virtual dimension 0, then the eta invariant extends to a map in bordism with values in $\mathbb{R}/\mathbb{Z}$.

In §3, we will give a different proof of Theorem 1.2 (b). We will use the eta invariant to show that if n is odd, then $\Omega_5^{spin}(B\mathbb{Z}/n)$ is generated by lens spaces; these all admit metrics of positive scalar curvature. We will then use Theorem 1.2 (a) to show that $\tilde\Omega_m^{spin}(B\mathbb{Z}/n)$ is generated by manifolds of positive scalar curvature for any $m \geq 5$; Theorem 1.2 (b) then follows from Theorem 1.3.

Once conjecture 1.1 is established for all the Sylow subgroups of a group G, it follows for G itself. If G is cyclic of odd order, Theorem 1.2 (b) applies. Thus only the case G is cyclic of 2 primary order or $G = Q$ is quaternion need to be dealt with to prove Theorem 1.2 (c) in full generality. The analysis of these cases is more delicate than that needed to prove Theorem 1.2 (b) owing to the presence of 2 torsion in the coefficient ring Ω^{spin}_*. The methods of [5] are quite topological in parts, but the eta invariant still plays an essential role.

If a spin manifold M admits a metric g of positive scalar curvature, then there are no harmonic spinors so we may lift the eta invariant from $\mathbb{R}/\mathbb{Z}$ to $\mathbb{R}$; see [9] for details. Let $\mathcal{R}^+(M)$ be the space of metrics of positive scalar curvature on M. In §4, we use the eta invariant to construct invariants of $\pi_0\mathcal{R}^+(M)$ if m is odd and of $\pi_1\mathcal{R}^+(M)$ if m is even; this second result uses the additivity of the eta invariant [6, 15, 19].

We say that M is a spherical space form if the Riemannian universal cover of M is the sphere; these are the Riemannian manifolds which have constant sectional curvature +1. We used a refinement of this invariant of $\pi_0(\mathcal{R}^+(M))$ in [4] to show that if M is a spherical space form of dimension $m \geq 5$ with non-trivial fundamental group, then $|\pi_0(\mathcal{R}^+(M))| = \infty$ so M admits a countable family of inequivalent metrics of positive scalar curvature. Thus this invariant has already proven very useful and we hope the analogous invariant of $\pi_1(\mathcal{R}^+)$ will prove equally useful in studying the topology of the moduli space.

We conclude §4 by using the eta invariant to show that if N is a spin manifold of dimension $n \equiv 0 \mod 4$ with $\hat{A}(N) = 0$, then $S^1 \times N$ does not admit a metric of positive scalar curvature; this is a special case of a theorem previously established by Gromov and Lawson [10] using other methods.

§2 The eta invariant.

We begin by establishing some notational conventions and recalling the basic analytical facts which we shall need. Let

$$\mathbb{Z}/n := \{\lambda \in \mathbf{C} : \lambda^n = 1\} \tag{2.1}$$

be the cyclic group of order n. Let $\rho_a(\lambda) := \lambda^a$ define a representation

$$\rho_a : \mathbb{Z}/n \to U(1). \tag{2.2}$$

If $\vec{a} = (a_1, \dots, a_k)$ is a collection of integers coprime to n, then

$$\rho_{\vec{a}} := \rho_{a_1} \oplus \dots \oplus \rho_{a_k} : \mathbb{Z}/n \to U(k) \tag{2.3}$$

defines a fixed point free action of $\mathbb{Z}/n$ on $S^{2k-1} \subset \mathbf{C}^k$ and we let

$$L(n; \vec{a}) = S^{2k-1}/\rho_{\vec{a}}(\mathbb{Z}/n) \tag{2.4}$$

be the resulting lens space. More generally, let G be a finite group and let

$$\tau : G \to U(k) \tag{2.5}$$

be a fixed point free representation of G for $k > 1$. We let

$$M(G,\tau) := S^{2k-1}/\tau(G). \tag{2.6}$$

We give $M(G,\tau)$ the inherited metric of constant sectional curvature $+1$; any odd dimensional complete Riemannian manifold with a metric of constant sectional curvature $+1$ is isometric to $M(G,\tau)$ for some (G,τ). These are the spherical space forms. The manifold $M(G,\tau)$ admits a spin structure if and only if $\det(\tau)^{1/2}$ extends to a representation of G and the choice of a square root defines the spin structure in question. We refer to Gilkey [8] for the proof of the following result which uses work of Donnelly [7].

Theorem 2.1: If $\rho \in R(G)$, then

$$\eta(M(G,\tau))(\rho) = \Sigma_{\lambda\in G,\lambda\neq 1}\mathrm{Tr}\,(\rho(g))\det(\tau(g))^{1/2}\det(I-\tau(g))^{-1}.$$

We can interpret the eta invariant in bordism. We refer to Atiyah et al [1, 2] for the proof of the following result; see also [9].

Theorem 2.2: Let m be odd, let $|G| < \infty$, and let $\rho \in R_0(G)$. Then $M \to \eta(M)(\rho)$ defines a group homomorphism from $\tilde{\Omega}_m^{spin}(BG)$ to $\mathbb{R}/\mathbb{Z}$.

It turns out that the eta invariant is additive with respect to splitting the manifold along a suitable hypersurface. This property is crucial to our discussion. Let X be a compact spin manifold and let W be a flat unitary bundle over X. Let Y be a hypersurface in X with trivial normal bundle. Assume the metric g on X is product near Y. Split X open along Y to define a new manifold X_Y with boundary; give X_Y the induced structures. Impose Atiyah-Patodi-Singer boundary conditions to define $\eta(X_Y,g,W)$.

Theorem 2.3: Let g be a metric of positive scalar curvature on X. Then

$$\eta(X,g,W) = \eta(X_Y,g,W).$$

Proof: We outline an argument of Wojciechowski; we refer to related work by Bunke [6] and Müller [15]. By [19, Theorem 0.1],

$$\eta(X,g,W) \equiv \eta(X_Y,g,W) \bmod \mathbb{Z}. \tag{2.7}$$

One reduces mod $\mathbb{Z}$ to eliminate the effect of the 0 spectrum and the corresponding small eigenvalues. The Lichnerowicz formula [13] generalizes to this setting to show there are no harmonic spinors, see [3] for details. Thus the spectrum of D is uniformly bounded away from zero and (2.7) extends to the desired equality in $\mathbb{R}$. ■

§3 Manifolds which admit metrics of positive scalar curvature.

We begin the proof of Theorem 1.2 (b) with the following Lemma.

Lemma 3.1: Let n be odd. Give the lens spaces the canonical spin structures and $\mathbb{Z}/n$ structures.

(a) $\tilde\Omega_5^{spin}(B\mathbb{Z}/n) = \text{span}_{\mathbb{Z}}\{[L(n;2,1,1)],[L(n;2,2,2)]\}$.

(b) The eta invariant completely detects $\tilde\Omega_5^{spin}(B\mathbb{Z}/n)$.

Proof: Let $M_1 = L(n;2,1,1)$ and $M_2 = L(n;2,2,2)$. Let

$$\mathcal{S}(n) = \text{span}_{\mathbb{Z}}\{[M_1],[M_2]\} \subseteq \tilde\Omega_5^{spin}(B\mathbb{Z}/n). \tag{3.1}$$

Define $\delta(M) = \eta(M)(\rho_{-3}(\rho_0 - \rho_2)^3)$. By Theorem 2.2, δ defines a homomorphism

$$\delta : \mathcal{S}(n) \to \mathbb{R}/\mathbb{Z}. \tag{3.2}$$

For $\lambda \neq 1$, define

$$f_1(\lambda) = \lambda^2(1-\lambda)^{-2}(1-\lambda^2)^{-1} \text{ and } f_2(\lambda) = \lambda^3(1-\lambda^2)^{-3}. \tag{3.3}$$

We use Theorem 2.1 to see:

$$\eta(M_i)(\rho) = n^{-1}\Sigma_{\lambda\in\mathbb{Z}/n,\lambda\neq 1}\text{Tr}\,(\rho(\lambda))f_i(\lambda). \tag{3.4}$$

Since $\rho_{-3}(\rho_0-\rho_2)^3 f_2(\lambda) = 1$ for $\lambda \neq 1$,

$$\delta(M_2) = n^{-1}\Sigma_{\lambda\in\mathbb{Z}/n,\lambda\neq 1}1 = (n-1)/n. \tag{3.5}$$

This shows that $|\text{Range}(\delta)| \geq n$. Since $(1-\lambda^2) = (1+\lambda)(1-\lambda)$,

$$f_1(\lambda) = \bar\lambda(1+\lambda)^2 f_2(\lambda). \tag{3.6}$$

Since $\bar\lambda(1+\lambda)^2 - 4 = (1-\lambda)(\bar\lambda - 1)$,

$$\eta(M_1 - 4M_2)(\rho) = \eta(M_2)(\rho(\rho_0-\rho_1)(\rho_{-1}-\rho_0)). \tag{3.7}$$

Consequently the same argument used to show (3.5) shows

$$\begin{aligned}\delta(M_1 - 4M_2) &= n^{-1}\Sigma_{\lambda\in\mathbb{Z}/n,\lambda\neq 1}(1-\lambda)(\bar\lambda-1) \\ &= n^{-1}\Sigma_{\lambda\in\mathbb{Z}/n}(1-\lambda)(\bar\lambda-1) = 2 \in \mathbb{Z}\end{aligned} \tag{3.8}$$

so $[M_1] \in \ker(\delta)$. Let $\gamma = \rho_{-2}(\rho_0 - \rho_2)$. Then

$$\eta(M_1 - 4M_2)(\gamma) = n^{-1}\Sigma_{\lambda\in\mathbb{Z}/n,\lambda\neq 1}(1-\lambda)^2/(1-\lambda^2)^2. \tag{3.9}$$

Let $n = 2q - 1$. Then $\lambda = \lambda^{2q}$ for $\lambda \in \mathbb{Z}/n$ so

$$\begin{aligned} \eta(M_1 - 4M_2)(\gamma) &= n^{-1}\Sigma_{\lambda\in\mathbb{Z}/n,\lambda\neq 1}(1-\lambda^{2q})^2/(1-\lambda^2)^2 \\ &= n^{-1}\Sigma_{\lambda\in\mathbb{Z}/n,\lambda\neq 1}(\lambda^{2q-2}+\lambda^{2q-4}+\ldots+\lambda^2+1)^2 \\ &= -q^2/n + n^{-1}\Sigma_{\lambda\in\mathbb{Z}/n}(\lambda^{2q-2}+\ldots+\lambda^2+1)^2 \\ &\equiv q^2/n \bmod \mathbb{Z}. \end{aligned} \tag{3.10}$$

This shows that $|\text{kernel}(\delta)| \geq n$ so

$$|\mathcal{S}(n)| \geq n^2. \tag{3.11}$$

The second term in the Atiyah-Hirzebruch bordism spectral sequence for the group $\tilde{\Omega}_*^{spin}(B\mathbb{Z}/n)$ is $\tilde{H}_p(B\mathbb{Z}/n;\Omega_q^{spin})$. We estimate

$$|\mathcal{S}(n)| \leq |\tilde{\Omega}_5^{spin}(B\mathbb{Z}/n)| \leq |\oplus_{p+q=5}\tilde{H}_p(B\mathbb{Z}/n;\Omega_q^{spin})| = n^2. \tag{3.12}$$

We compare (3.11) and (3.12) to see $\mathcal{S}(n) = \tilde{\Omega}_5^{spin}(B\mathbb{Z}/n)$. This proves (a); (b) now follows. ■

Remark: If $n = a \cdot b$ for a coprime to b, then

$$\tilde{\Omega}_m^{spin}(B\mathbb{Z}/n) = \tilde{\Omega}_m^{spin}(B\mathbb{Z}/a) \oplus \tilde{\Omega}_m^{spin}(B\mathbb{Z}/b). \tag{3.13}$$

Thus it suffices to study $n = p^\nu$ a prime power. We give below the additive structure of these groups; we omit further details as we shall not need this result.

$$\tilde{\Omega}_5^{spin}(B\mathbb{Z}/p^\nu) = \begin{cases} 0 & \text{if } p = 2 \text{ and } \nu = 1, \\ \mathbb{Z}/2^\nu \oplus \mathbb{Z}/2^{\nu-2} & \text{if } p = 2 \text{ and } \nu > 1, \\ \mathbb{Z}/3^{\nu+1} \oplus \mathbb{Z}/3^{\nu-1} & \text{if } p = 3, \\ \mathbb{Z}/p^\nu \oplus \mathbb{Z}/p^\nu & \text{if } p \geq 5. \end{cases} \tag{3.14}$$

We use Lemma 3.1 to establish the following result

Lemma 3.2: Let n be odd and let $m \geq 5$. We can choose generators which admit metrics of positive scalar curvature for $\tilde{\Omega}_m^{spin}(B\mathbb{Z}/n)$.

Proof: Note that the groups $E_{p,q}^2 = \tilde{H}_p(B\mathbb{Z}/n;\Omega_q^{spin})$ are finite groups of odd orders for each p, q, and n so the Atiyah-Hirzebruch spectral sequence implies that $\tilde{\Omega}_m^{spin}(B\mathbb{Z}/n)$ is a finite group of odd order for any m. We also note that all the torsion in Ω_*^{spin} is 2 primary. If $M \in \tilde{\Omega}_m^{spin}(B\mathbb{Z}/n)$ admits a metric of positive scalar curvature and if N is a simply connected spin manifold without boundary, then $M \times N$ also admits a metric of positive scalar curvature. Let $Y_1 = S^1$ with the canonical metric, $\mathbb{Z}/n$ structure, and spin structure; for $j > 1$, let

$$Y_{2j-1} := L(n;1,\ldots,1) \text{ and } y_{2j-1} := [Y_{2j-1}] \in \tilde{\Omega}_{2j-1}^{spin}(B\mathbb{Z}/n). \tag{3.15}$$

The Y_{2j-1} generate $\tilde{H}_*(B\mathbb{Z}/n)$ so $\tilde{\Omega}_*^{spin}(B\mathbb{Z}/n)$ is generated as a $\tilde{\Omega}_*^{spin}$ module by the y_{2j-1}. Since $\tilde{\Omega}_*^{spin}(B\mathbb{Z}/n)$ is a finite group of odd order, we may replace

Ω_*^{spin} by $\Omega_*^{spin} \otimes_{\mathbb{Z}} \mathbb{Z}[\frac{1}{2}]$. We can choose simply connected spin manifolds X_{4j} of dimension $4j$ so that if $x_{4j} := [X_{4j}] \in \Omega_*^{spin}$,

$$\Omega_*^{spin} \otimes_{\mathbb{Z}} \mathbb{Z}[\tfrac{1}{2}] = \mathbb{Z}[\tfrac{1}{2}][x_4, x_8, x_{12}, \ldots]. \tag{3.16}$$

This means that generators for $\tilde{\Omega}_m^{spin}(B\mathbb{Z}/n)$ can be taken of the form

$$y(j,I) := y_{2j-1} \cdot x_4^{i_1} \cdots x_{4k}^{i_k} \tag{3.17}$$

where $2j - 1 + 4(i_1 + \ldots + ki_k) = m$. If $j > 1$, then $Y(j,I)$ admits a metric of positive scalar curvature.

The manifold X_4 is a Kummer surface so $\hat{A}[X_4] = 2$. Thus by replacing X_{4k} by $X_{4k} \# n_k X_4^k$ where $n_k \in \mathbb{Z}[\frac{1}{2}]$ is chosen appropriately, we may assume $\hat{A}[X_{4k}] = 0$ for $k > 1$ so by Theorem 1.2 (a), these manifolds admit metrics of positive scalar curvature for $k > 1$. If some $i_\nu > 1$ for $\nu > 1$, then $Y(1,I)$ admits a metric of positive scalar curvature as desired. Finally, let $Y(1,I) = Y_1 \times X_4 \times X_4^{\nu-1}$ for $\nu \geq 1$. Since $Y_1 \times X_4 \in \tilde{\Omega}_5^{spin}(B\mathbb{Z}/n)$, we use Lemma 3.1 to see it can be represented by elements which admit metrics of positive scalar curvature. ■

Proof of Theorem 1.2 (b): Let M be a connected spin manifold of dimension $m \geq 5$ with $\pi_1(M) = \mathbb{Z}/n$ for n odd with $\alpha(M) = 0$ or equivalently $\hat{A}(M) = 0$. The argument given above shows that we may choose a simply connected manifold N_1 so that $[M] = [N_1]$ in the bordism group Ω_m^{spin}. Since $\hat{A}$ is a bordism invariant, $\hat{A}(N_1) = 0$ so N_1 admits a metric of positive scalar curvature by Theorem 1.2 (a). Since $[M] - [N_1] = 0$ in Ω_m^{spin}, the difference defines an element in $\tilde{\Omega}_m^{spin}(B\mathbb{Z}/n)$; by Lemma 3.2, this difference is bordant to a manifold N_2 which admits a metric of positive scalar curvature. Since $[M] = [N_1 + N_2]$ in $\Omega_m^{spin}(B\mathbb{Z}_n)$, we apply Theorem 1.3 to see M itself admits a metric of positive scalar curvature. ■

§4 The moduli space of metrics of positive scalar curvature.

Let M be a compact spin manifold without boundary. We take $G = \pi(M)$. In this section, we use the eta invariant to define real valued invariants of $\pi_0(\mathcal{R}^+(M))$ if m is odd and of $\pi_1(\mathcal{R}^+(M))$ if m is even. We begin with a result of Atiyah-Patodi-Singer [2]:

Theorem 4.1: Let M be a compact connected spin manifold without boundary of odd dimension m. Let $G = \pi_1(M)$. If $\rho \in R_0(G)$ and if $g \in \mathcal{R}^+(M)$, let $\eta(M,g)(\rho) \in \mathbb{R}$ be the eta invariant of the Dirac operator defined by the metric g with coefficients in ρ. Then $\eta(\cdot, \rho)$ is constant on arc components of $\mathcal{R}^+(M)$.

Proof: If g_t for $t \in [a,b]$ is a smooth 1-parameter family of metrics of positive scalar curvature, then $u^2 d\theta^2 + g_t$ is a metric of positive scalar curvature on $[a,b] \times M$ for u large. Consequently, by Theorem 2.2

$$\eta(M,g_a)(\rho) \equiv \eta(M,g_b)(\rho) \bmod \mathbb{Z}.$$

By the Lichnerowicz formula [13], there are no harmonic spinors since the metrics have positive scalar curvature. Since eigenvalues do not cross the origin, $\eta(M,g_t)(\rho)$ is a continuous function of t. It now follows that $\eta(M,g_t)(\rho)$ is independent of t. ∎

Remark: We have used a generalization of this invariant to prove that if M is a spherical space form of dimension $m \geq 5$, then $\mathcal{R}^+(M)$ has a countable number of arc components. See [3, 4] for details.

Next we suppose m is even. Fix some metric g_0 as the basepoint of $\mathcal{R}^+(M)$. Let $\Gamma : [0,2\pi] \to \mathcal{R}^+(M)$ be a smooth 1-parameter family of metrics of positive scalar curvature. Assume $\Gamma \equiv g_0$ near $\theta = 0$ and $\theta = 2\pi$. There exists a positive constant $u(\Gamma)$ so that if $u \geq u(\Gamma)$, the metric

$$g(\Gamma,u) := u^2 d\theta^2 + \Gamma(\theta) \in \mathcal{R}^+([0,2\pi] \times M). \tag{4.1}$$

The isotopy class of this metric in $\mathcal{R}^+$ is independent of u for $u \geq u(\Gamma)$ We use the trivial spin structure on S^1 to induce a spin structure on $S^1 \times M$; using the Mobius spin structure would not change the invariants which arise. The natural projection $\pi_1(S^1 \times M) \to \pi_1(M)$ defines a natural $G = \pi_1(M)$ structure on $S^1 \times M$. If $\rho \in R_0(G)$, let

$$\eta_1(\Gamma,\rho) := \eta(M \times S^1, u^2 d\theta^2 + \Gamma(\theta))(\rho); \tag{4.2}$$

this is independent of the parameter u by Theorem 4.1.

Theorem 4.2: Let M be a compact connected spin manifold without boundary of even dimension m. Then $\eta_1(\rho)$ defines group homomorphism

$$\eta_1(\rho) : \pi_1\mathcal{R}^+(M) \times R_0(\pi_1 M) \to \mathbb{Z}$$

for each $\rho \in R_0(\pi_1 M)$.

Proof: Let Γ_0 be the constant path, $\Gamma_0(\theta) = g_0$. Let $\rho = \rho_1 - \rho_0 \in R_0(G)$. By Theorem 2.2, the mod $\mathbb{Z}$ reduction of $\eta_1(\Gamma,\rho)$ is independent of Γ. To show $\eta_1(\Gamma,\rho) \in \mathbb{Z}$, it suffices to show $\eta_1(\Gamma_0,\rho) = 0$.

The metric $g(\Gamma_0)$ is a product metric on $S^1 \times M$. Let W be a unitary flat vector bundle on M; we extend W by pullback to $S^1 \times M$. Let $D = D(g(\Gamma_0),W)$ be the Dirac operator on $S^1 \times M$ with coefficients in W. We decompose:

$$D = \alpha_W(\partial_\theta + A_W) \tag{4.3}$$

where A_W is an operator of Dirac type on M and where α_W is a unitary bundle automorphism. Let

$$\eta(z,D) = \mathrm{Tr}_{L^2}(D(D^2)^{(-z-1)/2}). \tag{4.4}$$

Since α_W anti-commutes with A_W,

$$\eta(z,D) = \eta(z, \alpha_W(\partial_\theta - A_W)). \tag{4.5}$$

The map $\theta \to 2\pi - \theta$ is an isometry of $S^1 \times M$ which replaces $\alpha_W(\partial_\theta - A_W)$ by $\alpha_W(-\partial_\theta - A_W)$. Then

$$\eta(z,D) = \eta(z,-D) - \eta(z,\alpha(\partial_\theta + A_W)). \tag{4.6}$$

Since there are no harmonic spinors, $\eta(D) = 0$; this shows η_1 takes integer values.

We complete the proof by showing that η is a group homomorphism. Let $\Gamma_i : [0,2\pi] \to \mathcal{R}^+(M)$ and let $\Gamma_3 = \Gamma_1 \# \Gamma_2$ be the concatenation. Let W be a unitary bundle over M and let $\eta(\cdot,W,\Gamma)$ be the eta invariant of the Dirac operator for the metric defined by Γ with coefficients in W. We apply Theorem 2.3 to $X = S^1 \times M$ and $Y = \{0\} \times M$ to see

$$\eta(S^1 \times M,W,\Gamma_i) = \eta([0,2\pi] \times M,W,\Gamma_i). \tag{4.7}$$

We apply Theorem 2.3 with $X = S^1 \times M$ and $Y = \{0\} \times M \,\dot\sqcup\, \{\pi\} \times M$; we note that the manifold $X_Y = [0,\pi] \times M \,\dot\sqcup\, [\pi,2\pi] \times M$ is disconnected so

$$\begin{aligned} \eta(S^1 \times M,W,\Gamma_3)) &= \eta(X_Y,W,\Gamma_3) \\ &= \eta([0,\pi] \times M,W,\Gamma_1) + \eta([\pi,2\pi] \times M,W,\Gamma_2) \\ &= \eta(S^1 \times M,W,\Gamma_1) + \eta(S^1 \times M,W,\Gamma_2). \end{aligned} \tag{4.8}$$

We conclude by using the eta invariant to show certain manifolds do not admit metrics of positive scalar curvature. The following result also follows from Gromov and Lawson [10].

Theorem 4.3: Let N be a $4k$ dimensional spin manifold with $\hat{A}[N] \neq 0$. Then $S^1 \times N$ does not admit a metric of positive scalar curvature.

Proof: Give S^1 the canonical trivial spin structure. Let $\rho_\epsilon(n) = e^{2\pi i n\epsilon}$ for $\epsilon \in \mathbb{R}$ parametrize the irreducible unitary representations of $\pi_1(S^1) = \mathbb{Z}$. The following is well known; see [9] for example:

$$\begin{aligned} &\eta(S^1)(\rho_\epsilon - \rho_0) \equiv \epsilon \bmod \mathbb{Z}, \\ &\eta(S^1 \times N)(\rho) = \eta(S^1)(\rho)\hat{A}[N]. \end{aligned} \tag{4.9}$$

Consequently

$$\eta(S^1 \times N)(\rho_\epsilon - \rho_0) = \epsilon\hat{A}[N] \mod \mathbb{Z}. \tag{4.10}$$

Suppose $S^1 \times N$ admits a metric of positive scalar curvature. The bundles defined by $\rho(\epsilon)$ are all topologically trivial; the associated unitary flat connection on the trivial bundle is given by the connection 1-form $\epsilon d\theta$. Let

$$\eta(\epsilon) := \eta(S^1 \times N, \rho_\epsilon - \rho_0) \in \mathbb{R}. \tag{4.11}$$

Since there are no harmonic spinors, $\eta(\epsilon)$ is smooth in ϵ. By (4.10),

$$\partial_\epsilon \eta(\epsilon) = \hat{A}[N] \neq 0 \quad \text{so} \quad \eta(\epsilon + 1) = \eta(\epsilon) + \hat{A}[N] \neq \eta(\epsilon). \tag{4.12}$$

This is impossible since $\rho_\epsilon = \rho_{\epsilon+1}$. ■

References

[1] M. F. Atiyah, V. K. Patodi, and I. M. Singer, Spectral asymmetry and Riemannian geometry I, Math. Proc. Camb. Phil. Soc. **77** (1975), 43–69.

[2] —, II, **78** (1975) 405–432.

[3] B. Botvinnik and P. Gilkey, The eta invariant and metrics of positive scalar curvature (preprint).

[4] —, The eta invariant and metrics of positive scalar curvature on spherical space forms (preprint).

[5] B. Botvinnik, P. Gilkey, and S. Stoltz, The Gromov-Lawson-Rosenberg conjecture for gropus with periodic cohomology (preprint)

[6] U. Bunke, Relative index theory, J. Funct. Anal. 105 (1992) 63–76; see also On the gluing problem for the eta invariant (preprint).

[7] H. Donnelly, Eta invariants for G spaces, Indiana Univ. Math J 27 (1978), 889–918.

[8] P. Gilkey, The eta invariant and the K-theory of odd dimensional spherical space forms, Invent. Math. **76** (1984), 421–453.

[9] —, Invariance Theory, the heat equation, and the Atiyah-Singer index theorem 2nd Ed accepted for publication CRC press.

[10] M. Gromov and H. B. Lawson, Spin and scalar curvature in the presence of a fundamental group I, Annals of Math 111 (1980), 209–230.

[11] —, The classification of simply connected manifolds of positive scalar curvature, Ann. of Math. 111 (1980) 423–434.

[12] S. Kwasik and R. Schultz, Positive scalar curvature and periodic fundamental groups, Comment. Math. Helv. 65 (1990), 271–286.

[13] A. Lichnerowicz, Spineurs harmoniques, C. R. Acad. Sci. Paris 257 (1963) 7–9.

[14] T. Miyazaki, On the existence of positive curvature metrics on non simply connected manifolds, J. Fac. Sci. Univ. Tokyo Sect IA 30 (1984), 549–561.

[15] W. Müller, Eta invariants and manifolds with boundary, C. R. Acad. Sci. Paris Sér. I Math. 317 (1993) 181–186; Eta invariants and manifolds with boundary, preprint, MPI, Bonn, 1993.

[16] J. Rosenberg, C^* algebras, positive scalar curvature, and the Novikov conjecture, II. In Geometric Methods in Operator Algebras, Pitman Res. Notes 123, 341–374, Longman Sci. Techn., Harlow, 1986.; III Topology 25 (1986) 319–336.

[17] R. Schoen and S. T. Yau, The structure of manifolds with positive scalar curvature, Manuscripta Math. 28 (1979) 159–183.

[18] S. Stolz, Simply connected manifolds of positive scalar curvature, Ann. of Math. 196 (1992), 511–540.

[19] K. Wojciechowski, On the additivity of the eta invariant, preprint; see also: Spectral asymmetry, elliptic boundary value problem, and cutting and pasting of elliptic operators, preprint.

B. Botvinnik
Mathematics Department, University of Oregon
Eugene Oregon 97403 USA, botvinn@poincare.uoregon.edu

P. Gilkey
Mathematics Department, University of Oregon
Eugene Oregon 97403 USA, gilkey@math.uoregon.edu

Operator Theory:
Advances and Applications, Vol. 78

Generalized Strichartz Inequalities for the Wave Equation

J. Ginibre, G. Velo

Strichartz inequalities for the wave equation $\Box u = f$ are estimates of the solution u of the Cauchy problem for that equation, in the form of space time integral norms, in terms of similar norms of the inhomogeneity f and of suitable norms of the initial data. Following the seminal paper of Segal [18], the original result of Strichartz was obtained in the fundamental paper [19] and can be stated as follows. One is interested in the Cauchy problem for the wave equation

$$\begin{cases} \Box u = f \\ u(t=0) = u_0, \quad \partial_t u(t=0) = u_1 \end{cases} \tag{1}$$

in space time $\mathbb{R}^{n+1}$ with $n \geq 2$, with $\Box = \partial_t^2 - \Delta$. The initial data for the problem (1) are taken from the space

$$Y^\mu \equiv \dot{H}^\mu \oplus \dot{H}^{\mu-1} \tag{2}$$

with $\mu \in \mathbb{R}$, where $\dot{H}^\mu \equiv \dot{H}_2^\mu$ is the homogeneous Sobolev space defined more generally by

$$\dot{H}_r^\mu \equiv \dot{H}_r^\mu(I\!R^n) \equiv \Big\{ u; \| u; \dot{H}_r^\mu \| \equiv \| (-\Delta)^{\mu/2} u \|_r < \infty \Big\} \tag{3}$$

for $1 < r < \infty$, and $\| \cdot \|_r$ denotes the norm in $L^r \equiv L^r\ (I\!R^n)$. Then the solution u of the Cauchy problem (1) satisfies the estimate

$$\| u; L^{r_S}(\mathbb{R}^{n+1}) \| \leq C \Big\{ \| (u_0, u_1); Y^{1/2} \| + \| f; L^{\bar{r}_S}(\mathbb{R}^{n+1}) \| \Big\} \tag{4}$$

for any u_0, u_1, f for which the right hand side is finite, where $r_S = 2(n+1)/(n-1)$, and $(r, \bar{r})$ denote Hölder conjugate exponents $1/r + 1/\bar{r} = 1$. Subsequent generalisations of (4) consisted in extending that result to more general spaces than $L^{r_S}(\mathbb{R}^{n+1})$, typically to spaces $L^q(\mathbb{R}, \dot{H}_r^\rho)$ of L^q functions of the time variable with values in the Sobolev space $\dot{H}_r^\rho$ of the space variable or in similar spaces such as Besov spaces. The generalized version of the inequalities has now become an essential tool in the study of the Cauchy problem and of the theory of scattering for non linear wave equations. In particular they play an important role in recent studies of the Cauchy problem for critical nonlinearities and/or low regularity initial data.

The proof of the inequalities involves basically two types of ingredients. The first one consists of specific estimates, in particular stationary phase estimates, on the evolution group associated with the homogeneous equation. For the wave equation, those estimates are available from [2][16]. The second one consists of abstract arguments, not specific to the wave equation. Those arguments are mainly duality arguments and were first applied to the present problem in [19] in the context of the Fourier transform. Subsequent to the work of Strichartz, progress was made in several directions. Following work in [14] on the Klein Gordon equation, it was recognized in [17] that the space and time exponents that occur in the inequalities did not need to be equal, thereby providing most of the inequalities now available for the homogeneous equation. It then appeared on the related example of the Schrödinger equation that the duality arguments were more efficiently put to work in an operator form where the Fourier transform did not play an essential role [4]. That point of view was exploited partly in [4] and then fully in [21] to provide a large set of Strichartz inequalities for the Schrödinger equation. Those inequalities have become a fundamental tool in the study of the nonlinear Schrödinger equation (NLS) [12]. With the results of [2] [16] [17] [21] available, it was obvious how to write a large set of generalized Strichartz inequalities for the wave equation. That was immediately done and applied to the theory of scattering for the NLW equation [6] [7] and subsequently to the critical Cauchy problem [3]. Some of the inequalities were generalized to variable coefficient equations and equations on manifolds [10] [11] [15]. Similar inequalities have also been obtained in [22]. Finally in [13], the previous inequalities have been extended by the inclusion of some limiting cases, so far excluded, where one allows for the occurrence of the space L^2 in the time variable, an extension which is relevant in space dimension $n \geq 4$. That extension was made only in a codimension one subset of the natural range of the parameters, but it can be easily generalized to the full range thereof.

The purpose of the present lecture is to state and sketch the proof of the available inequalities, namely the bulk inequalities of [6] [7] extending those for the homogeneous equation obtained in [17], together with the limiting L_t^2 case obtained in [13], in the natural range of parameters. The (rather simple) proof is basically that given in [6], with an appropriate addition to cover the limiting L_t^2 case. The details will appear in [9].

In order to state the generalized inequalities, we need additional notation. We define the operators $\omega = (-\Delta)^{1/2}$ and $U(t) = \exp(i\omega t)$, $K(t) = \omega^{-1} \sin \omega t$, $\dot{K}(t) = \cos \omega t$. The Cauchy problem (1) is solved by $u = v + w$, where v is the solution of the homogeneous equation with the same data

$$\begin{cases} \square v = 0 \\ v(t=0) = u_0, \quad \partial_t v(t=0) = u_1 \end{cases} \tag{5}$$

namely

$$\begin{cases} v(t) &= \dot{K}(t)u_0 + K(t)u_1 \\ \partial_t v(t) &= K(t)\Delta u_0 + \dot{K}(t)u_1 \end{cases} , \tag{6}$$

and w is the solution of the inhomogeneous equation with zero data,

$$\begin{cases} \square w = f \\ w(t=0) = \partial_t w(t=0) = 0 \quad . \end{cases} \tag{7}$$

Let $L(t)$ be any of the operators $\omega^\lambda U(t)$, $\omega^\lambda K(t)$ or $\omega^\lambda \dot{K}(t)$ with $\lambda \in \mathbb{R}$ and let $\chi_\pm$ be the characteristic function of $\mathbb{R}^\pm$ in time. We define $L_R(t) = \chi_+(t)L(t)$ and $L_A(t) = \chi_-(t)L(t)$ where R and A stand for retarded and advanced. Then the Cauchy problem (7) is solved for positive time by

$$\begin{cases} w(t) = \int_0^t dt' K(t-t')f(t') \\ \qquad = (K_R *_t \chi_+ f)(t) \\ \partial_t w(t) = \int_0^t dt' \dot{K}(t-t')f(t') \\ \qquad = \left(\dot{K}_R *_t \chi_+ f\right)(t) \quad . \end{cases} \tag{8}$$

Similar formulas with advanced operators solve the Cauchy problem (7) for negative times. We restrict our attention from now on to positive times. The exponent r in various function spaces naturally occurs through positive multiples of the basic quantity $\alpha(r) \equiv 1/2 - 1/r$, namely $\beta(r) \equiv \frac{n+1}{2}\alpha(r)$, $\gamma(r) \equiv (n-1)\alpha(r)$ and $\delta(r) \equiv n\alpha(r)$. The exponent $\beta(r)$ (or rather $2\beta(r)$) is the loss of derivatives in the basic estimate (see (20) below), the exponent $\gamma(r)$ is the optimal time decay exponent of L^r solutions of the wave equation, and $\delta(r)$ appears naturally in Hölder and Sobolev inequalities. We denote by $\widehat{\ }$ the Fourier transform in $\mathbb{R}^n$ and by $*_x$ and $*_t$ the convolution in x or t, with the subscript x or t omitted when there is no risk of confusion. The use of Paley Littlewood dyadic decompositions is unavoidable in the present problem. They are defined in the following standard way. Let $\widehat{\psi} \in \mathcal{C}_0^\infty(\mathbb{R}^n)$ with $0 \leq \widehat{\psi} \leq 1$, $\widehat{\psi}(\xi) = 1$ for $|\xi| \leq 1$ and $\widehat{\psi}(\xi) = 0$ for $|\xi| \geq 2$. We define $\widehat{\varphi}_0(\xi) = \widehat{\psi}(\xi) - \widehat{\psi}(2\xi)$ and for any $j \in Z$, $\widehat{\varphi}_j(\xi) = \widehat{\varphi}_0(2^{-j}\xi)$ so that

$$Supp\ \widehat{\varphi}_j \subset \left\{\xi : 2^{j-1} \leq |\xi| \leq 2^{j+1}\right\}$$

and for any $\xi \in \mathbb{R}^n \setminus \{0\}$

$$\sum_j \widehat{\varphi}_j(\xi) = 1$$

with at most two non vanishing terms in the sum.

The proofs of the inequalities naturally yield them in terms of Besov spaces. With each tempered distribution u we associate the sequence of $\mathcal{C}^\infty$ functions $\varphi_j * u$, to be considered as a function of two variables j and x. The homogeneous Besov space $\dot{B}^\rho_{r,s}$ is defined for any $\rho \in \mathbb{R}$ and $1 \leq r,\ s \leq \infty$ by

$$\dot{B}^\rho_{r,s} = \left\{u : \|u; \dot{B}^\rho_{r,s}\| \equiv \|2^{\rho j}\varphi_j * u\ ;\ \ell^s_j(L^r_x)\| < \infty\right\} \tag{9}$$

where one takes first the L^r norm in the variable x and then the ℓ^s norm in the variable j ([20] p. 45 and p. 238). We shall need only the Besov spaces with $s = 2$

and in that case we shall omit that index. Comparison with the homogeneous Sobolev spaces $\dot{H}^\rho_r$ follows from the Mikhlin Hörmander theorem, which yields the inclusions [1] [20] :

$$\begin{aligned} \dot{B}^\rho_r &\subset \dot{H}^\rho_r \quad \text{for } 2 \le r < \infty \ , \\ \dot{B}^\rho_r &\supset \dot{H}^\rho_r \quad \text{for } 1 < r \le 2 \ . \end{aligned} \tag{10}$$

We shall state the generalized Strichartz inequalities in the Besov version, which is both the stronger one and the easier one to prove. The Sobolev version of the inequalities follows from the Besov version by the embeddings (10) and is obtained by replacing everywhere $\dot{B}^\rho_r$ by $\dot{H}^\rho_r$ and excluding the cases $r = 1$ and $r = \infty$. The main result can be stated as follows.

Proposition. Let ρ_1, ρ_2, $\mu \in \mathbb{R}$ and $2 \le q_1, q_2, r_1, r_2 \le \infty$ and let the following conditions be satisfied.

$$0 \le 2/q_i \le Min\,(\gamma(r_i), 1) \quad \text{for } i = 1, 2 \tag{11}$$

$$(2/q_i, \gamma(r_i)) \ne (1, 1) \text{ for } i = 1, 2 \tag{12}$$

$$\rho_1 + \delta(r_1) - 1/q_1 = \mu \tag{13}$$

$$\rho_1 + \delta(r_1) - 1/q_1 = 1 - (\rho_2 + \delta(r_2) - 1/q_2) \tag{14}$$

(1) Let $(u_0, u_1) \in Y^\mu$ (see (2)). Then v defined by (6) satisfies the estimates

$$\| v; L^{q_1}\left(\mathbb{R}, \dot{B}^{\rho_1}_{r_1}\right) \| + \| \partial_t v; L^{q_1}\left(\mathbb{R}, \dot{B}^{\rho_1 - 1}_{r_1}\right) \| \le C \| (u_0, u_1); Y^\mu \| \ . \tag{15}$$

(2) For any interval $I \subset \mathbb{R}$, possibly unbounded, the following estimates hold

$$\| K * f; L^{q_1}\left(I, \dot{B}^{\rho_1}_{r_1}\right) \| \le C \| f : L^{\bar{q}_2}\left(I, \dot{B}^{-\rho_2}_{\bar{r}_2}\right) \| \ . \tag{16}$$

(3) For any interval $I = [0, T)$, $0 < T \le \infty$, the function $w = K_R * \chi_+ f$ defined by (8) satisfies the estimates

$$\| w; L^{q_1}\left(I, \dot{B}^{\rho_1}_{r_1}\right) \| + \| \partial_t w; L^{q_1}\left(I, \dot{B}^{\rho_1 - 1}_{r_1}\right) \| \le C \| f : L^{\bar{q}_2}\left(I, \dot{B}^{-\rho_2}_{\bar{r}_2}\right) \| \ . \tag{17}$$

The same results hold with $\dot{B}^\rho_r$ replaced by $\dot{H}^\rho_r$ everywhere, under the additional assumption that $r_i < \infty$ $(i = 1, 2)$ wherever r_i occurs.

Because of the homogeneity of the wave equation and of the norms in the inequalities, the exponents ρ_1, $-\rho_2$ and μ are determined up to an irrelevant common translation, by the homogeneity conditions (13) (14). The allowed region (11) (12) for the remaining parameters (q_i, r_i) has a product structure and is best pictured in the plane of the variables $(1/q), 1/r)$. For $n \ge 4$, the allowed region is a quadrangle $ABCD$ with vertices $A = (0, 1/2)$, $B = (1/2, (n-3)/2(n-1))$, $C = (1/2, 0)$ and $D = (0, 0)$ corresponding to $(q = \infty, r = 2)$, $(q = 2, \gamma(r) = 1)$, $(q = 2, r = \infty)$ and $(q = \infty, r = \infty)$ respectively. For $n = 3$, it reduces to the triangle ACD

and for $n = 2$ to the smaller triangle AC_2D where $C_2 = (1/4, 0)$ corresponds to $(q = 4, r = \infty)$. The limiting case $q = 2$ occurs only for $n \geq 4$. The boundary of the quadrangle is entirely allowed except for the single point B for $n \geq 3$ (that point coincides with C for $n = 3$) which is excluded by (12). In addition, for the Sobolev version of the inequalities, the segment CD of the boundary is entirely excluded by the condition $r < \infty$. It is easy to locate the original Strichartz result in that picture. It corresponds to the case where $q = r$ and $2\beta(r) = 1$. The representative point S is the intersection of the diagonal $1/q = 1/r (= 1/r_S)$ with the segment AB.

Sketch of proof of the Proposition. By the definitions (6) (8) of v and w and the definitions of K and $\dot{K}$, the inequalities (15) (16) (17) follow from similar inequalities involving only U. In fact it is sufficient to prove the inequalities

$$\|U(.)u; L^{q_1}\left(\mathbb{R}, \dot{B}^{\rho_1}_{r_1}\right)\| \leq C\|u\|_2 \ , \tag{15'}$$

$$\|U * f : L^{q_1}\left(I, \dot{B}^{\rho_1}_{r_1}\right)\| \leq C\|f : L^{\bar{q}_2}\left(I, \dot{B}^{-\rho_2}_{\bar{r}_2}\right)\| \tag{16'}$$

for $I \subset \mathbb{R}$ and

$$\|U_R * f; L^{q_1}\left(I, \dot{B}^{\rho_1}_{r_1}\right)\| \leq C\|f; L^{\bar{q}_2}\left(I, \dot{B}^{-\rho_2}_{\bar{r}_2}\right)\| \tag{17'}$$

for $I = [0, T) \subset \mathbb{R}^+$ under the conditions (11) (12) and

$$\rho_i + \delta(r_i) - 1/q_i = 0 \qquad \text{for} \quad i = 1,2 \ . \tag{18}$$

A second reduction of the proof follows from the Sobolev inequalities and embeddings (which are elementary in Besov spaces) which imply that for fixed q_i and $\rho_i + \delta(r_i)$, the inequalities (15') (16') (17') are all the stronger as ρ_i or $1/r_i$ is larger, so that it suffices to prove them for the largest allowed values of $1/r_i$, namely to $\gamma(r_i) = 2/q_i$ in the bulk case $q_i > 2$.

From there on, the proof proceeds in three steps (we restrict our attention to the bulk case $q_i > 2$).

(1) Stationary phase estimates. The benefit of using dyadic decompositions is that it is sufficient to estimate the action of $U(t)$ on the generic building block φ_j of the decomposition. The basic estimate combines a stationary phase estimate and a trivial scaling to yield

$$\|U(t)\varphi_j\|_\infty \leq C \text{Min}\left\{2^{nj}, |t|^{-(n-1)/2}\, 2^{j(n+1)/2}\right\} \tag{19}$$

which by interpolation with unitarity of $U(t)$ in L^2 and substitution in the definition (9) directly yields the pointwise (in time) estimate of [2] [16], namely

$$\|U(t)f; \dot{B}^{-\beta(r)}_r\| \leq C|t|^{-\gamma(r)}\, \|f; \dot{B}^{\beta(r)}_{\bar{r}}\| \ . \tag{20}$$

(2) Disposal of the time variable. From (20) and the Hardy Littlewood inequality in time, one obtains

$$\| U_{(R)} *_t f : L^q \left(I, \dot{B}_r^{-\beta(r)} \right) \| \leq C \, \| f : L^{\bar{q}} \left(I, \dot{B}_{\bar{r}}^{\beta(r)} \right) \| \tag{21}$$

namely the diagonal ($r_1 = r_2, q_1 = q_2$) limiting ($2/q_i = \gamma(r_i)$) case of (16′) (17′).

(3) Abstract duality arguments. Those arguments are equation independent and basically state that if A is an operator and A^* its adjoint in a suitable context, it is equivalent for A or A^* or A^*A to be bounded. The relevant context has A as an operator from a dense subspace $\mathcal{D}$ (typically $\mathcal{S}(\mathbb{R}^{n+1})$) of a Banach space X to a Hilbert space $\mathcal{H}$ (typically $L^2(\mathbb{R}^n)$) so that A^* is an operator from $\mathcal{H}$ to the algebraic dual $\mathcal{D}_a$ of $\mathcal{D}$. Boundedness of A is to be understood as an operator from X to $\mathcal{H}$ (after suitable extension), and boundedness of A^* and A^*A as operators from $\mathcal{H}$ to X^* and from X to X^* (see [8] [9] for details). The relevant basic example is constructed from a one parameter unitary group $U(\cdot)$ in $\mathcal{H}$ with

$$Af = \int_I dt U(-t) f(t) \quad . \tag{22}$$

$$A^* v(t) = U(t) v \tag{23}$$

$$A^* \, Af = \int_I dt\,' \, U(t-t') f(t') \tag{24}$$

and the previous boundedness properties obviously hold with $X = L^1(I, \mathcal{H})$ and $X^* = L^\infty(I, \mathcal{H})$. In this framework and with obvious X, (15′) is the boundedness of A^* while the diagonal case of (16′) is the (equivalent) previously proved boundedness of A^*A. Furthermore an obvious factorisation argument shows that the off-diagonal cases of (16′) follow from the diagonal ones. Finally the off-diagonal (retarded) cases of (17′) also follow from the diagonal ones by an additional interpolation argument. This completes the sketch of the proof for the bulk cases $q_i > 2$. The limiting cases $q_i = 2$ mostly follow from a simple modification of the previous proof, whereby the Hardy-Littlewood inequality in time is replaced by the Young inequality at step (2) above.

References

[1] J. Bergh and J. Löfström, Interpolation spaces, Springer, Berlin, 1976.

[2] P. Brenner, On L_p-$L_{p'}$ estimates for the wave equation, Math. Z., **145** (1975), 251–254.

[3] J. Ginibre, A. Soffer and G. Velo, The global Cauchy problem for the critical non linear wave-equation, J. Funct. Anal., **110** (1992), 96–130.

[4] J. Ginibre and G. Velo, Scattering theory in the energy space for a class of non-linear Schrödinger equations, J. Math. Pures Appl., **64** (1984), 363–401.

[5] J. Ginibre and G. Velo, The global Cauchy problem for the nonlinear Klein-Gordon equation, Math. Z., **189** (1985), 487–505.

[6] J. Ginibre and G. Velo, Conformal invariance and time decay for nonlinear wave equations, II, Ann. Inst. Henri Poincaré (Physique Théorique), **47** (1987), 263–276.

[7] J. Ginibre and G. Velo, Scattering theory in the energy space for a class of nonlinear wave equations, Commun. Math. Phys., **123** (1989), 535–573.

[8] J. Ginibre and G. Velo, Smoothing properties and retarded estimates for some dispersive evolution equations, Commun. Math. Phys., **144** (1992),163–188.

[9] J. Ginibre and G. Velo, Generalized Strichartz inequalities for the wave equation, Preprint Orsay 94–60, submitted to J. Func. Anal.

[10] L. Kapitanski, Some generalisations of the Strichartz-Brenner inequality, Leningrad Math. J., **1** (1990), 693–726.

[11] L. Kapitanski, Cauchy problem for a semilinear wave equation, II, Jour. Sov. Math. **62** (1992), 2746–2776. III *ibid* 2619–2645.

[12] T. Kato, Nonlinear Schrödinger equations, Ann. Inst. Henri Poincaré (Physique Théorique), **46** (1987), 113–129.

[13] H. Lindblad and C. D. Sogge, On existence and scattering with minimal regularity for semilinear wave equations, preprint.

[14] B. Marshall, Mixed norm estimates for the Klein-Gordon equation, *in* Proc. Conf. Harm. Anal. in honor of A. Zygmund, Wadsworth, 1981, pp. 638–649.

[15] G. Mockenhaupt, A. Seeger and C. D. Sogge, Local smoothing of Fourier integrals and Carleson-Sjölin estimates, J. Amer. Math. Soc., **6** (1993), 65–130.

[16] H. Pecher, L^p-Abschätzungen und klassische Lösungen für nichtlineare Wellengleichungen, I, Math. Z. **150** (1976), 159–183.

[17] H. Pecher, Nonlinear small data scattering for the wave and Klein-Gordon equation, Math. Z., **185** (1984), 261–270.

[18] I. E. Segal, Space-time decay for solutions of wave equations, Adv. Math., **22** (1976), 304–311.

[19] R. S. Strichartz, Restriction of Fourier transform to quadratic surfaces and decay of solutions of wave equations, Duke Math. J., **44** (1977), 705–774.

[20] H. Triebel, Theory of function spaces, Birkhäuser, Basel, 1983.

[21] K. Yajima, Existence of solutions for Schrödinger evolution equations, Commun. Math. Phys., **110** (1987), 415–426.

[22] J. Harmse, On Lebesgue space estimates for the wave equation, Indiana Math. J., **39** (1990), 229–248.

J. Ginibre
Laboratoire de Physique Théorique et Hautes Energies Bâtiment 211,
Université de Paris-Sud, 91405 Orsay cedex, France

G. Velo
Dipartimento di Fisica, Università di Bologna
and INFN, Sezione di Bologna, Italy

Operator Theory:
Advances and Applications, Vol. 78

Around the Transfer Operator and the Trotter-Kato Formula

B. Helffer

Abstract

In preceding papers we have shown how the analysis of the transfer matrix method in statistical mechanics permits us to get a very natural result for the splitting for the transfer matrix. The purpose of this note is to analyze the link between this result and previous results obtained by J.Sjöstrand concerning the splitting between the two first eigenvalues of the Schrödinger operator. We present here improved results and analyze as a byproduct the convergence in the Trotter-Kato formula in a particular (non abstract but relatively general) case. As is known this is strongly related with the Feynman-Kac formula.

Acknowledgements : The material of this paper was partially prepared during stays of the author in St-Petersburg, Bologna and Holzhau. I want therefore to thank V. Buslaev, S. Graffi and V. Sordoni, M. Demuth and W. Schulze for their invitations. I thank also F. Daumer and S.T. Kuroda for useful discussions.

1 Introduction

In collaboration with J. Sjöstrand, we gave in [8] a rather general approach, mainly based on the Maximum Principle, for the estimate of the decay of correlations attached to "gaussian like" measures of the type

$$\exp -\Phi(X)dX \tag{1.1}$$

on $\mathbb{R}^n$ with Φ convex. More general and refined estimates were given by J.Sjöstrand in [18] and this was applied in [5] for the estimate of the splitting between the two largest eigenvalues of the Kac's operator. Another proof of this estimate (more correctly a variant) based on the analysis of the transfer matrix method (which is very standard at least for the study of the Ising model, see for example [19]) was proposed in [7]. In more general contexts, it is well known that similar ideas are still working and this is explained in Chapter 12 in the book by G. Parisi [10] or in our report [7]. Let us briefly recall the approach. We consider a particular case when $n = mp$ and we then take a matrix notation for the variable X:

$$X = x_j^i$$

where $i = 1,\dots,m$ corresponds to the row and $j = 1,\dots,p$ corresponds to the column. We use alternatively the notation: $X = (x_1,\dots,x_p)$ with $x_j \in \mathbb{R}^m$ or $X = (x^1,\dots,x^m)$ with $x^j \in \mathbb{R}^p$. We now consider the particular potential Φ :

$$\Phi^{(p)}(X) \equiv \Phi(X) \equiv \sum_{j=1}^{p} \left(V(x_j) + \frac{|x_j - x_{j+1}|^2}{4} \right) \tag{1.2}$$

where we take the convention that $x_{p+1} = x_1$. This example (1.2) appears naturally in quantum field theory when the so called "lattice approximation" is introduced. For this special class of potentials we can use as well the two different techniques mentioned before. The informations given by these two points of view are actually complementary. A dictionary was established in [7] between the properties of the measure $\exp -\Phi(X)dX$ and the spectral properties of the transfer matrix (which we call also Kac operator in former papers) :

$$K_V(x,y) = (2\pi h)^{-\frac{m}{2}} \exp -\frac{V(x)}{2} . \exp -\frac{|x-y|^2}{4h^2} . \exp -\frac{V(y)}{2} . \tag{1.3}$$

Playing with this dictionary intensively gives interesting connections between estimates of the quotient μ_2/μ_1 of the two first largest eigenvalues of the transfer matrix and corresponding estimates controlling the speed convergence of thermodynamic quantities or the decay of the correlations. This will be recalled in Section 2. But our main object in this article will be a more detailed analysis of the link between spectral properties of this Kac operator and analogous properties for the Schrödinger operator. This leads also to the analysis of the convergence problem in the Trotter-Kato formula where many still open problems seem to appear.

2 A link between statistical mechanics and quantum mechanics

We first recall some properties of the Kac operator (or transfer matrix), associated to a potential V via Formula 1.3. We impose (in this section) the following conditions for V

$$V(y) \geq (1/C)|y|^2 - C \tag{2.1}$$

for some strictly positive constant C, and

$$\forall \alpha \in \mathbb{Z}^m\ ,\ \exists C_\alpha\ s.t.\ |D_y^\alpha V(y)| \leq C_\alpha < y >^{(2-|\alpha|)_+}\ . \tag{2.2}$$

Under these assumptions, K_V is a compact strictly positive selfadjoint operator with strictly positive kernel. In particular it satisfies the assumptions of the extended Perron-Frobenius Theorem (see for example Theorem 3.3.2 in [3]) and K_V admits consequently a largest eigenvalue μ_1 equal to $||K_V||$ which is simple and corresponds to a unique strictly positive normalized eigenfunction which we denote by u_1. Let μ_j the sequence of eigenvalues that we order as a decreasing sequence tending to 0

$$0 \leq \mu_{j+1} \leq \mu_j \leq \cdots \leq \mu_2 < \mu_1 .$$

Thermodynamic limit

Let us first look at the thermodynamic limit. We start from

$$\int_{\mathbb{R}^{mp}} \exp -\Phi(X) dX = \mathrm{Tr}\,[(K_V)^p] = \sum_j \mu_j^p \tag{2.3}$$

In particular we get :

$$\lim_{p\to\infty} \frac{\ln \int_{\mathbb{R}^{mp}} \exp -\Phi(X) dX}{p} = \ln \mu_1 . \tag{2.4}$$

Mean value

We consider now the mean value with respect to our measure of some cylindrical functions. More precisely, if f is a function of x_1 in $C_0^\infty(\mathbb{R}^m)$, let us introduce

$$\langle f^{(1)} \rangle_p = \frac{\int f(x_1) \exp -\Phi(X) dX}{\int \exp -\Phi(X) dX} . \tag{2.5}$$

Easy computations give:

$$\lim_{p\to\infty} \langle f^{(1)} \rangle_p = \int f(y) u_1(y)^2 dy . \tag{2.6}$$

Pair correlation

Let us finally consider the correlation functions. If f is a function in $C_0^\infty(\mathbb{R}^m)$, we denote by $f^{(j)}$ the corresponding function on $\mathbb{R}^{mp}$:

$$X \to f(x_j) .$$

We shall now compute the pair correlation :

$$\mathrm{Cor}^{(p)}(f^{(i)}, g^{(j)}) = \langle (f^{(i)} - \langle f^{(i)} \rangle_p)(g^{(j)} - \langle g^{(j)} \rangle_p) \rangle_p .$$

Taking the limit as $p \to \infty$ we finally obtain when $f = g$ the following identity:

$$\lim_{p\to\infty} \mathrm{Cor}^{(p)}(f^{(1)}, f^{(j)}) = \sum_{k>1} \left(\frac{\mu_k}{\mu_1}\right)^{j-1} \left[\int f(y) u_k(y) u_1(y) dy\right]^2 , \tag{2.7}$$

(see (12.19) in [10]).

This explains why the control of the quotient μ_2/μ_1 is interesting. Using an indirect control of the correlation pairs we get in [7] the following

Theorem 2.1 If for some $\sigma > 0$

$$\text{Hess}\, V(x) \geq \sigma \,, \forall\, x \in \mathbb{R}^m \tag{2.8}$$

then

$$\frac{\mu_2}{\mu_1} \leq \exp - \left[\cosh^{-1}[h^2\sigma + 1]\right] \tag{2.9}$$

Let us recall also the result of J.Sjöstrand [17] which looks very classical but was obtained only recently using the Maximum Principle (in the same spirit as in [16]):

Theorem 2.2 If for some $\sigma > 0$

$$\text{Hess}\, V(x) \geq \sigma \;\;, \forall\, x \in \mathbb{R}^m \tag{2.10}$$

then the splitting between the two first eigenvalues λ_2 and $\lambda_1$of the Schrödinger operator $-h^2\Delta + V$ satisfies

$$\lambda_2 - \lambda_1 \geq \sqrt{2\sigma} h. \tag{2.11}$$

Remark 2.3 In the semiclassical limit, the semiclassical techniques permit to prove (see for example [1]) that:

$$\frac{\mu_2(h)}{\mu_1(h)} = \exp(\lambda_1(h) - \lambda_2(h)) + \mathcal{O}(h^2).$$

An easy computation shows that the two results are coherent. The link between the two statements is analyzed precisely in Section 4 in [7].

3 Around Trotter-Kato

Let us first recall a variant the Trotter-Kato Formula (see [20], [11](Theorem VIII.31)):

Theorem 3.1 If A and B are selfadjoint semi-bounded operators with domains $D(A)$ and $D(B)$ and $A + B$ is essentially selfadjoint on $D(A) \cap D(B)$, then

$$s - \lim_{N\to\infty} \left(\exp -\frac{tB}{2N} \circ \exp -\frac{tA}{N} \circ \exp -\frac{tB}{2N}\right)^N = \exp -t(A+B)\,. \tag{3.1}$$

In our situation, we can take $A = -\Delta$ and $B = V$.
This appears naturally when comparing the Kac operator and the Schrödinger operator as explained in [7]. For different problems, it is interesting to have a better control of the convergence and strong convergence appears in some case relatively weak. We shall prove in the next section the following

Proposition 3.2 Let V be a C^∞ potential satisfying

$$V(x) \geq 0 \tag{3.2}$$

and (2.2). Then, with $K(s) = \exp -\frac{s}{2}V \,.\, \exp s\Delta \,.\, \exp -\frac{s}{2}V$ for $s \in \mathbb{R}^+$, we have

$$||K(s) - \exp -s(-\Delta + V)||_{\mathcal{L}(L^2)} = \mathcal{O}(s^2) \tag{3.3}$$

as $s \to 0$.

Proposition 3.3 Under the assumptions of Proposition 3.2, let us denote by $\mu_1(s)$ and $\mu_2(s)$ the two largest eigenvalues of $K(s)$. Then

$$\ln \mu_1(s) - \ln \mu_2(s) = s(\lambda_2 - \lambda_1) + \mathcal{O}(s^2)\;; j = 1{,}2\;.$$

The inequality $\leq$ was proved in [7]) and the remainder term (which was $\mathcal{O}(s^{\frac{3}{2}})$ in this article) is improved here. The proof is just an application of the max-min principle (cf [11], Theorem XIII.1) for the two first eigenvalues and of Proposition 3.2.

Corollary 3.4 Theorem 2.2 is a direct consequence of Theorem 2.1.

Proposition 3.2 gives also a more precise result than the result deduced from the variant of the Trotter formula. We observe first the following lemma

Lemma 3.5 Let us assume the assumption of Trotter-Kato and that for some $\rho > 0$, the condition

$$||\exp -s(A+B) - \exp -\frac{sB}{2} \circ \exp -sA \circ \exp -\frac{sB}{2}||_{\mathcal{L}(L^2)} = \mathcal{O}(s^{1+\rho}) \tag{3.4}$$

is satisfied then

$$\lim_{N\to\infty} \left(\exp -\frac{tB}{2N} \circ \exp -\frac{tA}{N} \circ \exp -\frac{tB}{2N}\right)^N = \exp -t(A+B)\,, \tag{3.5}$$

in $\mathcal{L}(L^2)$.

We obtain indeed

$$||K(\frac{t}{N}) - \exp -\frac{t}{N}(-\Delta + V)||_{\mathcal{L}(L^2)} = \mathcal{O}((\frac{t}{N})^{1+\rho})\,, \tag{3.6}$$

and we get, observing that, as a consequence of the assumptions $A \geq 0$ and $B \geq 0$, the two operators $K(\frac{t}{N})$ and $\exp -\frac{t}{N}(A+B)$ are contractions:

$$||K(\frac{t}{N})^N - \exp -t(A+B)||_{\mathcal{L}(L^2)} = N\mathcal{O}((\frac{t}{N})^{1+\rho})\,. \tag{3.7}$$

We can easily get the result by recursion. If C and D are two contractions such that $||C - D|| \leq \epsilon$ we can indeed write:

$$C^n - D^n = C^{n-1}(C - D) + (C^{n-1} - D^{n-1})D\ ,$$

which gives

$$||C^n - D^n|| \leq \epsilon + ||C^{n-1} - D^{n-1}||\ .$$

We get finally from (3.7) for fixed t

$$||K(\frac{t}{N})^N - \exp -t(A + B)||_{\mathcal{L}(L^2)} = N^{-\rho}\mathcal{O}(1)\ . \tag{3.8}$$

and this proves the Lemma. Combining with Proposition 3.2, this gives finally the

Theorem 3.6 Under assumptions (3.2) and (2.2), we have

$$||K(\frac{t}{N})^N - \exp -t(-\Delta + V)||_{\mathcal{L}(L^2)} = N^{-1}\mathcal{O}(1)\ . \tag{3.9}$$

We shall see at the end of the next section how it is possible to relax a little the assumption (2.2).

4 Comparison between $\exp -t(-\Delta + V)$ and $\exp -\frac{t}{2}V \circ \exp t\Delta \circ \exp -\frac{t}{2}V$ through a pseudodifferential calculus

4.1 Introduction

This will be done by a more careful semi-classical analysis. Let V be a C^∞ potential satisfying

$$V(x) \geq 0 \tag{4.1}$$

and (2.2). It is a classical result as a consequence of (2.2) and of the positivity of V that the potential V satisfies for some constant C the following inequality:

$$|\nabla V(x)| \leq C V^{1/2}(x)\ \forall\, x \in \mathbb{R}^m\ . \tag{4.2}$$

4.2 Study of the Kac's operator

Let us consider now our operator

$$K(t) = \exp -\frac{t}{2}V \circ \exp t\Delta \circ \exp -\frac{t}{2}V \tag{4.3}$$

as a h-pseudodifferential operator $k^w(x,hD_x,t)$ (with $h = \sqrt{t}$) which is the operator associated by the Weyl quantification to the symbol: $(x,\xi,t) \to k(x,\xi,t)$ by the formula:

$$\begin{aligned} &k^w(x,hD_x,t)u(x) = \\ &\quad (2\pi)^{-m} \int\int \exp -i\,((x - y).\xi)\ .\ k((\tfrac{x+y}{2}), h\xi,t)u(y)dyd\xi \end{aligned} \tag{4.4}$$

Conversely k is called the h-symbol of $K(t)$ (with $h = \sqrt{t}$). The parameter $\sqrt{t}$ plays here the role of the semi-classical parameter h. Sometimes, it will be interesting to consider differently t and h by considering $-t\Delta$ as $-h^2\Delta$ with Weyl symbol ξ^2 and by considering $V_t = tV$ as a potential depending on a small parameter t. Let us introduce as in [7] the class:

$$\Sigma^s = \{p \in C^\infty(\mathbb{R}^m \times \mathbb{R}^m \times]0,h_0]) \ s.t \ |\partial_x^\alpha \partial_\xi^\beta p(x,\xi,h)| \leq C_{\alpha\beta} h^s\} \tag{4.5}$$

Here C^∞ is meant with respect with the (x,ξ) variables. For $s = 0$, this class Σ^0 is stable by composition, and the usual law of h-composition for the Weyl symbols can be used. We recall that the law of composition

$$r = p \sharp_h q$$

of two symbols p and q can be defined by:

$$p_h^w(x,hD_x,h) \ o \ q_h^w(x,hD_x,h) = r_h^w(x,hD_x,h) \ . \tag{4.6}$$

Of course it is well known that if p and q are in Σ^0 then $p \sharp_h q - p.q \in \Sigma^1$, but we shall need for our proof to do all the computations modulo $\mathcal{O}(h^4)$ and we use consequently the three first terms of the law of computation. Let us recall the formula (see [12] p. 80):

$$p \sharp_h q = \sum_{|\alpha|+|\beta|\leq 2} \frac{1}{\alpha!\beta!)} (\frac{1}{2})^{|\alpha|} (\frac{-1}{2})^{|\beta|} . h^{(|\alpha|+|\beta|)} (\partial_\xi^\alpha D_x^\beta p).(\partial_\xi^\beta D_x^\alpha q) + \mathcal{O}(h^3). \tag{4.7}$$

Although we want to work modulo $\mathcal{O}(h^4)$, we shall not need to know explicitely the coefficient of h^3 in the expansion. We observe indeed the following important properties.

Property P1

If we start from real symbols p and q, we observe that in the expansion the terms of $p \sharp_h q$ are alternatively real and imaginary. The same is true for $q \sharp_h p \sharp_h q$.

The other classical point is that

Property P2

The Weyl symbol of a selfadjoint operator is real.

Using the two arguments, one get the property that the symbol

$$q \sharp_h p \sharp_h q \text{ has a natural expansion in powers of } h^2 \ . \tag{4.8}$$

This remark permits us to avoid in the computation some coefficients of h^3 which will automatically vanish.

We now consider $K(t)$ as the composition of three h-pseudodifferential operators. The h-symbol $p(x,\xi)$ of $\exp h^2\Delta$ is clearly $\exp -\xi^2$:

$$p(x,\xi) = \exp -\xi^2 \ , \tag{4.9}$$

and the h-symbol of $\exp -\frac{t}{2}V$ is :

$$q(x,\xi) = \exp -\frac{}{t}2V(x) \; . \tag{4.10}$$

For the moment $q(x,\xi)$ is considered as depending on a parameter $t \in]0,1]$ which is independent of h, and we observe that q belongs to Σ^0 uniformly with respect to $t \in]0,1]$.

We want to compute the Weyl symbol of $K(t)$.

We now establish the estimate for the symbol. We prove that it belongs to Σ^0 and is actually in a better class. Inequality (4.2) gives, observing the identity,

$$\nabla_x(\exp -\frac{t}{2}V(x)) = -\frac{t}{2}(\nabla V)(x)(\exp -\frac{t}{2}V(x) \; ,$$

the following estimate

$$|\nabla_x(\exp -\frac{t}{2}V(x))| \leq C_0 t^{\frac{1}{2}} \; . \tag{4.11}$$

We now look at the second derivative. This is a sum of terms where the preceding argument can be used outside for one term which is:

$$t(\nabla^2 V)(x)(\exp -\frac{t}{2}V(x))$$

which is clearly $\mathcal{O}(t)$:

$$|\nabla_x^2(\exp -\frac{t}{2}V(x))| \leq C_2 t \; . \tag{4.12}$$

We now compute the Weyl symbol of $K(t)$ using the composition law of p.d.o. We shall first prove the:

Lemma 4.1

$$\sigma_h^w(K(t)) - \left[\exp -(\xi^2 + tV(x))\right] \in \Sigma^4 \; . \tag{4.13}$$

for $t = h^2$.

We first study:

$$\begin{aligned}(\exp -\tfrac{t}{2}V(x)) \sharp_h (\exp -\xi^2) \sim_3 \;& \exp -(\xi^2 + \tfrac{t}{2}V(x)) \\ & + h(\tfrac{t}{4i})[\xi.\nabla V(x)] \exp -(\xi^2 + \tfrac{t}{2}V(x)) \\ & + h^2 r_2(t,x,\xi) \exp -(\xi^2 + \tfrac{t}{2}V(x))\end{aligned} \tag{4.14}$$

Here $\sim_3$ means modulo $\mathcal{O}(h^3)$. It is sufficient to apply the law of composition modulo $\mathcal{O}(h^3)$ because we know a priori (cf (4.8)) that the coefficient of h^3 will vanish. We know also that the coefficient of h will vanish. Here we obtain that $r_2(t,x,\xi)$ is a linear combination of terms of the type $t^2(\nabla_x^\alpha V)(\nabla_x^\beta V)$ with $|\alpha| = |\beta| = 1$ and $t(\nabla_x^\alpha V)$ with $|\alpha| = 2$, multiplied by polynomials in the ξ variable. One then gets easily

$$r_2(t,x,\xi) \exp -(\xi^2 + \frac{t}{2}V(x)) \in \Sigma^2 \tag{4.15}$$

if $t = h^2$.

Composing again with $\exp -\frac{t}{2}V(x)$ on the right we observe that the second term in the expansion disappears and we have now to prove that:

$$(\frac{t}{4i})[\xi.\nabla V(x)] \exp -(\xi^2 + \frac{t}{2}V(x)) \in \Sigma^1 \tag{4.16}$$

for $t = h^2$ and

$$r_2(t,x,\xi) \exp -(\xi^2 + \frac{t}{2}V(x)) \in \Sigma^2 \text{ for } t = h^2 . \tag{4.17}$$

These estimates say that the symbol behaves a little better are a consequence of:

$$\nabla(\exp -\frac{t}{2}V(x)) \in \Sigma^1 \tag{4.18}$$

and

$$\nabla^2(\exp -\frac{t}{2}V(x)) \in \Sigma^2. \tag{4.19}$$

We have proved the lemma.

4.3 Study of $\exp -t(-\Delta + V)$:

Let us now consider $\tilde{K}(t) = \exp -t(-\Delta + V)$, that we again prefer to consider as : $\tilde{K}(t) = \exp -(-h^2\Delta + tV)$. Using the functional calculus of Helffer-Robert [12], we prove the

Lemma 4.2

$$\sigma_h^w(\tilde{K}(t)) - \left[\exp -(\xi^2 + tV(x))\right] \in \Sigma^4 \tag{4.20}$$

If f is a good real function and if $p^w(x,hD_x,h)$ is in a good family of h-pseudo-differential selfadjoint operators then the functional calculus gives (see [12] p. 143) that $f(p^w(x,hD_x,h))$ is also a pseudodifferential operator and give explicit formulas for the symbol. Here we take $f(s) = \exp -s$ and $p^w(x,hD_x,h) = -h^2\Delta + tV$. The Lemma III-7 in [12] and the fact that a_t satisfies the estimates :

$$|\partial_x^\alpha \partial_\xi^\beta a_t(x,\xi)| \leq C_{(\alpha,\beta)}(a_t(x,\xi) + 1) , \tag{4.21}$$

where the constants $C_{(\alpha,\beta)}$ are independent of $t \in]0,1]$. This estimate permits to apply the functional calculus uniformly with respect to t. We then get with $a_t(x,\xi) = \xi^2 + tV(x)$:

$$\sigma_t^w(\tilde{K}(t)) \sim_3 f(a_t(x,\xi)) + h^2 \left[\sum_{k=1}^{3}(-1)^k(k!)^{-1}d_{2,k}f^{(k)}(a_t(x,\xi))\right] \tag{4.22}$$

Again we know a priori that the coefficient of h^3 in the expansion is zero and that consequently the above-written formula is indeed true modulo $\mathcal{O}(h^4)$. In order to

verify this last statement, we verify that this is true for any polynomial of an operator $P = \text{ op }_h^w(p)$ with p real and independent of h. Then, it is clear by density, that it is also true for any C^∞- function for which the functional calculus is well defined.
It is then sufficient to analyze the coefficient of h^2 and we have to compute explicitely the symbol $d_{2,k}$. The general theory says that it is a polynomial function of derivatives of a_t. But we have to compute these coefficients more precisely. We want indeed to prove that

$$(x,\xi) \to \left[\sum_{k=1}^{3}(-1)^k(k!)^{-1}d_{2,k}(x,\xi)f^{(k)}(a_t(x,\xi))\right] \in \Sigma^4 \tag{4.23}$$

for $t = h^2$, using the improvements (4.11) and (4.12). In order to perform this computation we come back to the proof of the functional calculus (formal part). Let us first recall the Lemma III.7.

Lemma 4.3 For $P(h) = Op_h^w(a)$ and $z \in \mathbb{C} \setminus [0,\infty[$, let us denote by

$$\sum_{j\geq 0} b_{z,j}h^j$$

the formal Weyl symbol of the resolvent $R(z,h) = (P(h) - z)^{-1}$. Then we have:

$$b_{z,0} = (a - z)^{-1} \tag{4.24}$$

and, for any integer $j \geq 1$,

$$b_{z,j} = \sum_{k=1}^{2j-1} d_{j,k} \, . \, b_{z,0}^{k+1} \, . \tag{4.25}$$

where the $d_{j,k}$ are polynomial functions of a and its derivatives.

In order to compute the different coefficients $d_{2,k}$, it is sufficient to follow the computation of the symbol of the resolvent in the Weyl calculus. We have the following recursion formula (cf. (31) p. 135 in [12]):

$$b_{z,j+1} = -b_{z,0} \sum_{\left\{\begin{array}{l} \ell+|\alpha|+|\beta| = j+1 \\ 0\leq \ell \leq j \end{array}\right\}} \Gamma(\alpha,\beta)(\nabla_\xi^\alpha\nabla_x^\beta a)(\nabla_x^\alpha\nabla_\xi^\beta b_{z,\ell}) \tag{4.26}$$

with

$$\Gamma(\alpha,\beta) = \left(\alpha! \, . \, \beta! \, . \, 2^{|\alpha|}(-2)^{|\beta|}\right)^{-1} .$$

We first find that

$$b_{z,1} = 0 \tag{4.27}$$

because $\{b_{z,0},a\} = 0$.

We now analyze the crucial:

$$\begin{array}{rl} b_{z,2} & = -b_{z,0} \sum_{|\alpha|+|\beta|=2} \Gamma(\alpha,\beta)(\nabla_\xi^\alpha \nabla_x^\beta a)(\nabla_x^\alpha \nabla_\xi^\beta b_{z,0}) \\ & = -b_{z,0} \sum_{|\alpha|=2} \Gamma(\alpha,0)(\nabla_\xi^\alpha a)(\nabla_x^\alpha b_{z,0}) \\ & -b_{z,0} \sum_{|\beta|=2} \Gamma(0,\beta)(\nabla_x^\beta a)(\nabla_\xi^\beta b_{z,0}) . \end{array} \tag{4.28}$$

Let us first look at the first term in the r.h.s. of the last equality. We have just to analyze $\nabla_x^\alpha b_{z,0}$ for $|\alpha| = 2$. But this term is the sum of terms of the type:

$$tb_{z,0}^2 \nabla_x^\beta V \text{ , with } |\beta| = 2$$

and of the type

$$t^2 b_{z,0}^3 \nabla_x^\beta \cdot V \nabla_x^\gamma V \text{ , with } |\beta| = |\gamma| = 1 .$$

Let us now look at the second term in the r.h.s. of the last equality. We have now to analyze the product of $t\nabla_x^\beta V$, with $|\beta| = 2$ by $\nabla_\xi^\beta b_{z,0}$ which can be written as a sum of $b_{z,0}^2$ and of $\xi^\beta b_{z,0}^3$. This proves that $d_{2,3}$ is a sum of terms like $t^2 \nabla_x^\beta V \nabla_x^\gamma V$, with $|\beta| = |\gamma| = 1$, and like $t\xi^\beta \nabla_x^\beta V$, with $|\beta| = 2$. One then get easily (using (4.2) that (for $t = h^2$)

$$d_{2,3} f^{(3)}(a) \in \Sigma^2 .$$

Similarly $d_{2,2}$ is a sum of terms like $t\nabla_x^\beta V$, with $|\beta| = 2$. One get then immediately that

$$d_{2,2} f^{(2)}(a) \in \Sigma^2 .$$

This ends essentially our proof.

4.4 Conclusion

From this we get that

$$K(t) - \tilde{K}(t) = O(t^2) \text{ in } \mathcal{L}(L^2), \tag{4.29}$$

where we have used the Calderon-Vaillancourt Theorem for the L^2-continuity of the pseudo-differential operators in order to control the remainder terms.
This permits to improve the result of the preceding section as previously announced. As an extension F.Daumer [2] has obtained the following

Theorem 4.4 Let V be a C^∞ potential satisfying (4.1)

$$V(x) \geq 1$$

and for some $0 < \rho \leq 1$

$$|D_x^\alpha V(x)| \leq C_\alpha V^{(1-\rho|\alpha|)_+} \text{ , } \forall \alpha , \tag{4.30}$$

then we get similarly

$$K(t) - \tilde{K}(t) = O(t^{1+\inf(\frac{1}{2}, \rho)}) \text{ in } \mathcal{L}(L^2) . \tag{4.31}$$

This permits to prove to extend our preceding results for more general V's (in particular potentials with quartic terms).

5 Around Campbell-Hausdorff Formula

Let us analyze more in detail the formal Campbell-Hausdorff formula. We refer to Goodman ([4] p. 13). We start from

$$\exp -tX \,.\, \exp -tY = \exp -F(X,Y;t) \tag{5.1}$$

where

$$F(X,Y;t) = t(X+Y) - t^2(\frac{1}{2}[X,Y]) + t^3\left(\frac{1}{12}[X,[X,Y]] + \frac{1}{12}[Y,[Y,X]]\right) + \mathcal{O}(t^4), \tag{5.2}$$

and has to be considered as a formal series in t. From this formula, we get immediately that :

$$F(X,Y;t) - t(X+Y) = \mathcal{O}(t^2) \tag{5.3}$$

but we hope to prove, like in the case of the pseudodifferential calculus, that the approximation is much better if we consider now:

$$\exp -\frac{t}{2}X \,.\, \exp -tY \,.\, \exp -\frac{t}{2}X = \exp -G(X,Y;t) \tag{5.4}$$

and we want consequently to compute $G(X,Y;t) - t(X+Y)$. Using (5.2) we get after tedious computations

$$G(X,Y;t) = t(X+Y) + t^3\left(-\frac{1}{24}[X+2Y,[X,Y]]\right) + \mathcal{O}(t^4)\,. \tag{5.5}$$

The coefficient of the coefficient of t^3 for $Y = -\Delta$ and $X = V$ does not seem to vanish in general. But the vanishing of t^2 explain probably the improvement in the convergence observed in the preceding section.

6 Conclusion

In my talk in the conference I was asking for similar results at the abstract level. One week after this interesting conference, Professor S.T.Kuroda kindly indicated to me that Professor Berezanski mentioned the work of Dzh.L.Rogava [13] which intersects with our discussion. The following theorem is indeed announced :

Theorem 6.1 Let A and B be positive definite selfadjoint operators in a Hilbert space H. Suppose that $D(B) \subset D(A)$. Suppose also that $A+B$ is selfadjoint on $D(B)$. Then for any $N \in \mathbb{N}$

$$|| \left(\exp -\frac{tB}{2N} \circ \exp -\frac{tA}{N} \circ \exp -\frac{tB}{2N}\right)^N - \exp -t(A+B) ||_{\mathcal{L}(H)} = \mathcal{O}(\ln N/\sqrt{N})\,.$$

It is difficult to see from the sketch of the proof if the condition $D(B) \subset D(A)$ is important or not in the proof.

References

[1] M. Brunaud and B. Helffer: Un problème de double puits provenant de la théorie statistico-mécanique des changements de phase (ou relecture d'un cours de M. Kac), *preprint LMENS*, March 1991.

[2] F. Daumer: Personal communication, June 1994.

[3] J. Glimm, A. Jaffe: Quantum physics (a functional integral point of view), Springer Verlag, Second edition (1987).

[4] R. Goodman: Nilpotent Lie groups: structure and applications to analysis. Lecture Notes in Mathematics, n^0 562, Springer-Verlag (1966).

[5] B. Helffer: Universal estimate of the gap for the Kac operator in the convex case, *Comm. in Math. Physics*, Vol.161, n° 3, (1994), p. 631–642.

[6] B. Helffer: Spectral properties of the Kac operator in large dimension Proceeding of the Summer School in Vancouver, August 1993.

[7] B. Helffer: Correlation decay and gap of the transfer operator, Proceedings of the conference in the Euler Institut in St Petersbourg (submitted in April 1994).

[8] B. Helffer, J. Sjöstrand: On the correlation for Kac like models in the convex case, Report No. 9, 1992–93, Institut Mittag-Leffler, March 1993. To appear in *Journal of statistical physics.*

[9] M. Kac: Mathematical mechanisms of phase transitions, Brandeis lectures (1966), *Gordon and Breach.*

[10] G. Parisi: Statistical field theory, Frontiers in Physics, Addison-Wesley Publishing Company (1984)

[11] M. Reed, B. Simon: Methods of modern mathematical physics, Tome I-IV, Functional Analysis, Academic Press 1972-79.

[12] D. Robert: Autour de l'approximation semi-classique, *Birkhäuser*, Vol.68, Progress in Mathematics, (1987)

[13] Dzh. L. Rogava: Error bounds for Trotter-type formulas for Self-adjoint operators, Functional analysis and its applications (1993) p. 217–219.

[14] B. Simon: The $P(\phi)_2$ Euclidean quantum field theory, Princeton Series in Physics (1974).

[15] B. Simon: Functional integration and quantum physics, *Pure and Applied mathematics* $n^0$86, Academic press , New York, (1979).

[16] I.M. Singer, B. Wong, S.T. Yau and S.S.T. Yau: An estimate of the gap of the first two eigenvalues of the Schrödinger operator, *Ann. Scuola Norm. Sup. Pisa* (4), 12 (1985), p. 319–333.

[17] J. Sjöstrand: Potential wells in high dimensions II, more about the one well case. preprint March 1991, *Ann. Inst. Poincaré*, Section Physique théorique, 58 (1)(1993).

[18] J. Sjöstrand: Ferromagnetic integrals, correlations and maximum principles, Preprint Université Paris-Sud (1993).

[19] C.J. Thompson: Mathematical Statistical Mechanics, Princeton University Press (1972).

[20] H. Trotter: On the product of semigroups of operators, Proc.Amer. Math.Soc. 10 (1959) p. 549–551.

Bernard Helffer
DMI-ENS, 45 rue d'Ulm
F-75230 Paris Cédex, France

Operator Theory:
Advances and Applications, Vol. 78

Bands and Gaps for Periodic Magnetic Hamiltonians

R. Hempel[1], I. Herbst[2]

1. Introduction.

In this contribution, we wish to present some recent results and to discuss an open problem in the spectral theory of magnetic Hamiltonians

$$H(\vec{a}) = (-i\nabla - \vec{a}(x))^2 ,$$

acting in $L_2(\mathbb{R}^n)$, with $n \geq 2$, and $\vec{a} \in C^1(\mathbb{R}^n; \mathbb{R}^n)$. Here we focus on the cases where

(1) the field $B = d\vec{a}$ is periodic with respect to some discrete lattice Γ spanning $\mathbb{R}^n$, e. g., $\Gamma = \mathbf{Z}^n$, or where, more strongly,

(2) the vector potential $\vec{a}$ is periodic with respect to the lattice Γ.

In the second case, $H(\vec{a})$ commutes with translations $\in \Gamma$, and we may apply Floquet theory to obtain a direct fiber integral decomposition. In the first case, after going through any translation $\in \Gamma \setminus \{0\}$, the operator will in general differ from the original one by a non-trivial gauge transformation.

Periodicity of the vector potential $\vec{a}$ implies periodicity of the magnetic field, plus certain conditions on the magnetic flux. For example, it is easy to see that in dimension $n = 2$, the total flux through a period parallelogram vanishes. In general, it follows from some basic results in cohomology theory that periodicity of B plus certain flux conditions are equivalent to the existence of an associated periodic vector potential; cf. Proposition 1.

Let us now first give some basic results on bands and gaps; Section 2 will be devoted to the problem of absolute continuity of $H(\vec{a})$.

It is well known that the magnetic Hamiltonian with constant field $B \neq 0$ in dimension $n = 2$ has pure point spectrum consisting of isolated eigenvalues of infinite multiplicity. One may think of this type of spectrum as a version of band – gap structure, with bands degenerated into points. However, constant fields in odd dimensions do not produce gaps.

There is a remarkable recent result due to Brüning and Sunada [2] (cf. also [1]) which says that for periodic field B the spectrum of $H(\vec{a})$ has *band structure* in the sense that there are at most a finite number of gaps in any given bounded interval $[0,R]$.

---

[1] Research partially supported by USNSF grant DMS-9401417.
[2] Research partially supported by USNSF grant DMS-9307147.

In a recent paper, it was shown by Hempel and Herbst ([6; Corollary 3.10]) that gaps can be produced in any dimension $n \geq 2$ by a periodic array of "magnetic barriers", with periodic vector potential or just periodic field. In order to describe this result (plus some generalizations) we again define for $\vec{a} \in C^1(\mathbb{R}^n;\mathbb{R}^n)$ the closed sets

$$M = \{x \in \mathbb{R}^n; B(x) = 0\}, \quad M_{\vec{a}} = \{x \in \mathbb{R}^n; \vec{a}(x) = 0\};$$

note that $M_{\vec{a}} \setminus M$ has measure zero while $M \setminus M_{\vec{a}}$ can be large. The condition that $M \setminus M_{\vec{a}}$ have measure zero is essential for most of the results in [6]; Section 4 in [6] gives a discussion of what this condition means.

Theorem 1. Let $\vec{a}$ be periodic with respect to the lattice Γ and assume that $M \setminus M_{\vec{a}}$ has measure zero. Suppose there exists a compact set M_0 with non-empty interior such that, writing $M_j = M_0 + j$ for $j \in \Gamma$, we have $M_j \cap M_k = \emptyset$, for all $k \neq j$ and $M = \cup_{j \in \Gamma} M_j$, up to a set of measure zero. Let $E_1 < E_2 < \ldots < E_k < \ldots$ denote the eigenvalues of the (Dirichlet) Laplacian $-\Delta_{M_0}$ on the closed set M_0, as defined in [6].

Finally, let $\varepsilon > 0$ and $E > 0$, and suppose that $E_K < E < E_{K+1}$. Then there exists $m = m(\varepsilon,E)$ such that for $|\mu| \geq m$ the following holds:

(1) $\sigma(H(\mu\vec{a})) \cap (E_k - \varepsilon, E_k + \varepsilon) \neq \emptyset$, for $k = 1, \ldots, K$.

(2) $\sigma(H(\mu\vec{a})) \cap [0,E] \subset \cup_{k=1}^{K}(E_k - \varepsilon, E_k + \varepsilon)$.

(3) The spectrum of $H(\mu\vec{a})$ inside any of the intervals $(E_k - \varepsilon, E_k + \varepsilon)$, $k = 1, \ldots, K$, consists of a finite number of disjoint closed intervals (which may degenerate into points).

(4) $H(\mu\vec{a})$ has no singular continuous part.

(5) For all but a discrete set of μ's, $H(\mu\vec{a})$ is absolutely continuous in the interval $[0,E]$. In particular, there are no eigenvalues inside $[0,E]$ and the intervals of spectrum occuring in (3) are non-degenerate, for all but a discrete set of μ's.

Proof. Since we assume that $M \setminus M_{\vec{a}}$ has measure zero, it follows from [6] that $H(\mu\vec{a})$ converges in norm resolvent sense to the (Dirichlet) Laplacian $-\Delta_M$ on the set M, as $|\mu| \to \infty$. By periodicity, $-\Delta_M$ is nothing but an infinite direct sum of copies of $-\Delta_{M_0}$ and the spectrum of $-\Delta_M$ consists of the sequence of eigenvalues E_k, each eigenvalue having infinite multiplicity. Now, since norm resolvent convergence for self-adjoint operators implies convergence of spectra, statements (1) and (2) follow.

By Floquet theory, the spectrum of $H(\mu\vec{a})$ is given by the union of the ranges of the band functions, which are continuous functions on a fundamental cell of the dual lattice, and (3) follows (note that, by compactness, only a finite number of band functions have values in a given bounded interval, for any fixed μ).

Claims (4) and (5) are a direct consequence of Theorems 2 and 4 in Section 2. ∎

As a corollary, we obtain a version of Theorem 1 where the assumptions are made in terms of the field B. Let $g_1, \dots, g_n$ denote a set of linearly independent vectors in $\mathbb{R}^n$ spanning the lattice Γ. We first prepare an auxiliary result.

Proposition 1. Suppose B is a 2-form on $\mathbb{R}^n$ which is C^1 (or C^∞) and Γ-periodic in the sense that $B(x+g_j) = B(x)$ for $j = 1, \dots, n$ and for all $x \in \mathbb{R}^n$. Suppose $dB = 0$ and that the flux conditions

$$\int_{C_{ij}} B = 0, \quad 1 \le i < j \le n, \tag{1}$$

are satisfied, where $C_{ij} : [0,1] \times [0,1] \to \mathbb{R}^n$ is given by

$$C_{ij}(t,s) = tg_i + sg_j. \tag{2}$$

Then there is a Γ-periodic 1-form $\vec{a}$ of class C^1 (or of class C^∞) such that $B = d\vec{a}$.

This result can be proved without using cohomology theory, but the proof is somewhat lengthy and we refrain from reproducing it here. On the other hand, an application of de Rham's theorem to the case of a smooth closed 2-form on the n-torus T^n reduces the proof of Proposition 1 to the problem of determining an explicit basis for the module of 2-cycles; it is easy to see that such a basis is conveniently given by the 2-tori defined via eqn. (2) (cf., e.g., [12; pp. 334 and 239]).

Combining Theorem 1 and Proposition 1, we now obtain the following corollary:

Corollary 1. Let B be a smooth Γ-periodic magnetic field (more precisely, a smooth, Γ-periodic, exact two-form on $\mathbb{R}^n$) satisfying the flux conditions described in eqns. (1) and (2). Furthermore, suppose that there exists a compact set M_0 as in Theorem 1 such that, in addition, the interior $\operatorname{int}(M_0)$ is a simply connected Lipschitz domain and M_0 is the closure of its interior.

Let $-\Delta_M$ and E_k be as above. Then, for any given $\varepsilon > 0$, $E > 0$ and K as in Theorem 1, there exists $m = m(\varepsilon, E)$ such that the statements (1) – (5) hold true for $|\mu| \ge m$.

Proof. We have to produce a smooth vector potential $\vec{a}$ such that $d\vec{a} = B$, satisfying the conditions of Theorem 1. As a first step, Proposition 1 yields the existence of a periodic vector potential $\vec{a}_0 \in C^\infty(\mathbb{R}^n; \mathbb{R}^n)$ such that $d\vec{a}_0 = B$. In general, $\vec{a}_0$ will not vanish on M_0, however. Following the construction at the end of Section 4 in [6] and using Stein [11; p. 181], we next find a function $f \in C^\infty(\mathbb{R}^n)$ such that $df(x) = \vec{a}_0(x)$ for $x \in \operatorname{int}(M_0)$. Pick a (real-valued) function $\varphi \in C_0^\infty(\mathbb{R}^n)$ which is 1 in a neighborhood of M_0, and which vanishes in a neighborhhood of each M_j, for all $j \in \Gamma$, $j \ne 0$. Define $F \in C^\infty(\mathbb{R}^n)$ by

$$F(x) = \sum_{j \in \Gamma} (\varphi f)(x-j), \quad x \in \mathbb{R}^n.$$

Then $\vec{a}(x) = \vec{a}_0(x) - dF(x)$ is periodic and satisfies $d\vec{a} = B$; furthermore $\vec{a}(x) = 0$ for almost every $x \in M$. ∎

2. On the absolute continuity of periodic Hamiltonians.

The central question of this note is to determine the nature of the spectrum separating gaps of $H(\vec{a})$ and of $H(\mu\vec{a})$, for $\mu \in \mathbb{R}$. Here the strongest result one might head for is to show that for periodic $\vec{a}$ the spectrum of $H(\vec{a})$ is absolutely continuous:

Conjecture. Suppose $\vec{a} \in C^1(\mathbb{R}^n;\mathbb{R}^n)$ is periodic with respect to the lattice Γ. Then $H(\vec{a})$ is absolutely continuous.

Below, we will present some evidence in support of this conjecture. We shall not persue the case of periodic field B here; note, however, that the constant field case in 2-dimensions provides only a weak example, and one might still ask for the nature of the spectrum in cases where B is periodic, but non-constant. In this direction, Iwatsuka [7] has announced some results on line-broadening.

Let us now sketch how far one can get in the question of absolute continuity by using the approach of L. Thomas [13] (reproduced in [10]) or some more abstract results on the structure of the "Bloch-variety", the union of the graphs of the band-functions. We find that $H(\vec{a})$ has no singular continuous part, so the only obstacle in proving absolute contiunity is the possibility of eigenvalues. Furthermore, $H(\vec{a})$ is absolutely continuous for small $\vec{a}$, and $H(\mu\vec{a})$ is a. c. in any given interval $[0,R]$ for all but a discrete set of μ's (which we expect to be empty). We begin with results which exploit the fact that the Bloch variety associated with $H(\vec{a})$ can we viewed as the zero set of a suitable real-analytic function of several variables; cf. Gérard [4], Wilcox [15], Kuchment [9] for related methods and results in the case of periodic $-\Delta + V$.

The Floquet decomposition of $H(\vec{a})$ is fundamental to our analysis (cf., e. g., [10]): Let Q denote a fundamental cell of the dual lattice spanned by the vectors K_j (note that our Q corresponds to $\tilde{Q}$ in [10; p. 305]). Defining a new Hilbert space $\mathcal{H}$ via a direct fiber integral of ℓ_2-spaces,

$$\mathcal{H} = \int_Q^{\oplus} \ell_2(\mathbf{Z}^n)dk,$$

there exists a unitary operator $U : L_2(\mathbb{R}^n) \to \mathcal{H}$ such that

$$UH(\vec{a})U^{-1} = \int_Q^{\oplus} H_k(\vec{a})dk, \tag{3}$$

with suitable operators $H_k(\vec{a})$, acting in $\ell_2(\mathbb{Z}^n)$. We are now ready to state the following theorem:

Theorem 2. Suppose that $\vec{a} \in C^1(\mathbb{R}^n;\mathbb{R}^n)$ is periodic with respect to the lattice Γ. Then:

(a) The singular continuous spectrum of $H(\vec{a})$ is empty.

(b) The eigenvalues of $H(\vec{a})$ form a discrete subset of $\mathbb{R}$.

(c) If λ is an eigenvalue of $H(\vec{a})$, then λ is an eigenvalue of $H_k(\vec{a})$ for all $k \in Q$.

Proof. We start from the direct fiber decomposition (3). Using regularized determinants, one can construct a real analytic function $F(\lambda,k)$ on $\mathbb{R} \times Q$ with the property that E is an eigenvalue of $H_k(\vec{a})$ iff $F(E,k) = 0$; cf. [4], [9], [15].

We first prove (b) and (c). By the direct fiber decomposition of $H(\vec{a})$, λ_0 is an eigenvalue of $H(\vec{a})$ iff there exists a set $M_0 \subset Q$ of positive measure such that λ_0 is an eigenvalue of $H_k(\vec{a})$ for all $k \in M_0$ iff $F(\lambda_0,k) = 0$ for all $k \in M_0$. It now follows from Theorem A that the set of eigenvalues is discrete.

Furthermore, if λ_0 is an eigenvalue of $H(\vec{a})$, then $F(\lambda_0,k) = 0$ for all $k \in Q$ (by Lemma A.1), so λ_0 is an eigenvalue of $H_k(\vec{a})$ for all $k \in Q$.

In order to prove that the singular continuous spectrum of $H(\vec{a})$ is empty, it is enough to show that for any Borel set $\Lambda \subset \mathbb{R} \setminus \sigma_{pp}(H(\vec{a}))$ of measure zero, and any $f \in L_2(\mathbb{R}^n)$, we have $\|P_\Lambda(H(\vec{a}))f\|^2 = 0$, where

$$P_\Lambda(H(\vec{a})) = \chi_\Lambda(H(\vec{a}))$$

is the spectral projection associated with $H(\vec{a})$ and Λ. By [10; Thm. XIII.85 (c)], we see that

$$\|P_\Lambda(H(\vec{a}))f\|^2 = \int_Q \|P_\Lambda(H_k(\vec{a}))f_k\|_{\ell_2}^2 \, dk.$$

The integrand on the RHS will vanish at k unless Λ contains an eigenvalue of $H_k(\vec{a})$. But the set

$$\{k \in Q; \Lambda \cap \sigma(H_k(\vec{a})) \neq \emptyset\} = \{k \in Q; F(\lambda,k) = 0 \text{ for some } \lambda \in \Lambda\}$$

has measure zero by Theorem A, and we are done. ∎

Remark. There is not much new in Theorem 2 beyond what can already be found in Kuchment [9], Gérard [4], or in Wilcox [15]. In particular, it is shown by Gérard [4] that the resolvent of a periodic Schrödinger operator $H = -\Delta + V$, with V relatively bounded with respect to the Laplacian with bound < 1, can be analytically continued through the real axis (in a suitable sense), outside a discrete set of points. This establishes the absence of singular continuous spectrum, but leaves the possibility of a discrete set of eigenvalues. Below and in the Appendix, we give a self-contained introduction into the required machinery, keeping the use of algebra to a strict minimum. Note that Gérard seems to share our belief that there actually shouldn't be any such eigenvalues, for relatively bounded perturbations [5; p. 48].

L. Thomas' celebrated proof of the absolute continuity of periodic Schrödinger operators $-\Delta + V$ can be adapted to the magnetic case as long as the vector potential is not too large.

Theorem 3. ([6; Thm. 3.12]) For any given lattice Γ there exists a constant $c > 0$ such that $H(\vec{a})$ is absolutely continuous for all Γ-periodic vector potentials $\vec{a} \in C^1(\mathbb{R}^n; \mathbb{R}^n)$ satisfying

$$\sup_x |\vec{a}(x)| < c.$$

Proof. By Theorem 2(a), it is enough to find a constant $c > 0$ such that $H(\vec{a})$ has no eigenvalues if $\sup |\vec{a}(x)| < c$. By Theorem 2(c), λ is an eigenvalue of $H(\vec{a})$ if and only if λ is an eigenvalue of $H_k(\vec{a})$, for all $k = (k_1, \ldots, k_n) \in Q$ if and only if λ is an eigenvalue of $H_k(\vec{a})$ for all $k \in \mathbf{C}^n$. We now take take Im k_1 to infinity, along suitable lines. As in the original proof of Thomas [13], we find by a direct calculation that

$$\liminf_{\mathrm{Im}\, k_1 \to \infty} \left\| (H_k(\vec{a}) + 1)^{-1} \right\| = 0,$$

provided $\sup |\vec{a}|$ is small. But then no eigenvalue of the family $H_k(\vec{a})$ can be constant. ∎

Remark. In the the case of periodic Hamiltonians $-\Delta + V$, with mild assumptions on V ([10; p. 305 ff]), a similar calculation reveals that $\|(H_k + 1)^{-1}\|$ tends to zero as $\mathrm{Im}\, k_1 \to \infty$. It is interesting to point out why a smallness conditions seems to be unavoidable in the magnetic case:

Let $P(k) = k + \sum m_j K_j$, acting in $\ell_2(\mathbb{Z}^n)$, where $m_j \in \mathbb{Z}$ and the vectors K_j, $j = 1, \ldots, n$ span the dual lattice. We then find that controlling the term coming from $\vec{a} \cdot \nabla$ requires control of $\|P(k)(P(k)^2 + 1)^{-1}\|$, for $\mathrm{Im}\, k_1 \to \infty$ on some line. It turns out that $\liminf_{\mathrm{Im}\, k_1 \to \infty} \|P(k)(P(k)^2 + 1)^{-1}\|$ is a certain finite number, which is independent of the lines along which k_1 tends to ∞, but, alas, this number is not zero.

We finally introduce a coupling $\mu \in \mathbb{R}$ and ask for the absolute continuity of $H(\mu\vec{a})$.

Theorem 4. Suppose that $\vec{a} \in C^1(\mathbb{R}^n; \mathbb{R}^n)$ is periodic with respect to the lattice Γ. Then:

(a) For any fixed $R > 0$, the set of coupling constants $\mu \in \mathbb{R}$ for which $H(\mu\vec{a})$ is not absolutely continuous in the interval $[0,R]$ is discrete.

(b) The set of coupling constants $\mu \in \mathbb{R}$ for which $H(\mu\vec{a})$ is not absolutely continuous is at most countable.

Proof. Let $R > 0$, and suppose for a contradiction that there exist sequences $\{\mu_j\} \subset \mathbb{R}$ with $\mu_j \to \mu_0 \in \mathbb{R}$, $\mu_j \neq \mu_k$ for $j \neq k$, and $\{\lambda_j\} \subset [0,R]$ such that λ_j is an eigenvalue of $H(\mu_j\vec{a})$, for all $j \in \mathbf{N}$. Without restriction we may assume that the sequence $\{\lambda_j\}$ converges. Let $M \geq 0$ be such that $|\mu_j| \leq M$. We first conclude from Theorem 2(c) that λ_j is an eigenvalue of $H_k(\mu_j\vec{a})$, for all $k \in Q$.

For each fixed $k \in Q$, $(H_k(\mu\vec{a}), \mu \in \mathbf{C})$ is a self-adjoint holomorphic family of type (A) in the sense of Kato. Furthermore, $H_k(\mu\vec{a})$ has compact resolvent for all μ. By [8; Thm. VII-3.9], there exists a countable family of functions $\Lambda_m(k,\mu)$, $m \in \mathbf{N}$, real-analytic in $\mu \in \mathbb{R}$, which describe the eigenvalues of $H_k(\mu\vec{a})$. In addition, an easy compactness argument (using the standard bound given in [8;p. 391] for the derivatives $\frac{\partial}{\partial\mu}\Lambda(k,\mu)$) implies that for each fixed $k \in Q$ there exists at most a finite number of indices m such that $\Lambda_m(k,\mu) \in [0,R]$ for some $|\mu| \leq M$.

We now pick any $k_0 \in Q$. Since λ_j is an eigenvalue of $H_{k_0}(\mu_j\vec{a})$, for all $j \in \mathbf{N}$, and since only finitely many functions $\Lambda_m(k_0,\mu)$ meet the interval $[0,R]$, for $|\mu| \leq M$, there exists an index $m(k_0)$ such that

$$\lambda_j = \Lambda_{m(k_0)}(k_0,\mu_j) \tag{4}$$

for an infinite number of j's. Without restriction, we may assume that (4) holds for all $j \in \mathbf{N}$. Write $\Lambda_0 = \Lambda_{m(k_0)}(k_0,0)$. Suppose $k \in Q$. As above there exists an index $m(k)$ such that

$$\lambda_j = \Lambda_{m(k)}(k,\mu_j)$$

for infinitely many j. By unique continuation, $\Lambda_{m(k)}(k,\mu) = \Lambda_{m(k_0)}(k_0,\mu)$ for all μ, in particular for $\mu = 0$. Thus for each $k \in Q$

$$\Lambda_{m(k)}(k,0) = \Lambda_0$$

so that Λ_0 is an eigenvalue of $H_k(0)$ for all $k \in Q$. But this implies that $H(0) = -\Delta$ has an eigenvalue, in contradiction with the absolute continuity of $-\Delta$. ∎

Appendix.

Theorem A. Suppose $I \subset \mathbb{R}$ is an open interval and $\mathcal{O} \subset \mathbb{R}^n$ is an open connected set. Suppose $f : I \times \mathcal{O} \to \mathbf{C}$ is real analytic and not identically zero. Define the measure μ on the Borel sets Λ of $\mathbb{R}$ by

$$\mu(\Lambda) = \text{meas}\,\{k \in \mathcal{O}; f(\lambda,k) = 0 \quad \text{for some } \lambda \in \Lambda\}).$$

Then μ has no singular continuous component and, in addition, the set of pure points of μ has no accumulation point in I.

We first prove two lemmas.

Lemma A.1. Suppose $\mathcal{O} \subset \mathbb{R}^n$ is an open connected set and $h : \mathcal{O} \to \mathbf{C}$ is real analytic. Then if meas $(h^{-1}(0)) > 0$ it follows that $h \equiv 0$.

Proof. Let k_0 be a point of density of the zero set of h. Then it easily follows by induction that all derivatives of h vanish at k_0. By unique continuation, h is zero. ∎

Lemma A.2. Suppose $\mathcal{O} \subset \mathbb{R}^n$ is open and connected and the map $f : \mathcal{O} \to \mathbb{R}$ is real analytic. Then, if f is not constant, it follows that if $\Lambda \subset \mathbb{R}$ is a Borel set with meas $(\Lambda) = 0$ we have meas $(f^{-1}(\Lambda)) = 0$.

Remark. Lemma A.2 is a special case of Theorem A with $f(\lambda,k) = \lambda - f(k)$.

Proof. We can assume $\mathcal{O} = J_1 \times \cdots \times J_n$ where J_ℓ is an open interval. The lemma is easy to prove if $n = 1$ so we assume $n > 1$. Let

$$\Gamma_1 = \{(k_2, \ldots, k_n) \in J_2 \times \cdots \times J_n; f(k_1, k_2, \ldots, k_n) \text{ is constant in the variable } k_1\}.$$

It is easy to see that Γ_1 is measurable. By Lemma A.1, if meas $(\Gamma_1) > 0$ it follows that for any k_1, $k_1' \in J_1$

$$f(k_1,k_2,\ldots,k_n) - f(k_1',k_2,\ldots,k_n) = 0$$

for all $(k_2,\ldots,k_n) \in J_2 \times \cdots \times J_n$, in other words, $f(\cdot, k_2, \ldots, k_n)$ is constant for all $(k_2,\ldots,k_n) \in J_2 \times \cdots \times J_n$. Similarly, we define Γ_j, $j = 2,\ldots,n$. If meas $(\Gamma_j) > 0$ for all j, it then easily follows that f is constant. Without loss of generality we can therefore assume meas $(\Gamma_1) = 0$. Write $k = (k_1,k')$ and $f_{k'}(k_1) = f(k_1,k')$ with $k_1 \in J_1$, $k' \in J_2 \times \cdots \times J_n$. By Fubini's theorem we have

$$\begin{aligned} \text{meas}\,(f^{-1}(\Lambda)) &= \int_{f^{-1}(\Lambda)} dk = \int_{J_2\times\cdots\times J_n} \left(\int_{k'=a,\ k\in f^{-1}(\Lambda)} dk_1\right) da \\ &= \int_{J_2\times\cdots\times J_n} \left(\int_{f_a^{-1}(\Lambda)} dk_1\right) da. \end{aligned}$$

But for almost every $a \in J_2 \times \cdots \times J_n$, f_a is non-constant so by the result for $n = 1$, $f_a^{-1}(\Lambda)$ has measure zero for almost every a. The result follows. ∎

Proof of Theorem A. If $\mu(\{\lambda_j\}) > 0$ for a sequence $\lambda_j \to \lambda \in I$ then by Lemma A.1, $f(\lambda_j,k) = 0$, for all $k \in \mathcal{O}$, and thus $f = 0$. This contradiction proves the second statement.

Let

$$M = \{k \in \mathcal{O}; f(\lambda,k) = 0 \text{ for all } \lambda \in I\}.$$

Then by Lemma A.1, M is a set of measure zero, closed in the relative topology. Suppose $\Lambda \subset \mathbb{R}$ has Lebesgue measure zero and contains no pure points of μ. It is enough to show that every point $(\lambda_0,k_0) \in I \times (\mathcal{O} \setminus M)$ has an open neighborhood $I_0 \times \mathcal{O}_0$ with $\mathcal{O}_0 \subset \mathcal{O} \setminus M$ such that

$$\{k \in \mathcal{O}_0; f(\lambda,k) = 0 \ \text{ for some } \lambda \in \Lambda \cap I_0\}$$

has measure zero. Clearly we can assume $f(\lambda_0,k_0) = 0$. For notational simplicity in what follows we take $(\lambda_0,k_0) = 0$. Then there is a neighborhood U_1 of 0 contained in $I \times (\mathcal{O} \setminus M)$ and functions p_j, $j = 1,\ldots,\ell$, real analytic on U_1, and integers $r_1,\ldots,r_\ell \geq 1$ so that on U_1

$$f = p_1^{r_1} \cdots p_\ell^{r_\ell}.$$

In addition, p_j is prime ($j = 1, \ldots, \ell$) in the sense that if on any neighborhood of 0 we have $p_j = h_1 h_2$ with h_1 and h_2 real analytic and $h_1(0) = 0$ then $h_2(0) \neq 0$. This follows from the unique factorization property of the germs of analytic functions at 0 ([F; p. 116]). We can assume $U_1 = I_1 \times \mathcal{O}_1$ where I_1 is an interval of $\mathbb{R}$ and $\mathcal{O}_1$ is an open subset of $\mathbb{R}^n$. By the Weierstrass Preparation Theorem ([F; Thm. 3.23]) and the fact that $f(\lambda, 0) \not\equiv 0$, we can write (after shrinking U_1 perhaps)

$$f = h\tilde{p}_1^{r_1} \cdots \tilde{p}_\ell^{r_\ell} \qquad \text{on } U_1 = I_1 \times \mathcal{O}_1\,,$$

where h is non-zero on U_1 and $\tilde{p}_j$ is a *Weierstrass polynomial*:

$$\tilde{p}_j(\lambda, k) = \lambda^{n_j} + a_{j1}(k)\lambda^{n_j - 1} + \ldots + a_{jn_j}(k),$$

with $a_{jm}(k)$ real analytic on $\mathcal{O}_1$, $a_{jm}(0) = 0$. In addition, $\tilde{p}_j$ is prime.

The discriminant $D_j(k)$ of $\tilde{p}_j$ is real analytic on $\mathcal{O}_1$ and is not identically zero ([14], [5; Thm. 6.11]) because $\tilde{p}_j$ is prime. The set M_j where D_j is zero is a closed set of measure zero and $\tilde{p}_j$ has distinct roots on $\mathcal{O}_1 \setminus M_j$. It is enough to find for each j and each point $(\lambda_0', k_0') \in I_1 \times (\mathcal{O}_1 \setminus M_j)$ a neighborhood $I_2 \times \mathcal{O}_2$ of (λ_0', k_0') such that

$$\{k \in \mathcal{O}_2; \tilde{p}_j(\lambda, k) = 0 \ \text{ for some } \lambda \in I_2 \cap \Lambda\}$$

has measure zero. We can assume $\tilde{p}_j(\lambda_0', k_0') = 0$. By the Implicit Function Theorem there exists a neighborhood $I_2 \times \mathcal{O}_2$ of (λ_0', k_0') and n_j functions $\lambda_1, \ldots, \lambda_{n_j}$, real analytic on $\mathcal{O}_2$, such that for $(\lambda, k) \in I_2 \times \mathcal{O}_2$

$$\tilde{p}_j(\lambda, k) = 0 \iff \lambda = \lambda_i(k), \qquad \text{for some } i \in \{1, \ldots, n_j\}.$$

We can assume that $\mathcal{O}_2$ is an open rectangle. We have

$$\begin{aligned} &\{k \in \mathcal{O}_2; \tilde{p}_j(\lambda, k) = 0 \ \text{ for some } \lambda \in I_2 \cap \Lambda\} \\ &\quad = \cup_{i=1}^{n_j} \{k \in \mathcal{O}_2; \lambda = \lambda_i(k) \text{ for some } \lambda \in I_2 \cap \Lambda\}. \end{aligned}$$

The set

$$\{k \in \mathcal{O}_2; \lambda = \lambda_i(k) \text{ for some } \lambda \in I_2 \cap \Lambda\} = \lambda_i^{-1}(I_2 \cap \Lambda)$$

will have non-zero measure only if $\operatorname{Im} \lambda_i(k) = 0$, for all $k \in \mathcal{O}_2$ (Lemma A.1) and if for some $\lambda_0 \in I_2 \cap \Lambda$, $\lambda_i(k) = \lambda_0$ for all $k \in \mathcal{O}_2$ (Lemma A.2). But the latter condition implies $f(\lambda_0, k) = 0$ for all $k \in \mathcal{O}_2$ and thus for all $k \in \mathcal{O}$. But Λ was chosen to be free of pure points of μ, so this is a contradiction. ∎

Acknowledgements. The authors are most grateful to S. Agmon for bringing Chr. Gérard's paper [4] to their attention. They would also like to thank P. Michor (Vienna) for several useful conversations.

References

[1] J. Brüning and T. Sunada, On the spectrum of gauge-periodic elliptic operators. Soc. Math. France Astérisque **210** (1992), 65–74.

[2] J. Brüning and T. Sunada, in preparation.

[3] M. Field, Several complex variables and complex manifolds I. Cambridge University press, Cambridge 1982.

[4] Chr. Gérard, Resonance theory for periodic Schrödinger operators. Bull. Soc. Math. France **118** (1990), 27–54.

[5] H. Grauert and K. Fritzsche, Several complex variables. Springer Verlag, 1976.

[6] R. Hempel and I. Herbst, Strong magnetic fields, Dirichlet boundaries, and spectral gaps. Preprint 1994. To appear in Commun. Math. Phys.

[7] A. Iwatsuka, On Schrödinger operators with magnetic fields. In: Lecture Notes in Mathematics, vol. **1450** (Conf. Proc., ed. by H. Fujita, T. Ikebe, S. T. Kuroda), pp. 157–172. Springer Verlag, Berlin 1990.

[8] T. Kato, Perturbation theory for linear operators. Springer Verlag, New York 1966.

[9] P. Kuchment, Floquet theory for partial differential equations. Birkhäuser, Basel 1993.

[10] M. Reed and B. Simon, Methods of Modern Mathematical Physics, Vol. IV. Analysis of Operators. Academic Press, New York 1978.

[11] E. Stein, Singular Integrals and differentiability properties of functions. Princeton University Press, Princeton 1986.

[12] R. Stöcker und H. Zieschang, Algebraische Topologie. Teubner, Stuttgart 1989.

[13] L. E. Thomas, Time dependent approach to scattering from impurities in a crystal. Commun. Math. Phys. **33** (1973), 335–343.

[14] B. L. van der Waerden, Modern Algebra, vol. I. Frederick Unger 1966

[15] C. H. Wilcox, Theory of Bloch waves. J. Analyse Math. **33** (1978), 146–167.

Rainer Hempel
On leave from Univ. of Alabama, Birmingham, AL 35294, USA.
Address after August 1, 1994:
Institut für Analysis, TU Braunschweig, Pockelsstr. 14, D-38106 Braunschweig, Germany.
Ira Herbst
On leave from Univ. of Virginia, Charlottesville, VA 22903, USA.
Erwin Schrödinger International Institute for Mathematical Physics
Pasteurgasse 4/7, A-1090 Vienna, Austria

Operator Theory:
Advances and Applications, Vol. 78

H^∞-Calculus for Second Order Elliptic Operators in Divergence Form

M. Hieber

1. Introduction

Heat kernel estimates play an important role in the investigation of the question whether or not a given sectorial operator admits bounded imaginary powers or an H^∞-calculus on an L^p space. Indeed, it was shown in [7] and [10] that, roughly speaking, the negative of a given generator $-A$ of a holomorphic semigroup T on $L^2(\Omega)$ admits an H^∞-calculus on $L^p(\Omega)$, $1 < p < \infty$, with optimal angle, provided the following conditions hold : A admits an H^∞-calculus on $L^2(\Omega)$, the operators $T(t)$ are integral operator for all $t > 0$ and Ω as well as the kernel of $T(t)$ or the kernel of the resolvent of $-A$ satisfy certain regularity conditions. These conditions are in particular satisfied whenever Ω is a bounded Lipschitz domain in $\mathbb{R}^n$, $(n \geq 3)$, T satisfies an upper Gaussian estimate and the kernel of the resolvent of $-A$ satisfies a certain Hölder regularity assumption.

For the time being, let $n \geq 3$ and consider an elliptic differential operator $-A^D$ in divergence form in $L^2(\Omega)$ with symmetric L^∞-coefficients and subject to Dirichtlet boundary conditions. Then it was shown by Davies [4] that the semigroup generated by $-A^D$ satisfies an upper Gaussian estimate and it was proved by Grüter and Widmann [9] that the regularity condition on the kernel of the resolvent of $-A^D$ is satisfied in this case. Hence, as an application of the above cited result, it is proved in [7] and [10] that, given $\theta > 0$, there exists $\mu > 0$ such that $A^D + \mu$ admits a bounded H^∞-calculus on $L^p(\Omega)$, $1 < p < \infty$, of angle θ. For the case of Neumann boundary conditions we refer to [7]. Very recently it was shown by Duong and Robinson [8] that the above mentioned result remains true without assuming the additional Hölder regularity condition on the kernel of the resolvent of $-A$.

In this note we use the Duong-Robinson result to obtain the existence of an H^∞-calculus on $L^p(\Omega)$ for second order elliptic operators described as above but subject to more general boundary conditions than those of Dirichlet or Neumann type. Considering in particular the function $\lambda \mapsto \lambda^{it} (t \in \mathbb{R})$, it follows that A admits bounded imaginary powers.

Our interest in this problem is twofold : First it is known [15] that the domain $D(A^\alpha)$ of the fractional powers A^α of A $(0 < \alpha < 1)$ coincides with the complex interpolation space $[E,D(A)]_\alpha$ provided A is a densely defined operator on a Banach space E possessing bounded imaginary powers. Second, by a result of Dore and Venni [5], results on "maximal regularity" for equations of the form

$$u'(t) + Au(t) = f(t), \qquad u(0) = u_0$$

are intimately connected with the theory of bounded imaginary powers. Both of these results are of great importance in the functional analytic approach to quasi-linear, parabolic problems (cf.[1]). In this context, we are in particular interested in H^∞-calculus results on $L^p(\Omega)$ for elliptic operators with non-smooth coefficients. For more information on this subject we refer to [2], [3], [6], [11], [13] and [14].

2. Gaussian estimates and H^∞-calculus on L^p-spaces

As usual, we denote by $D(A)$, $\sigma(A)$, $\rho(A)$ the domain, spectrum and resolvent set of a closed operator A, respectively. Moreover, $\mathcal{L}(E)$ denotes the space of bounded linear operators on a Banach space E. If $G \subset \mathbb{C}$ is open, then $H^\infty(G)$ means the space of all functions $f : G \to \mathbb{C}$ which are holomorphic and bounded; equipped with the sup-norm $||\cdot||_\infty$, $H^\infty(G)$ becomes a Banach algebra. Given $G \subset \mathbb{C}$, A is said to *admit an H^∞-calculus for G*, if $\sigma(A) \subset \overline{G}$ and there is a continuous algebra homomorphism $\Phi : H^\infty(G) \to \mathcal{L}(E)$ such that the functions $f_\mu(\lambda) := (\mu-\lambda)^{-1}$, $\mu \notin \overline{G}$, are mapped to $\Phi f_\mu = (\mu - A)^{-1}$.

Let $\Omega \subset \mathbb{R}^n$ be an open set and let T be a C_0-semigroup on $L^2(\Omega)$ with generator A. In the following we always identify $L^p(\Omega)$ with a subspace of $L^p(\mathbb{R}^n)$ by extending functions by zero. Let G_p be the Gaussian semigroup on $L^p(\mathbb{R}^n)$, i.e. G_p is given by $G_p(t)f := k_t * f$, where

$$k_t(x) := (\frac{1}{4\pi t})^{n/2} e^{-\frac{|x|^2}{4t}}, \qquad (t > 0, x \in \mathbb{R}^n).$$

We say that the semigroup T satisfies an *upper Gaussian estimate* if there exist constants $a \geq 0, M,b > 0$ such that

$$|T(t)f| \leq Me^{at}G_2(bt)|f| \qquad (t \geq 0)$$

for all $f \in L^2(\Omega)$. Assume now that T is a C_0-semigroup on $L^2(\Omega)$ which satisfies an upper Gaussian estimate. Then there exist consistent semigroups T_p on $L^p(\Omega)$, $(1 \leq p < \infty)$, such that $T = T_2$ and

$$|T_p(t)f| \leq Me^{at}G_p(bt)|f| \qquad (f \in L^p(\Omega), t \geq 0).$$

By *consistent* we mean that $T_p(t)f = T_2(t)f$ for all $t \geq 0$ and all $f \in L^2(\Omega) \cap L^p(\Omega)$. We denote by A_p the generator of T_p. Finally, for $\theta \in [0,\pi)$ we define open and closed sectors in the complex plane by

$$S_\theta^0 := \{z \in \mathbb{C}\backslash\{0\}; |arg z| < \theta\} \quad \text{and} \quad S_\theta := \{z \in \mathbb{C}\backslash\{0\}; |arg z| \leq \theta\} \cup \{0\}.$$

The following result is proved in [8;Cor.3.4].

Theorem 2.1. (Duong, Robinson). Let $1 < p < \infty$ and let $\Omega \subset \mathbb{R}^n$ be a bounded Lipschitz domain. Suppose that $-A$ generates a holomorphic C_0-semigroup T on $L^2(\Omega)$ of angle $\varphi \in (0,\pi/2]$. Assume that T satisfies an upper Gaussian estimate and that A admits an H^∞-calculus for S_α^0 with $\alpha > \pi/2 - \varphi$. Then, given $\theta > \pi/2 - \varphi$, there exists a constant $\mu > 0$ such that $A_p + \mu$ admits an H^∞-calculus on $L^p(\Omega)$ for every sector S_θ^0.

In the following section we use Theorem 2.1 to obtain the existence of an H^∞-calculus on $L^p(\Omega)$ for elliptic operators in divergence form subject to rather general boundary conditions.

3. Elliptic operators in divergence form with L^∞-coefficients

Let $\Omega \subset \mathbb{R}^n$ be a bounded Lipschitz domain and let $a_{ij} \in L^\infty(\Omega)$ such that $a_{ij} = a_{ji}$ for all $1 \le i,j \le n$. Suppose that there exist constants $c_1,c_2 > 0$ such that

$$c_1|\xi|^2 \le \sum_{i,j=1}^{n} a_{ij}(x)\xi_i\xi_j \le c_2|\xi|^2 \qquad x-a.e.$$

for all $\xi \in \mathbb{R}^n$. Define the bilinear form a by

$$a(u,v) := -\sum_{i,j=1}^{n} \int a_{ij}(x)\partial_i u \partial_j v dv.$$

We then distinguish between three cases :

a) Set $D(a) := H_0^1(\Omega)$. Then a is a closed form and we denote by A^D the operator in $L^2(\Omega)$ associated with a. We call A^D a uniformly elliptic operator subject to Dirichlet boundary conditions.

b) Set $D(a) := H^1(\Omega)$. Again a is a closed form and we denote by A^N the operator in $L^2(\Omega)$ associated with a. We call A^N a uniformly elliptic operator subject to Neumann boundary conditions.

c) Set $D(a) := V$, where V is a closed ideal of $H^1(\Omega)$ containing $H_0^1(\Omega)$. Denote by A^V the operator in $L^2(\Omega)$ associated with a.

Observe that $-A^D, -A^N$ and $-A^V$ are self-adjoint operators and that they generate bounded analytic C_0-semigroups T^D, T^N and T^V, respectively, of angle $\pi/2$ on $L^2(\Omega)$. Furthermore it follows from results due to Davies [4; Cor. 3.2.8, Thm. 3.2.9] that the semigroups T^D and T^N satisfy an upper Gaussian estimate. Denote by $-A_p^D, -A_p^N$ and $-A_p^V$ the generators of the consistent semigroups T_p^D, T_p^N and T_p^V on $L^p(\Omega)$, respectively. Theorem 2.1 implies that, given $\theta > 0$, there exists a constant $\mu > 0$ such that $A_p^D + \mu$ and $A_p^N + \mu$ admit an H^∞-calculus on $L^p(\Omega)$ for every sector S_θ^0 (see also [7] and [10]).
For the case c) we obtain the following result.

Theorem 3.1. Let $1 < p < \infty$ and let $\Omega \subset \mathbb{R}^n$ be a bounded Lipschitz domain. Let $-A_p^V$ be defined as above. Then, given $\theta > 0$, there exists $\mu > 0$ such that $A_p^V + \mu$ admits an H^∞-calculus on $L^p(\Omega)$ for every sector S_θ^0.

Proof. It was proved by Ouhabaz [12] that in the given situation we have $T^V(t) \leq T^N(t)$ for all $t \geq 0$. Therefore, by the Dunford-Pettis theorem, for each $t > 0$, $T^V(t)$ can be represented as an integral operator whose kernel is dominated by the kernel associated to $T^N(t)$. Hence, T^V satisfies an upper Gaussian estimate and the assertion follows from Theorem 2.1. □

Remark 3.2. Let $\Omega \subset \mathbb{R}^n$ be bounded with C^1 boundary $\partial\Omega$. Let $V := \{u \in H^1(\Omega); \gamma(u) = 0 \text{ on } \Gamma_0\}$, where Γ_0 is a closed subset of $\partial\Omega$ and γ is the trace operator from $H^1(\Omega)$ into $H^{1/2}(\Omega)$. Then V is an ideal of $H^1(\Omega)$. We do, however, not know whether all closed ideals of $H^1(\Omega)$ containing $H_0^1(\Omega)$ are of this form. For the special case $n = 1$ and $\Omega = (a,b)$ one can prove that the only closed ideals of $H^1(a,b)$ containing $H_0^1(a,b)$ are $V_1 := H_0^1(a,b)$, $V_2 := \{u \in H^1(a,b); u(a) = 0\}$, $V_3 := \{u \in H^1(a,b); u(b) = 0\}$ and $V_4 := H^1(a,b)$.

References

[1] H. Amann, Linear and Quasilinear Parabolic Problems. Book in preparation, 1994.

[2] H. Amann, M. Hieber, G. Simonett, Bounded H_∞-calculus for elliptic operators. Differential Integral Equations 7, (1994), 613–653.

[3] M. Cowling, I. Doust, A. McIntosh, A. Yagi, Banach space operators with a bounded H^∞ functional calculus. preprint (1993).

[4] E.B. Davies, Heat kernels and spectral theory. Cambridge University Press, Vol. 92, Cambridge, 1989.

[5] G. Dore, A. Venni, On the closedness of the sum of two closed operators. Math. Z. 196, (1987), 189–201.

[6] X.T. Duong, H^∞ functional calculus of elliptic operators with C^∞ coefficients on L^p spaces of smooth domains. J. Austral. Math. Soc. 48, (1990), 113–123.

[7] X.T. Duong, A. McIntosh, Functional calculi of second order elliptic partial differential operators with bounded measurable coefficients. Preprint (1993).

[8] X.T. Duong, D.W. Robinson, Gaussian bounds, Brownian estimates and H_∞-functional calculus. Preprint (1994).

[9] M. Grüter, K.O. Widmann, The Green function for uniformly elliptic equtions. Manuscprita Math. 37, (1982), 303–343.

[10] M. Hieber, Heat kernel estimates and bounded H^∞-calculus on L^p spaces. In: Partial Differential Equations; Models in Physics and Biology, S. Nicaise, G. Lumer, B.-W. Schulze (eds.), Akademie Verlag, Berlin, (1994), to appear.

[11] M. Hieber, J. Prüss, H^∞-calculus for generators of bounded C_0-groups and positive contraction semigroups on L^p-spaces. Preprint (1994).

[12] E. Ouhabaz, Propriétés d'ordre et de contractivité des semi-groupes avec applications aux opérateurs elliptiques. Ph.D. Thesis, Besançon, (1992).

[13] J. Prüss, H. Sohr, Imaginary powers of elliptic second order differential operators in L^p-spaces. Hiroshima Math. J. 23, (1993), 161–192.

[14] R.T. Seeley, The resolvent of an elliptic boundary value problem. Amer. J. Math. 91, (1969), 889–920.

[15] H. Triebel, Interpolation Theory, Function Spaces, Differential Operators. North-Holland, Amsterdam, 1978.

Matthias Hieber,
Mathematisches Institut, Universität Zürich
Winterthurerstr. 190, CH-8057 Zürich
Switzerland

Operator Theory:
Advances and Applications, Vol. 78

Path Integral for the Relativistic Schrödinger Semigroup

T. Ichinose

1. Introduction.

We consider the Weyl quantized relativistic Hamiltonian $H = H_A \dot{+} \Phi$ defined through the quadratic form corresponding to the classical symbol

$$\sqrt{(p - A(x))^2 + m^2} + \Phi(x), \qquad (p,x) \in \mathbb{R}^d \times \mathbb{R}^d, \tag{1.1}$$

of a spinless particle with mass $m \geq 0$ in an electromagnetic field (e.g. Landau-Lifschitz [9]), where the light velocity c is taken to be equal to 1. For the vector and scalar potentials $A : \mathbb{R}^d \to \mathbb{R}^d$ and $\Phi : \mathbb{R}^d \to \mathbb{R}$ of the field we shall assume that

$$A \in L^{1+\delta}_{\text{loc}}(\mathbb{R}^d) \text{ for some } \delta > 0 \text{ and } \Phi \in L^1_{\text{loc}}(\mathbb{R}^d),\ \Phi(x) \geq 0 \text{ a.e.} \tag{1.2}$$

The quadaratic form concerned is the maximal symmetric form $h = h_{A,\Phi}$ given by

$$h[u] \equiv h[u,u] = m\|u\|^2 + \frac{1}{2}\iint_{|x-y|>0} |e^{-i(x-y)\cdot A(\frac{1}{2}(x+y))}u(x) - u(y)|^2 n(x-y)dxdy + \int \Phi(x)|u(x)|^2 dx, \tag{1.3a}$$

where

$$n(y) = \begin{cases} 2(2\pi)^{-(d+1)/2} m^{d+1}(m|y|)^{-(d+1)/2} K_{(d+1)/2}(m|y|), & m > 0, \\ \pi^{-(d+1)/2}\Gamma(\frac{d+1}{2})|y|^{-(d+1)}, & m = 0, \end{cases} \tag{1.3b}$$

with form domain $Q(h) = \{u \in L^2(\mathbb{R}^d); h[u] < \infty\}$.

It was shown in Ichinose-Tsuchida [6] that this form h is a closed form with form domain $Q(h)$ including $C_0^\infty(\mathbb{R}^d)$ as a form core, so that the minimal symmetric form $h_{\min}$ defined as the form closure of $h|C_0^\infty(\mathbb{R}^d) \times C_0^\infty(\mathbb{R}^d)$ coincides with h. Therefore there exists a unique selfadjoint operator $H = H_A \dot{+} \Phi$ with domain $D(H)$ associated with the form h such that $h[u,v] = (Hu,v)$ for $u \in D(H)$, $v \in Q(h)$. We expect that the condition (1.2) is minimal to assure that h defines a quadratic form in $L^2(\mathbb{R}^d)$ with form domain including $C_0^\infty(\mathbb{R}^d)$, so long as $\Phi(x)$ is nonnegative.

If $A(x)$ is sufficiently smooth and has bounded derivatives $\partial_x^\alpha A(x)$ up to sufficiently high order and $\Phi(x)$ locally square-integrable with $\Phi(x) \geq 0$ a.e., then H coincides with $\mathcal{H}_A^w + \Phi$ on $C_0^\infty(\mathbb{R}^d)$, where $\mathcal{H}_A^w$ is the Weyl pseudo-differential operator $\mathcal{H}_A^w$:

$$(\mathcal{H}_A^w u)(x) = (2\pi)^{-d} \iint_{\mathbb{R}^d \times \mathbb{R}^d} e^{i(x-y)p} \sqrt{(p - A(\frac{x+y}{2}))^2 + m^2} u(y)dydp, \quad u \in C_0^\infty(\mathbb{R}^d), \tag{1.4}$$

and Φ is the multiplication operator by the function $\Phi(x)$.

It was shown in Nagase-Umeda [11] (cf. Ichinose [3]) that $\mathcal{H}_A^w$ is essentially selfadjoint on $C_0^\infty(\mathbb{R}^d)$.

It is this selfadjoint operator H that we will consider as the *Weyl quantized relativistic Hamiltonian* corresponding to the classical symbol (1.1).

In this note we want to prove a path integral representation for the semigroup $e^{-t[H-m]}$, completing our previous results in Ichinose-Tamura [5] and Ichinose [4].

2. Path Integral Representation.

When $A \in L^4_{\text{loc}}(\mathbb{R}^d)$ with $\operatorname{div} A \in L^2_{\text{loc}}(\mathbb{R}^d)$ and $\Phi \in L^2_{\text{loc}}(\mathbb{R}^d)$ with $\Phi(x) \geq 0$ a.e., the Feynman-Kac-Itô formula [12, pp. 159–163] holds for the nonrelativistic Schrödinger operator $(2m)^{-1}(-i\nabla - A(x))^2 + \Phi(x)$. This condition is minimal to assure it to define a linear operator in $L^2(\mathbb{R}^d)$ with domain including $C_0^\infty(\mathbb{R}^d)$, and in this case it is in fact essentially selfadjoint on $C_0^\infty(\mathbb{R}^d)$, as was proved by Leinfelder-Simader [10]. However, since $\operatorname{div} A$ is a distribution, the Feynman-Kac-Itô formula does not necessarily make sense for vectors A belonging to $L^2_{\text{loc}}(\mathbb{R}^d)$ and for functions Φ in $L^1_{\text{loc}}(\mathbb{R}^d)$, with $\Phi(x) \geq 0$ for almost all x. Therefore the nonrelativistic Schrödinger operator is defined through the *quadratic form* (see[1, pp. 8–10], as long as the Coulomb gauge condition $\operatorname{div} A = 0$ (in distributional sense) is not imposed. In fact the Feynman-Kac-Itô formula can be given a meaning, provided the divergence term is left out and the stochastic integral is interpreted in Stratonowich sense.

In this section we give a path integral formula for the semigroup $e^{-t[H-m]}$. Emphasized is that here H is defined *through the quadratic form*. This strongly contrasts with the Feynman-Kac-Itô formula as mentioned above.

To state the result we need some notations about the Lévy process. λ_x is the probability measure on the space $D_x \equiv D_x([0,\infty) \to \mathbb{R}^d)$ of the right-continuous paths $X : [0,\infty) \to \mathbb{R}^d$ with left-hand limits and $X(0) = x$ such that

$$\exp[-t(\sqrt{p^2+m^2} - m)] = \int_{D_x} e^{ip(X(t)-x)} d\lambda_x(X), \qquad t \geq 0, \quad p \in \mathbb{R}^d. \tag{2.1}$$

By the Lévy-Khinchin formula (e.g. Ikeda-Watanabe [7])

$$\sqrt{p^2+m^2} - m = -\int_{|y|>0} [e^{ipy} - 1 - ipyI_{\{|y|<1\}}]n(dy), \qquad p \in \mathbb{R}^d, \tag{2.2}$$

where $I_{\{|y|<1\}}$ is the indicator function for the set $\{|y|<1\}$ and $n(dy) = n(y)dy$ is the Lévy measure with density $n(y)$ in (1.3b). For each $X \in D_x$, $\widetilde{N}_X(dsdy) = N_X(dsdy) - dsn(dy)$ by definition, where $N_X(dsdy)$ is a counting measure on $(0,\infty)\times(\mathbb{R}^d\backslash\{0\})$ defined by $N_X((t,t']\times U) = \#\{s \in (t,t']; 0 \neq X(s)-X(s-) \in U\}$ with $0 < t < t'$ and U a Borel set in $\mathbb{R}^d\backslash\{0\}$. We have $\int_{D_x} N_X(dsdy)d\lambda_x(X) = dsn(dy)$ (e.g. Ikeda-Watanabe [7]). The Lévy-Itô theorem gives the representation of X in D_x:

$$X(t) = x + \int_0^{t+}\int_{|y|\geq 1} yN_X(dsdy) + \int_0^{t+}\int_{0<|y|<1} y\widetilde{N}_X(dsdy). \tag{2.3}$$

$X(s)$ has at most finitely many points s at which the jump $|X(s)-X(s-)|$ exceeds a given positive constant. In particular, $X(s)$ has at most countably many discontinuities. It is bounded on every finite interval $[0,t]$, but not of bounded variation there.

Then we have the following path integral representation for the semigroup $e^{-t[H-m]}$.

Theorem. For $g \in L^2(\mathbb{R}^d)$,

$$(e^{-t[H-m]}g)(x) = \int_{D_x} e^{-S(t,X)}g(X(t))d\lambda_x(X), \tag{2.4a}$$

with

$$\begin{aligned} S(t,X) &= i\int_0^{t+}\int_{|y|\geq 1} A(X(s-)+y/2)\cdot yN_X(dsdy) \\ &\quad + i\int_0^{t+}\int_{0<|y|<1} A(X(s-)+y/2)\cdot y\widetilde{N}_X(dsdy) \\ &\quad + i\int_0^t ds\,\text{p.v.}\int_{0<|y|<1} A(X(s)+y/2)\cdot y\,n(dy) + \int_0^t \Phi(X(s))ds. \end{aligned} \tag{2.4b}$$

Remark. It can be seen that $S(t,X)$ can also be written as

$$\begin{aligned} S(t,X) &= \lim_{R\to\infty}\Big[i\int_0^{t+}\int_{0<|y|<R} A(X(s-)+y/2)\cdot y\widetilde{N}_X(dsdy) \\ &\quad + i\int_0^t ds\,\text{p.v.}\int_{0<|y|<R} A(X(s)+y/2)\cdot y\,n(dy)\Big] + \int_0^t \Phi(X(s))ds. \end{aligned} \tag{2.4b$'$}$$

Thus the representation (2.4a) with $S(t,X)$ given by (2.4') has a close analogy with the Feynman-Kac-Itô formula for the nonrelativistic Schrödinger semigroup (see Simon [12]).

The zero-mass limit (as the mass m tends to zero) as well as the nonrelativistic limit (as the light velicity c tends to ∞) (cf. [2]) may be discussed with this path integral formula.

3. Proof of Theorem.

For $A(x)$ and $\Phi(x)$ in $\mathcal{B}^\infty(\mathbb{R}^d)$ we have proved in [5] for the semigroup $e^{-t[H-m]}$ the path integral representation (2.4a) with

$$\begin{aligned} S(t,X) &= i\int_0^{t+}\int_{|y|\geq 1} A(X(s-)+y/2)\cdot yN_X(dsdy) \\ &+i\int_0^{t+}\int_{0<|y|<1} A(X(s-)+y/2)\cdot y\widetilde{N}_X(dsdy) \\ &+i\int_0^t ds\int_{0<|y|<1} (A(X(s)+y/2)-A(X(s)))\cdot y\,n(dy) \\ &+\int_0^t \Phi(X(s))ds, \end{aligned} \tag{3.1}$$

which is of course equal to (2.4b).

So we assume (1.2) for $A(x)$ and $\Phi(x)$. Choose a sequence $\{A_k\}$ in $\mathcal{B}^\infty(\mathbb{R}^d)$ with $|A_k(x)| \leq |A(x)|$ which is convergent to $A(x)$ in $L^{1+\delta}_{\text{loc}}$ and pointwise a.e., and a sequence $\{\Phi_k\}$ in $\mathcal{B}^\infty(\mathbb{R}^d)$ with $0 \leq \Phi_k(x) \leq \Phi(x)$ which is convergent to $\Phi(x)$ in L^1_{loc} and pointwise a.e., as $k \to \infty$. Then we have

$$(e^{-t[H_k-m]}g)(x) = \int_{D_x} e^{-S_k(t,X)}g(X(t))d\lambda_x(X), \tag{3.2}$$

where $S_k(t,X)$ is the $S(t,X)$ in (3.1) with A_k and Φ_k in place of A and Φ, and H_k is the selfadjoint operator associated with the form $h_k \equiv h_{A_k,\Phi_k}$ in (1.3a,b). We shall show both sides of (3.2) converge to those of (2.4a) as $k \to \infty$.

As far as the left-hand side is concerned, by [6, Lemma 3.6], H_k converges to H in the strong resolvent sense, and by [8, Chap. IX, Theorem 2.16, p.504], $\{\exp[-t(H_k-m)]g\}_{k=1}^\infty$ converges to $\exp[-t(H-m)]g$, uniformly on each bounded t-interval in $[0,\infty)$, in L^2 and, if a subsequence is taken, pointwise a.e.

To see convergence of the right-hand side, we show that $\{\exp[-S_k(t,X)]\}_{k=1}^\infty$ converges for a.e. x and that λ_x-a.e. X, as $k \to \infty$, and its limit can be written as $e^{-S(t,X)}$. Put $S_k(t,X) = \sum_{j=1}^4 S_k^{(j)}(t,X)$ and $S(t,X) = \sum_{j=1}^4 S^{(j)}(t,X)$. We need to show for each j that for λ_x-a.e. X, $\exp[-S_k^{(j)}(t,X)]$ converges as $k \to \infty$.

(i) For $j = 1$ the same proof as in [4] is valid, by taking a Borel set K_1 in $\mathbb{R}^d$ of Lebesgue measure zero such that $|A(x)|$ is finite and $A_k(x) \to A(x)$ for $x \notin K_1$.

(ii) For $n > 0$ let $\sigma_n(X) = \inf\{s > 0;\ |X(s-)| > n\}$. Then for λ_x-a.e. X, $\lim_{n\to\infty}\sigma_n(X) = \infty$. For k, ℓ integers put $A_{k\ell}(x) = A_k(x) - A_\ell(x)$, and $G_2^{k\ell} = \{(X,s,y) \in D_x\times(0,t]\times\{0 < |y| < 1\};\ |A_{k\ell}(X(s-)+y/2)\cdot y| > 1\}$ and for each $X \in D_x$ $G_2^{k\ell}(X) = \{(s,y) \in (0,t]\times\{0 < |y| < 1\};\ |A_{k\ell}(X(s-)+y/2)\cdot y| > 1\}$. The complements of $G_2^{k\ell}$ in $D_x\times(0,t]\times\{0 < |y| < 1\}$ and $G_2^{k\ell}(X)$ in $(0,t]\times\{0 < |y| < 1\}$ are denoted by $\mathcal{C}G_2^{k\ell}$ and $\mathcal{C}G_2^{k\ell}(X)$, respectively.

Then we have, for n fixed and for an arbitrary compact subset K of $\mathbb{R}^d$ with Lebesgue measure $|K|$,

$$\begin{aligned}
&\int_K dx \int_{D_x} |S_k^{(2)}(t\wedge\sigma_n(X),X) - S_\ell^{(2)}(t\wedge\sigma_n(X),X)|d\lambda_x(X)\\
&\le \int_K dx \int_{D_x} \left| \iint_{G_2^{k\ell}(X)} I_{[0,\sigma_n(X)]}(s) A_{k\ell}(X(s-)+y/2)\cdot y\widetilde{N}_X(dsdy)\right| d\lambda_x(X)\\
&+ \int_K dx \int_{D_x} \left| \iint_{CG_2^{k\ell}(X)} I_{[0,\sigma_n(X)]}(s) A_{k\ell}(X(s-)+y/2)\cdot y\widetilde{N}_X(dsdy)\right| d\lambda_x(X)\\
&\equiv \int_K I_1^{k\ell}dx + \int_K I_2^{k\ell}dx.
\end{aligned}$$

For $I_1^{k\ell}$ we have

$$\begin{aligned}
&\int_K I_1^{k\ell}dx \le \int_K dx \int_{D_x} d\lambda_x(X)\\
&\quad\times \iint_{G_2^{k\ell}(X)} I_{[0,\sigma_n(X)]}(s)|A_{k\ell}(X(s-)+y/2)\cdot y|(N_X(dsdy)+dsn(dy))\\
&\le 2\int_K dx \int_{D_x} d\lambda_x(X) \iint_{G_2^{k\ell}(X)} I_{[0,\sigma_n(X)]}(s)|A_{k\ell}(X(s)+y/2)\cdot y|^{1+\delta}dsn(dy)\\
&\le 2\int dx \int_0^t \int_{0<|y|<1} dsn(dy) \int_{|z|\le n} |A_{k\ell}(z+y/2)\cdot y|^{1+\delta}k_0(s,x-z)dz\\
&\le 2n_\delta t \int_{|z|\le n+1} |A_{k\ell}(z)|^{1+\delta}dz,
\end{aligned}$$

with $n_\delta = \int_{0<|y|<1}|y|^{1+\delta}n(dy)$, where in the second inequality we have used that $|A_{k\ell}(X(s)+y/2)\cdot y| > 1$ on $G_2^{k\ell}(X)$. For $I_2^{k\ell}$ we have

$$\begin{aligned}
(I_2^{k\ell})^2 &= \int_{D_x} d\lambda_x(X) \iint_{CG_2^{k\ell}(X)} I_{[0,\sigma_n(X)]}(s)|A_{k\ell}(X(s)+y/2)\cdot y|^2 dsn(dy)\\
&\le \int_{D_x} d\lambda_x(X) \iint_{CG_2^{k\ell}(X)} I_{[0,\sigma_n(X)]}(s)|A_{k\ell}(X(s)+y/2)\cdot y|^{1+\delta} dsn(dy)\\
&= \int_0^t \int_{0<|y|<1} dsn(dy) \int_{|z|\le n} |A_{k\ell}(z+y/2)\cdot y|^{1+\delta}k_0(s,x-z)dz,
\end{aligned}$$

where the inequality is due to that $|A_{k\ell}(X(s)+y/2)\cdot y| \le 1$ on $CG_2^{k\ell}(X)$. Hence

$$\begin{aligned}
&\left(\int_K I_2^{k\ell}dx\right)^2 \le |K| \int (I_2^{k\ell})^2 dx\\
&= |K| \int dx \int_0^t \int_{0<|y|<1} dsn(dy) \int_{|z|\le n} |A_{k\ell}(z+y/2)\cdot y|^{1+\delta}k_0(s,x-z)dz\\
&\le |K| n_\delta t \int_{|z|\le n+1} |A_{k\ell}(z)|^{1+\delta}dz.
\end{aligned}$$

Thus we have

$$\begin{aligned}&\int_K dx \int_{D_x} |S_k^{(2)}(t\wedge\sigma_n(X),X) - S_\ell^{(2)}(t\wedge\sigma_n(X),X)|d\lambda_x(X)\\ &\le \int_K I_1^{k\ell}dx + \int_K I_2^{k\ell}dx\\ &\le 2n_\delta t\int_{|z|\le n+1}|A_{k\ell}(z)|^{1+\delta}dz + (|K|n_\delta t)^{1/2}\Big(\int_{|z|\le n+1}|A_{k\ell}(z)|^{1+\delta}dz\Big)^{1/2},\end{aligned}$$

which tends to zero as $k,\ell\to\infty$. Since K is arbitrary, it can be seen, by passing to a subsequence, for a.e. x, that as $k\to\infty$, $\{S_k^{(2)}(t\wedge\sigma_n(X),X)\}_{k=1}^\infty$ converges to a limit in L^1 with respect to λ_x and, by passing to a subsequence, for λ_x-a.e. X. This limit is what is to be denoted by $S^{(2)}(t\wedge\sigma_n(X),X)$. Further, since $\lim_{n\to\infty}\sigma_n(X)=\infty$ for λ_x-a.e. X, we see that $\{S_k^{(2)}(t,X)\}_{k=1}^\infty$ converges to a limit for λ_x-a.e. X, which is to be denoted by $S^{(2)}(t,X)$, and hence $\exp[-S_k^{(2)}(t,X)]\to\exp[-S^{(2)}(t,X)]$.

(iii) We can show with the theory of singular integrals that $a(x)\equiv$ p.v. $\int_{0<|y|<1} A(x+y/2)\cdot y\,n(dy)$ exists pointwise a.e. in x, while $a_k(x)\equiv\int_{0<|y|<1}(A_k(x+y/2)-A_k(x))\cdot y\,n(dy) =$ p.v. $\int_{0<|y|<1}A_k(x+y/2)\cdot y\,n(dy)$ exists for every x, and that as $k\to\infty$, $a_k(x)$ converges to $a(x)$ in $L^{1+\delta}_{\rm loc}$. With the same $\sigma_n(X)$ as in (ii), we have, for n fixed,

$$\begin{aligned}&\int_{\mathbb{R}^d} dx \int_{D_x} |S_k^{(3)}(t\wedge\sigma_n(X),X) - S_\ell^{(3)}(t\wedge\sigma_n(X),X)|^{1+\delta}d\lambda_x(X)\\ &=\int dx\int d\lambda_x(X)\Big|\int_0^t I_{[0,\sigma_n(X)]}(s)[a_k(X(s))-a_\ell(X(s))]ds\Big|^{1+\delta}\\ &\le\int dx\int d\lambda_x(X)\Big(\int_0^t ds\Big)^\delta\int_0^t I_{[0,\sigma_n(X)]}(s)|a_k(X(s))-a_\ell(X(s))|^{1+\delta}ds\\ &\le\ t^\delta\int dx\int_0^t ds\int_{|z|\le n}|(a_k(z)-a_\ell(z))|^{1+\delta}k_0(s,x-z)dz\\ &\le\ t^{1+\delta}\int_{|x|\le n}|(a_k(x)-a_\ell(x)|^{1+\delta}dx\to 0,\qquad k,\ell\to\infty,\end{aligned}$$

where in the first inequality we have used the Hölder inequality. Similarly to (ii), we can see, by passing to a subsequence, for a.e. x, that as $k\to\infty$, $\{S_k^{(3)}(t\wedge\sigma_n(X),X)\}_{k=1}^\infty$ converges to a limit in $L^{1+\delta}$ with respect to λ_x and, by passing to a subsequence, for λ_x-a.e. X. This limit is what is to be denoted by $S^{(3)}(t\wedge\sigma_n(X),X)$. Further, since $\lim_{n\to\infty}\sigma_n(X)=\infty$ for λ_x-a.e. X, we see that $\{S_k^{(3)}(t,X)\}_{k=1}^\infty$ converges to a limit for λ_x-a.e. X, which is to be denoted by $S^{(3)}(t,X)$, and hence $\exp[-S_k^{(3)}(t,X)]\to\exp[-S^{(3)}(t,X)]$.

(iv) The proof of the $j=4$ case is the same as in [4].

Acknowledgements. It is a great pleasure to thank Professors M. Demuth and B.-W. Schulze for the kind invitation to this Conference.

References

[1] H.C. Cycon, R.G. Froese, W. Kirsch and B. Simon, Schrödinger Operators, Springer, Berlin – Heidelberg – New York, 1987.

[2] T. Ichinose, The nonrelativistic limit problem for a relativistic spinless particle in an electromagnetic field, J. Functional Analysis, **73**, 233–257(1987).

[3] T. Ichinose, Essential selfadjointness of the Weyl quantized relativistic Hamiltonian, Ann. Inst. H. Poincaré, Phys. Théor. **51**, 265–298(1989).

[4] T. Ichinose, On the Weyl quantized relativistic Hamiltonian, Operator Theory: Advances and Applications, **57**, 127–137(1992).

[5] T. Ichinose and H. Tamura, Imaginary-time path integral for a relativistic spinless particle in an electromagnetic field, Commun. Math. Phys. **105**, 239–257(1986).

[6] T. Ichinose and T. Tsuchida, On essential self-adjointness of the Weyl quantized relativistic Hamiltonian, Forum Math. **5**, 539–559(1993).

[7] N. Ikeda and S. Watanabe, Stochastic Differential Equations and Diffusion Processes, North-Holland / Kodansha, Amsterdam, Tokyo, 1981.

[8] T. Kato, Perturbation Theory for Linear Operators, 2nd ed., Springer, Berlin – Heidelberg – New York, 1976.

[9] L.D. Landau and E.M. Lifschitz, Course of Theoretical Physics Vol.2, The Classical Theory of Fields, 4th revised English ed., Pergamon Press, Oxford, 1975.

[10] H. Leinfelder and C.G. Simader, Schrödinger operators with singular magnetic vector potentials, Math. Z. **176**, 1–19(1981).

[11] M. Nagase and T. Umeda, Weyl quantized Hamiltonians of relativistic spinless particles in magnetic fields, J. Functional Analysis **92**, 136–154(1990).

[12] B. Simon, Functional Integration and Quantum Mechanics, Academic Press, New York, 1979.

Takashi Ichinose
Department of Mathematics,
Faculty of Science, Kanazawa University
Kanazawa 920-11, Japan

Operator Theory:
Advances and Applications, Vol. 78

Semiclassical Spectral Asymptotics and Multiparticle Quantum Theory

V. Ivrii[1]

Abstract

In this talk I would like to present the very first results of my intervention to the *Multiparticle Quantum Theory* (MQT). A lot of nice results were obtained by mathematical physicists but the *Theory of Semiclassical Spectral Asymptotics* (TSSA) was applied only in a few papers and I believe that the systematic application will provide some progress. This is my long term project. However, it is vital to combine methods of TSSA and MQT. The first talk with this subject was given in École Polytechnique $2\frac{1}{2}$ months ago but I managed to advance after and this address is improved, corrected and advanced version of that talk.

1. General.

Let us consider the following operator (quantum Hamiltonian)

$$H = \sum_{1\le j\le N} (\Delta_{x_j} + V(x_j)) + \sum_{1\le j<k\le N} |x_j - x_k|^{-1} \tag{1}$$

describing N same type particles in the external field with the potential V and repulsing one another according the Coulomb law. Here $x_j \in \mathbb{R}^d$ and $(x_1,\dots,x_N) \in \mathbb{R}^{Nd}$, Δ is a positive Laplacian in $\mathbb{R}^d$, function $V(x)$ is assumed to be real-valued. Mass is equal to $\frac{1}{2}$ and Planck constant and a charge are equal to 1 here. The crucial question is the quantum statistics. We assume that the particles (electrons) are fermions. That means that the Hamiltonian should be considered on the Fock space $\mathcal{H} = \bigwedge_{1\le j\le N} L^2(\mathbb{R}^d)$ of the functions antisymmetric with respect to all variables $x_1,\dots,z_N$ while for bosons one should consider a space symmetric functions or (what is the same in our problem) on the space $\otimes_{1\le j\le N} L^2(\mathbb{R}^d)$ of all functions. We neglect the fact that one particle is described by the wave function $\phi \in L^2(\mathbb{R}^d,\mathbb{C}^q)$ rather than by the wave function $\phi \in L^2(\mathbb{R}^d,\mathbb{C})$. One can easily adjust our arguments to the case $q > 1$. Let us assume that

$$\text{Operator } H \text{ is self-adjoint on } \mathcal{H}. \tag{2}$$

We will never discuss this assumption. We are interested in the *ground state energy* $E = E(N)$ of our system i.e. in the lowest eigenvalue of the operator

[1]Work was partially supported by NSERC grant OGP0138277.

H on $\mathcal{H}$. The first approximation is the Hartree-Fock (or Thomas-Fermi) theory. Namely, let us introduce the space density of the particle with the state $\psi \in \mathcal{H}$:

$$\rho(x) = N \int |\psi(x,x_2,\dots,x_N)|^2 dx_2 \cdots dx_N \tag{3}$$

(antisymmetricity of ψ implies that it doesn't matter what variable x_j is replaced by x). Let us write the Hamiltonian, describing the corresponding "quantum liquid":

$$\mathcal{E}(\rho) = \kappa \int \rho^{1+\frac{2}{d}} dx + \int V\rho dx + \frac{1}{2} \iint |x-y|^{-1} \rho(x)\rho(y) dxdy \tag{4}$$

where the numerical value of κ is $4\pi^2 \omega_d^{-\frac{2}{d}} d(2+d)^{-1}$, ω_d is the volume of the unit ball in $\mathbb{R}^d$. The classical sense of the second and the third terms is clear and why the density of the kinetic energy is $\kappa\rho^{1+\frac{2}{d}}$ in the semiclassical approximation will be explained later. So, the problem is to minimize this functional under restrictions

$$\rho \geq 0, \qquad \int \rho dx = N. \tag{5}$$

The solution if exists is unique because functional $\mathcal{E}(\rho)$ is strictly convex)[1]. The existence and the property of this solution denoted further by ρ_{TF} is known in the series of physically important cases.

To justify the heuristic formula $E \sim \mathcal{E}_{TF} = \mathcal{E}(\rho_{TF})$ and to find the error estimate let us deduce the lower and upper estimates of E. For lower estimate we apply the *electrostatic inequality* due to E.H. Lieb (later we replace it by more precise bound due to Bach-Graf-Solovej)[2]):

$$\begin{aligned} \sum_{1\leq j<k\leq N} \int |x_j - x_k|^{-1} |\psi(x_1,\dots,x_N)|^2 dx_1 \cdots dx_N \geq \\ \frac{1}{2} \iint |x-y|^{-1} \rho(x)\rho(y) dxdy - C \int \rho^{1+\frac{1}{d}} dx. \end{aligned} \tag{6}$$

This inequality holds for all (not necessarily antisymmetric) functions ψ with $\|\psi\|_{L^2(\mathbb{R}^{dN})} = 1$. Actually it was proven for $d = 3$ but generalization for $d \geq 2$ is rather trivial. So,

$$\begin{aligned} \langle H\psi,\psi\rangle \geq \sum_{1\leq j\leq N} \langle A_{x_j}\psi,\psi\rangle + \\ \frac{1}{2} \iint |x-y|^{-1} (\rho - \rho_{TF})(x)(\rho - \rho_{TF})(y) dxdy - \\ \frac{1}{2} \iint |x-y|^{-1} \rho_{TF}(x)\rho_{TF}(y) dxdy - C \int \rho^{1+\frac{1}{d}} dx \end{aligned} \tag{7}$$

[1]Operator with the Schwartz kernel $|x-y|^{-1}$ is positive due to Fefferman-La Llave decomposition for $d \geq 2$.

[2]I hate to confess that in [I2] I failed to mention V. Bach who did the most pioneering work.

where $\langle , \rangle$ means the inner product in $\mathcal{H}$ and A is one-particle Schrödinger operator with the Thomas-Fermi potential W

$$A = \Delta + W, \qquad W = V + |x|^{-1} * \rho_{TF}. \tag{8}$$

The physical sense of the second term in W is transparent. Skipping the positive second term in the right-hand expression of (7) and believing that the third term is not very important for the ground state function ψ we see that we need to estimate from below the first term. Here assumption that ψ is antisymmetric is crucial. Namely, for general ψ the best possible estimate is $N\lambda_1$ where λ_1 is the lowest eigenvalue of A (we always assume that there is sufficiently many eigenvalues under the bottom of the essential spectrum of A) and we cannot go further. However, for antisymmetric ψ we get instead

$$\sum_{1\le j\le N} \lambda_j = \mathbf{N}_1(A-\bar\lambda) + \bar\lambda N \tag{9}$$

where $\mathbf{N}(B)$, $\mathbf{N}_1(B)$ are the number and the sum of all the negative eigenvalues of operator B (such that $\operatorname{Spec}_{\mathrm{ess}}(B) \subset \bar{\mathbb{R}}^+$) and $\bar\lambda = \lambda_N < 0$. Applying the semiclassical approximation (need to be justified!) one gets

$$\mathbf{N}_1(A-\bar\lambda) \sim -(2\pi)^{-d}\omega_d d(d+2)^{-1}\int (W-\bar\lambda)_-^{\frac{d+2}{2}}\,dx$$

and therefore the below estimate for the ground state energy is

$$\begin{gathered} E \ge -(2\pi)^{-d}\omega_d d(d+2)^{-1}\int (W-\bar\lambda)_-^{\frac{d+2}{2}}\,dx + \bar\lambda N - \\ \frac{1}{2}\iint |x-y|^{-1}\rho_{TF}(x)\rho_{TF}(y)\,dxdy - \text{error}. \end{gathered} \tag{10}$$

Furthermore, applying semiclassical approximation for the number $\mathbf{N}(A-\bar\lambda)$ of eigenvalues below $\bar\lambda$ (and this number should be approximately N) one gets an equality

$$N = (2\pi)^{-d}\omega_d\int (W-\bar\lambda)_-^{\frac{d}{2}}\,dx + \text{error}. \tag{11}$$

On the other hand, let us consider the Euler-Lagrange equation for $\rho = \rho_{TF}$:

$$(1+\frac{2}{d})\kappa\rho^{\frac{2}{d}} + W = \nu \quad (\rho > 0), \qquad W = V + |x|^{-1} * \rho \tag{12}$$

with the Lagrange factor ν. Expressing ρ and integrating we get

$$N = (2\pi)^{-d}\omega_d\int (W-\nu)_-^{\frac{d}{2}}\,dx. \tag{13}$$

Comparing (11) and (13) we get that with some error $\bar\lambda = \nu$. Substituting to the first term in (10) $\bar\lambda = \nu$ and $\nu - W = \kappa(1+\frac{d}{2})\rho^{\frac{2}{d}}$ we get the lower estimate

$E \geq \mathcal{E}_{TF}$-error. This construction is well-known in MQT and even more advanced below estimates don't use TSSA arguments.

To get the upper estimate one takes a test function $\psi(x_1,\dots,x_n)$ which is an antisymmetrization with respect to $(x_1,\dots,x_n)$ of the product $\phi_1(x_1)\cdots\phi_N(x_N)$ where $\phi_1,\dots,\phi_N$ are orthonormal eigenfunctions of A corresponding to eigenvalues $\lambda_1,\dots,\lambda_N$. Namely this function minimized the expression

$$\sum_{1\leq j\leq N} \langle A_{x_j}\psi,\psi\rangle$$

in $\mathcal{H}$. One can write $\psi = \frac{1}{N!}\det\bigl(\phi_i(x_j)\bigr)_{i,j=1,\dots,N}$ and it is called *Slater determinant*. Obviously, $\|\psi\| = 1$ and $\rho(x) = e_N(x,x)$ where $e_N(x,y) = \sum_j \phi_j(x)\bar{\phi}_j(y)$ is the Schwartz kernel of the projector to the subspace spanned on $\{\phi_j\}_{1\leq j\leq N}$. Pretty easy calculations show that

$$\langle H\psi,\psi\rangle = \sum_{1\leq j\leq N}\lambda_j + \frac{1}{2}\iint |x-y|^{-1}\bigl(\rho-\rho_{TF}\bigr)(x)\bigl(\rho-\rho_{TF}\bigr)(y)dxdy - \\ \frac{1}{2}\iint |x-y|^{-1}\rho_{TF}(x)\rho_{TF}(y)dxdy - \frac{1}{2}\iint |x-y|^{-1}|e_N(x,y)|^2dxdy.$$

Let us replace in the first term N by $\mathbf{N}(A-\bar{\lambda})$. Then we get $\mathbf{N}_1(A-\bar{\lambda})+\bar{\lambda}\mathbf{N}(A-\bar{\lambda})$ with an error $\bar{\lambda}\bigl(N-\mathbf{N}(A-\bar{\lambda})\bigr)$ linked with the possible degeneration of the eigenvalue $\bar{\lambda}=\lambda_N$. Then we get an upper estimate

$$\begin{aligned}&E \leq \mathbf{N}_1(A-\bar{\lambda}) + \bar{\lambda}\mathbf{N}(A-\bar{\lambda}) + \\ &\frac{1}{2}\iint |x-y|^{-1}\bigl(\rho-\rho_{TF}\bigr)(x)\bigl(\rho-\rho_{TF}\bigr)(y)dxdy - \\ &\frac{1}{2}\iint |x-y|^{-1}\rho_{TF}(x)\rho_{TF}(y)dxdy - \frac{1}{2}\iint |x-y|^{-1}|e_N(x,y)|^2dxdy.\end{aligned} \tag{14}$$

Treating the third and fifth terms in the right-hand expression as errors and applying the semiclassical approximation to the first one we get the upper estimate $E \leq \mathcal{E}_{TF} + \text{error}$.

So, there are two types of errors. The first is due to replacement of E by

$$\mathbf{N}_1(A-\bar{\lambda}) + \bar{\lambda}\mathbf{N}(A-\bar{\lambda}) - \frac{1}{2}\iint |x-y|^{-1}\rho_{TF}(x)\rho_{TF}(y)dxdy$$

and the second is an error of the semiclassical approximation of two first terms including may be the replacement $\bar{\lambda}$ by ν. Actually, negligence of the third term in (14) is the semiclassical approximation either.

2. Large molecule. I.

The most known example is the case of n large nuclei with the charges $Z_1,\dots,Z_n$ located at fixed points $y_1,\dots,y_n$. Then

$$V(x) = -\sum_{1\le j\le n} \frac{Z_j}{|x-y_j|} \tag{15}$$

in (1). Let us assume that

$$Z = Z_1 + \ldots + Z_N = N, \qquad Z_j \ge \epsilon N \quad \forall j(\epsilon > 0). \tag{16}$$

The first assumptions means that the system is neutral and it yields that $\nu = 0$. Let us assume that $d = 3$. The properties of ρ_{TF} are well-known. In particular, $\rho_{TF} = N^2\bar\rho(xN^{\frac13})$ where $\bar\rho$ doesn't depend on N explicitly and it is analytic (excluding points $\bar y_j = y_jN^{\frac13}$), positive and vanishing at infinity as $\bigl(\min_j|x-y_j|\bigr)^{-4}$. Then

$$\mathcal{E}(\rho_{TF}) = \kappa_0 N^{\frac73} \tag{17}$$

with constant $\kappa_0 = \bar{\mathcal{E}} < 0$ (depending on $\bar y_j$). It can be proven by the methods of mathematical physics solely that

$$E \ge \mathbf{N}_1(A) - CN^{\frac53}. \tag{18}$$

Let us consider operator A. Rescaling as before we get in the new coordinates

$$A = N^{\frac43}\bar A_h, \qquad \bar A_h = h^2\Delta + \bar W, \qquad h = N^{-\frac13} \tag{19}$$

with potential $\bar W$ which doesn't depend on N explicitly. Then $\mathbf{N}_1(A) = N^{\frac43}\mathbf{N}_1(\bar A_h)$. Weylian semiclassical approximation for $\mathbf{N}_1(\bar A_h)$ is $\kappa_0 h^{-3}$ and it yields to approximation $\kappa_0 N^{\frac73}$ for E. The error is $O(h^{-2})$ and it yields to the error $O(N^2)$ in the final answer. If potential $\bar W$ was smooth the error would be $O(h^{-1})$ for the rescaled one-particle problem and the final error would be $O(N^{\frac53})$. However, the Coulomb singularity of $\bar W$ at $\bar y_j$ produces additional term $\kappa_1 h^{-2}$ with $\kappa_1 = \sum_j z_j^2$, $z_j = Z_jN^{-1}$ in the semiclassical approximation for $\mathbf{N}_1(\bar A_h)$ and it yields the *Scott correction term*

$$\sum_j \mathrm{const}\, Z_j^2 \tag{20}$$

in the final answer. The error $O(h^{-1})$ (and even $O(h^{\delta-1})$) was recovered in one atom case by Fefferman and Seco (see [2, 3] for references to earlier papers) and this and all the previous proofs heavily use the spherical symmetry of the problem. Actually, in all these papers variables were separated and the ODE was carefully investigated and the semiclassical approximation for the final answer was obtained by the WKB method. This proof is very sophisticated and very long and it leads to the record-beating estimate but it works only for spherically symmetric system.

So, the final error was $O(N^{\frac{5}{3}})$. On the other hand, by methods of the semiclassical spectral asymptotics for PDE Ivrii and Sigal [8] got an error estimate $O(h^{-\frac{4}{3}})$ for semiclassical approximation (with Scott correction term) for $\mathbf{N}_1(\bar{A}_h)$ and it leads to the final error $O(N^{\frac{16}{9}})$. Later the error was improved up to $O(h^{-1})$. So, the final error is $O(N^{\frac{5}{3}})$ again. I should note that both results were obtained under (physically reasonable) assumption $|y_j - y_k| \ge \epsilon N^{-\frac{1}{3}} \quad \forall j \ne k$ with an arbitrarily small constant $\epsilon > 0$. It is very important that $\bar{W} - z_j|x+\bar{y}_j|^{-1} = O(1)$ as $x \to \bar{y}_j$.

However, one can get the remainder estimate $O(N^2)$ without these assumptions. Then applying Teller's lemma)[3] one gets immediately that the the excess (ground state energy for molecule with the added potential energy of repulsion between nuclei minus the sum of ground state energies for atoms) is larger than

$$\epsilon_1 N^{\frac{7}{3}}(N^{\frac{1}{3}}\ell + 1)^{-7} - CN^2$$

where ℓ is the minimal distance between nuclei. Here the first term is the lower bound for excess in the Thomas-Fermi approximation and Teller lemma deals with it and second is the remainder estimate. Therefore, assuming that nuclei occupy positions minimizing the ground state energy one gets estimate $\ell \ge N^{-\frac{2}{7}}$. So, our assumption is justified in this model. Applying now remainder estimate $O(N^{\frac{5}{3}})$ one improves this estimate to $\ell \ge N^{-\frac{5}{21}}$)[4]. The last thing I want to explain is why one can replace $\bar{\lambda}$ by $\nu = 0$. In the lower it is trivial: $\sum_{\lambda_j \le \bar{\lambda}} \lambda_j$ doesn't increase if one replaces arbitrary $\bar{\lambda}$ by 0. In the upper bound one can apply all the above arguments with N replaced by $\bar{N} = \max_{\lambda_j \le 0} j$ and then one can replace $\bar{\lambda} = \lambda_{\bar{N}}$ by 0. Then one gets the proper upper bound for the ground state energy $E(Z,\bar{N})$ of the same system but with $\bar{N}$ electrons. However, according to Zhislin's theorem $E(Z,\bar{N}) \ge E(Z,N)$ if $Z = N \ge \bar{N}$. I recall that Z is the total charge. Note that these arguments belong to MQT. On the other hand, if $\bar{N} > N$ (i.e. $\lambda_{N+1} \le 0$) one needs to estimate the sum of the eigenvalues laying between λ_N and 0. Due to semiclassical arguments their number is $O(N^{\frac{2}{3}})$ and $O(N)$ (if $\lambda_N \le 0$) and their contribution is $O(N^{\frac{5}{3}})$. Moreover, under standard condition to the Hamiltonian trajectories (see below) one can replace "O" by "o" everywhere in the previous sentence.

3. Large molecule. II.

In order to improve the remainder estimate $O(N^{\frac{5}{3}})$ in the final answer one needs to improve the remainder estimate $O(h^{-1})$ in the semiclassical asymptotics for $\mathbf{N}_1(\bar{A}_h)$ and the remainder estimate in the original reduction.

Let us consider first the semiclassical asymptotics for $\mathbf{N}_1(\bar{A}_h)$. It is standard that the remainder estimate $O(h^{-1})$ cannot be improved without global assumption: one needs to consider closed Hamiltonian trajectories at the energy level 0 for

[3] In the Thomas-Fermi approximation excess $\asymp N^{\frac{7}{3}}(N^{\frac{1}{3}}\ell + 1)^{-7}$

[4] Actually, estimate $\ell \ge N^{-\frac{1}{3}}$ can be recovered from MQT arguments solely

the corresponding classical Hamiltonian $a(x,\xi) = |\xi|^2 + W(x)$. It is known in the theory of the semiclassical spectral asymptotics and the same conclusion is made in [2, 3] on the ground of their calculations. So, let us consider $\Sigma = \{(x,\xi), a(x,\xi) = 0\}$ with the natural measure $\mu = dxd\xi : da$ and Hamiltonian flow $\Phi_a(t)$ preserving it. The first very basic conclusion is the following: *if $\mu(P_a) = 0$ where P_a denotes the set of points of Σ, periodic with respect to Φ_a, then*

$$\mathbf{N}_1(\bar{A}_h) = \kappa_0 h^{-3} + \kappa_1 h^{-2} + \bigl(\kappa_2 + o(1)\bigr)h^{-1} \qquad \text{as } h \to 0 \tag{21}$$

where κ_0,κ_1 are the leading and Scott coefficients and

$$\kappa_2 = \kappa_1 \int W_-^{\frac{1}{2}} \Delta W dx = -\kappa_2 \int \rho_{TF}^{\frac{4}{3}} dx \tag{22}$$

is *Schwinger correction*; κ_1,κ_2 are known positive numbers. One can easily justify the basic assumption $\mu(P_a) = 0$ for the atom using the known properties of W. Moreover, one can prove this for molecule as well assuming that $|y_j - y_k|N^{\frac{1}{3}} \to \infty \quad \forall j \neq k$ (physically reasonable assumption). In order to improve this remainder estimate one should assume some properties of the Hamiltonian flow. Namely, to get $O\bigl(h^{\delta-1}\bigr)$ with $\delta > 0$ one should assume no more than power growth of the Jacobian matrix of $\Phi(t)$ and some estimates for the measure of "almost periodic" points. However, for atom it is completely integrable and one can check this properties. For molecule with

$$|y_j - y_k| \geq N^{-\frac{1}{3}+\delta'} \quad \forall j \neq k \tag{23}$$

the necessary properties are due to the properties of atoms (one should take $|t| \leq h^{-\delta''}$ with arbitrarily small $\delta'' > 0$). The assumption to the nuclear distances is physically reasonable. So, in the physically interesting case this part of the job is done.

For atoms the arguments of Fefferman-Seco go further: they got the same remainder estimate with the small but *specified* exponent $\delta > 0$. This is the advantage of the explicit calculation over general methods of PDE in the (pretty rare) case when these calculations are possible. Moreover, using the methods of the semiclassical spectral asymptotics one can prove that under assumption (23)

$$\mathbf{N}_1(\bar{A}_h) = \sum_{1 \leq j \leq n} \mathbf{N}_1(\bar{A}_{jh}) + O\bigl(h^{-1+\delta}\bigr)$$

where $\bar{A}_{jh}$ mean reduced "one-atomic" operators and $\delta = 12\delta' - 1$ for $\delta' \leq \frac{1}{2}$. Therefore, applying Fefferman-Seco result (asymptotics of $\mathbf{N}_1$) for atom one extends it with the same (pretty small) exponent δ for molecules under this assumption. In the "optimal positions of nuclei model" we already got $\delta' = \frac{2}{21}$ and it is sufficient for this extension. However, I don't know how to extent Fefferman-Seco results for term (26) (see below). But it is not necessary, if the lower bound for ℓ (in the original model) is all we want to know. One can now improve it further.

4. Large molecule. III. Correlations.

Only now I pass to the main topic of my talk. Now we need to improve remainder estimate in the reduction. First of all, Bach and Graf-Solovej improved the electrostatic inequality and applying it to the problem in question improved the lower estimate:

$$E \ge \mathbf{N}_1(\bar{A}) - \frac{1}{2}\iint |x-y|^{-1}\rho(x)\rho(y)dxdy - \kappa_3\int \rho_{TF}^{\frac{4}{3}}dx - O(N^{\frac{5}{3}-\delta}) \qquad (24)$$

with some (known) constant $\kappa_3 > 0$. One can rewrite the third *(Dirac correction)* term in the form $\kappa_3 N^{\frac{5}{3}}$. The proof is pretty complicated but completely in frames of MQT (no PDE!). To get an upper estimate we try again the Slater determinant test function. Then we need to consider asymptotics of

$$-\frac{1}{2}\int |e(x,y)|^2|x-y|^{-1}dxdy \qquad (25)$$

and

$$\frac{1}{2}\int (\rho-\rho_{TF})(x)(\rho-\rho_{TF})(y)|x-y|^{-1}dxdy. \qquad (26)$$

Let us start from the first one. We call it *correlation.*

Theorem 1. In our conditions

$$\int |e(x,y)|^2|x-y|^{-1}dxdy = 2\kappa_3\int \bar{\rho}^{\frac{4}{3}}dxN^{\frac{5}{3}} - O(N^{\frac{4}{3}}). \qquad (27)$$

This theorem is a corollary of the following local version and some functional analytical arguments (nevertheless not arguments of mathematical physics):

Theorem 2. Let us consider Schrödinger operator $A_h = h^2\Delta + W$. Let A_h be self-adjoint in $L^2(X)$ where X is a domain in $\mathbb{R}^d, d\ge 2$ containing the unit ball $B(0,1)$ and let

$$|D^\alpha W(x)| \le c \qquad \forall\alpha : |\alpha| \le K. \qquad (28)$$

Let $\zeta(x,y)$ be supported in $B(0,\frac{1}{2})\times B(0,\frac{1}{2})$ and

$$|D^{\alpha,\beta}_{x+y,x-y}\zeta| \le c|x-y|^{-p-|\beta|} \qquad \forall\alpha,\beta : |\alpha|+|\beta| \le K \qquad (29)$$

with $0<p<d$. Here $K=K(d)$. Then

$$\mathcal{R} = \Big|\int |e(x,y)|^2\zeta(x,y)dxdy - \mathrm{Weyl}\Big| \le Ch^{1-d-p} \qquad (30)$$

with

$$\mathrm{Weyl} = (2\pi)^{-2d}\int dx \iiint_{|\xi|^2+W(x)\le 0, |\eta|^2+W(x)\le 0} e^{ih^{-1}\langle z,\xi-\eta\rangle}\zeta(x,z)dzd\xi d\eta$$

where $e(x,y)$ is the Schwartz kernel of the spectral projector $\Pi(0)$. Here $\mathrm{Weyl} = O(h^{-d-p})$ and moreover $\mathrm{Weyl} \asymp h^{-d-p}$ under natural conditions.

The last theorem is due to more general theorem:

Theorem 3. Let

$$A = \sum_{|\alpha|\le m} a_\alpha(x)(hd)^\alpha$$

be a self-adjoint)[5] in $L^2(X,\mathbb{H})$ operator ($\mathbb{H} = \mathbb{C}^D$), $X \supset B(0,1)$ and

$$|D^\beta a_\alpha| \le c \qquad \forall \alpha,\beta : |\alpha| \le m, |\beta| \le K \tag{31}$$

in $B(0,1)$ with $K = K(d,p)$. Let $Q'(x,hD), Q''(x,hD)$ be h-pseudo-differential operators with symbols supported in $B(0,\frac{1}{2}) \times B(0,c)$ and satisfying (31) (with derivatives on x,ξ) and ψ', ψ'' be functions supported in $B(0,\frac{1}{2})$ and satisfying (31). Then

(i) For any $x \in B(0,1)$, $\gamma \in (h,\frac{1}{2}])$ estimate

$$\int_{|x-y|\ge\gamma} |Q'(x,hD_x)\psi'(x)e(x,y,0)|^2 dy \le Ch^{1-d}\gamma^{-1} \tag{32}$$

holds.

(ii) Let $A(x,\xi)$ be microhyperbolic at the energy level 0 in some ξ-directions i.e. for any $(x,\xi) \in \operatorname{supp} Q'$ let there exist $\ell \in \mathbb{S}^d$ such that

$$\langle (l \cdot \partial_\xi A(x,\xi))v,v\rangle \ge \epsilon|v|^2 - C|A(x,\xi)^2|^2 \qquad \forall v \in \mathbb{H}. \tag{33}$$

Then estimate

$$\mathcal{R} \le Ch^{1-d-p} + Ch^{\frac{1}{2}-d-\frac{p}{2}-\delta} \tag{34}$$

holds with arbitrarily small $\delta > 0$.

(iii) Moreover, under conditions (33) and $\mathbb{H} = \mathbb{C}$ estimate (30) holds for $d > p > 0$. Moreover, in this case for ζ satisfying (31) estimate (30) holds with $p = 0$.

(iv) Finally, if

$$\langle a(x,\xi)v,v\rangle \ge \epsilon|\xi|^m|v|^2 \qquad \forall x \in B(0,1), \xi \in \mathbb{R}^d, v \in \mathbb{H} \tag{35}$$

then all the above estimates hold with $Q' = I$.

Remark 4.

(i) Condition (33) for the Schrödinger operator means that $|W| \ge \epsilon$. Then applying rescaling-partition method we replace it first by more pleasant condition $|W| + |\nabla W| \ge \epsilon$ and then we get rid of the last condition for $d \ge 2$.

---

[5] With some boundary conditions; we don't care how A is defined out of $B(0,1)$, where $B(x,r)$ means a ball of radius r with a center at x.

(ii) To improve estimate (30) one needs condition to Hamiltonian trajectories (and "offenders" are periodic trajectories and not loops as I wrongly indicated in [7]. It is much more pleasant to deal with periodic trajectories). However, the remainder estimate better than $O(h^{1-d})$ requires the more complicated principal part of the asymptotics taking in account Hamiltonian rays from x to y and $\zeta(x,y)$ with disjoint x,y instead of germs at point x.

(iii) Returning to original MQT problem I should notice that the principal part of the correlation energy asymptotics provides term $\kappa N^{\frac{5}{3}}$ in the final answer and remainder estimate (30) produces contribution $O(N^{\frac{4}{3}})$ to the final remainder estimate. This is much smaller than contribution from other terms.

Conjecture Estimate (30) holds in frames of theorem 3(i).

I have proved it recently but I need to check my arguments.

5. Large molecule. IV.

Let us consider term (26). I recall that $\rho(x) = e(x,x)$. The local theorem is now

Theorem 5.

(i) In frames of theorem 2

$$\mathcal{R}_2 = \left|\rho(x) - h^{-d}\rho_0(x)\right| \le Ch^{1-d} \qquad \forall x \in B(0,\frac{1}{2}) \tag{36}$$

provided either $d \ge 3$ or

$$W(x) \le -\epsilon \qquad \forall x \in B(0,1); \tag{37}$$

here and below $\rho_0(x) = (2\pi)^{-d}\omega_d W_-^{\frac{d}{2}}$.

(ii) Moreover, for $d = 2$

$$\mathcal{R}_3 = \iint (\rho - h^{-d}\rho_0)(x)(\rho - h^{-d}\rho_0)(y)|x-y|^{-p} \le Ch^{2-2d} \tag{38}$$

$$\mathcal{R}_4 = \int \left|(\rho - h^{-d}\rho_0)(x)\right| dx \le Ch^{1-d} \tag{39}$$

without condition (31); I recall $0 \le p < 2$.

Estimate (36) is rather known under condition (37). Other estimates follow from it and applying rescaling method one gets rid of (37). Applying this local theorem one can recover only estimate $O(N^{\frac{5}{3}})$ for (26). In order to improve it one should consider Hamiltonian trajectories. However, there is a difference between (36), (39) and (38): there is no spatial mollification in (36), there is rather averaging in (39) and only (38) contains some mollification. Due to this in (36), (39) not only closed trajectories seem to cause troubles but loops as well.

Definition 6. Point $(x,\xi) \in \Sigma$ is a loop point if $\Phi_t(x,\xi) = (x,\xi')$ for some $t \neq 0, \xi'$. Note, that $\xi' = \xi$ means exactly that (x,ξ) is periodic.

One can introduce measure $\mu_x = d\mu : d\xi$ on Σ_x, the layer of Σ over x and one can prove easily that left-hand expression in (31) is $o(h^{1-d})$ provided μ_x-measures of the loop points over x is 0 and we can control long-term trajectories. Moreover, if μ-measure of all the loop trajectories is 0 then $\mathcal{R} = o(h^{2-d})$ and $\mathcal{R}_1 = o(h^{1-d})$. *These results with the possible adjustments for well-controlled Hamiltonian flow permit to recover estimate $o(N^{\frac{5}{3}})$ and even $O(N^{\frac{5}{3}-\delta})$ for (26) under condition (23).*

I would like to notice that the μ-measure of the set of loop points which are not periodic is 0 and therefore these points are of little importance in $\mathcal{R}, \mathcal{R}_1$. Really, due to condition $\xi' \neq \xi$ the loop points form a set of the (natural) measure 0 along any non-periodic trajectory. Due to this one can improve a bit remainder estimates in theorem 5 under standard condition to closed trajectories. However, there is a special treatment for (38) which is most important for us: due to Fefferman-La Llave (or similar) decomposition one can rewrite the left-hand expression ($\mathcal{R}_3$) in a form

$$\int_{t\geq h^{1-\delta}} \int \Bigl(\int \psi(t^{-1}(x-z))\bigl(\rho - h^{-d}\rho_0\bigr)(x)dx\Bigr)^2 t^{-d-p-1}dtdz \tag{40}$$

with C_0^∞ function ψ plus the same integral restricted to $|x-y| \leq \bar\gamma = h^{\frac{1}{2}-\delta}$. The last integral doesn't exceed $C\bar\gamma^{d-p}h^{1-d}$ while in (40) only closed trajectories but not loops are important for sure.

6. Jellium model on sphere.

To understand the link of the periodic trajectories and (25) let us consider the high-density neutral jellium on the sphere instead of standard $\mathbb{R}^d$. The neutral jellium model means that the potential V is created by the charges uniformly distributed on the manifold X with the density $\bar\rho = N(\mathrm{vol}\, X)^{-1}$. Then $\rho_{TF}(x) = \bar\rho$ and $W = 0$. So, we get $A = -\Delta$. I will discuss only upper bound. There is a very little doubt that the lower bound can be obtained by Bach-Graf-Solovej method. In this case $\lambda = \lambda_N \sim (2\pi)^{\frac{1}{2}}\omega_d^{-\frac{1}{d}}\bar\rho$. In the general model one can get easily upper bound (substituting the Slater determinant as a test function)

$$\mathbf{N}_1(A-\lambda) + \lambda N + \kappa_3 N^{\frac{4}{3}} + \frac{1}{2}\iint \bigl(\rho(x)-\bar\rho\bigr)\bigl(\rho(y)-\bar\rho\bigr)|x-y|^{-1}dxdy + O(N) \tag{41}$$

and going to semiclassical expression one gets easily the answer in the form $\kappa_0 N^{\frac{5}{3}} + O(N^{\frac{4}{3}})$ with well-established constants κ_j. Moreover, under condition $\mu(P_a) = 0$ one gets easily the upper bound in the form $\kappa_0 N^{\frac{5}{3}} + \kappa_1 N^{\frac{4}{3}} + o(N^{\frac{4}{3}})$.

Let us consider ultimately $X = \mathbb{S}^d$, $d \geq 2$. There is an explicit formula for $\mathbf{N}_1(A-\lambda_N) + \lambda_N N$ (nevertheless it contains non-Weylian term of the magnitude

$N^{\frac{4}{3}}$) and we need to estimate only (26). Note, that nevertheless eigenfunctions corresponding to eigenvalues less than λ_N are not chosen uniquely, their total contribution to $\rho(x)$ is defined uniquely. However, it is not true for eigenfunctions corresponding to eigenvalue λ_N because only some but not all of them should be taken (the number of these eigenfunctions should be $N - \mathbf{N}(A - \lambda_{N-1})$). This arbitrary choice doesn't affect other terms in (41). One can note easily that (26) doesn't change if one takes $\rho(x)$ corresponding only to these last M eigenfunctions and $\bar{\rho} = (\operatorname{vol} X)^{-1} M$. Let $M_N \asymp N^{1-\frac{1}{d}}$ be the dimension of the eigenspace corresponding to λ_N; formula is known but out of importance. It is not true that for any N and $M = 1, \dots, M_N$ one can pick eigenfunctions with $\rho(x) = \bar{\rho}$. However, $|\rho(x) - \bar{\rho}| \le CM_N$ for any choice of these eigenfunctions and $\operatorname{mes}\{x, |\rho(x) - \bar{\rho}| \ge \epsilon M_N\} \le \varepsilon(N,\epsilon)$ for an appropriate choice of these eigenfunctions where $\epsilon > 0$ is arbitrarily small and $\varepsilon(N,\epsilon) \to 0$ as $N \to \infty$. Thus, *upper estimate*

$$\mathbf{N}_1(A - \lambda) + \lambda N + \kappa_3 N^{\frac{4}{3}} + o\big(N^{\frac{4}{3}}\big)$$

holds and I am sure that one can replace $o\big(N^{\frac{4}{3}}\big)$ by $O\big(N^{\frac{4}{3}-\delta}\big)$ using the same arguments.

Note, that in this model to reach the proper upper estimate one should choose carefully eigenfunctions from the upper level. Moreover, adding some potential we replace the degenerate eigenvalue by the dense eigenvalue cluster and choosing eigenfunctions one should minimize (26) rather than simply picking up corresponding to the lowest eigenvalues (surely, the real ground state function hardly is the Slater determinant).

Moreover, it is not very difficult to treat jellium model for manifold X with the boundary and even with singularities of some type.

7. Large molecule in strong magnetic field.

The ends of my talk is highly speculative but I hope to justify my conjectures in the future. In this model one uses operator (1) with operator Δ replaced by $(D_1 - Bx_2)^2 + D_2^2 + D_3^2 - Bh$ ($h = 1$ *here*); so the physical dimension is 3 and magnetic field is constant and directed along axe x_3. I believe that the arguments which lead to the lower bound before will work in this case as well (at least if B isn't large enough) and we get one-particle problem as before with D_j replaced by hD_j ($h = N^{-\frac{1}{3}}$) and B replaced by $\beta = N^{-1}B$. Nothing should change for $\beta \le c$. In the case of $\beta > c$ asymptotics of $\mathbf{N}_1$ was treated in [5], section 12.5 and [11]. The formula is

$$-\kappa \int \sum_j (W + 2\beta hj - \bar{\lambda})_-^{\frac{3}{2}} 2\beta h^{-2} dx$$

($j \in \mathbb{Z}^+$) and under some assumptions the remainder estimate is $O(\beta^{\frac{1}{3}} h^{-1})$ for $\beta h \le c$ (for smooth potential one can recover even $O(h^{-1})$). I believe that asymptotics of (25) will cause very little trouble. However, term (26) is really dangerous and the first reason is very simple: the correct semiclassical formula for ρ is now

$$\rho = \kappa' \sum_j (W + 2\beta hj - \bar{\lambda})_-^{\frac{1}{2}} 2\beta h \tag{42}$$

instead of $\rho = \kappa''(W - \bar{\lambda})_-^{\frac{3}{2}}$ and the difference is at least βh and this leads to (26) larger than $\beta^2 h^{-4}$ for sure (and this leads finally to $\beta^2 N^{\frac{5}{3}}$). That means that we need to change Thomas-Fermi equation: now it is a system (42) and

$$W = V + |x|^{-1} * \rho, \qquad V = \sum_k -z_k|x - y_k| \tag{43}$$

where dilatation is already done and ρ, V, W are divided by N^2, N,N respectively. One can note that (41) is not smooth and due to this solution cannot be very smooth at points with $W = \bar{\lambda} - 2\beta hj$. It looks like W has Lipshitz 4-th order derivatives and it is pretty flat this points but in order to apply the strongest SSA arguments one needs to have W really smooth! However, it is not so bad. I have a good reason to believe that for $\beta \leq h^{-1+\delta}$ the following procedure should work well: one replaces (42) by

$$\rho = \kappa''(W - \bar{\lambda})_-^{\frac{3}{2}} h^{-3} \sum_{0 \leq \ell < L} c_\ell \big(\beta h (W - \bar{\lambda})^{-1}\big)^\ell \tag{44}$$

where c_ℓ are known numbers such that the difference between (42) and (43) is $O\big((\beta h)^L\big)$ the sense of distributions, $c_0 = 1$ and L is pretty large. Then we get smooth solution (sure, there are some problems with points where $W = \bar{\lambda}$ but they are pretty remote). I think that for $\beta \geq h^{-1+\delta}$ some other type of smoothing near "bad surfaces" is possible.

Remark (March, 95). During last year my efforts were directed on this problem. Now I can claim that it is almost done for $B \ll N^3$: I got very accurate remainder estimate under some conjecture, which is definitely true for one atom; in the general case I got a less precise remainder estimate. The main trouble is that the solution of the magnetic Thomas-Fermi equation is not very smooth and one needs to make appropriate mollifications to engage the machinery of the semiclassical analysis.

References

[1] V. Bach. Error bound for the Hartree-Fock energy of atoms and molecules, Commun. Math. Phys., **147** (1992), 527–548.

[2] C. Fefferman, L. Seco. The ground state energy of a large atom, Bull. A.M.S **23** (1990), 525–530.

[3] C. Fefferman, L. Seco. On the Dirac and Schwinger corrections to the ground state energy of an atom, Adv. Math. (to appear).

[4] G.M. Graf, P. Solovej. A correlation estimate with applications to quantum systems with Coulomb interactions (to appear).

[5] V. Ivrii. Semiclassical microlocal analysis and precise spectral asymptotics. Chapters 4, 10, 12 (Preprints 1, 7, 8), École Polytechnique, (1990–92). Revised version is available by e-mail.

[6] V. Ivrii. Semi-classical asymptotics for exchange energy. Sém. EDP, École Polytechnique, 1993/94.

[7] V. Ivrii. Around Scott correction term(s). Travaux Conf. EDP Sant Jean de Monts (1994) (to appear).

[8] V. Ivrii, I.M. Sigal. Asymptotics of the ground state energies of large Coulomb systems. Ann. Math **138** (1993), 243–335.

[9] E.H. Lieb. The stability of matter: from atoms to stars (Selecta). Springer-Verlag (1991).

[10] I.M. Sigal. Lectures on large Coulomb systems (to appear).

[11] A. Sobolev. The quasi-classical asymptotics of local Riesz means for the Schrödinger operator in a strong homogeneous magnetic field, Duke Math. J., (1994).

Victor Ivrii
Department of Mathematics, University of Toronto, 100, St.George Str.,
Toronto, Ontario M5S 1A1, CANADA
e-mail:ivrii@math.toronto.edu
Fax:(416)978-4107

Operator Theory:
Advances and Applications, Vol. 78

Product Formulas and Error Estimates

S. T. Kuroda, K. Kurata

1. Introduction

Recently there has been considerable interest in the use of Trotter type product formulas, in theoretical physics (see [15] and literature cited there) as well as in numerical computation (see, e.g., [6] and the literature cited there). The basic formula is the Trotter product formula $e^{A+B} = \lim_{n\to\infty}(e^{A/n}e^{B/n})^n$, which is a two-factor product formula. It is also widely recognized that the symmetrized three factor formula $e^{A+B} = \lim_{n\to\infty}(e^{A/2n}e^{B/n}e^{A/2n})^n$ is effective in numerical computations. In fact the order of error formally derived from the Taylor expansion is $O(n^{-1})$ and $O(n^{-2})$, respectively. We also mention that a systematic investigation for constructing multi-factor product formulas with higher order of error has been done by M. Suzuki ([14], [15]) (for bounded generators). The convergence of multi-factor formulas as the number of factors becomes large is also proved in [15]. (For numerical applications of multi-factor formulas, see also [5].)

This note is a preliminary report of our attempt to derive some error estimates for these product formulas with unbounded generators and apply them to Schrödinger operators. When the generators A and B are bounded, error estimates can be given in the operator norm and are well established ([13]). When A and B are not bounded, estimates will generally be of strong type like $e^{(A+B)}u - (e^{A/n}e^{B/n})^n u = O(n^{-\alpha})$. (See, however, Remark at the end of this section.) The exponent α will depend on u. For differential operators it is expected that the order of error will be better as the regularity of u improves. The purpose of this note, which is partly a review, is to comfirm it in a simple situation. Most of the proofs will be omitted. The details with some generalizations will be published elsewhere.

In this note we only consider the case of skew-selfadjoint generators. Replacing A etc. by $-itA$ etc., we assume that A and B are selfadjoint operators in a Hilbert space X and put

$$U(t) = e^{-it(A+B)}, \quad U_A(t) = e^{-itA}, \quad U_B(t) = e^{-itB}, \tag{1.1}$$

$$W_2(t) = e^{-itA}e^{-itB} = U_A(t)U_B(t), \tag{1.2}$$

$$W_3(t) = e^{-itA/2}e^{-itB}e^{-itA/2} = U_A(t/2)U_B(t)U_A(t/2). \tag{1.3}$$

We call an estimate of $U(t) - W_j(t/n)^n$ with large n a product estimate and an estimate of $U(t) - W_j(t)$ with small $|t|$ a small time estimate. The general principle is that $U(t) - W_j(t) = O(t^p)$, $t \to 0$, implies $U(t) - W_j(t/n)^n = t^p O(n^{-(p-1)})$, $n \to \infty$, with t fixed (see, e.g., [9]).

Remark. During the conference Professor Y. Berezansky reminded the first author of a recent work of Dzh. L. Rogava ([11]) concerning the norm convergence of product formulas with unbounded generators. However, [11] is concerned with the parabolic cases ($U(t) = e^{-t(A+B)}$ with A, B positive) and may not be applicable to the Schrödinger case.

2. Bounded generators

We reproduce some relevant estimates from [13] for the purpose of comparison . In [13] they are given for more general Banach algebra situations and for $A_1 + \cdots + A_p$ instead of $A + B$. Here, we write down only the following two formulas with $t > 0$ for simplicity:

$$\|U(t) - W_2(t)\| \leq \frac{1}{2}t^2 \|[A,B]\|, \tag{2.1}$$

$$\|U(t) - W_3(t)\| \leq \frac{1}{12}t^3 \left(\|[\,[A,B],B]\| + \frac{1}{2}\|[\,[A,B],A]\| \right), \tag{2.2}$$

where $[A,B] = AB - BA$.

Remarks

1. For skew-selfadjoint generators we do not need the factor $e^{t(\|A\|+\|B\|)}$ which is present in the general case treated in [13].

2. Using computations similar to the proof of (2.2) one derives the following error estimates for an unsymmetrized three-factor formula. Let $0 \leq \theta \leq 1$. Then,

$$\begin{aligned} &\|U(t) - U_A(\theta t)U_B(t)U_A((1-\theta)t)\| \\ &\leq \frac{t^2}{2}\left\{ |1-2\theta|\,\|[A,B]\| + \frac{t}{3}\left(\theta\|[[A,B],B]\| + (1-\theta)^2\|[[A,B],A]\|\right)\right\}. \end{aligned} \tag{2.3}$$

This reveals computationally that $O(t^3)$ approximation would be achieved only when $\theta = 1/2$. This is a special example of a general theorem ([14], Theorem 2) stating that the symmetry property like $W_3(-t) = W_3(t)^{-1}$ improves the order of approximation by 1.

3. Unbounded generators

3.1. For unbounded generators the process from a small time estimate to a product estimate is not so straightforward and given as follows.

Proposition 3.1. Let C be a selfadjoint operator in a Hilbert space X and let $\{V(t)\}_{t\in\mathbb{R}}$ be a family of unitary operators in X. Let $K \subset X$ and assume that there exists a function $c(t) \geq 0$ such that

$$\left\|\left(e^{-itC} - V(t)\right)u\right\| \leq c(t), \quad \forall t \in \mathbb{R}, \quad \forall u \in K. \tag{3.1}$$

Fix $t > 0$ and assume that $e^{-isC}u \in K$ for all $s \in [0,t]$. Then

$$\left\|\left(e^{-itC} - V(t/n)^n\right)u\right\| \leq nc(t/n). \tag{3.2}$$

The proof is a repetition of the proof of Theorem VIII.29 of [9]. However, the assumption $e^{-isC}u \in K$, which facilitates the proof, makes the application of this proposition harder. Thus, for unbounded generators a small time estimate of smart form need not lead to an equally nice product estimate.

3.2. In this subsection we assmue that

$$A,\ B, \text{ and } A+B \text{ are selfadjoint} \tag{3.3}$$

and restrict our attention to the case that A and B are semi-bounded from below. We first make the simple but useful observation that adding constants to A and B multiplies $U(t)$ and $(U_A(t/n)U_B(t/n))^n$ by the same constant of absolute value 1 so that it does not change the error estimates. Thus we assume without loss of generality that

$$A,\ B,\ A+B \geq 1. \tag{3.4}$$

For $\alpha > 0$ we define A^α by the spectral theorem and denote by $D(A^\alpha)$ the domain of A^α with the graph norm $(\|u\|^2 + \|A^\alpha u\|^2)^{1/2}$. T is A^α-bounded if $D(T) \supset D(A^\alpha)$ and T is a bounded operator from $D(A^\alpha)$ to X. These notations apply to B^α and $(A+B)^\alpha$ as well. In particular, $\|u\|_\alpha$ denotes the (graph) norm of $D((A+B)^\alpha)$. We shall state theorems only for $t > 0$. For $t < 0$ similar statements hold.

Theorem 3.1. In addition to (3.3) assume that

$$A^\alpha,\ B^\alpha,\ A^{\alpha-1}B \text{ are } (A+B)^\alpha\text{-bounded} \tag{3.5}$$

for some α with $1 \leq \alpha \leq 2$. Then

$$\|(U(t) - U_A(t)U_B(t))u\| \leq d_\alpha t^\alpha \|u\|_\alpha, \quad u \in D((A+B)^\alpha). \tag{3.6}$$

If $1 \leq \alpha < 2$, then

$$\|(U(t) - U_A(t)U_B(t))u\| \leq t^\alpha o(1), \quad t \to \infty, \quad u \in D((A+B)^\alpha), \tag{3.7}$$

uniformly on any compact set of $D((A+B)^\alpha)$.

Theorem 3.2. Assume (3.3) and (3.5) and let $1 \le \alpha \le 2$. Then

$$\|U(t)u - (U_A(t/n)U_B(t/n))^n u\| \le \frac{d_\alpha t^\alpha}{n^{\alpha-1}} \|u\|_\alpha, \quad u \in D((A+B)^\alpha), \tag{3.8}$$

If $1 \le \alpha < 2$, then

$$\|U(t)u - (U_A(t/n)U_B(t/n))^n u\| = \frac{|t|^\alpha}{n^{\alpha-1}} o(1), \quad n \to \infty, \quad u \in D((A+B)^\alpha) \tag{3.9}$$

uniformly for u in a compact set of $D((A+B)^\alpha)$ and for t in a compact interval.

The proof of these theorems is based on simple estimates of the error term

$$R_n(t,A)u = U_A(t)u - \sum_{k=0}^{n-1} \frac{(-it)^k}{k!} A^k u, \quad u \in D(A^{n-1}), \tag{3.10}$$

of the Taylor expansion. The estimates are

$$\|R_n(t,A)u\| \le c_\alpha t^\alpha \|A^\alpha u\|, \quad u \in D(A^\alpha), \quad n-1 \le \alpha \le n, \tag{3.11}$$

$$R_n(t,A)u = t^\alpha o(1), \quad t \to 0, \quad u \in D(A^\alpha), \quad n-1 \le \alpha < n. \tag{3.12}$$

Details need not be written. We only remark that the proof is simple because we can base it on the spectral theorem for selfadjoint operators.

In applications it is convenient to have some sufficient conditions for (3.5) to hold.

Theorem 3.3. In addition to (3.3) and (3.4) assume that

$$(A+B)^2 = A^2 + AB + BA + B^2. \tag{3.13}$$

Then (3.5) holds for all α with $1 \le \alpha \le 2$.

The proof is an easy application of the interpolation theorem between $D(A+B)$ and $D((A+B)^\alpha)$.

Remarks

1. Theorems 3.1, 3.2, and 3.3 will remain valid without the assumption of semi-boundedness of A and B if the following changes are made: (i) $(A+B)^\alpha$ etc. are replaced by $|A+B|^\alpha$ etc.; (ii) AB is assumed to be closable. Note that with assumption (3.4), i.e. with the existence of bounded A^{-1}, AB is automatically closed.
2. Similar considerations can be made for the three factor formula $W_3(t)$. The range of α is $2 \le \alpha \le 3$ and instead of (3.13) we assume
$$(A+B)^3 = A^3 + A^2B + \cdots + B^3. \tag{3.14}$$
One needs a long computation to carry through. This was done in [7], but we do not elaborate further details here.
3. It is also possible to obtain similar estimates when generators are sectorial operators. These will be investigated in future publications.
4. During the conference Professor V. Enss made a remark that if $[A,B]$ is bounded, estimates similar to (2.1) and (2.2) would be derived by a limiting procedure. In fact under assumptions (3.4) and (3.13) AB and BA are $(A+B)^2$-bounded so that one has
$$\|(U(t)-W_2(t))(A+B)^{-2}\| \le \frac{1}{2}t^2\|[A,B](A+B)^{-2}\|. \qquad (3.15),$$

4. Applications to Schrödinger operators

4.1. Schrödinger operators. In this section we consider the following situation:

$$X = L^2(\mathbb{R}^n), \quad A = -\Delta, \quad B = V(x)\cdot. \tag{4.1}$$

Here $V(x)$ is a real measurable function. A is selfadjoint with the domain $D(A) = H^2(\mathbb{R}^n)$, the Sobolev space of order 2, and B with $D(V) = \{u \in L^2 \mid Vu \in L^2\}$.

What we intend to do is to find some sufficient conditions for (3.3) and (3.13) to hold. For (3.3) only the selfadjointness of $A+B$ matters.

4.2. Selfadjointness of $A+B$. Some general criteria for the selfadjointness of $A+B$ on $D(A)\cap D(B)$ were given by H. Sohr [12] and N. Okazawa[8]. Sohr's assumption, in our notation, is that $\mathrm{Re}(Au,B^{-1}u)+a\|u\|^2 \ge 0$, $\forall u \in D(A)$, with $0 \le a < 1$, and Okazawa's is that $\mathrm{Re}(Au,B_nu)+a\|B_nu\| \ge -c\|u\|^2 - b\|B_nu\|\,\|u\|$, where B_n is the Yosida approximation of B. The difficult part is that this is the only assumption. If one assumes also the essential selfadjointness of $A+B$, the problem becomes very simple. This is noticed essentially by N. Yamazaki[16]. Though stated only for some concrete operators, the following proposition is in [16].

Proposition 4.1. Let A and B be selfadjoint in X. Suppose that $D(A) \cap D(B)$ is dense in X and that there exists $D \subset D(A) \cap D(B)$, dense in X, such that
(i) $A+B$ is essentially selfadjoint on D, i.e. the restriction $(A+B)_{|D}$ of $A+B$ to D is essentially selfadjoint,
(ii) there exists $c > 0$ such that

$$\|(A+B)u\| + \|u\| \geq c(\|Au\| + \|Bu\|), \quad u \in D. \tag{4.2}$$

Then $A+B$ is selfadjoint on $D(A) \cap D(B)$.

Condition (4.2) may be replaced by a conditon of Sohr-Okazawa type.

Proposition 4.1′. Assumption (ii) in Proposition 4.1 can be replaced by
(ii′) there exists a, $0 \leq a < 1$, such that

$$\mathrm{Re}(Au,Bu) + a\|Bu\|^2 \geq 0, \quad u \in D. \tag{4.2′}$$

Proof. (4.2′) implies $\|(A+B)u\|\,\|Bu\| \geq \mathrm{Re}(\,(A+B)u,Bu) \geq (1-a)\|Bu\|^2$. Denoting the closure of T by T^a, we then see that $(A+B)_{|D}{}^a \subset A+B$. The same relation can be derived more directly from (4.2). Since the LHS is selfadjoint by assumption (ii) and the RHS is symmetric, they are equal. This finishes the proof.

Remark. (4.2′) with $0 \leq a < 1/2$ implies (4.2), but not vice versa.

We now apply Proposition 4.1 to the Schrödinger operator. We take $D = C_0^\infty(\mathbb{R}^n)$. The essential selfadjointness is a consequence of Kato's inequality if $V \in L^1_{loc}$ and is, say, bounded below (see [3], [10]). By computations similar to those in [8] one sees easily that condition (4.2′) is satisfied if $V \in C^1$, $V(x) \geq V_0 \in \mathbb{R}$, and there exist α with $0 \leq \alpha < 4$ and $\beta \geq 0$ such that

$$|\nabla V(x)|^2 \leq \alpha |V(x)|^3 + \beta. \tag{4.3}$$

This is the same condition as given by Okazawa ([8], (6.3)).

For later convenience we reproduce a sketch of the proof. By adding a large constant to V, which does not affect the conclusion, we may assume that $V(x) \geq V_0 > 1$ and $\gamma \equiv (\beta/V_0 + \alpha) < 4$. By partial integration we see that $(Au,Bu) = (\nabla u, \nabla(Vu)) = \|V^{1/2}\nabla u\|^2 + (V^{1/2}\nabla u, V^{-1/2}\nabla V \cdot u)$. Applying the Schwarz inequality to the second term on the RHS and using assumption (4.3), we obtain $\mathrm{Re}(Au,Bu) \geq (1-\varepsilon)\|V^{1/2}\nabla u\|^2 - (1/4\varepsilon)\gamma\|Vu\|^2$. Since $0 \leq \gamma < 4$, we can choose ε such that $0 < \varepsilon < 1$ and $a \equiv (1/4\varepsilon)\gamma < 1$, which proves the assertion.

4.3. Verification of (3.13). An abstract criterion corresponding to Proposition 4.1 is the following proposition, the proof of which is also straightforward.

Proposition 4.2. Let A, B, and $A+B$ be selfadjoint in X and satisfy (3.4). Suppose that $\tilde{D} \equiv D(A^2) \cap D(AB) \cap D(BA) \cap D(B^2)$ is dense in X and that there exists $D \subset \tilde{D}$, dense in X, such that

(i) $(A+B)^2$ is essentially selfadjoint on D,
(ii) there exists $c>0$ such that

$$\|(A+B)^2u\| + \|u\| \geq c(\|A^2u\| + \|ABu\| + \|BAu\| + \|B^2u\|), \quad u \in D. \tag{4.4}$$

Then (3.13) holds.

We now apply Proposition 4.2 to Schrödinger operators. We take $D = C_0^\infty$. The essential selfadjointness of the powers of the Schrödinger operator as we need it has been proved by P. R. Chernoff([1]) for a semi-bounded potential. (See also [4] for a generalization and a concise exposition.) We also quote [2]. Thus, (i) is all right if V with $V(x) \geq 1$ is sufficiently smooth.

To deal with (ii) we need to do a lengthy computation. The computation is along the same line as exemplified in the last part of §4.2, but six cross terms like $AB \cdot A^2$ arising from the expansion of $(A^2 + AB + BA + B^2)^2$ have to be dealt with in a systematic way. As a result of these computations we have the following proposition.

Proposition 4.3. Suppose that $V \in C^2$ and $V(x) \geq 1$. Assume that there exists δ_1 and δ_2 sufficiently small and K_1, $K_2 > 0$ such that

$$|\nabla V(x)|^2 \leq K_1 + \delta_1 V(x)^3, \tag{4.5}$$

$$|\triangle V(x)|^2 \leq K_2 + \delta_2 V(x)^4, \tag{4.6}$$

then (4.4) holds. (K_1 and K_2 may depend on δ_1 and δ_2, respectively.)

Remark. In principle we can estimate how small δ_1 and δ_2 should be. But it is complicated and we do not elaborate it here.

Since (4.5) covers (4.3) if $\delta_1 < 4$, we finally obtain the following theorem.

Theorem 4.4. Assume that $V(x)$ satisfies all the assumptions of Proposition 4.3. Then the conclusions of Theorem 3.2 hold for the Schrödinger operator $A = -\triangle$, $B = V$, and $A + B = -\triangle + V$.

Acknowledgements. The first author (STK) would like to express his thanks to Professor Y. Berezansky and Professor V. Enss for their comments given during the conference.

References

[1] P. R. Chernoff, Essential self-adjointness of powers of generators of hyperbolic equations, J. Funct. Anal. **12**(1973), 401–414.

[2] Cordes, H. O., Spectral Theory of Linear Operators and Comparison Algebras, London Math. Soc. Lecture Note Ser. **76**, Cambridge Univ. Press, 1987.

[3] Kato, T., Schrödinger operators with singular potentials, Israel J. Math. **13**(1973), 135–148.

[4] Kato, T., A remark to the preceding paper by Chernoff, J. Funct. Anal. **12**(1973), 415–417.

[5] Kuroda, S. T., Topics in the spectral methods in numerical computation – Product formulas, Differential Equations with Applications to Mathematical Physics, W. F. Ames et al ed., Academic Press 1993, 213–219.

[6] Kuroda, S. T., and T. Suzuki, A time-dependent method for computing eigenfunctions and eigenvalues of Schrödinger operators, Japan J. Appl. Math. **7**(1990), 231–253.

[7] Nakaya, Y., Error estimates for Lie-Trotter product formula (in Japanese), Master Thesis, Gakushuin Univ., 1994.

[8] Okazawa, N., On the perturbation of linear operators in Banach and Hilbert spaces, J. Math. Soc. Japan **34**(1982), 677–701.

[9] Reed, M., and B. Simon, Methods of Mathematical Physics, I: Functional Analysis, Academic Press, 1972.

[10] Reed, M., and B. Simon, Methods of Mathematical Physics, II: Fourier Analysis and Self-Adjointness, Academic Press, 1975.

[11] Rogava, Dzh. L., Error bounds for Trotter-type formulas for self-adjoint operators, Functional Anal. Appl. **27**(1993), 217–219.

[12] Sohr, H., Über die Selbstadjungiertheit von Schrödinger-Operatoren, Math. Z. **160**(1978), 255–261.

[13] Suzuki, M., Decomposition formulas of exponential operators and Lie exponentials with some applications to quantum mechanics and statistical physics, J. Math. Phys. **26**(1985), 601–612.

[14] Suzuki, M., General theory of fractal path integrals with applications to many-body theories and statistical physics, J. Math. Phys. **32**(1991), 400–407.

[15] Suzuki, M., Convergence of general decompositions of exponential operators, Comm. Math. Phys. **163**(1994), 491–508.

[16] Yamazaki, S., On the perturbation of eigenvalues of Schrödinger operators of unharmonic osscillator type (in Japanese), Master Thesis, Univ. Tokyo, 1981.

S. T. Kuroda
Departmant of Mathematics, Gakushuin University
1-5-1 Mejiro, Toshima-ku, Tokyo, 171 Japan

Kazuhiro Kurata
Department of Mathematics, Tokyo Metropolitan University
1-1 Minami-Osawa, Hachioji, 192-03, Japan

Operator Theory:
Advances and Applications, Vol. 78

On Inequalities for the Bound States of Schrödinger Operators

A. Laptev

Abstract

We improve the Lieb constant in the Cwikel-Lieb-Rozenblum inequality for the number of bound states of Schrödinger operators whose potential equals the characteristic function of a measurable set. Similar idea also gives better constants in the Lieb-Thirring inequalities for the respective moments of eigenvalues.

1. Let H be a Schrödinger operator in $\mathbb{R}^n$

$$H = -\Delta - V. \tag{1}$$

In what follows we can assume without loss of generality that the potential V is non-negative, $V \geq 0$. Let $\lambda_1 \leq \lambda_2 \leq \ldots \leq \lambda_j \leq \ldots \leq 0$ be the non-positive eigenvalues of H. We shall study the quantities

$$S_{\gamma,n}(V) = \sum_j |\lambda_j|^\gamma, \qquad \gamma \geq 0. \tag{2}$$

It is known (see [8]) that there exist universal constants $L_{\gamma,n}$ such that

$$S_{\gamma,n}(V) \leq L_{\gamma,n} \int V^{\frac{n}{2}+\gamma} dx, \tag{3}$$

where $\gamma > 1/2$ $(n = 1)$, $\gamma > 0$ $(n = 2)$ and $\gamma \geq 0$ $(n \geq 3)$.

Let $N(\mu,V)$ be the number of bound states (non-positive eigenvalues) of the operator (1) not exceeding $-\mu$, $\mu \geq 0$. It is clear that $S_{0,n}(V) = N(0,V)$. Lieb and Thirring [10] proved the inequality (3) under the condition $\gamma > \max(0,1 - n/2)$ and obtained some estimates on the constants $L_{\gamma,n}$ (see also [8], [12], [4] for related papers). In the case $\gamma = 0$, $n \geq 3$ the inequality (3) was first obtained by Rozenblum [14]. Then independently of Rozenblum and of each other it was proved by Cwikel [3] and Lieb [6]. Later it was reproved again in [11]. All these proofs are different and give different estimates for the constants $L_{0,n}$. The best estimate is due to Lieb and all the attempts to improve it were unsuccessful.

2. In this paper we give a simple proof of the Cwikel-Lieb-Rozenblum and Lieb-Thirring inequalities for Schrödinger operators whose potentials equal the characteristic function of sets of the finite Lebesgue measure. Thus we study the eigenvalue problem for the Schrödinger operator in $\mathbb{R}^n$

$$H = -\Delta - \lambda\chi_\Omega, \qquad \lambda > 0, \tag{4}$$

where χ_Ω is the characteristic function of a measurable set $\Omega \subset \mathbb{R}^n$. Denote by v_n the volume of the unit ball $\mathbb{B}^n$.

Theorem 1. Let $\Omega \subset \mathbb{R}^n$, $n \geq 3$, be a finite measure set, $|\Omega| \leq \infty$. Then

$$N(0,\lambda V) \leq M_{0,n}\lambda^{n/2}|\Omega|,$$

where

$$M_{0,n} = (2\pi)^{-n} v_{n-1} \frac{n^{n/2}}{(n-2)^{n/2}}.$$

Remark 1. If $n = 3$ we have $0.0780 < M_{0,3} = \frac{1}{8\pi^3}\frac{4\pi}{3}3^{3/2} = \frac{\sqrt{3}}{2\pi^2} \approx 0.0877 < 0.1156$ which gives a better upper bound compared with [6].

Remark 2. Let $L^{cl}_{0,n} = (2\pi)^{-n}v_n$ be the constant appearing in the classical Weyl asymptotic formula for the number of negative eigenvalues of the operator (4) when $\lambda \to \infty$. Then

$$\lim_{n\to\infty} (L^{cl}_{0,n})^{-1} M_{0,n} = e.$$

Denote by λ_k^D the eigenvalues of the Dirichlet boundary value problem

$$-\Delta u = \lambda u, \quad u \in L_2(\Omega),$$

$$u|_{\partial\Omega} = 0.$$

Let $N_\Omega^D(\lambda)$ now equal the number of eigenvalues λ_k^D not exceeding λ. Applying Theorem 1 combined with the well-known variational principle (see [13]) we immediately obtain the following statement.

Corollary 2. If $|\Omega| < \infty$ and $n \geq 3$ then

$$N_\Omega^D(\lambda) \leq M_{0,n}\lambda^{n/2}|\Omega|. \tag{5}$$

Remark 3. Li and Yau [11] proved that for $n \geq 1$

$$\sum_{k=1}^N \lambda_k^D \geq \frac{n}{n+2}\frac{N^{\frac{n+2}{n}}}{(L^{cl}_{0,n}|\Omega|)^{2/n}}.$$

Since $N\lambda_N^D \geq \sum_{k=1}^N \lambda_k^D$ this implies that

$$N_\Omega^D(\lambda) \leq C_n\lambda^{n/2}|\Omega|, \tag{6}$$

where $C_n = L^{cl}_{0,n}\frac{(n+2)^{n/2}}{n^{n/2}}$. The last estimate is sharper than (5). However $\lim_{n\to\infty} M_{0,n}C_n^{-1} = 1$.

Proof of Theorem 1. Let $n(\alpha)$, $\alpha > 0$, be the number of eigenvalues of the operator

$$K = \lambda\chi_\Omega(-\Delta)^{-1}\chi_\Omega$$

which are greater or equal than α. Applying the Birman-Schwinger principle [1], [15] we obtain that $N(0,\lambda V) = n(1)$, and this therefore reduces the proof of Theorem 1 to the estimate of $n(1)$.

Now let $\varphi_a(t) = (t-a)_+$, $0 < a < 1$, $t \in \mathbb{R}^1$, be a convex function such that $\varphi_a(t) = t - a$ if $t > a$ and $\varphi_a(t) = 0$ if $t \le a$. Using the Berezin - Lieb trace inequality (see [2], [7] and also [5]) we have

$$n(1) \le \frac{1}{1-a}\operatorname{Tr}\varphi_a(\lambda\chi_\Omega(-\Delta)^{-1}\chi_\Omega) \le \frac{1}{1-a}\operatorname{Tr}\left(\chi_\Omega\varphi_a(\lambda(-\Delta)^{-1})\chi_\Omega\right). \tag{7}$$

This implies that

$$\begin{aligned} n(1) &\le \frac{|\Omega|}{(1-a)(2\pi)^n}\int(\lambda|\xi|^{-2}-a)_+\,d\xi \\ &= \frac{\lambda^{n/2}a^{1-n/2}|\Omega|}{(1-a)(2\pi)^n}\int_{\mathbb{S}^{n-1}}\int_0^1(r^{n-3}-r^{n-1})\,dr\,d\theta \\ &= \lambda^{n/2}\frac{a^{1-n/2}|\Omega|}{(1-a)(2\pi)^n}v_n\frac{2}{(n-2)}. \end{aligned}$$

The minimum value of the function $u(a) = (1-a)^{-1}a^{1-n/2}$ is reached at the point $a = 1 - 2/n$, and so we finally obtain

$$\begin{aligned} N(\lambda,0) &= n(1) \le \lambda^{n/2}\frac{(1-2/n)^{1-n/2}|\Omega|}{(1-(1-2/n))(2\pi)^n}v_n\frac{2}{(n-2)} \\ &= (2\pi)^{-n}\frac{n^{n/2}}{(n-2)^{n/2}}v_n\lambda^{n/2}|\Omega|. \end{aligned}$$

The proof is complete.

3. Using a similar approach we can find an upper bound on the quantity $S_{\gamma,n}(V)$, $\gamma > 0$, introduced in (2) when $V = \lambda\chi_\Omega$.

Theorem 3. If $|\Omega| < \infty$ and $n/2 + \gamma - 1 > 0$, $\gamma > 0$, then

$$S_{\gamma,n}(\lambda V) \le M_{\gamma,n}\lambda^{n/2+\gamma}|\Omega|,$$

where

$$M_{\gamma,n} = (2\pi)^{-n}v_n\left(\frac{n/2+\gamma}{n/2+\gamma-1}\right)^{n/2+\gamma}\frac{\Gamma(\frac{n}{2}+1)\Gamma(\gamma+1)}{\Gamma(\gamma+\frac{n}{2}+1)}.$$

Proof. It is clear that

$$S_{\gamma,n}(\lambda V) = \gamma \int_0^\infty \mu^{\gamma-1} N(\mu,\lambda\chi_\Omega)\, d\mu.$$

Therefore by analogy with (7) we obtain for any $0 \le a \le 1$ that

$$S_{\gamma,n}(\lambda V) \le \frac{\gamma}{1-a} \int \mu^{\gamma-1} \mathrm{Tr}\left(\chi_\Omega \varphi_a(\lambda(-\Delta+\mu)^{-1})\chi_\Omega\right) d\mu. \tag{8}$$

The trace in the right hand side of (8) can be calculated explicitly and we have

$$\begin{aligned} S_{\gamma,n}(\lambda V) &\le \frac{|\Omega|\gamma}{(2\pi)^n(1-a)} \int_0^\infty \int \mu^{\gamma-1}(\lambda(|\xi|^2+\mu)^{-1} - a)_+ \, d\xi\, d\mu \\ &= \frac{\lambda^{n/2+\gamma}|\Omega|}{(2\pi)^n} \frac{a^{1-n/2-\gamma}}{(1-a)} v_n \frac{\Gamma(n/2+1)\Gamma(\gamma+1)}{(\gamma+n/2-1)\Gamma(n/2+\gamma)}. \end{aligned}$$

Minimizing $a^{1-n/2-\gamma}(1-a)^{-1}$ we obtain that $a = 1-(n/2+\gamma)^{-1}$ and this completes the proof of the theorem.

Remark 4. If $n=3$ and $\gamma=1$ then $M_{1,3} \approx 0.02422$, which is better than 0.04030 obtained in [8]. However this still does not prove the conjecture (see [8], page 476) that $L_{1,3} < L_{0,3}^{cl} = 0.01689$. The inequality $M_{1,n} < L_{0,n}^{cl}$ becomes true only begining from $n \ge 5$.

Remark 5. By analogy with Remark 2 we obtain that

$$\lim_{n\to\infty} (L_{\gamma,n}^{cl})^{-1} M_{\gamma,n} = e,$$

where $L_{\gamma,n}^{cl}$ is the constant appearing in the classical Weyl asymptotic formula for the value $S_{\gamma,n}(\lambda\chi_\Omega)$ when $\lambda \to \infty$.

Acknowledgement. I am grateful to D.Vassiliev who pointed out that the estimate (6) following from [11] is sharper than (5).

References

[1] M. Sh. Birman, The spectrum of singular boundary problems. Mat. sb. **55(97)**, (1961), 268. (Russian).

[2] F. Berezin, Convex functions of operators, Mat. sb., **88**, (1972), 268–276. (Russian).

[3] M. Cwikel, Weak type estimates for singular values and the number of bound states of Schrödinger operators. Ann. Math. **106**, (1977), 93–100.

[4] B. Helffer and D. Robert, Riesz means of bounded states and semi-classical limit connected with a Lieb-Thirring conjecture II. Ann.Inst.Henri Poincaré **53**, (1990), 139–147.

[5] A. Laptev and Yu. Safarov, A generalization of the Berezin-Lieb inequality. Preprint, Mittag-Leffler Inst., (1993).

[6] E.H. Lieb, Bounds on the eigenvalues of the Laplace and Schrödinger operators. Bull of the AMS. **82**, (1976), 751–753.

[7] E.H. Lieb, The classical limit of quantum spin systems, Comm. Math. Phys., **31**, (1973), 327–340.

[8] E.H. Lieb, On Characteristic exponents in turbulence. Comm. Math. Phys. **92**, (1984), 473–480.

[9] V. Guillemin, Some classical theorems in spectral theory revisited, seminar on sing. of sol. of diff. equations, Princeton University Press, Princeton, 1979, 219–259.

[10] E.H. Lieb and W.E. Thirring, Inequalities for the moments of the eigenvalues of Schrödinger equation and their relation to Sobolev inequalites. Studies in Mathematical Physics (E.H. Lieb, B. Simon and A. Wightman, eds.), Princeton Univ. Press, Princeton, NJ, 1976.

[11] P. Li and S.T. Yau, On the Schrödinger equation and the eigenvalue problem. Comm. Math. Phys. **88**, (1983), 309–318.

[12] A. Martin, New results on the moments of the eigenvalues of the Schrödinger Hamiltonian and applications. Comm. Math. Phys. **129**, (1990), 161–168.

[13] M. Reed and B. Simon, Methods of mordern mathematical physics v. 4., Academic Press, New York, San Francisco, London, 1978.

[14] G.V. Rozenblum, On the eigenvalues of the first boundary problem in unbounded domains. Mat. Sb. **89(2)**, (1972), 234–247.

[15] J. Schwinger, On the bound states of a given potential. Proc. Nat. Acad. Sci. **47**, (1961), 122–129.

A. Laptev
Department of Mathematics, Royal Institute of Technology
S-10044 Stockholm, Sweden
E-mail address: laptev@math.kth.se

Operator Theory:
Advances and Applications, Vol. 78

Some Examples of two-term Spectral Asymptotics for Sets with Fractal Boundary

M. Levitin[1], D. Vassiliev

Abstract

We construct a multiparametric family of sets in a Euclidean space which have fractional Minkowski dimension of the boundary. Using the renewal theory, we explicitly construct two-term spectral asymptotics for these sets.

1. Introduction.

Let G be an arbitrary bounded open set in $\mathbb{R}^m$. We consider a (variational) spectral problem for the Dirichlet Laplacian $-\Delta_G$ on G; such a problem always has a purely discrete spectrum. We define the *spectral counting function* $N(\lambda; G)$ as the number of eigenvalues of $-\Delta_G$ which are less than a given number λ.

It is well known that for any bounded open set $G \subset \mathbb{R}^m$ the spectral counting function $N(\lambda; G)$ has a one-term Weyl asymptotics:

$$N(\lambda; G) = \gamma_m |G|_m \lambda^{m/2} + o\left(\lambda^{m/2}\right) \qquad \text{as} \qquad \lambda \to +\infty, \tag{1.1}$$

where $|\cdot|_m$ denotes the m-dimensional Lebesgue volume, and

$$\gamma_m = \frac{1}{(4\pi)^{m/2}\Gamma(1+m/2)}$$

is a constant depending only upon m.

Moreover, in many situations the one-term Weyl asymptotic formula (1.1) can be improved. In particular, if the boundary ∂G is piecewise smooth and under some additional assumptions on the geodesic flow on G we have [6, 10]

$$N(\lambda; G) = \gamma_m |G|_m \lambda^{m/2} - \widetilde{\gamma}_m |\partial G|_{m-1} \lambda^{(m-1)/2} + o\left(\lambda^{(m-1)/2}\right) \qquad \text{as} \quad \lambda \to +\infty, \tag{1.2}$$

where

$$\widetilde{\gamma}_m = \frac{1}{4(4\pi)^{(m-1)/2}\Gamma((m+1)/2)}.$$

Note that it is still an open question whether (1.2) holds for *all* bounded open sets with smooth boundary.

[1]Research supported by a SERV Analysis grant # H-55567

In the case when ∂G is non-smooth or fractal, the asymptotic behaviour of the spectral counting function depends, in particular, upon the so-called *interior Minkowski* (or *box-counting*) *dimension* of the boundary. If the interior Minkowski dimension of the boundary $d = \dim_{\mathcal{M}}(\partial G)$ lies in the interval $(m-1,m)$, and the corresponding upper Minkowski content is finite, then one has, using variational technique, an *estimate* of the order of the remainder term in Weyl's formula (1.1) (see [8]):

$$N(\lambda; G) = \gamma_m |G|_m \lambda^{m/2} + O\left(\lambda^{d/2}\right) \qquad \text{as} \qquad \lambda \to +\infty \,. \tag{1.3}$$

However, it is highly unlikely that a general formula exists for the remainder term in (1.3) because of the complex interplay between the asymptotic behaviour of $N(\lambda; G)$, the asymptotic behaviour as $\epsilon \to +0$ of the volume of the Minkowski sausage

$$\mu(\epsilon; G) = |\{x \in G : \mathrm{dist}(x,\partial G) < \epsilon\}|_m \,,$$

and the capacitary density function of the boundary (see, e. g., [1]). By this reason, no further improvement can be achieved for arbitrary sets G. Moreover, even if we restrict ourselves to special types of sets, it is highly unlikely that any improvement can be achieved by using solely the standard variational technique (Dirichlet–Neumann bracketing).

It is also well known that there is no direct analogy between two-term spectral asymptotics in the smooth and fractal cases. In particular, the remainder term in (1.3) does not necessarily behave as const $\lambda^{d/2}$, see [4].

In this paper we construct a multiparametric family of sets with fractal boundary for which we can explicitly compute the two-term asymptotics of the spectral counting function. The description of the sets is given in Section 2, and the main results are stated and explained in Sections 3–4.

2. Construction of the iterated set $\mathcal{G}$.

Let $G_1 \subset \mathbb{R}^m$ be a bounded open set; for simplicity we assume that it has a smooth boundary ∂G_1. We shall denote by G_r an r times scaled copy of G_1 (so that $|G_r|^m = r^m |G_1|_m$) placed somewhere in $\mathbb{R}^m$. The set G_1 is the first component (iteration) of the set $\mathcal{G}$ we want to construct. Now, let us fix M coefficients $r_1, \ldots, r_M$ from the interval (0,1) (not necessarily different) and include in $\mathcal{G}$ scaled copies $G_{r_1}, \ldots, G_{r_M}$ placed in such way that they do not intersect G_1 and each other. We repeat the same procedure for all the M sets of the second generation, obtaining M^2 sets of the third generation, and so on. We include in $\mathcal{G}$ all the sets of all the generations:

$$\mathcal{G} = \left(\bigcup_{n=1}^{+\infty} \bigcup_{\{i_1,\ldots,i_n\}} G_{i_1 \cdots i_n} \right) \cup G_1 \,. \tag{2.1}$$

Here n is the number of the generation and each of the indices i_k takes the values from 1 to M. Note that the indices $i_1, \ldots, i_n$ are not necessarily different.

We impose the only technical condition on the coefficients r_j:

$$\sum_{j=1}^{M} (r_j)^m < 1 < \sum_{j=1}^{M} (r_j)^{m-1} . \tag{2.2}$$

It can be shown that the left inequality (2.2) guarantees that we can place all the sets $G_{i_1 \ldots i_n}$ in $\mathbb{R}^m$ in such a way that they do not intersect each other and that $\mathcal{G}$ is bounded. Elementary calculations give

$$|\mathcal{G}|_m = \frac{|G_1|_m}{1 - \sum_{j=1}^{M} {r_j}^m} . \tag{2.3}$$

Let us define the number d as the (unique) positive real root of the equation

$${r_1}^d + {r_2}^d + \ldots + {r_M}^d = 1 . \tag{2.4}$$

Such a root exists and lies in the interval $(m-1, m)$ because ${r_1}^d + \ldots + {r_M}^d - 1$ is a monotone function of d which takes values of opposite sign on the ends of this interval according to (2.4).

One can show that d is precisely the interior Minkowski dimension of the boundary $\partial\mathcal{G}$. Therefore, $\mathcal{G}$ is indeed a set with a fractal boundary. Thus, by (1.3), its spectral counting function has the asymptotics

$$N(\lambda; \mathcal{G}) = \gamma_m |\mathcal{G}|_m \lambda^{m/2} + O\left(\lambda^{d/2}\right) \quad \text{as} \quad \lambda \to +\infty . \tag{2.5}$$

Our main aim is to find an explicit formula for the remainder term in (2.5). To simplify notation, we shall work further on with the function

$$\mathcal{N}(\lambda; \mathcal{G}) = \gamma_m |\mathcal{G}|_m \lambda^{m/2} - N(\lambda; \mathcal{G}) . \tag{2.6}$$

3. Spectral asymptotics for the set $\mathcal{G}$.

Before stating the result we have to recall the following definition. A finite set of positive real numbers $\{\rho_1, \rho_2, \ldots, \rho_M\}$ is called *arithmetic* if all the ratios ρ_j/ρ_k, $j,k = 1, \ldots, M$, are rational numbers. The maximal number ρ such that ρ_j/ρ are integer for all $j = 1, \ldots, M$, is called the *span* of the set $\{\rho_1, \rho_2, \ldots, \rho_M\}$. If at least one ratio ρ_j/ρ_k is irrational, then the set $\{\rho_1, \rho_2, \ldots, \rho_M\}$ is called *non-arithmetic*.

Theorem 3. If the set $\{-\log r_1, -\log r_2, \ldots, -\log r_M\}$ is non-arithmetic, then $\mathcal{N}(\lambda; \mathcal{G})$ has a power-type asymptotics

$$\mathcal{N}(\lambda; \mathcal{G}) = \mathbf{q} \lambda^{d/2} + o\left(\lambda^{d/2}\right) \quad \text{as} \quad \lambda \to +\infty , \tag{3.1}$$

where

$$\mathbf{q} = \mathbf{q}(m,r_1,\ldots,r_M,G_1) = \left(2\sum_{j=1}^{M} r_j^d(-\log r_j)\right)^{-1} \int_0^{+\infty} \mathcal{N}(\eta;G_1)\,\eta^{-(1+d/2)}\,d\eta \tag{3.2}$$

is a positive constant. If the set $\{-\log r_1, -\log r_2, \ldots, -\log r_M\}$ is arithmetic with span ρ, then $\mathcal{N}(\lambda;\mathcal{G})$ has the asymptotics

$$\mathcal{N}(\lambda;\mathcal{G}) = \mathbf{p}(\log\lambda)\lambda^{d/2} + o\left(\lambda^{d/2}\right) \qquad \text{as} \qquad \lambda \to +\infty\,, \tag{3.3}$$

where

$$\mathbf{p}(z) = \rho\left(\sum_{j=1}^{M} r_j^d(-\log r_j)\right)^{-1} \sum_{k=-\infty}^{+\infty} \mathcal{N}\left(e^{z-2k\rho};G_1\right) e^{-d(z-2k\rho)/2} \tag{3.4}$$

is a uniformly bounded left-continuous 2ρ-periodic function; the set of points of discontinuity of the function $\mathbf{p}$ is dense in $\mathbb{R}$.

Proof. Let us consider a "branch" of the set $\mathcal{G}$ starting from one of the sets G_{r_j} of the second generation (i. e., we include in this branch all the sets $G_{r_j\cdots}$). This "branch" is similar to the whole set $\mathcal{G}$ with coefficient r_j (to within an exact positioning of its components, which is unessential for the purpose of computing the spectral counting function as far as these components are mutually disjoint). Therefore, we can rewrite (2.1) in an implicit form

$$\mathcal{G} = \left(\bigcup_{j=1}^{M} \mathcal{G}_{r_j}\right) \cup G_1\,, \tag{3.5}$$

where $\mathcal{G}_r$ is the r times scaled copy of $\mathcal{G}$. This immediately implies

$$\mathcal{N}(\lambda;\mathcal{G}) = \left(\sum_{j=1}^{M} \mathcal{N}(\lambda;\mathcal{G}_{r_j})\right) + \mathcal{N}(\lambda;G_1)\,,$$

and, since by change of variables $\mathcal{N}(\lambda;\mathcal{G}_a) = \mathcal{N}(a^2\lambda;\mathcal{G})$ for any $a > 0$,

$$\mathcal{N}(\lambda;\mathcal{G}) = \left(\sum_{j=1}^{M} \mathcal{N}({r_j}^2\lambda;\mathcal{G})\right) + \mathcal{N}(\lambda;G_1)\,. \tag{3.6}$$

We shall seek the solution of the functional equation (3.6) in the form

$$\mathcal{N}(\lambda;\mathcal{G}) = f(\log\lambda)\lambda^{d/2}\,. \tag{3.7}$$

Substituting (3.7) into (3.6), dividing the resulting equation by $\lambda^{d/2}$, using (2.4), and switching to the new variable $z = \log\lambda$, we obtain the following functional equation for $f(z)$:

$$f(z) - \sum_{j=1}^{M} a_j f(z-\alpha_j) = g(z)\,, \tag{3.8}$$

where we have denoted

$$\alpha_j = -2\log r_j\,, \qquad a_j = {r_j}^d\,,$$

$$g(z) = \mathcal{N}(e^z; G_1)\, e^{-dz/2}\,.$$

We immediately see that by (2.4)

$$\sum_{j=1}^{M} a_j = 1\,. \tag{3.9}$$

Note also that

$$\mathcal{N}(\lambda, \mathcal{G}) \equiv \gamma_m |\mathcal{G}|_m \lambda^{m/2}$$

for sufficiently small λ, therefore with the account of the inequality $d < m$ we have

$$f(-\infty) = 0\,. \tag{3.10}$$

Similar arguments show that

$$g(-\infty) = g(+\infty) = 0\,;$$

moreover, $g(z)$ vanishes at $\pm\infty$ exponentially.

Functional equations of the type (3.8) are well known in probability theory; they are called *renewal equations*. It is easy to show that for any exponentially decreasing at infinity piecewise continuous right-hand side $g(z)$ the problem (3.8)-(3.10) has a unique solution $f(z)$. The asymptotic behaviour of this solution as $z \to +\infty$ is described by the so-called renewal theorems (see, for example, [2]) which state that

$$f(z) \to \text{const} = \left(\sum_{j=1}^{M} a_j\alpha_j\right)^{-1} \int_{-\infty}^{+\infty} g(z)\, dz \tag{3.11}$$

if the set $\{\alpha_1, \dots, \alpha_M\}$ is non-arithmetic, and

$$f(z) \sim \alpha\left(\sum_{j=1}^{M} a_j\alpha_j\right)^{-1} \sum_{k=-\infty}^{+\infty} g(z-\alpha k) \qquad \text{as} \qquad z \to +\infty \tag{3.12}$$

if the set $\{\alpha_1, \dots, \alpha_M\}$ is arithmetic with span α; it is easily seen that the function in the right-hand side of (3.12) is α-periodic.

It must be noted, however, that one should apply (3.11) and (3.12) in our case with great care. The problem is that in our case the function $g(z)$ is discontinuous, and Feller's proofs need adaptation to this situation.

Now, using (3.7), (3.11), (3.12), and returning to the initial parameter λ, we obtain (3.1)–(3.4). The proof of the fact that $\mathbf{p}(z)$ has a dense set of discontinuities is a modification of a similar proof in [4].

4. Discussion.

The same technique as in Section 3 allows us to obtain multi-term asymptotic expansions for such characteristics of the set $\mathcal{G}$ like the heat content and the partition function, see the details in [9]. It is also possible to extend the results of the Theorem to the case $d \geq m$. As easily seen, in this case the set $\mathcal{G}$ does no longer have a finite volume, so even Weyl's asymptotics (1.1) is no longer valid. In addition, in many cases one can obtain further terms in the asymptotic expansions for $N(\lambda, \mathcal{G})$ using careful analysis of the renewal equation; these results will appear shortly in [12]. Note that a different approach to a similar problem (operators associated with self-similar measures) also using the renewal theory is developed in [11]. The renewal theory also helps to obtain asymptotics for the heat content and the partition function for some *connected* domains with fractal self-similar boundary, see [3]. Some spectral problems *on* (one-dimensional) self-similar fractals were considered in [7].

Finally, our construction of the set $\mathcal{G}$ allows us to give once again a (negative) answer to the famous question of Mark Kac "Can one hear the shape of a drum?" While it is well known that this is not true, there is no direct analogy between the cases of drums with piecewise smooth and fractal boundary. In particular, in the piecewise smooth case one can always "hear" the $(m-1)$-dimensional volume of the boundary. A natural analogue in the fractal case would be the interior Minkowski content of the boundary. Nevertheless, we can easily construct two different isospectral sets with fractal boundary which have different Minkowski content. For example, one can construct two sets $\mathcal{G}$ using the same coefficients $r_1, \dots, r_M$ and taking as basic sets G_1 two isospectral planar domains of [5]. Certainly, the spectra of these two sets will be the same, however explicit computations using the renewal theory show that the Minkowski contents can be made to be different.

References

[1] J. Brossard and R. Carmona, Can one hear the dimension of a fractal?, Commun. Math. Phys. **104** (1986), 103–122.

[2] W. Feller, An introduction to probability theory and its applications, Vol. 2, 2nd rev. ed., John Wiley & Sons, N. Y., 1966.

[3] J. Fleckinger, M. Levitin and D. Vassiliev, Heat equation on the triadic von Koch snowflake: asymptotic and numerical analysis, Proc. London Math. Soc. (to appear).

[4] J. Fleckinger-Pelle and D. Vassiliev, An example of a two-term asymptotics for the "counting function" of a fractal drum, Trans. Amer. Math. Soc. **337** (1993), 99–116.

[5] C. Gordon, David L. Webb, S. Wolpert, One cannot hear the shape of a drum, Bull. Amer. Math. Soc. **27** (1992), 134–138.

[6] V. Ja. Ivrii, Second term of the spectral asymptotic expansion of the Laplace-Beltrami operator on manifolds with boundary, Functional. Anal. Appl. **14** (1980), 98–106.

[7] J. Kigami and M. L. Lapidus, Weyl's problem for the spectral distribution of Laplacians on p.s.f. self-similar fractals, Commun. Math. Phys. **158** (1993), 93–125.

[8] M. L. Lapidus, Fractal drum, inverse spectral problems for elliptic operators and a partial resolution of the Weyl-Berry conjecture, Trans. Amer. Math. Soc. **325** (1991), 465–529.

[9] M. Levitin and D. Vassiliev, Spectral asymptotics, renewal theorem, and the Berry conjecture for a class of fractals, preprint (1994).

[10] R. B. Melrose, Weyl's conjecture for manifolds with concave boundary, Geometry of the Laplace Operator, Proc. Sympos. Pure Math., vol. 36, Amer. Math. Soc., Providence, R. I., 1980, 257–273.

[11] K. Naimark and M. Solomyak, On the eigenvalue behaviour for a class of operators related to self-similar measures on $\mathbb{R}^d$, C. R. Acad. Sci. Paris (1994) (to appear).

[12] M. van den Berg and M. Levitin, in preparation.

M Levitin
Department of Mathematics, Heriot-Watt University,
Riccarton, Edingburgh EG 14 4AS, U.K.
e-mail: levitin@cara.ma.hw.ac.uk

D. Vassiliev
Centre for Mathematical Analysis and Its Applications, University of Sussex,
Falmer, Brighton BN1 9QH, U.K.
e-mail: d.vassiliev@sussex.ac.uk

Operator Theory:
Advances and Applications, Vol. 78

Estimates for Fourier Transforms of Surface-Carried Densities on Surfaces with Singular Points

O. Liess

1. We study asymptotic estimates for a class of Fourier integrals which are defined on surfaces with singular points and show how these estimates can be applied to obtain decay estimates for solutions of Cauchy problems of constant coefficient hyperbolic equations. Integrals of the type which we consider appear e.g. in the study of solutions of the system of elasticity for cubic crystals. Related estimates for smooth surfaces or for more general oscillatory integrals in a fully C^∞-setting have a long tradition in analysis and have proved useful in many situations.

2. Let $S \subset R^{n+1}$ be a surface defined in a neighborhood U of $0 \in R^{n+1}$, $n \geq 2$, $0 \in S$. We assume that $S \setminus 0$ is C^∞ and that S has a "node" at 0. By this we mean that we can find a C^∞- function $g : U \to R$ such that $S = \{z \in U; g(z) = 0\}$, grad $g(0) = 0$, but so that grad $g(z) \neq 0$ for any point $z \neq 0$ in S. The coordinates in R^{n+1} will be denoted by z or by (t,x), where $t \in R$ and $x \in R^n$.

We shall then assume that in these coordinates $(\partial/\partial t)^2 g(0) \neq 0$, so in particular the rank of the Hessian matrix $H_{zz}g(0)$ of g calculated at 0 is at least one. Two particular cases seem here of special interest for applications in crystal elasticity: when rank $H_{zz}g(0)$ is $n+1$ and when rank $H_{zz}g(0)$ is one. In the first case we shall call the singularity at zero "conical" and in the second the "node" will be called "uniplanar", provided some natural additional geometric conditions are satisfied.

Our main aim is to estimate the Fourier transform $I(\lambda)$,

$$I(\lambda) = \int_S e^{-i\langle z,\lambda\rangle} u(z)\, d\sigma(z), \quad \lambda \in R^n, \tag{1}$$

of the distribution $u\, d\sigma$, where d σ is the surface element on S and u is a compactly supported C^∞-density on S. By the latter we mean that we can find a C^∞-function u' defined on U so that u' coincides with u on S. Slightly more general situations can also be considered and are in fact necessary in the applications. If S were everywhere smooth and with nowhere vanishing total curvature, it would follow from a result of Hlawka (cf. [4]; also see [3], [5] or [17]) that

$$|I(\lambda)| \leq c|\lambda|^{-n/2}. \tag{2}$$

Here the situation will be somewhat more complicated in that we will have a singularity at 0 but our main result (cf. Theorem 1 below) will state that under some additional geometric assumptions the estimate (2) still remains valid. Actually the conditions which we impose in this paper are not the most general ones under which one can study the integral in (1) and the main motivation to work with the present conditions is that they appear in our initial example in crystal elasticity.

3. To describe our conditions, let us first apply the Malgrange preparation theorem. We can thus find a neighborhood U' of 0, a function $G \in C^\infty(U')$ with $G(0) \neq 0$ and C^∞-functions a and b such that $g(t,x) = G(t,x)(t^2 + a(x)t + b(x))$ on U'. Since our result will be local near 0 it will be no loss of generality then to assume from the very beginning that g has the form $g(t,x) = t^2 + a(x)t + b(x)$. We shall further assume for convenience that grad $a(0) = 0$. Actually, this condition is not strictly speaking essential for what follows, but it can always be achieved after a linear change of variables and it is easier to understand the invariant meaning of the assumptions made later on if it is satisfied.

Let us now denote by $\Delta(x) = a^2(x) - 4b(x)$ the discriminant of the function g in the variable t. We shall assume that Δ is nonnegative and that it vanishes only at $x = 0$. Geometrically, the condition on Δ just means that we can split S near 0 into the two sheets S^+ and S^-, where

$$S^\pm = \{z \in U; g^\pm(z) = 0\}, \quad g^\pm(z) = t - [-a(x) \pm \sqrt{\Delta(x)}\ ]/2. \tag{3}$$

It is also clear that Δ must vanish of even order at zero and it is not difficult to see that the assumption that rank $H_{zz}g(0)$ is one, implies that this order of vanishing is at least four. It follows from this that the two functions $g^\pm$ are both C^1 and we have $\text{grad}_z g^+(0) = \text{grad}_z g^-(0) = (1,0,\dots,0)$. In particular tangent spaces to $S^\pm$ at 0 are well-defined and coincide there. It is the latter property which is called "uniplanarity" in differential geometry (cf.[16]). We shall assume in most of this report that we can find constants c' and c'' for which

$$c'|x|^4 \leq \Delta(x) \leq c''|x|^4 \text{ for small } x. \tag{4}$$

It can be shown that the "slowness surface" of crystal-elasticity for cubic crystals is smooth except for a finite number of points which are either of conic or of uniplanar type and that for the uniplanar case the condition (4) is satisfied. (Cf. e.g. [1] or [9].)

The last geometric assumptions we shall impose are now that $\varphi^\pm = -H_{xx}a(x) \pm \sqrt{\tilde{\Delta}(x)} \neq 0$ if $x \neq 0$, where $H_{xx}a$ is the Hessian of a at 0, and $\tilde{\Delta}(x)$ is the expression $\sum_{|\alpha|=4} \partial_x^\alpha \Delta(0)x^\alpha/\alpha!$, and that the total curvature of the surfaces $S'^\pm = \{x \in R^n; \varphi^\pm/2 = 1\}$ does never vanish. At this point we should observe that all conditions have been formulated in a fixed special choice of linear coordinates in R^{n+1}. The extent to which they have an invariant meaning will be studied in the extended version of this report. Let us mention already here however the fact that it follows from our conditions that the surfaces $S'^\pm$ are bounded and strictly convex.

(The second assertion is a consequence of a classical result of Hadamard which says that a surface which has nowhere vanishing total curvature is necessarily convex: cf. e.g. [2].) It is moreover a rather straightforward calculation that if the total curvature of $S'^{\pm}$ has nonvanishing total curvature, then the same is true for all points of S sufficiently close to zero.

4. We have now introduced all assumptions on the geometric shape of S and can state the main result.

Theorem 1 Let $S = S^+ \cup S^-$ be a surface of form (3), which has an uniplanar node at 0 and satisfies the conditions above for some choice of linear coordinates in R^{n+1}. Also consider a compactly supported C^∞- density u on S. If the support of u lies in a sufficiently small neighborhood of 0, then the Fourier transform $I(\lambda)$ of $u\,d\sigma$ defined in (1) will satisfy the estimate (2).

Remark 2 As in any result of this type, only finite regularity is needed in the argument. What is more important however is that it is not necessary to require u to be regular in the sense of extendibility near 0 on S. Indeed, if u is C^∞ (or just C^k for some sufficiently high k) away from 0 and has a singularity at 0 which is "related" to the type of singularity the functions $g^\pm$ have there, then the conclusion of the theorem will still remain valid.

We also mention for completeness the following result analoguous to Theorem 1 for the case of a conical singularity.

Theorem 3 Let S be given by (3) and assume that $\langle H_{zz}g(0)v,v\rangle$ is a nondegenerate quadratic form of positivity one. Also let $I(\lambda)$ be defined by (1). Then we have the estimate $|I(\lambda)| \le c|\lambda|^{-(n-1)/2}$.

5. Let us now also describe briefly what kind of applications one can obtain from Theorem 1 when one wants to study decay properties for solutions of global Cauchy problems for constant coefficient hyperbolic linear partial differential operators in a half-space. We recall here that decay properties of the type in question are useful when one wants to understand the long-time (or global-in-time) existence of small classical solutions of nonlinear perturbations of the equations under consideration. (Cf. e.g. [6], [7], [9], [11], [12], [13], [14], [18], among others.)

To describe the general setting, let us start from a homogeneous polynomial $(\tau,\xi) \to P(\tau,\xi)$ in the variables $(\tau,\xi) \in R^{n+1}$, which is hyperbolic with respect to the direction $\tau = 0$. By this we mean that $P(1,0,\dots,0) \neq 0$ and that the roots τ of the equation $P(\tau,\xi) = 0$ are real for any fixed $\xi \in R^n$. We assume that P is of degree m, call the roots "$\tau_i(\xi)$" and label them in such a way that $\tau_1(\xi) \le \tau_2(\xi) \le \dots \le \tau_m(\xi)$ for any ξ. Multiple roots are written here according to their multiplicities. To simplify notations we shall assume that the discriminant $\Delta(\xi)$ of $P(\tau,\xi)$ in the variable τ does not vanish identically, i.e. that the roots $\tau_i(\xi)$ are distinct at generic points. We shall assume moreover that for $\xi \neq 0$ the multiplicity of the roots is at most two. Next we fix some $\xi^0 \neq 0$ and some i^0 for which $\tau_{i^0}(\xi^0) = \tau_{i^0+1}(\xi^0)$. (We assume of course that such ξ^0, i^0 exist.) Since $\tau_{i^0}(\theta\xi^0) = \tau_{i^0+1}(\theta\xi^0)$ for any positive θ, we call ξ^0 a singular direction then. Let us also denote by

$$S = \{\xi \in R^n; \tau_{i^0}(\xi) = 1 \text{ or } \tau_{i^0+1}(\xi) = 1\}.$$

It follows that $\xi^0/\tau_{i^0}(\xi^0) \in S$ and that S is part of what is called sometimes the "slowness surface" of P, viz. $\{\xi \in R^n; P(1,\xi) = 0\}$, in mathematical physics. It is no loss of generality to assume henceforth that $\tau_{i^0}(\xi^0) = 1$ and we shall denote $S^+ = \{\xi \in R^n; \tau_{i^0}(\xi) = 1\}$, $S^- = \{\xi \in R^n; \tau_{i^0+1}(\xi) = 1\}$. In particular, $\xi^0 \in S^+ \cap S^-$. Our main assumption is now that S is a surface of the type from Theorem 1 in a neighborhood of ξ^0. (In Theorem 1 we actually had $\xi^0 = 0$, but it is immediate how to reformulate our conditions near some arbitrary "ξ^0".) In particular $\xi \to \tau_{i^0}(\xi)$ is a C^1-function near ξ^0. As a motivation, we should state that the slowness surface for the system of elasticity for cubic crystals satisfies the above conditions in the nearly isotropic case.

Associated with $P(D)$ we shall now consider integrals of form

$$u(t,x) = \int_\Gamma e^{it\tau_{i^0}(\xi) + i\langle x, \xi\rangle} \kappa(\xi) q(\xi) g(\xi)\, d\xi, \tag{5}$$

where g is in $S(R^n)$, i.e. in the Schwartz space of functions which decay rapidly with all their derivatives at infinity, q is positively homogeneous in ξ of some degree $\mu > -n+1$ which satisfies an estimate of form $|q(\xi)| \leq c|\xi|^\mu$ in a conic neigborhood of ξ^0 and κ is a function which is $C^\infty(R^n \setminus \{0\})$, positively homogeneous of degree zero, and vanishes outside some conic neighborhood Γ of ξ^0. We do not specify explicitly the regularity properties which q must have.

The reason why we insert such a κ is of course that we want to localize our estimates to a conic neighborhood of ξ^0. An integral very similar to the right hand side of (5) is the one obtained when we replace $\tau_{i^0}(\xi)$ by $\tau_{i^0+1}(\xi)$ there. Integrals of the type (5) appear if we solve the Cauchy problem for $P(D)$ with Cauchy data in $S(R^n)$ by Fourier transformation. (Cf. e.g. [15]. Actually we will obtain then also integrals similar to (5) with τ_{i^0} replaced by any other root function τ_i.) The g corresponds then to the Cauchy data and the q is a rational expression in terms of the root functions τ_i. Of course, in that case, we will have to consider integrals of form (5) extended over all of R^n . Using partitions of unity built from functions of type κ we can then cut down to small conic sets and thus localize the difficulties. What we can then do with the aid of the theorems 1 and 3 is estimate the contribution to the solutions of the Cauchy problem which come from a small conic neighborhood of a singular direction of uniplanar, respectively of conic, type. It is a fact that for scalar equations the functions q which appear in this way have the regularity properties which one needs in our arguments. However these q will blow up near the singular directions for such equations and will not satisfy the condition $|q(\xi)| \leq c|\xi|^{\mu}$. It is therefore interesting to observe that the $q's$ which one obtains for the system of elasticity for cubic crystals do not blow up near singular directions.

What we want to do next is to see how one can obtain estimates on the decay of $u(t,x)$ from Theorem 1. Let us then chose linear coordinates in R^n_ξ so that $(\partial/\partial\xi_i)\tau_{i^0}(\xi^0) = 0$, $i = 1,\dots,n-1$, $(\partial/\partial\xi_n)\tau_{i^0}(\xi^0) \neq 0$, so that in particular $S^\pm$ can locally near ξ^0 be written as $\{\xi; |\xi' - \xi'^0| < \delta, \xi_n = h^\pm(\xi')\}$ for some functions $\xi' \to h^\pm(\xi')$, $\xi' = (\xi_1, \xi_2, \dots, \xi_{n-1})$. The assumptions imply that the functions $h^\pm(\xi')$ are C^1 in a neighborhood U of ξ'^0 and that they lie in $C^\infty(U' \setminus \{\xi'^0\})$ for some perhaps smaller neighborhood U' of ξ'^0. It also follows that $\xi_n^0 \neq 0$, since otherwise we had $\tau_{i^0}(\xi^0) = \langle \xi^0, \nabla\tau_{i^0}(\xi^0)\rangle = 0$ in view of Euler's relation for homogeneous functions. To make a choice, we shall assume that $\xi_n^0 > 0$. The relation between the functions $h^\pm$ and τ_{i^0}, τ_{i^0+1}, is of course $\tau_{i^0}(\xi', h^+(\xi')) \equiv \tau_{i^0+1}(\xi', h^-(\xi')) \equiv 1$. It follows from this that $\nabla_{\xi'} h^+(\xi') = -\nabla_{\xi'}\tau_{i^0}(\xi', h^+(\xi'))/[(\partial/\partial\xi_n)\tau_{i^0}(\xi', h^+(\xi'))]$. We can conclude therefore applying once more Euler's relation for homogeneous functions that

$$L(\xi') \stackrel{\text{def}}{=} h^+(\xi') - \langle \xi', \nabla_{\xi'} h^+(\xi')\rangle = \tau_{i^0}(\xi', h^+(\xi'))/[(\partial/\partial\xi_n)\tau_{i^0}(\xi', h^+(\xi'))],$$

so we get $L(\xi') = 1/(\partial/\partial\xi_n)\tau_{i^0}(\xi', h^+(\xi'))$. In particular we have $L(\xi'^0) \neq 0$, and can therefore assume that $L(\xi') \neq 0$ for $|\xi' - \xi'^0| \leq \varepsilon$, if we shrink ε.

We need also consider in this context the map $\xi' \to \eta(\xi')$, $\eta(\xi') = \begin{pmatrix} \xi' \\ h^+(\xi') \end{pmatrix}$, respectively the map $(\nu, \theta') \to \psi(\nu, \theta') = \nu\eta(\theta')$, if $|\theta' - \xi'^0| < \varepsilon$ and $\nu > 0$. Both maps are C^1 and ψ is a C^1-diffeomorphism from $\{\nu > 0, |\theta' - \xi'^0| < \delta\}$, δ small, to a conic neighborhood of ξ^0 since we have that $\det\left(\dfrac{\partial\psi(\nu,\theta')}{\partial(\theta',\nu)}\right)(\nu, \xi'^0) = \nu^{n-1}L(\xi'^0) \neq 0$. Indeed,

$$\det\left(\frac{\partial\psi(\nu,\theta')}{\partial(\nu,\theta')}\right) = \det\begin{pmatrix} \nu & 0 & \dots & 0 & \theta_1 \\ 0 & \nu & \dots & 0 & \theta_2 \\ 0 & 0 & \dots & 0 & \theta_3 \\ \cdot & \cdot & \dots & \cdot & \cdot \\ \cdot & \cdot & \dots & \cdot & \cdot \\ \cdot & \cdot & \dots & \cdot & \cdot \\ 0 & 0 & \dots & \nu & \theta_{n-1} \\ \nu\dfrac{\partial h^+}{\partial\theta_1}(\theta') & \nu\dfrac{\partial h^+}{\partial\theta_2}(\theta') & \dots & \nu\dfrac{\partial h^+}{\partial\theta_{n-1}}(\theta') & h^+(\theta') \end{pmatrix},$$

and the last determinant is $[h^+(\theta') - \langle\theta', \nabla_{\theta'} h^+(\theta')\rangle]\nu^{n-1}$, when calculated explicitly. We also have that $|\eta(\theta')| \geq c$ if $|\theta' - \xi'^0| < \varepsilon$ since $\eta(\xi'^0) = \xi^0 \neq 0$.

To study the integral in (5) we can therefore make the change of variables $(\nu,\theta') \to \psi(\nu,\theta') = \xi$. It follows from the above calculations and notations that we have

$$\int_0^\infty e^{it\nu}\nu^{n-1+\mu}\,d\nu \int_{|\theta'-\xi'^0|<\varepsilon} e^{i\nu x_n h^+(\theta')+i\langle \nu x',\theta'\rangle} q(\eta(\theta'))(\kappa g)(\nu\eta(\theta'))L(\theta')\,d\theta',$$

provided the conical support of κ has been chosen small enough initially. We want here to apply Theorem 1 to the inner integral which is of the type considered in (1), although it is of course extended only over S^+, rather than S. The role of λ in Theorem 1 will be played by νx and the dimension of the problem is by one less than there. Of course this can at most give an estimate of form $|u(t,x)| \leq c|x|^{-(n-1)/2}$, which will only be interesting when $|t|$ can be dominated by $|x|$. It should be therefore observed at this point that the estimate of u when $|t|$ dominates $|x|$ is in general easier to obtain. In any case, when we want to apply Theorem 1, than there are no problems with the geometry of the surface S, since we have assumed just that it is as required in the theorem. The only problems which remain still to be checked are

A) that the regularity of the function $\theta' \to q(\eta(\theta'))(\kappa g)(\nu\eta(\theta'))L(\theta')$ is as required in the proof of the theorem and

B) that the estimates depend on ν in a way which makes it possible to integrate in ν without loosing decay.

Both requirements can be checked in particular situations and are in fact true e.g. when u is a component of a solution of the system of elasticity for cubic crystals near a singular direction of uniplanar type.

Details of the arguments will be given elsewhere.

References

[1] Duff, G.F.D., The Cauchy problem for elastic waves in an anisotropic medium, Phil.Transactions Royal Soc. London, Ser. A Nr. 1010, 252, (1960), 249–273.

[2] Gallot S., Hubin D., Lafontaine J., Riemannian Geometry, Springer Universitext. Second Edition, Springer Verlag, 1990.

[3] Herz, C.S., Fourier transforms related to convex sets, Ann. Math. 75 (1962), 81–92.

[4] Hlawka, E., Über Integrale auf konvexen Körpern, Monatshefte der Mathematik, Bd. 54:1 (1950), 1–36.

[5] Hörmander, L., The Analysis of Linear Partial Differential Operators I, Springer Verlag Grundlehren, vol. 256, 1983.

[6] Klainerman, S., Global existence for nonlinear wave equations, Comm. Pure Appl. Math. 33 (1980), 43–101.

[7] Klainerman, S., Ponce, G., Global, small amplitude solutions to nonlinear evolution equations, Comm. Pure Appl. Math. 36 (1983), 133–141.

[8] Liess, O., Decay estimates for the solutions of the system of crystal optics, Asymptotic Analysis 4 (1991), 61–95.

[9] Liess, O., Global existence for the nonlinear equations of crystal optics. Proc. conf. "Journees equations aux der. partielles", Saint-Jean-de-Monts, 1989.

[10] Liess, O., Conical refraction and higher microlocalization, Springer Lecture Notes in Mathematics, vol. 1555, 1993.

[11] Racke, R., Lectures on nonlinear evolution equations: initial value problems, Vieweg Verlag, Wiesbaden, Aspects of mathematics, E 19, 1992.

[12] Segal, I.E., Dispersion for nonlinear nonrelativistic equations, Ann. E.N.S. 4 (1968) 459–497.

[13] Segal, I.E., Space time decay for solutions of the wave equation, Adv.Math. 22 (1976) 305–311.

[14] Shatah, J., Global existence of small solutions to nonlinear evolution equations, J.D.E., 46 (1982) 409–425.

[15] Shilov, G.E., Generalized functions and partial differential equations, Gordon and Breach, New York, 1968. (Translated from Russian)

[16] Sommerville, D.M.Y., Analytical Geometry of three Dimensions, Cambridge, at the University Press, 1934.

[17] Stein, E.M., Harmonic analysis: real variable methods, orthogonality, and oscillatory integrals, Princeton University Press, Princeton, 1993.

[18] Strauss, W., Decay and asymptotics for $\Box u = F(u)$, J. Funct. Anal. 2 (1968), 409–457.

[19] von Wahl, W., L^p-decay rates for homogeneous wave equations, Math. Z. 120(1971), 93–106.

Otto Liess
University of Bologna, Department of Mathematics
Bologna, Italy.

Operator Theory:
Advances and Applications, Vol. 78

On Adiabatic Reduction Theory

A. Martinez, G. Nenciu

1. Introduction.

Consider, in a Hilbert space $\mathcal{H}$, the evolution:

$$i\varepsilon \frac{d}{dt} U_\varepsilon(t,s) = H(t) U_\varepsilon(t,s) \quad ; \quad U_\varepsilon(s,s) = 1 \quad ; \quad t,s \in \mathbb{R} \tag{1.1}$$

in the limit $\varepsilon \to 0$, where $H(t)$ is self-adjoint, approaches limits $H_\pm$ sufficiently fast as $t \to \pm\infty$, and satisfies the gap condition:

$$(H1) \qquad \begin{array}{l} \sigma := \sigma(H(t)) = \sigma_1(t) \cup \sigma_2(t)\ ,\ t \in \mathbb{R}. \\ \mathrm{Inf}_{t\in\mathbb{R}}\ \mathrm{dist}\big(\sigma_1(t)\ ,\ \sigma_2(t)\big) = d > 0. \end{array} \tag{1.2}$$

Let $P_j(t)$ $(j = 1,2)$ denote the spectral projections of $H(t)$ corresponding to $\sigma_j(t)$, $P_{j,\pm} = \lim_{t\to\pm\infty} P_j(t)$, and $\mathbf{S}$ be the scattering operator associated to (1.1):

$$\mathbf{S} = \lim_{\substack{t\to+\infty \\ s\to-\infty}} e^{itH_+/\varepsilon} U_\varepsilon(t,s) e^{-isH_-/\varepsilon}. \tag{1.3}$$

The matrix elements of $\mathbf{S}$ are defined by

$$S_{j,k} = P_{j,+} \mathbf{S} P_{k,-} \tag{1.4}$$

and we set

$$\mathcal{P}_{j,k} = \|S_{j,k}\|^2 \quad ; \quad \mu = \mathcal{P}_{1,2}^{1/2} + \mathcal{P}_{2,1}^{1/2}. \tag{1.5}$$

The adiabatic reduction theory [12, 6, 10] (see also [8, 2, 11, 13, 1, 3] for earlier and related results) is concerned with the following two problems:

(i) The proof of the fact that the transition probabilities $\mathcal{P}_{j,k}$ $(j \neq k)$ are small in the limit $\varepsilon \to 0$;

(ii) The approximation of $S_{j,j}$ $(j = 1,2)$ by scattering operators $\tilde{S}_j$ of some effective self-adjoint families $h_{\varepsilon,j}(t)$ of Hamiltonians in reduced Hilbert spaces $\mathcal{H}_j$ of dimensions $\dim \mathcal{H}_j = \mathrm{Rank} P_j(0)$.

In the analytic case (i.e. when $H(t)$ has a holomorphic continuation in a complex strip around real axis), it has been shown that $\mathcal{P}_{j,k} = \mathcal{O}(e^{-k_1/\varepsilon})$, and that $S_{j,j}$ can be computed via an effective evolution up to errors of order $\mathcal{O}(e^{-k_2/\varepsilon})$ $(k_1, k_2 > 0)$.

In [12, 6, 10] these results are proved by constructing approximate $U_\varepsilon(t,s)$-invariant subspaces, i.e. families of projections $P_\varepsilon(t)$ satisfying for some positive k:

$$\left\| i\varepsilon \frac{d}{dt} P_\varepsilon(t) - [H(t), P_\varepsilon(t)] \right\| = \mathcal{O}(e^{-k/\varepsilon}). \tag{1.6}$$

which gives $k_1 = k_2 = k$. The projections P_ε were obtained in [12] (see also [14]) by resumation of an asymptotic series

$$P_\varepsilon(t) \sim \sum_{j=0}^{\infty} E_j(t)\varepsilon^j. \tag{1.7}$$

which does not provide a very good control on k, which can be quite small (the same is true for the iteration scheme used in [6]). On the other hand it is proved in [9], using a priori microlocal exponential weighted estimates, that $\mathcal{P}_{1,2} + \mathcal{P}_{2,1} = \mathcal{O}(e^{-2\Sigma/\varepsilon})$, where Σ is arbitrarily close to some positive constant Σ_0, which has a nice geometrical interpretation. Then a natural question is whether one can also improve the value of k_2. Actually, at the heuristic level, one expects to be possible to construct effective Hamiltonians such that $k_2 \sim -\varepsilon \ln(\mathcal{P}_{1,2}^{1/2} + \mathcal{P}_{2,1}^{1/2})$, i.e. the effective evolution approximates the actual one (after an appropriate identification) up to errors of the order of the square roots of transition probabilities.

We prove below that it is indeed not hard to construct a general reduction scheme with the above properties. The hard problem is to prove that the effective Hamiltonian $h_\varepsilon(t)$ we obtain in this way has the right analytic properties. More precisely, assuming that $H(t)$ is holomorphic inside a strip, one would like that $h_\varepsilon(t)$ also has a holomorphic continuation in a (possibly narrower) complex strip with uniform estimates there as $\varepsilon \longrightarrow 0$. This is an important issue since the analytic properties of $h_\varepsilon(t)$ are crucial in obtaining the Dykhne-Landau-Zener formulae for the transition probabilities (see e.g. [4, 10, 3] and references therein). Up to now, our results in this respect are incomplete: we are able to prove that $h_\varepsilon(t)$ has the required analytic properties only when $H(t)$ is a bounded operator. The optimal result (if true) would be the holomorphy of $h_\varepsilon(t)$ in the largest strip for which the gap condition for $H(t)$ still holds true.

Section 2 contains the result and Section 3 the proofs. In the appendix, taking advantage of the simplicity of the iteration scheme given by Sjöstrand in [14], we simplify and improve the estimates of [12] on the E_j's.

2. The result.

In addition to the gap condition (H1), we assume that $H(t)$ is bounded and satisfies:

(H2)
$H(t)$ has a holomorphic bounded continuation $H(z)$ in $\mathcal{I}_\lambda = \{|\mathrm{Im}\, z| < \lambda\}$, and there exist $B < \infty$, $\eta > 2$ and operators $H_\pm$ such that
$\mathrm{Sup}_{z\in\mathcal{I}_\lambda} \langle \mathrm{Re}\, z\rangle^\eta \|\frac{d}{dz}H(z)\| \le B$; $\|H(t) - H_\pm\| = \mathcal{O}(\langle t\rangle^{-\eta})$ $(t \in \mathbb{R}$, $\pm t > 0)$. (2.1)

where $\langle t\rangle = (1+t^2)^{1/2}$. In what follows, $A_0(t,s)$ is the parallel transport of $P_1(t)$, i.e. the solution of

$$i\frac{d}{dt}A_0(t,s) = i[\dot{P}_1(t), P_1(t)]A_0(t,s) \quad ; \quad A_0(s,s) = \mathbb{1}. \tag{2.2}$$

and in particular $A_0(t,s)$ has limits $A_0(\pm\infty, s)$ and $A_0(t,\pm\infty)$ as t or s tends to $\pm\infty$.

Theorem. Assume (H1)–(H2), and suppose $\mu < 1/8$ for $\varepsilon \in (0, \varepsilon_0]$, $\varepsilon_0 > 0$ fixed. Then there exist $\lambda_1 > 0$, and a family $h_\varepsilon(z) \in \mathcal{L}(\text{Ran } P_1(0))$ such that for $0 < \varepsilon \le \varepsilon_0$:

(i) $h_\varepsilon(z)$ depends holomorphically on z in $\mathcal{I}_{\lambda_1}$;

(ii) There exists $b < \infty$ such that for all $z \in \mathcal{I}_{\lambda_1}$

$$\begin{aligned} &\|\frac{d}{dz} h_\varepsilon(z)\| \le b\langle \text{Re } z\rangle^{1-\eta}, \\ &\|h_\varepsilon(t) - h_\pm\| = \mathcal{O}(\langle t\rangle^{1-\eta}), \quad (t \in \mathbb{R}\,,\ \pm t > 0), \end{aligned} \tag{2.3}$$

where

$$h_\pm = P_1(0)A_0(0,\pm\infty)H_\pm A_0(\pm\infty,0)P_1(0), \tag{2.4}$$

(iii) $h_\varepsilon(t)$ has uniform asymptotic expansion in $\mathcal{I}_{\lambda_1}$ as $\varepsilon \to 0$. In particular

$$h_\varepsilon(z) = P_1(0)A_0(0,z)H(z)A_0(z,0)P_1(0) + \mathcal{O}(\varepsilon), \tag{2.5}$$

(iv) If

$$\tilde{S} = \lim_{\substack{t\to+\infty \\ s\to-\infty}} e^{ith_+/\varepsilon}\tilde{U}_\varepsilon(t,s)e^{-ish_-/\varepsilon}, \tag{2.6}$$

where

$$i\varepsilon\frac{d}{dt}\tilde{U}_\varepsilon(t,s) = h_\varepsilon(t)\tilde{U}_\varepsilon(t,s)\ ;\ \tilde{U}_\varepsilon(s,s) = \mathbb{1}, \tag{2.7}$$

then

$$\|S_{1,1} - A_0(+\infty,0)\tilde{S}A_0(0,-\infty)\| \le \frac{2\mu}{(1-2\sqrt{2\mu})^2}. \tag{2.8}$$

Remark 1. Suppose in particular that

$$\sigma_1(t) = e_1(t) \cup e_2(t) \quad ; \quad \inf_{t\in\mathbb{R}} \text{dist}\big(e_1(t), e_2(t)\big) > 0.$$

Let $p_j(t)$ be the spectral projection of $H(t)$ corresponding to $e_j(t)$, $p_{j,\pm} = \lim_{t\to\pm\infty} p_j(t)$, and $s_{1,2} = p_{1,+}\mathbf{S}p_{2,-}$. Notice that $\sigma\big(h_\varepsilon(t)\big) = \sigma_1(t) + \mathcal{O}(\varepsilon)$ and let (for ε sufficiently small) $\hat{p}_{j,\varepsilon}(t)$ be the spectral projection of $h_\varepsilon(t)$ associated to $e_j(t) + \mathcal{O}(\varepsilon)$. If $\tilde{s}_{1,2} = \hat{p}_{1,+}\tilde{S}\hat{p}_{2,-}$, we get from (2.8):

$$\|s_{1,2} - A_0(+\infty,0)\hat{s}_{1,2}A_0(0,-\infty)\| \le \frac{2\mu}{(1-2\sqrt{2\mu})^2}.$$

Therefore

$$|\ \|s_{1,2}\| - \|\hat{s}_{1,2}\|\ | \le \frac{2\mu}{(1-2\sqrt{2\mu})^2},$$

i.e. the square root of the transition probabilities computed from the effective two-level system gives the square root of the transition probabilities between the two levels of the initial system up to errors of order μ. This is in accordance with the heuristic intuition.

Remark 2. If in addition to (2.1) we also assume

$$\operatorname*{Sup}_{\substack{z\in\mathcal{I}_\lambda\\ \pm\mathrm{Re}\ z\geq 0}} \langle \mathrm{Re}\ z\rangle^\eta \|H(z)-H_\pm\| \leq B,$$

then our proof also gives

$$\operatorname*{Sup}_{\substack{z\in\mathcal{I}_{\lambda_1}\\ \pm\mathrm{Re} z\geq 0}} \langle \mathrm{Re} z\rangle^{\eta-1} \|h_\varepsilon(z)-h_{\varepsilon,\pm}\| \leq b.$$

In particular, the analytic assumptions made in [5, 6] are verified by $h_\varepsilon(t)$, and therefore their results concerning the Dykhne formula can in principle be applied to h_ε under suitable geometrical assumptions.

Remark 3. Our proof gives the following value for λ_1:

$$\lambda_1 = \frac{\lambda d}{16eR(1+2B\lambda/d)},$$

where $R \leq \operatorname*{Sup}_{z\in\mathcal{I}_\lambda} \|H(z)\|$.

3. The proof.

For $z \in \mathcal{I}_\lambda$ define

$$Q_\pm(z) = \lim_{s\to\pm\infty} U_\varepsilon(z,s)P_1(s)U_\varepsilon(s,z), \tag{3.1}$$

where the existence of the limits comes from a standard "Cook's method" argument. Then at any fixed ε, $Q_\pm(z)$ is holomorphic in $\mathcal{I}_\lambda$, and it satisfies the Heisenberg equation:

$$i\varepsilon\frac{d}{dz}Q_\pm(z) = [H(z), Q_\pm(z)]. \tag{3.2}$$

Now consider for t real the standard adiabatic evolution (see [ASY]) defined by

$$\begin{aligned} H_A(t) &= H(t) + i\varepsilon[\dot{P}_1(t), P_1(t)], \\ i\varepsilon\frac{d}{dt}U_A(t,s) &= H_A(t)U_A(t,s) \quad ; \quad U_A(s,s) = \mathbb{1}, \end{aligned}$$

and denote

$$\mathcal{U} = \lim_{\substack{t\to+\infty\\ s\to-\infty}} U_A(0,t)U_\varepsilon(t,s)U_A(s,0),$$

where the limit exists by same reasons as before. Then by the definitions of $\mathcal{P}_{j,k}$ and μ, and the fact that $P_1(t)U_A(t,s) = U_A(t,s)P_1(s)$ for any (t,s), we get easily:

$$\|[\mathcal{U}, P_1(0)]\| \leq \mu. \tag{3.3}$$

Since also, for t real, $\|Q_+(t) - Q_-(t)\|$ does not depend on t, we get by conjugating with $U_A(0,t)$ and taking the limit $t \to +\infty$:

$$\|Q_+(t) - Q_-(t)\| = \|\mathcal{U} P_1(0)\mathcal{U}^* - P_1(0)\|,$$

and therefore, by (3.3):

$$\|Q_+(t) - Q_-(t)\| \leq \mu. \tag{3.4}$$

Define

$$Q(z) = \chi(z)Q_+(z) + \big(1 - \chi(z)\big)Q_-(z), \tag{3.5}$$

where $\chi(z) = e^{\pi z/4\lambda}/(e^{\pi z/4\lambda} + e^{-\pi z/4\lambda})$. Using (3.4), we get easily for t real:

$$\|Q^2(t) - Q(t)\| \leq 2\mu,$$

which proves that the distance between the spectrum of $Q(t)$ and $\{0,1\}$ is less than $\sqrt{2\mu}$. Therefore, for $\mu < 1/8$ we can set

$$Q_1(t) = \frac{1}{2i\pi}\int_{|\zeta-1|=1/2} \big(\zeta - Q(t)\big)^{-1} d\zeta,$$

which is a projection. Since by (3.2) and (3.4), $Q(t)$ satisfies

$$\|i\varepsilon\dot{Q}(t) - [H(t), Q(t)]\| \leq \pi\varepsilon\mu/2\lambda(e^{\pi t/4\lambda} + e^{-\pi t/4\lambda})^2,$$

we obtain for Q_1:

$$\|i\varepsilon\dot{Q}_1(t) - [H(t), Q_1(t)]\| \leq \frac{\pi\varepsilon\mu}{\lambda(1-2\sqrt{2\mu})^2(e^{\pi t/4\lambda} + e^{-\pi t/4\lambda})^2} =: \delta(\varepsilon, t). \tag{3.6}$$

Applying in our situation the method of [1], we also get:

$$\|Q_1(t) - P_1(t)\| = \mathcal{O}(\varepsilon\langle t\rangle^{1-\eta}). \tag{3.7}$$

From this point, one can adapt the reduction theory of [11, 10]. Suppose that ε is sufficiently small so that $\|Q_1(t) - P_1(t)\| \leq a < 1$ for all $t \in \mathbb{R}$. Define

$$M(t) = \Big(1 - \big(Q_1(t) - P_1(t)\big)^2\Big)^{-1/2}\Big(Q_1(t)P_1(t) + \big(1 - Q_1(t)\big)\big(1 - P_1(t)\big)\Big). \tag{3.8}$$

Then (see [7] Ch.I.4) $M(t)^{-1}$ exists, $M(t)^{-1} = M(t)^*$, and

$$Q_1(t) = M(t)P_1(t)M(t)^{-1}.$$

Moreover, $M(t)$ satisfies:

$$\begin{aligned} &M(t) = \mathbb{1} + \mathcal{O}(\varepsilon\langle t\rangle^{1-\eta}),\\ &\lim_{t\to\pm\infty}\frac{d}{dt}M(t) = \lim_{t\to\pm\infty}\frac{d}{dt}M(t)^{-1} = 0. \end{aligned} \tag{3.9}$$

Now, consider

$$H_1(t) = H(t) + \big(1 - 2Q_1(t)\big)\big(i\varepsilon \dot{Q}_1(t) - [H(t), Q_1(t)]\big)$$

and the associated evolution $U_1(t,s)$:

$$i\varepsilon \frac{d}{dt} U_1(t,s) = H_1(t) U_1(t,s) \quad ; \quad U_1(s,s) = 1.$$

It satisfies (see e.g. [11]): $Q_1(t)U_1(t,s) = U_1(t,s)Q_1(s)$. Define

$$\Phi(t,s) = A_0(0,t) M(t)^{-1} U_1(t,s) M(s) A_0(s,0). \tag{3.10}$$

We then have:

$$\begin{aligned} &[\Phi(t,s), P_1(0)] = 0, \\ &i\varepsilon \frac{d}{dt} P_1(0)\Phi(t,s)P_1(0) = h_\varepsilon(t) P_1(0)\Phi(t,s)P_1(0), \end{aligned} \tag{3.11}$$

with

$$h_\varepsilon(t) = P_1(0)A_0(0,t)M(t)^* \left(H(t) + i\varepsilon M(t)\dot{M}(t)^*\right) M(t)A_0(t,0)P_1(0). \tag{3.12}$$

We claim that $h_\varepsilon(t)$ has the required properties announced in the theorem. First of all, using (3.9) we see that (2.4) is true, as well as (2.5) for real z. Also, denoting $\tilde{U}_\varepsilon(t,s) = P_1(0)\Phi(t,s)P_1(0)$ and defining $\tilde{S}$ as in (2.6), we can prove (2.8) in the following way. Let $\Omega(t,s) = U_1(s,t)U_\varepsilon(t,s)$. Then using (3.6) we get

$$\|\Omega(t,s) - \mathbb{1}\| \leq \frac{1}{\varepsilon} \left| \int_s^t \delta(\varepsilon,u)du \right| \leq \frac{2\mu}{(1-2\sqrt{2\mu})^2},$$

and therefore, using (3.9)-(3.10):

$$\|\tilde{S} - \lim_{\substack{t\to+\infty \\ s\to-\infty}} e^{ith_+/\varepsilon} P_1(0)A_0(0,t)U_\varepsilon(t,s)A_0(s,0)P_1(0)e^{-ish_-/\varepsilon}\| \leq \frac{2\mu}{(1-2\sqrt{2\mu})^2}.$$

This estimate and (2.4) permit to deduce (2.8) easily. Now, it remains to show the analytic properties of $h_\varepsilon(t)$. By construction, it is sufficient to investigate those of $Q_\pm(z)$, and we restrict our attention to $Q_-(z)$ (the other one being similar). For t real and ζ outside $\sigma(t)$, we define recursively $q_j(t,\zeta)$ by:

$$q_0(t,\zeta) = \big(\zeta - H(t)\big)^{-1} \quad ; \quad q_j(t,\zeta) = i\big(\zeta - H(t)\big)^{-1} \frac{d}{dt} q_{j-1}(t,\zeta), \tag{3.13}$$

and we set

$$E_j(t) = \frac{1}{2i\pi} \int_{\gamma(t)} q_j(t,\zeta) d\zeta, \tag{3.14}$$

where $\gamma(t)$ is a complex simple loop enclosing $\sigma_1(t)$, leaving $\sigma_2(t)$ in its exterior, and keeping a distance larger than $d/2$ from $\sigma(t)$. Then, at the level of formal power-series in ε, we have (see [14]):

$$\begin{aligned} i\varepsilon\frac{d}{dt}\sum_{j\geq 0} E_j(t)\varepsilon^j &= [H(t), \sum_{j\geq 0} E_j(t)\varepsilon^j], \\ \Big(\sum_{j\geq 0} E_j(t)\varepsilon^j\Big)^2 &= \sum_{j\geq 0} E_j(t)\varepsilon^j. \end{aligned} \tag{3.15}$$

It follows from Lemma 1 in [12] that E_j given by (3.14) coincide with those constructed in [12]. Taking advantage of the fact that (3.13)-(3.14) is simpler than the induction formula in [12], one can improve and simplify the estimations given there. More precisely, the results of the appendix and a convenient choice of γ show that for $k+j \geq 1$:

$$\left\| \frac{d^k}{dt^k} E_j(t) \right\| \leq C_0 A_1^j B_1^k (j+k)! \langle t \rangle^{-\eta},$$

with $C_0 = 1 + (2/\pi d)\,\mathrm{Sup}_{t\in\mathbb{R}}\,\mathrm{length}(\sigma_1(t))$, $A_1 = 4B_1/d$, and $B_1 = 4Bd^{-1} + 2\lambda^{-1}$. Using that $(j+k)! \leq e^{j+k} j! k!$, it follows that for any $j \geq 1$, $E_j(t)$ has a holomorphic continuation in $\mathcal{I}_{(1/eB_1)}$, and satisfies there:

$$\|E_j(z)\| + \|\dot{E}_j(z)\| \leq \mathrm{const.}\langle t \rangle^{-\eta} (eA_1)^j j!. \tag{3.16}$$

Now consider for $N \geq 1$:

$$T_N(z) = \sum_{j=0}^{N} E_j(z)\varepsilon^j \quad ; \quad \Delta_N(z) = Q_-(z) - T_N(z), \tag{3.17}$$

which are holomorphic in $\mathcal{I}_\lambda$ (but not uniformly bounded as $\varepsilon \to 0$). We wish to prove that for a suitable choice of $N = N(\varepsilon)$, both $T_N(z)$ and $\Delta_N(z)$ (and thus also $Q_-(z)$) satisfy uniform estimates in $\mathcal{I}_{\lambda_1}$ for some positive λ_1. Using (3.2) and (3.15), we get

$$i\varepsilon\dot{\Delta}_N = [H, \Delta_N] - i\dot{E}_N\varepsilon^{N+1},$$

and therefore

$$\Delta_N(z) = -\varepsilon^N \int_{-\infty}^{z} U_\varepsilon(z,s)\dot{E}_N(s)U_\varepsilon(s,z)ds, \tag{3.18}$$

where the integration is made along any path $\Gamma(z)$ inside $\mathcal{I}_\lambda$. In particular, denoting $z = t + i\tau$, $(t, \tau$ real) and taking $\Gamma(z) = (-\infty, t] \cup [t, t+i\tau]$, we get:

$$\Delta_N(z) = U_\varepsilon(z,t)\Delta_N(t)U_\varepsilon(t,z) - i\varepsilon^N \int_0^{\tau} U_\varepsilon(z,t+is)\dot{E}_N(t+is)U_\varepsilon(t+is,z)ds. \tag{3.19}$$

Let $D(z)$ be the smallest complex disc containing $\sigma(H(z))$, and denote $c(z)$ its centrum, $R(z)$ its radius, and $R = \sup_{z\in\mathcal{I}_{(1/eB_1)}} R(z)$. For z, z' in $\mathcal{I}_{(1/eB_1)}$, consider

$$\hat{U}_\varepsilon(z,z') = U_\varepsilon(z,z')\exp\Big(-\frac{i}{\varepsilon}\int_z^{z'} c(s)ds\Big).$$

Then

$$i\varepsilon \frac{d}{dz}\hat{U}_\varepsilon(z,z') = \big(H(z) - c(z)\big)\hat{U}_\varepsilon(z,z'),$$

and the Gronwall inequality gives for t real, $|\tau| < 1/eB_1$ and $|s| < 1/eB_1$:

$$\|\hat{U}_\varepsilon(t+i\tau, t+is)\| \le e^{R|\tau - s|/\varepsilon}. \tag{3.20}$$

Then choose $N = N_\varepsilon = [1/eA_1\varepsilon]$. From the Stirling formula and (3.16), we get for $|\mathrm{Im} z| < 1/eB_1$:

$$\|\varepsilon^{N_\varepsilon}\dot{E}_{N_\varepsilon}(z)\| \le \mathrm{const.}\varepsilon^{-1/2}e^{-1/eA_1\varepsilon}\langle t\rangle^{-\eta}. \tag{3.21}$$

For $t \in \mathbb{R}$, one has from (3.18) and (3.21):

$$\|\Delta_{N_\varepsilon}\| \le \mathrm{const.}\varepsilon^{-1/2}e^{-1/eA_1\varepsilon}\langle t\rangle^{1-\eta}. \tag{3.22}$$

Since in (3.19) one can replace U_ε by $\hat{U}_\varepsilon$, one has from (3.20)-(3.22), and for $z = t + i\tau \in \mathcal{I}_{(1/eB_1)}$:

$$\|\Delta_{N_\varepsilon}(z)\| \le \mathrm{const.}\varepsilon^{-1/2}e^{(2R|\tau| - 1/eA_1)/\varepsilon}\langle t\rangle^{1-\eta}. \tag{3.23}$$

Take

$$0 < \lambda_1 < \frac{1}{2eRA_1}.$$

Then for $z \in \mathcal{I}_{\lambda_1}$, we get from (3.23):

$$\Delta_{N_\varepsilon}(z) \le \mathrm{const.}\langle t\rangle^{1-\eta}e^{-\delta/\varepsilon} \tag{3.24}$$

with some $\delta > 0$. Also, from the Stirling formula and (3.16), we see that $T_{N_\varepsilon}(z)$ is uniformly bounded in $\mathcal{I}_{\lambda_1}$, so that from (3.24) and (3.17), we obtain finally that, for $z \in \mathcal{I}_{\lambda_1}$, $Q_-(z)$ is uniformly bounded as $\varepsilon \to 0$, and moreover:

$$Q_-(z) = T_{N_\varepsilon}(z) + \mathcal{O}\left(\langle t\rangle^{1-\eta}e^{-\delta/\varepsilon}\right). \tag{3.25}$$

The same result holds for $Q_+(z)$, and in view of (3.5), (3.8) and (3.12), this finishes the proof of the theorem.

Appendix

Lemma A.1. Under the assumptions of the theorem, and for $\zeta \in \mathbb{C}$ such that $\mathrm{dist}\big(\zeta, \sigma(t)\big) \ge g > 0$, one has for any real t and any $k \ge 1$:

$$\left\| \frac{d^k}{dt^k}\big(\zeta - H(t)\big)^{-1} \right\| \le \frac{1}{g}A_0^k k!\langle t\rangle^{-\eta},$$

where $A_0 = \frac{B}{g} + \frac{1}{\lambda}$.

Proof. The assumtions and the Cauchy formulae imply for $k \geq 1$:

$$\left\| \frac{d^k}{dt^k} H(t) \right\| \leq B\lambda^{1-k} k! \langle t \rangle^{-\eta}. \tag{A.1}$$

Denoting $q_0(t,\zeta) = \left(\zeta - H(t)\right)^{-1}$, we then get by Leibniz formula:

$$\frac{d^k}{dt^k} q_0 = q_0 \sum_{\ell=1}^{k} \frac{k!}{\ell!(k-\ell)!} \frac{d^\ell}{dt^\ell} H. \frac{d^{k-\ell}}{dt^{k-\ell}} q_0. \tag{A.2}$$

We notice that the result is obvious for $k = 1$, and we proceed by induction. We then get from (A.2):

$$\begin{aligned} \left\| \frac{d^k}{dt^k} q_0 \right\| &\leq \frac{Bk!}{g^2} \langle t \rangle^{-\eta} \left(\lambda^{1-k} + \sum_{\ell=1}^{k-1} A_0^{k-\ell} \lambda^{1-\ell} \langle t \rangle^{-\eta} \right) \\ &\leq \frac{\lambda B k!}{g^2} A_0^k \langle t \rangle^{-\eta} \sum_{\ell=1}^{\infty} (A_0 \lambda)^{-\ell} \quad \leq \quad \frac{\lambda B}{g(\lambda A_0 - 1)} \frac{1}{g} A_0^k k! \langle t \rangle^{-\eta}. \end{aligned}$$

from which the result follows since $\frac{\lambda B}{g(\lambda A_0 - 1)} = 1$.

Lemma A.2. Defining $q_j(t,\zeta)$ as in (3.13), one has for any ζ such that dist $\left(\zeta, \sigma(t)\right) \geq g > 0$ and any $j, k \geq 0$, $j + k \geq 1$:

$$\left\| \frac{d^k}{dt^k} q_j(t,\zeta) \right\| \leq \frac{1}{g} A_1^j B_1^k (j+k)! \langle t \rangle^{-\eta},$$

where A_1 and B_1 are any positive numbers satisfying:

$$B_1 > A_0 \quad ; \quad A_1 \left(1 - \frac{A_0}{B_1} \right) > \frac{B_1}{g}.$$

Proof. By Lemma A.1, the result is true for $j = 0$. Since

$$\frac{d^k}{dt^k} q_j = \sum_{\ell=0}^{k} \frac{k!}{\ell!(k-\ell)!} \frac{d^\ell}{dt^\ell} q_0 . \frac{d^{k-\ell}}{dt^{k-\ell}} q_{j-1},$$

we get by an induction argument:

$$\left\| \frac{d^k}{dt^k} q_j \right\| \leq \frac{1}{g} A_1^j B_1^k (j+k)! \langle t \rangle^{-\eta} \sum_{\ell=0}^{k} \frac{k!(j+k-\ell)!}{(k-\ell)!(j+k)!} \frac{B_1 A_0^\ell}{g A_1 B_1^\ell},$$

and thus, since $\frac{k!(j+k-\ell)!}{(k-\ell)!(j+k)!} \leq 1$:

$$\left\| \frac{d^k}{dt^k} q_j \right\| \leq \frac{1}{g} A_1^j B_1^k (j+k)! \langle t \rangle^{-\eta} \frac{B_1}{g A_1} \left(1 - \frac{A_0}{B_1} \right)^{-1},$$

from which the result follows easily.

References

[1] J.E. Avron, R. Seiler, L.G. Yaffe: Adiabatic Theorems and Applications to the Quantum Hall Effect, Comm. Math. Phys. 110, 33–89 (1987).

[2] L.M. Garrido: Generalized adiabatic invariance, J. Math. Phys. 5, 355–362 (1964).

[3] V. Jaksic, J. Segert: Exponential approach to the adiabatic limit and the Landau-Zener formula, Rev. Math. Phys., Vol.4 (4), 529–574 (1992).

[4] A. Joye: Proof of the Landau-Zener formula, Preprint CPT Marseille (1992).

[5] A. Joye, H. Kunz, C.E. Pfister: Exponential decay and geometric aspects of transition probabilities in the adiabatic limit, Annals of Physics 208, 299–332 (1991).

[6] A. Joye, C.E. Pfister: Superadiabatic evolution and transition probability between two non-degenerate levels isolated in the spectrum, J. Math. Phys. 34, 454–479 (1993).

[7] T. Kato: Perturbation theory for linear operators, Springer Berlin, 2nd ed. (1976).

[8] A. Lenard: Adiabatic invariants to all orders, Ann. Phys. 6, 261–276 (1959).

[9] A. Martinez: Precise exponential estimates in adiabatic theory, to appear in J. Math. Phys. (1994).

[10] Ph.A. Martin, G. Nenciu: Semiclassical inelastic S-matrix for one-dimensional N-states systems, Preprint ESI 89 (1994).

[11] G. Nenciu: Adiabatic theorem and spectral concentration, Comm. Math. Phys. 82, 121–135 (1981).

[12] G. Nenciu: Linear adiabatic theory. Exponential estimates, Comm. Math. Phys. 152, 479–496 (1993).

[13] S.J. Sancho: m-th order adiabatic invariance for quantum systems, Proc. Phys. Soc. Lond. 89, 1–5 (1966).

[14] J. Sjöstrand: Projecteurs adiabatiques du point de vue pseudo-différentiel, C. R. Acad. Sci. Paris, t. 317, Série I, 217–220 (1993).

André Martinez
Université Paris-Nord
Département de Mathématiques
CNRS URA n. 742, F-93430 Villetaneuse (France)

Gheorghe Nenciu
University of Bucharest
Department of Theoretical Physics
PO Box MG 11, R-76900 Bucharest (Romania)

Operator Theory:
Advances and Applications, Vol. 78

Recurrence for Fractional Powers of Diffusion Operators in Terms of Volume-Growth

I. McGillivray

1 Introduction.

This contribution summarizes the contents of [11]. There are now sharp conditions available for determining the recurrence of the Laplace-Beltrami operator on a complete Riemannian manifold [2][10] which carry over to the setting of strongly local Dirichlet forms [14]; roughly speaking, quadratic volume growth or less implies recurrence. Let L be the negative-definite self-adjoint diffusion operator associated to a strongly local Dirichlet form. The subordinated semigroup of index α, $0 < \alpha < 2$ is the strongly continuous submarkovian semigroup generated by $-(-L)^{\alpha/2}$. The problem of finding recurrence conditions for subordinated diffusion operators has not received much attention — although see [13], [12]. If L is the Laplace-Beltrami operator on a complete Riemannian manifold with non-negative Ricci curvature, the heat kernel lower bounds [3] entail that $-(-L)^{\alpha/2}$ is recurrent if the volume growth function $v(r)$ satisfies $v(r) \leq cr^{\alpha}$ for large r. We present here a recurrence condition for subordinated diffusion operators in terms of the volume growth function. For one-dimensional diffusion operators our hypotheses are weak, but in the general case we need a strong "radial symmetry" condition. An outcome of our result is that if the derivative $\dot{v}(r)$ of the volume growth function satisfies $\dot{v}(r) \leq cr^{\alpha-1-\epsilon}$ for some $0 < \epsilon \leq \alpha - 1$ and $1 < \alpha < 2$ then $-(-L)^{\alpha/2}$ is recurrent.

2 Some general results.

We start by recalling some definitions and facts from the theory of Dirichlet forms, for which we refer to [5] for more details.

Let $(X,\mathcal{B},m)$ be a σ-finite measure space and $\mathcal{E}$ a symmetric Dirichlet form on $L^2(X,m)$ with domain $D(\mathcal{E})$. The standard inner product in $L^2(X,m)$ is written $(\cdot,\cdot)$. The associated resolvent, semigroup and negative definite self adjoint operator are denoted $(G_\lambda)_{\lambda>0}$, $(T_t)_{t>0}$ and L respectively. The form $\mathcal{E}$ is said to be irreducible if for $B \in \mathcal{B}$, whenever $T_t(1_B u) = 1_B T_t u$ for all $u \in L^2(X,m)$ and $t > 0$ then either $m(B) = 0$ or $m(X - B) = 0$. A strongly continuous submarkovian semigroup is induced on $L^1(X,m)$ by $(T_t)_{t>0}$. Let $\mathcal{B}_+$ stand for the collection of m-a.e. non-negative measurable functions on X and $L^1_+(X,m)$ for $L^1(X,m) \cap \mathcal{B}_+$. We say that $\mathcal{E}$ is transient if $Gf := \lim_{t\to\infty} \int_0^t T_s f\, ds < \infty$ m-a.e. for all $f \in L^1_+(X,m)$, and recurrent if irreducible and non-transient. In this case $Gf = \infty$ m-a.e. for all $f \in \mathcal{B}_+$, $f \not\equiv 0$. The operator G appearing above is called the Green operator.

We assume that the form $\mathcal{E}$ is regular. Thus there exists a properly associated Hunt process $M = (\Omega,\mathcal{F},\mathcal{F}_t,X_t,P_x)$. The form $\mathcal{E}$ is said to be Harris recurrent if for $\mathcal{E}$-q.e. $x \in X$, P_x-a.e. $\int_0^\infty f(X_t)dt = \infty$ for all $f \in \mathcal{B}_+$, $f \not\equiv 0$.

The main object of study in this article is a class of subordinated semigroups analogous to the stable symmetric semigroups of index α, $0 < \alpha < 2$. Let $\gamma_t^{(\alpha)}(s)$, $s > 0$ be the unique continuous function satisfying

$$e^{-ta^{\alpha/2}} = \int_0^\infty e^{-as}\gamma_t^{(\alpha)}(s)ds, \quad a,t > 0,$$

Define

$$T_t^{(\alpha)} := \int_0^\infty T_s\gamma_t^{(\alpha)}(s)ds,\ t > 0. \tag{1}$$

Then $T_t^{(\alpha)}$ is a strongly continuous submarkovian semigroup on $L^2(X,m)$, with corresponding Dirichlet from $\mathcal{E}^{(\alpha)}$. Let $\{E_\lambda : 0 \le \lambda < \infty\}$ be a spectral family for $-L$ in $\mathcal{H} := L^2(X,m)$. By the spectral theorem,

$$D(\mathcal{E}^{(\alpha)}) = \{u \in \mathcal{H} : \int_{[0,\infty)} \lambda^\alpha d(E_\lambda u,u) < \infty\}, \tag{2}$$

$$\mathcal{E}^{(\alpha)}(u,u) = \int_{[0,\infty)} \lambda^{\alpha/2} d(E_\lambda u,u),\ u \in D(\mathcal{E}^{(\alpha)}). \tag{3}$$

One can show that $\mathcal{E}^{(\alpha)}$ regular if $\mathcal{E}$ is.

We now give a recurrence condition at the level of the sample paths of the subordinated process.

Proposition 2.1 $\mathcal{E}^{(\alpha)}$ is recurrent if and only if for m-a.e. $x \in X$,

$$\int_0^\infty t^{\alpha/2-1} f(X_t)dt = \infty\ P_x\text{-a.e.} \tag{4}$$

for all $f \in \mathcal{B}_+$, $f \not\equiv 0$. In this case (4) holds $\mathcal{E}^{(\alpha)}$-q.e.

From the construction of the subordinated process it is clear that any set N of zero $\mathcal{E}$-capacity also has zero $\mathcal{E}^{(\alpha)}$-capacity. The converse however is false, see [4], theorem 6.4.
We finish this section with a generally applicable condition for transience of the subordinated form.

Proposition 2.2 Suppose that $\mathcal{E}$ is transient. Then $\mathcal{E}^{(\alpha)}$ is transient for $0 < \alpha < 2$.

Remark. The converse of this result is false. According to [13] the stable symmetric semigroup of index α, $0 < \alpha < 2$ in dimension $d = 2$ is transient, whereas the Brownian semigroup is recurrent.

3 One-dimensional results.

The aim of this section is to give a recurrence condition for subordinated one-dimensional diffusions.

Let $\mathcal{B}$ stand for the Borel σ-algebra on $\mathbb{R}$ and dx the Lebesgue measure. Let ρ be a locally integrable Borel measurable function satisfying $ess\text{-}\inf_K \rho > 0$ for all compacts K, and m be a strictly positive continuous function on $\mathbb{R}$. For ease of reference we make the following definitions. Define

$$\mathbb{D}_\rho(u,v) := \frac{1}{2}\int_{\mathbb{R}} u'v'\rho \cdot dx \; u,v \in C_0^\infty(\mathbb{R})$$

and $D(\mathbb{D}_\rho^m)$ to be the closure of $C_0^\infty(\mathbb{R})$ in $L^2(\mathbb{R},m\rho^{-1}\cdot dx)$ w.r.t. $\mathbb{D}_{\rho,1}^{1/2}$; the extension of $\mathbb{D}_\rho$ to $D(\mathbb{D}_\rho^m)$ is written $\mathbb{D}_\rho^m$. The above conditions on ρ and m guarantee that $\mathbb{D}_\rho^m$ is an irreducible Dirichlet form on $L^2(\mathbb{R},m\rho^{-1}\cdot dx)$. We also use the following abbreviations: $\mathbb{D}_\rho := \mathbb{D}_\rho^1$, $\mathbb{D}^m := \mathbb{D}_1^m$ and $\mathbb{D} := \mathbb{D}_1^1$. The function $s(x) := \int_0^x \rho^{-1}(y)dy$ is continuous and strictly increasing, so that the inverse function $\bar{s}$ defined by $\bar{s}(s(x)) = x$ is well-defined, continuous, strictly increasing and maps the range $R(s)$ of s onto $\mathbb{R}$. Define a $\rho^{-1}\cdot dx$–symmetric strongly continuous submarkovian semigroup by

$$\bar{T}_t u := T_t(u\circ\bar{s})\circ s, \tag{1}$$

where T_t is the Brownian semigroup on $R(s)$ with Dirichlet boundary conditions. It is easily seen that

$$(u,u-\bar{T}_t u)_{L^2(\mathbb{R},\rho^{-1}\cdot dx)} = (u\circ\bar{s},u\circ\bar{s}-T_t(u\circ\bar{s}))_{L^2(R(s),dx)},$$

thus the Dirichlet form corresponding to $\bar{T}_t$ is $\mathbb{D}_\rho$. Let $M = (\Omega,\mathcal{F},\mathcal{F}_t,X_t,P_x,x\in\mathbb{R})$ be the Brownian motion on $\mathbb{R}$ (corresponding to the form $\mathbb{D}$) in canonical form.

We include the proof of the following result to show how the sample path methods of the last section are applied.

Proposition 3.1 Let ϕ be a finite monotonically increasing function defined on $[0,\infty)$ such that

$$m(x) \le \phi(|\,x\,|) \text{ for all } x\in\mathbb{R},. \tag{2}$$

$$\int_1^\infty r^{-\frac{\alpha}{2-\alpha}+\epsilon}\phi(r)dr < \infty \text{ for some } \epsilon > 0. \tag{3}$$

Then $\mathbb{D}^{m,(\alpha)}$ is recurrent.

Proof. Let ν be the speed measure $m\cdot dx$ whose associated positive continuous additive functional is $A_t = \int_0^t m(X_s)ds$ (w.r.t. M) and with inverse functional τ_t. It suffices to show by 2.1 that for m-a.e. $x\in\mathbb{R}$, $\int_0^\infty t^{\alpha/2-1}f(X_{\tau_t})dt = \infty$ P_x-a.e. for all $f\in\mathcal{B}_+$, $f\not\equiv 0$.

A change of variables leads to

$$\int_0^\infty t^{\alpha/2-1} f(X_{\tau_t})dt = \int_0^\infty (A_t)^{\alpha/2-1} f(X_t)m(X_t)dt.$$

From (2), A_t is dominated by $\int_0^t \phi(| X_s |)ds$. According to the law of the iterated logarithm [9] for all $x \in \mathbb{R}$, P_x-a.e. there exists a positive constant c such that $| X_t |\leq c(2t \log\log t)^{1/2}$ for t sufficiently large. Observe that given $\epsilon > 0$, $c > 0$ there exists a constant c' such that

$$r^{-\frac{1}{2-\alpha}} \int_1^r \phi(c(s\log\log s)^{1/2})ds \leq c' \int_1^\infty u^{-\frac{\alpha}{2-\alpha}+\epsilon}\phi(u)du.$$

Thus by (3), P_x-a.e. there exist constants c_1,c_2,c_3 such that

$$\int_0^t \phi(| X_s |)ds \leq c_1 + \int_0^t \phi(c_2(s\log\log s)^{1/2})ds \leq c_1 + c_3 t^{\frac{1}{2-\alpha}}$$

for t large. It follows that $A_t^{\alpha/2-1} \geq ct^{-1/2}$ for large t and consequently

$$\int_0^\infty t^{\alpha/2-1} f(X_{\tau_t})dt \geq c\int_1^\infty t^{-1/2} f(X_t)m(X_t)dt.$$

Using the fact that $\mathbb{D}^{(1)}$ is recurrent [13] and 3.4 the last expression is P_x-a.s. infinite, as desired. □

With the help of this result, it is possible to give a recurrence condition for subordinated one-dimensional diffusions.

Theorem 3.2 Assume that $| \int_0^{\pm\infty} \rho^{-1}(y)dy |= \infty$. Assume further that there exists a finite monotonically increasing function ϕ defined on $[0,\infty)$ such that

$$m \circ \bar{s}(x) \leq \phi(| x |) \text{ for all } x \in \mathbb{R}, \tag{4}$$

$$\int_1^\infty r^{-\frac{\alpha}{2-\alpha}+\epsilon}\phi(r)dr < \infty \text{ for some } \epsilon > 0. \tag{5}$$

Then $\mathbb{D}_\rho^{m,(\alpha)}$ is recurrent.

4 Strongly local Dirichlet spaces.

In Theorem 4.1 we give a recurrence condition for some subordinated strongly local Dirichlet forms.

Let X be a locally compact separable Hausdorff space, $\mathcal{B}$ the Borel σ-algebra on X, and m an everywhere positive Radon measure on $\mathcal{B}$. Let $\mathcal{E}$ be an irreducible regular Dirichlet form on $L^2(X,m)$ having the strong local property i.e. whenever $u,v \in D(\mathcal{E})$ with $u = v$ on $Supp[v]$, then $\mathcal{E}(u,v) = 0$. In this situation there exists a Radon measure valued positive semi definite symmetric bilinear mapping $(u,v) \to \Gamma(u,v)$ on $D(\mathcal{E}) \times D(\mathcal{E})$ such that $\mathcal{E}(u,v) = \int_X d\Gamma(u,v)$. Set

$$d(x,y) := \sup\{u(x) - u(y) : u \in C_0(X) \cap D(\mathcal{E}), d\Gamma(u,u) \le dm\},\ x,y \in X,$$

and

$$B_r(x) := \{y \in X : d(x,y) < r\},\ x \in X, r > 0.$$

We assume

$$d \text{ is a metric on } X \text{ compatible with the original topology and} \tag{1}$$

$$\text{each ball } B_r(x)\ (x \in X, r > 0) \text{ is relatively compact.} \tag{2}$$

The metric d is called the diffusion metric. Fix $b \in X$ and define $F(x) := d(x,b)$. We write B_r for $B_r(b)$. The volume growth function v is defined by $v(r) := m(B_r)$. Let $\sigma(F)$ be the sub-σ-algebra of $\mathcal{B}$ generated by F and $\mathcal{F}$ the projection from $L^2(X,\mathcal{B},m)$ onto $L^2(X,\sigma(F),m)$. We shall also make use of the following additional conditions.

$$d\Gamma(F,F) << dm \text{ and } \Gamma(F,F) := \frac{d\Gamma(F,F)}{dm} = 1 \ m\text{-a.e.,} \tag{3}$$

$$dv << dr \text{ and } \dot{v} := \frac{dv}{dr} \text{ is continuous and strictly positive.} \tag{4}$$

Let m^* be the Borel measure on the open real half-line $\mathbb{R}_+$ given by $\dot{v}(r)dr$ and let $\mathcal{E}^*$ be the closure of

$$\mathcal{E}^*(u,v) := \int_0^\infty u'v'\dot{v}(r)dr \ u,v \in C_0^\infty(\mathbb{R})$$

in $L^2(\mathbb{R}_+,m^*)$ (which makes sense by (4)). For $u^*,v^* \in C_0^\infty(\mathbb{R}_+)$ the transformation rule [5], together with (3) and (4) above entail

$$\mathcal{E}(u^* \circ F, v^* \circ F) = \int_X (u' \circ F)(v' \circ F) d\Gamma(F,F) = \mathcal{E}^*(u^*,v^*).$$

Thus by a density argument $u^* \circ F$ belongs to $D(\mathcal{E})$ whenever u^* belongs to $D(\mathcal{E}^*)$ and moreover $\mathcal{E}(u^* \circ F, v^* \circ F) = \mathcal{E}^*(u^*,v^*)$ for all u^*,v^* in $D(\mathcal{E}^*)$. We denote the resolvent associated to $\mathcal{E}^*$ by G^*_λ. The standard inner product in $L^2(\mathbb{R}_+,m^*)$ is written $(\cdot,\cdot)_{m^*}$. Our last assumption is a strong "radial symmetry" condition: $\mathcal{E}$ possesses a core $\mathcal{C}$ left invariant by $\mathcal{F}$ such that $\mathcal{F}u = u^* \circ F$ for some $u^* \in D(\mathcal{E}^*)$ whenever $u \in \mathcal{C}$ and

$$\mathcal{E}(u \circ F, v) = \mathcal{E}(u \circ F, \mathcal{F}v) \tag{5}$$

for all $u^* \in D(\mathcal{E}^*)$ and $v \in \mathcal{C}$.

Theorem 4.1 Assume (1)–(5) and that $\int_0^\infty \dot{v}^{-1}(r)dr = \infty$.
Define $s(r) := \int_0^r \dot{v}(s)^{-1}ds$ and $\bar{s}$ by $\bar{s}(s(r)) = r$. Assume further that there exists a finite monotonically increasing function ϕ defined on $[0,\infty)$ such that

$$\dot{v}^2 \circ \bar{s}(r) \leq \phi(r) \text{ for all } r \in \mathbb{R}_+, \tag{6}$$

$$\int_1^\infty r^{-\frac{\alpha}{2-\alpha}+\epsilon}\phi(r)dr < \infty \text{ for some } \epsilon > 0. \tag{7}$$

Then $\mathcal{E}^{(\alpha)}$ is recurrent.

Remark. Suppose that $\dot{v}(r) \leq cr^\gamma$ (c constant, $0 \leq \gamma < 1$). Then $\mathcal{E}^{(\alpha)}$ is recurrent if $\gamma < \alpha - 1$.

5 Examples.

We give here two examples for which the conditions (1)–(5) are fulfilled.

Riemannian manifolds. We use the notation of [8]. Let M be a geodesically complete connected n-dimensional Riemannian manifold. The Riemannian inner product, gradient and volume measure on the Borel σ-algebra $\mathcal{B}$ of M are denoted $< \cdot,\cdot >$, ∇, m respectively. We consider the strongly local Dirichlet form $\mathcal{E}$ with domain $D(\mathcal{E})$ given by the closure of

$$\mathcal{E}(u,v) := \int_M < \nabla u, \nabla v > dm \quad u,v \in C_0^\infty(M)$$

in $L^2(M,m)$ corresponding to the Laplace-Beltrami operator on M. The diffusion metric d is the geodesic distance. The geodesic ball of radius r centred on a fixed point b in M is denoted B_r and the growth function is $v(r) := m(B_r)$. Assumptions (1) and (2) are fulfilled by [8], 2.91 and 2.105.

Let $F(x) = d(x,b)$. If $\sum_{i,j} g_{ij}dx_i dx_j$ is a local expression for the Riemannian inner product $< \cdot,\cdot >$ in a chart (U,ψ) then the function defined locally by $det(g_{ij})$ is denoted g.

Proposition 5.1 Assume that $F \in C^\infty(M - \{b\})$, $v \in C^1(\mathbb{R}_+)$ and

$$L_X((\dot{v} \circ F)^{-1} g^{1/2}) = 0 \text{ on } M - \{b\}, \tag{1}$$

where $X := \nabla F$ and $\dot{v} := \frac{dv}{dr}$. If the conditions of 5.2 hold then $\mathcal{E}^{(\alpha)}$ is recurrent.

The revolution surfaces [8], 5.11 provide concrete examples for which (1) hold.

Gradient Dirichlet forms. Let w be a strictly positive continuous radially symmetric function defined on $\mathbb{R}^d$, $d \geq 1$. Let $\mathcal{E}$ be the closure of

$$\mathcal{E}(u,v) := \int_{\mathbb{R}^d} \nabla u \cdot \nabla v w \cdot dx \quad u,v \in C_0^\infty(\mathbb{R}^d)$$

in $L^2(\mathbb{R}^d, w \cdot dx)$. It is not difficult to see that (1)–(5) hold.

References

[1] Carmona, R., Masters, W., Simon, B., Relativistic Schrödinger operators: asymptotic behaviour of the eigenfunctions, J. Funct. Anal., 91, No. 1, (1990), 117–142.

[2] Cheng, S.Y., Yau, S.T., Differential equations on Riemannian manifolds and their geometric applications, Comm. Pure Appl. Math., 28, (1975), 333–354.

[3] Davies, E.B., Heat kernels and spectral theory, Cambridge University Press, 1989.

[4] Falconer, K.J., The geometry of fractal sets, Cambridge University Press, 1985.

[5] Fukushima, M., Dirichlet forms and Markov processes, Kodansha and North-Holland, 1980.

[6] Fukushima, M., On recurrence criteria in the Dirichlet space theory, in *Local time to global property, Control and Physics*, ed. Elworthy, Research Notes in Math., 150 Longman, 1987.

[7] Fukushima, M., Oshima, Y., Takeda, M., Dirichlet forms and symmetric Markov processes, Walter de Gruyter & Co., Berlin – New York, 1994.

[8] Gallot, S., Hulin, D., Lafontaine, J., Riemannian geometry, Springer-Verlag, 1993.

[9] Karatzas, I., Shreve, S.E., Brownian motion and stochastic calculus, Springer-Verlag, 1988.

[10] Karp, L., Subharmonic functions, harmonic mappings and isometric immersions, in *Seminar on differential geometry*, ed. Yau, Ann. Math. Studies, 102, Princeton, 1982.

[11] McGillivray, I., A recurrence condition for some subordinated strongly local Dirichlet forms. Preprint, 1994.

[12] Ôkura, H., Capacitary inequalities and global properties of symmetric Dirichlet forms, preprint, 1994.

[13] Port, S.C., Stone, C.J., Infinitely divisible processes and their potential theory (II part), Ann. Inst. Fourier, 21, 4, (1971), 179–265.

[14] Sturm, K.T., Analysis on local Dirichlet spaces, Preprint 1994.

I. McGillivray
Inst. of Mathematics, TU Clausthal
Erzstr. 1, D-38678 Clausthal-Zellerfeld, Germany

Operator Theory:
Advances and Applications, Vol. 78

Band Spectrum for Schrödinger Operators with Strong Periodic Magnetic Fields

S. Nakamura

1 Introduction

In a recent paper [5], Hempel and Herbst considered the magnetic Schrödinger operator (without scalar potential): $H(\lambda A) = (p - \lambda A(x))^2$, where $p = -\mathrm{i}\,\partial_x$ and $A(x)$ is a vector potential. They studied the behavior of $H(\lambda A)$, or its resolvent as $\lambda \to \infty$. They showed that if $\{x \mid A(x) = 0\} = \{x \mid dA(x) = 0\}$ up to a null set, then $H(\lambda A)$ converges to the Dirichlet Laplacian on $\{x|A(x) = 0\}$ as $\lambda \to \infty$ in the sense of strong resolvent convergence. With additional conditions, e.g., if $A(x)$ is periodic, they also showed that they converges in the norm resolvent sense, and hence the spectrum converges. Thus, in such cases (with additional conditions), if λ is sufficiently large then $H(\lambda A)$ has spectrum bands, and the gaps are in fact open.

In this paper we study the band structure of the spectrum using a semi-classical method. In particular, we prove the existence of the spectrum gaps by different method for a larger class of operators. Moreover, we show that the widths of the bands (or clusters in the spectrum) are bounded by $C\exp(-\alpha\sqrt{\lambda})$ with some $\alpha > 0$.

We consider the Schrödinger operator:

$$H(\lambda A,V) = (p - \lambda A(x))^2 + V(x) \quad \text{on } L^2(\mathbb{R}^n),$$

where $A(x)$ is a vector potential, $V(x)$ is a scalar potential, and $\lambda \geq 0$ is a field strength parameter. For simplicity, we suppose $n = 3$. We always suppose $A(x)$ is C^1-class and $V(x)$ is bounded. Then $H(\lambda A,V)$ is essentially self-adjoint on $C_0^\infty(\mathbb{R}^3)$ and is bounded from below(cf. [4], Chapter 1). The magnetic field is given by

$$B_i(x) = \partial_j A_k(x) - \partial_k A_j(x), \quad x \in \mathbb{R}^3,$$

where $(i,j,k) = (1,2,3), (2,3,1)$ or $(3,1,2)$.

We suppose that $B(x)$ is periodic with respect to some lattice. Let G be a lattice in $\mathbb{R}^3$, i.e., $G = \left\{\sum_{j=1}^3 n_j\alpha_j \,\middle|\, n_j \in \mathbb{Z}, j = 1,2,3\right\}$, where $\{\alpha_1,\alpha_2,\alpha_3\}$ is a set of independent vectors in $\mathbb{R}^3$. G acts on $\mathbb{R}^3$ by the natural additive action:

$$g(x) = x + g, \quad g\Omega = \{x + g \,|\, x \in \Omega\} \quad \text{for } x \in \mathbb{R}^3, \Omega \subset \mathbb{R}^3, g \in G.$$

$\Omega \subset \mathbb{R}^3$ is called a fundamental domain with respect to G, if

$$g\Omega \cap h\Omega = \emptyset \quad \text{for } g \neq h,\ g,h \in G,$$

and

$$\bigcup_{g\in G} g\overline{\Omega} = \mathbb{R}^3.$$

We always suppose Ω is bounded and the boundary $\partial\Omega$ is piecewise C^1-class.

Assumption (A): $B(x)$ and $V(x)$ are periodic with respect to G, i.e.,

$$B(g(x)) = B(x), \quad V(g(x)) = V(x), \quad \text{for } x \in \mathbb{R}^3,\ g \in G. \tag{1.1}$$

Moreover, there exists a fundamental domain Ω such that

$$|B(x)| \geq \gamma > 0 \quad \text{for } x \in \partial\Omega. \tag{1.2}$$

Now we fix a fundamental domain Ω satisfying the above assumption. We let H^Ω be the Dirichlet Hamiltonian on Ω, i.e.,

$$H^\Omega = H^\Omega(\lambda A, V) = (p - \lambda A(x))^2 + V(x) \quad \text{on } L^2(\Omega)$$

with $D(H^\Omega) = H^2(\Omega) \cap H_0^1(\Omega)$. Then H^Ω has discrete spectrum, and we denote the spectrum by $\Sigma = \Sigma(\lambda) = \sigma(H^\Omega(\lambda A, V))$.

The following is our main result:

Theorem 1.1 Suppose Assumption (A). Then for any $\beta < \gamma$, there are $\alpha > 0$ and $C > 0$ such that

$$\sigma(H) \cap (-\infty, \beta\lambda] \subset \left\{ s \in (-\infty, \beta\lambda] \;\middle|\; \mathrm{dist}(s, \Sigma) < C\exp(-\alpha\sqrt{\lambda}) \right\} \tag{1.3}$$

for any $\lambda > 0$, where $\mathrm{dist}(\cdot,\cdot)$ denotes the distance in $\mathbb{R}$.

Remark 1: If $Z = \{x \in \Omega \mid B(x) = 0\}$ has nonempty interior and $\{x \mid B(x) = 0\} = \{x \mid A(x) = 0\}$ up to a null set, then $\Sigma(\lambda)$ converges to the spectrum of the Dirichlet Hamiltonian on Z (cf. [5]). Thus $\Sigma \cap (-\infty, \beta\lambda]$ is nonempty for large λ. Even if Z is discrete, we can show the infimum of $\Sigma(\lambda)$ is $o(\lambda)$ (smaller order than λ, in fact), and the above result is meaningful.

Remark 2: We will give an explicit estimate for the constant α in Section 3.

Remark 3: We can also prove analogous theorem for the case $n \neq 3$. If $n = 2$, the result holds without change. For $n \geq 4$, the field strength $|B(x)|$ is replaced by the largest absolute value of the eigenvalues of $(B_{ij})_{ij}$. We omit the detail here.

Remark 4: We expect that our estimate is *not* optimal, but the estimate in (1.3) should be "$< C\exp(-\alpha\lambda)$", which is expected by an observation of WKB solutions. This can be showed for special cases, but we have not been able to prove it in general.

Actually, we can show that $\sigma(H)$ has nonempty components in the small neighborhood of each point of Σ, and the corresponding spectral projection is infinite dimensional.

Corollary 1.2 In addition to the conditions of Theorem 1.1, if $A(x)$ is periodic, then $H(\lambda A,V)$ has band spectrum (possibly degenerate) and the width of each band is at most $O(\exp(-\alpha\lambda))$ as $\lambda \to \infty$.

The corollary is a simple consequence of Theorem 1.1 and the Floquet-Bloch theory (cf. [13], Section XIII-16). If $A(x)$ is not periodic, then the spectrum may be very complicated, and it is expected that H has Cantor spectrum under certain conditions ([2], [7], etc.).

The idea of the proof is fairly simple. We recall the Avron-Herbst-Simon estimate: $(p - \lambda A(x))^2 \geq \pm\lambda B_j(x)$ for any j ([1]). This implies, roughly speaking, $(p - \lambda A(x))^2$ is bounded from below by $\lambda|B(x)|$. Then we construct a method of tunneling estimates analogous to the Agmon method, where $\lambda|B(x)| + V(x)$ plays the role of the potential for the Schrödinger operator without magnetic field (see, e.g., [3], [11], [6], [8]). Combining this with a cut-and-paste technique (cf. [9]), we construct an approximation of $H(\lambda A,V)$ in terms of $H^{\Omega}(\lambda A,V)$ to prove Theorem 1.1.

Acknowledgement: This work was motivated by a seminar talk by Professor Rainer Hempel at the Holzhau conference. The author thanks Professor Hempel and Professor Ira Herbst for valuable discussion and encouragement. He also thanks Professor Michael Demuth for organizing the wonderful conference.

2 Tunneling estimates for the magnetic barrier

In this section, we consider $H = H(\lambda A,V)$ on an arbitrary bounded domain $\Omega \subset \mathbb{R}^3$ and suppose that $A(x)$ is C^1-class on $\overline{\Omega}$ with bounded derivatives and that $V(x)$ is bounded. First we recall the Avron-Herbst-Simon estimate, and prove it for the completeness.

Lemma 2.1 *([1])* For any $i \neq j$,

$$H(\lambda A,V) \geq \lambda\,(\partial_i A_j(x) - \partial_j A_i(x)) + V(x). \tag{2.1}$$

Proof: If we write

$$k_j = p_j - \lambda A_j = -\,\mathrm{i}\,\partial_j - \lambda A_j(x), \tag{2.2}$$

then it is easy to see

$$[k_i, k_j] = \mathrm{i}\,\lambda\,(\partial_i A_j - \partial_j A_i)\,, \tag{2.3}$$

and

$$H(\lambda A, V) = \sum_{j=1}^{3} k_j^2 + V(x).$$

We also note

$$-\,\mathrm{i}\;[k_i,k_j] \leq k_i^2 + k_j^2$$

since $(k_i + \mathrm{i}\, k_j)^*(k_i + \mathrm{i}\, k_j) \geq 0$. Combining them, we have

$$H(\lambda A, V) \geq k_i^2 + k_j^2 + V \geq -\mathrm{i}\,[k_i,k_j] + V = \lambda\,(\partial_i A_j - \partial_j A_i) + V.$$

∎

We use the following version of the above estimate:

Lemma 2.2 For any $\varepsilon > 0$, there is $C > 0$ such that

$$H(\lambda A,V) \geq \lambda(|B(x)| - \varepsilon) - C, \quad \lambda \geq 0. \tag{2.4}$$

Proof: Since V is bounded, it suffices to consider $H(\lambda A,0)$. For each $x \in \overline{\Omega}$, we can find a coordinate system such that $B(x) = (|B(x)|,0,0)$. Then we choose a neighborhood $E(x)$ of x in $\overline{\Omega}$ such that

$$|B(x) - B(y)| \leq \varepsilon/2 \quad \text{for } y \in E(x). \tag{2.5}$$

We take a finite point $x_1, \dots, x_m$ in $\overline{\Omega}$ so that $\overline{\Omega} = \bigcup_{j=1}^m E(x_j)$ by the compactness of $\overline{\Omega}$. We let $J_i(x) \in C^\infty(\overline{\Omega})$ such that

$$\begin{aligned}
&J_i(x) \geq 0, \quad x \in \overline{\Omega}, i = 1, \dots, m;\\
&\operatorname{supp} J_i \subset E(x_i), \quad i = 1, \dots, m;\\
&\sum_{i=1}^m J_i(x)^2 = 1, \quad x \in \overline{\Omega}.
\end{aligned}$$

By Lemma 2.1 and (2.5), we have for each i:

$$\begin{aligned}
J_i H J_i &\geq \lambda J_i B_1 J_i = \lambda J_i^2 B_1\\
&\geq \lambda J_i^2\,(B_1(x_i) - \varepsilon/2)\\
&\geq \lambda J_i^2(|B(x)| - \varepsilon).
\end{aligned}$$

On the other hand, by the standard argument (cf. [4]), we have

$$H = \sum_i J_i H J_i - \frac{1}{2}\sum_i |\nabla J_i|^2.$$

Combining these, we conclude

$$H \geq \lambda \sum_i J_i^2(|B| - \varepsilon) - C = \lambda(|B| - \varepsilon) - C.$$

∎

Thus H is bounded from below by $|B(x)|$ up to small or lower order error. Motivated by this observation, we define an analogue of the Agmon metric at an energy $\beta \in \mathbb{R}$ by

$$ds_\beta^2 = (|B(x)| - \beta)_+ dx^2$$

and let $\mathbf{d}_\beta(\cdot,\cdot)$ be the pseudodistance generated by the metric ds_β^2. Here we have used the notation $(*)_+ = \max(*,0)$. The following resolvent estimate is the main result of this section:

Theorem 2.3 Let $\beta > 0$ and let $D \subset \Omega$ be a closed subset such that

$$\inf\{|B(x)|\,|x \in D\} > \beta.$$

Then for any $\varepsilon > 0$ and $M > 0$, there is $C > 0$ such that

$$\left\|\chi_D(H-z)^{-1}\right\| \le C\lambda^{-1}, \quad \lambda \ge 0, \tag{2.6}$$

if $\operatorname{Re} z \le \beta\lambda$ and

$$\left\|(H-z)^{-1}\right\| \le M \exp\left((\mathbf{d}_\beta(D,\Omega_\beta) - \varepsilon)\sqrt{\lambda}\right), \tag{2.7}$$

where χ_D is the indicator function of D and

$$\Omega_\beta = \{\xi \in \Omega \mid |B(x)| \le \beta\}.$$

Proof: Since $D \subset \Omega_\beta^c$, $\mathbf{d}_\beta(D,\Omega_\beta) > 0$. Thus we may suppose ε is so small that $\mathbf{d}_\beta(D,\Omega_\beta) - \varepsilon > 0$ without loss of generality.

We first choose $\delta > 0$ so small that

$$\mathbf{d}_{(\beta+2\delta)}(D,\Omega_{\beta+2\delta})) > \mathbf{d}_\beta(D,\Omega_\beta) - \varepsilon/3. \tag{2.8}$$

Then naturally $D \cap \Omega_{(\beta+2\delta)} = \emptyset$. We set

$$W(x) = (\beta + 2\delta - |B(x)|)_+,$$

and set

$$\tilde{H}(\lambda) = H(\lambda A,V) + \lambda W(x).$$

Then by Lemma 2.2 with $\varepsilon = \delta$, we have

$$\begin{aligned}
\tilde{H}(\lambda) - \lambda\beta &\ge \lambda(|B(x)| - \delta) - C + \lambda(\beta + 2\delta - |B(x)|)_+ - \lambda\beta \\
&= \lambda\Big[(\beta + 2\delta - |B(x)|)_+ - (\beta + 2\delta - |B(x)|)\Big] + \lambda\delta - C \\
&\ge \lambda\delta - C,
\end{aligned}$$

and the right hand side is positive if λ is sufficiently large. In particular, $\sigma(\tilde{H}(\lambda)) \subset [\lambda(\beta+\delta) - C,\infty)$. Moreover, if we set

$$\tilde{V}(x) = \begin{cases} |B(x)| - \delta & \text{if} \quad |B(x)| \ge \beta + 2\delta, \\ \beta + \delta & \text{if} \quad |B(x)| < \beta + 2\delta, \end{cases}$$

then we also have

$$\tilde{H} \geq \lambda \tilde{V} - C. \tag{2.9}$$

Now we let $\alpha = \beta + 2\delta$.

Lemma 2.4 There is $C > 0$ such that

$$\left\| \chi_D(\tilde{H} - z)^{-1} \chi_{\Omega_\alpha} \right\| \leq (\lambda\delta - C)^{-1} \exp\left[-(\mathbf{d}_\beta(D,\Omega_\beta) - (2/3)\varepsilon)\sqrt{\lambda} \right] \tag{2.10}$$

if $\operatorname{Re} z \leq \beta$ and λ is sufficiently large.

Proof: By the standard computations, we have

$$\begin{aligned} e^{\sqrt{\lambda}\rho(x)} \tilde{H}(\lambda) e^{-\sqrt{\lambda}\rho(x)} &= \sum_j (k_j + \mathrm{i}\sqrt{\lambda}\partial_j \rho(x))^2 + V(x) + \lambda W(x) \\ &= \tilde{H}(\lambda) + \mathrm{i}\sqrt{\lambda} \sum_j (k_j(\partial_j \rho) + (\partial_j \rho) k_j) - \lambda |\nabla \rho(x)|^2 \end{aligned}$$

for any C^1-function $\rho(x)$ on $\overline{\Omega}$. This and (2.9) imply

$$\operatorname{Re}\left[e^{\sqrt{\lambda}\rho} (\tilde{H}(\lambda) - z) e^{-\sqrt{\lambda}\rho} \right] \geq \lambda(\tilde{V} - |\nabla\rho|^2) - \operatorname{Re} z - C. \tag{2.11}$$

By the choice of δ: (2.8), we can find $\rho(x) \geq 0$ such that

$$\begin{aligned} &\rho(x) = 0 \quad \text{on } \Omega_\alpha; \\ &\rho(x) \geq \mathbf{d}_\beta(D,\Omega_\beta) - (2/3)\varepsilon \quad \text{on } D; \\ &|\nabla\rho(x)| \leq \sqrt{(|B(x)| - \beta - 2\delta)_+} = \sqrt{\tilde{V}(x) - \beta - \delta}. \end{aligned}$$

Then by (2.11), we obtain

$$\begin{aligned} \operatorname{Re}\left[e^{\sqrt{\lambda}\rho} (\tilde{H}(\lambda) - z) e^{-\sqrt{\lambda}\rho} \right] &\geq \lambda \left(\tilde{V} - (\tilde{V} - \beta - \delta) \right) - \operatorname{Re} z - C \\ &= \lambda(\beta + \delta) - \operatorname{Re} x - C \\ &\geq \delta\lambda - C \quad \text{if } \operatorname{Re} z \leq \lambda\beta, \end{aligned}$$

and hence

$$\left\| \left[e^{\sqrt{\lambda}\rho} (\tilde{H} - z) e^{-\sqrt{\lambda}\rho} \right]^{-1} \right\| = \left\| e^{\sqrt{\lambda}\rho} (\tilde{H} - z)^{-1} e^{-\sqrt{\lambda}\rho} \right\| \leq (\delta\lambda - C)^{-1}$$

for large λ. By this and the conditions on ρ, we learn

$$\begin{aligned} \left\| \chi_D(\tilde{H} - z)^{-1} \chi_{\Omega_\alpha} \right\| &= \left\| \left(\chi_D\, e^{-\sqrt{\lambda}\rho} \right) \left(e^{\sqrt{\lambda}\rho} (\tilde{H} - z)^{-1} e^{-\sqrt{\lambda}\rho} \right) \left(e^{\sqrt{\lambda}\rho} \chi_{\Omega_\alpha} \right) \right\| \\ &\leq \left\| \chi_D\, e^{-\sqrt{\lambda}\rho} \right\| (\lambda\delta - C)^{-1} \\ &\leq (\lambda\delta - C)^{-1} \exp\left[-(\mathbf{d}_\beta(D,\Omega_\beta) - (2/3)\varepsilon)\sqrt{\lambda} \right]. \end{aligned}$$

∎

We now use the resolvent equation:

$$\chi_D(H-z)^{-1} = \chi_D(\tilde{H}-z)^{-1} - \chi_D(\tilde{H}-z)^{-1}\lambda W(H-z)^{-1},$$

and the assumption (2.7). Since $\operatorname{supp} W \subset \Omega_\alpha$, we can apply Lemma 2.4 to obtain

$$\begin{aligned}\left\|\chi_D(H-z)^{-1}\right\| &\le \left\|(\tilde{H}-z)^{-1}\right\| + \left\|\chi_D(\tilde{H}-z)^{-1}\chi_{\Omega_\alpha}\right\| \|\lambda W\| \left\|(H-z)^{-1}\right\| \\ &\le (\lambda\delta - C)^{-1} + C\lambda(\lambda\delta - C)^{-1}\exp\left[-(\varepsilon/3)\sqrt{\lambda}\right] \\ &\le C_1\lambda^{-1}\end{aligned}$$

with some $C_1 > 0$ if λ is sufficiently large. The claim for bounded λ is obvious and this completes the proof of Theorem 2.3. ∎

3 Periodic magnetic Schrödinger operators

Throughout this section, we suppose Assumption (A), and let $\gamma > 0$ be the constant in it. We fix a constant β so that $0 < \beta < \gamma$. We set

$$\mathbf{d} = \mathbf{d}_\beta(\partial\Omega,\Omega_\beta), \quad \text{with } \Omega_\beta = \{x \in \Omega \mid |B(x)| \le \beta\},$$

where $\mathbf{d}_\beta(\cdot,\cdot)$ is the Agmon distance defined in the last section. We will show that Theorem 1.1 holds for any $\alpha < \mathbf{d}$. We fix such α and then choose $\varepsilon > 0$ so small that

$$\mathbf{d}_\beta((\partial\Omega)_\varepsilon,\Omega_\beta) > (\mathbf{d}+\alpha)/2, \tag{3.1}$$

where

$$(\partial\Omega)_\varepsilon = \left\{x \in \mathbb{R}^3 \;\middle|\; \operatorname{dist}(x,\partial\Omega) < \varepsilon\right\}.$$

We set

$$\Gamma = \Omega \cup (\partial\Omega)_\varepsilon = \left\{x \in \mathbb{R}^3 \;\middle|\; \operatorname{dist}(x,\Omega) < \varepsilon\right\}.$$

and apply Theorem 2.3 for Γ. Let $j(x)$ be a C_0^∞-function such that $\operatorname{supp} j \subset \Gamma$, $j(x) \ge 0$ for any x, and

$$\sum_{g\in G} j_g(x) = 1 \quad \text{for any } x \in \mathbb{R}^3, \text{ where } j_g(x) = j(g^{-1}(x)).$$

We now let

$$\mathcal{H}_0 = \bigoplus_{g\in G} L^2(g\Gamma)$$

and let

$$H_0 = \bigoplus_{g\in G} H^g \quad \text{on } \mathcal{H}_0,$$

where

$$H^g = H(\lambda A(g^{-1}(x)),V(g^{-1}(x))) \quad \text{on } L^2(g\Gamma)$$

with the Dirichlet boundary condition. H^g is the same operator as H^0 for each $g \in G$ up to a translation. H_0 is our approximation system of $H(\lambda A,V)$. The identification map is constructed as follows: Since $B(x)$ is G-periodic, there exists a gauge transform $T_g = \exp[\mathrm{i}\,\lambda\Phi_g(x)]$ for each $g \in G$ such that

$$T_g H^g T_g^* = H(\lambda A, V) \quad \text{on } L^2(g\Gamma).$$

Then we define an operator J by

$$J(\oplus_g \varphi_g) = \sum_{g\in G} j_g T_g \varphi_g, \quad \text{for } (\oplus_g \varphi_g) \in \mathcal{H}_0.$$

By the definition of $j_g(x)$, J is a bounded map from $\mathcal{H}_0$ to $\mathcal{H} = L^2(\mathbb{R}^3)$, and the adjoint J^* is an isometry from $\mathcal{H}$ to $\mathcal{H}_0$. We approximate H by JH_0J^* and consider the error. This is analogous to the discussion in Sections 4 and 7 of Nakamura-Bellissard [9], and we omit the detail.

The following formula follows by direct computations: For $z \in \rho(H)\cap\rho(H_0)$,

$$(H-z)^{-1} = J(H_0-z)^{-1}J^* - (H-z)^{-1}M(H_0-z)^{-1}J^*,$$

where $M = HJ - JH_0 \in B(D(H_0),\mathcal{H})$. Moreover, M is decomposed as $M = \oplus_{g\in G}\, m_g$ and

$$\begin{aligned} m_g\varphi_g &= Hj_gT_g\varphi_g - j_gT_gH^g\varphi_g \\ &= -\mathrm{i}\,T_g\left[(p-\lambda A(g^{-1}(x)))\cdot(\nabla j_g) + (\nabla j_g)\cdot(p-\lambda A(g^{-1}(x)))\right]\varphi_g. \end{aligned}$$

Thus the differential operator m_g is supported in $\operatorname{supp}(\nabla j_g)$ which is included in $g[(\partial\Omega)_\varepsilon]$. By Theorem 2.3, we learn

$$\left\|(\nabla j_g)(H^g-z)^{-1}\right\| \le C\lambda^{-1}$$

if $\operatorname{Re} z \le \beta\lambda$ and

$$\left\|(H^g-z)^{-1}\right\| \le C\exp\left[(\mathbf{d}_\beta((\partial\Omega)_\varepsilon,\Omega_\beta)-\varepsilon_1)\sqrt{\lambda}\right] \tag{3.2}$$

with some $\varepsilon_1 > 0$. By (3.1), we can take $\varepsilon_1 > 0$ so small that

$$\mathbf{d}_\beta((\partial\Omega)_\varepsilon,\Omega_\beta) - \varepsilon_1 \ge \alpha.$$

Under the same assumption, we can also show

$$\left\|(p-\lambda A(g^{-1}(x)))\cdot(\nabla j_g)(H^g-z)^{-1}\right\| \le C\lambda^{-1/2}$$

by the standard argument using resolvent equations. These imply

$$\left\|M(H_0-z)^{-1}\right\| \le C\lambda^{-1/2}$$

under the condition (3.2). We note that in order to prove the above estimates, we use the fact that the multiplicity of the supports of $\{m_g\}$ are at most $2^3 = 8$, and a simplified version of Cotler's lemma. Then we can construct the resolvent of H by

$$(H - z)^{-1} = J(H_0 - z)^{-1} J^* \left[1 + M(H_0 - z)^{-1} J^*\right]^{-1}$$

for sufficiently large λ and the above z. The above equality is verified easily for z far away from $\sigma(H)$, and then for any $z \in \mathbb{C}$ such that $[\cdots]$ is invertible by the analytic continuation. Thus we have proved:

Lemma 3.1 Let $0 < \beta < \gamma$ and let $0 < \alpha < \mathbf{d}$. If $\operatorname{Re} z \le \beta\lambda$ and if

$$\left\|(H^g - z)^{-1}\right\| \le C \exp(\alpha\sqrt{\lambda}),$$

then $z \in \rho(H)$. In particular,

$$\sigma(H) \cap (-\infty,\beta\lambda] \subset \left\{ s \in (-\infty,\beta\lambda] \;\middle|\; \operatorname{dist}(s,\sigma(H^g)) < C \exp(-\alpha\sqrt{\lambda}) \right\}. \quad (3.3)$$

Mimicking the above argument to approximate H^0 in terms of H^Ω, we can also prove

Lemma 3.2 There is $C > 0$ such that

$$\sigma(H^g) \cap (-\infty,\beta\lambda] \subset \left\{ s \in (-\infty,\beta\lambda] \;\middle|\; \operatorname{dist}(s,\Sigma) < C \exp(-\alpha\sqrt{\lambda}) \right\}. \quad (3.4)$$

These complete the proof of Theorem 1.1.

References

[1] Avron, J., Herbst, I., Simon, B.: Schrödinger operators with magnetic fields. I. General interactions. Duke Math. J. **45**, 847–882 (1978).

[2] Bellissard, J.: C^*-algebras in solid state physics: 2D-electrons in a uniform magnetic field. Operator algebras and Applications (eds. Evans, E. V., Takesaki, M.), Cambridge University Press 1988.

[3] Briet, Ph., Combes, J. M., Duclos, P.: Spectral stability under tunneling. Commun. Math. Phys. **126**, 133–156 (1989).

[4] Cycon, H, L., Froese, R. G., Kirsch, W., Simon, B.: Schrödinger Operators, with Applications to Quantum Mechanics and Global Geometry, Springer Verlag, Berlin 1987.

[5] Hempel, R., Herbst, I.: Strong magnetic fields, Dirichlet boundaries, and spectral gaps, Preprint 1994, to appear in Commun. Math. Phys.

[6] Helffer, B., Sjöstrand, J.: Multiple wells in the semiclassical limit. I., Comm. P. D. E. **9**, 337–408 (1984).

[7] Helffer, B., Sjöstrand, J.: Semiclassical analysis for the Harper's equation III. Cantor structure of the spectrum, Mem. Math. Soc. France **39** (1989).

[8] Helffer, B., Sjöstrand, J.: Effect tunnel pour l'equation de Schrödinger avec champ magnétique, Ann. Scuola Norm. Sup. Pisa **14**, 625–657 (1987).

[9] Nakamura, S., Bellissard, J.: Low energy bands do not contribute to quantum Hall effect. Commun. Math. Phys. **131**, 283–305 (1990).

[10] Reed, M., Simon, B.: Methods of Modern Mathematical Physics. Vol. I–IV. Academic Press 1972–1979.

[11] Simon, B.: Semiclassical analysis of low-lying eigenvalues, II. Tunneling, Ann. Math. **120**, 89–118 (1984).

Shu Nakamura
Department of Mathematical Sciences, University of Tokyo
3-8-1, Komaba, Meguro-ku, Tokyo, Japan 153

Operator Theory:
Advances and Applications, Vol. 78

Mellin Pseudodifferential Operators with Operator Symbols and its Applications

V. S. Rabinovich

1. Introduction.

Mellin pseudodifferential operators based on Mellin transform naturally arise in the theory of boundary value problems on manifolds with singularities (see the pioneering paper [1], the recent monographs [2]–[3] and papers [4]–[8]).

It should be noted that the class of Mellin pseudodifferential operators ($M\Psi DO$) with operator symbols $a(t,\lambda)$ tending to indepent from $t \in \mathbb{R}_+$ symbols $a(\lambda)$ for $t \to +0$ is enough class for investigation of boundary value problems for differential operators with coefficients having first kind discontinuities at the conical points of the boundary.

An application of $M\Psi DO$ with non-stabilized symbols for $t \to +0$ gives a good possibility to investigate Fredholm properties of a boundary value problem for differential operators with point second kind discontinuties in domains with such singularities of boundary in which the tangent cone does not exist.

In this paper we introduce a class of $M\Psi DO$ with slowly varying operator symbols when $t \to +0$. We concentrate our attention at the problem of local invertibility of $M\Psi DO$ at the point zero because the local principle reduces the question for a boundary value problem operator to be Fredholm to the problem of its local invertibility.

The problem of local invertibility of the boundary value problem at conical points of the boundary is reduced to the problem of local invertibility $M\Psi DO$ at the point 0.

The main results of the paper are criterium of local invertibility of $M\Psi DO$ at the point 0 and theorem about local improving of solution behavior of Mellin pseudodifferential equations are main results of the paper.

(We note that scalar ΨDO on $\mathbb{R}^n$ with slowly varying symbols at infinity from L.Hörmander's class $S^m_{1,0}$ have been studied in [8] and ones from $S^m_{\varrho,\delta}$ $(0 \leq \delta \leq \varrho \leq 1, \delta < 1)$ have been studied in [9].)

We applied the mentioned results to operators from Boutet-de-Monvel's algebra in a cone. Our construction of these operators is similar to a construction of the paper [4], but we use the calculus of $M\Psi DO$ with non-stabilized operator symbols when $t \to +0$).

The problem of the local invertibility of a boundary value problem at a singular point of boundary in which the tangent cone does not exist is reduced to a problem in conical domain for differential operator with coefficients having second kind discontinuities by means of special type of change of variables.

As an example we consider Dirichlet and Neumann problems in domain without tangent cone at the point 0.

2. Definitions and preleminary results.

Let $\mathbb{H}$, $\mathbb{H}_j$, $\mathbb{H}'_j$ denote complex Hilbert spaces which are assumed to be separable, $\mathbb{B}(\mathbb{H}_1,\mathbb{H}_2)$ denotes the space of bounded linear operators mapping $\mathbb{H}_1$ into $\mathbb{H}_2$, $\mathbb{R}_\mu = \{\lambda \in \mathbb{C} : \mathrm{Im}\ \lambda = \mu\}$.

Definition 2.1. We say that operator-function $Q(\lambda) \in \Lambda_\mu(\mathbb{H},\mathbb{H}')$ if $Q(\lambda) \in C^\infty(\mathbb{R}_\mu,\mathbb{B}(\mathbb{H},\mathbb{H}'))$ and for some $\varepsilon \in \mathbb{R}$

$$\|Q(\lambda)\|_{\mathbb{B}(\mathbb{H},\mathbb{H}')} \leq C < \lambda >^{\varepsilon}\ , \quad \lambda \in \mathbb{R}_\mu \quad (< \lambda >= (1+|\lambda|^2)^{\frac{1}{2}}).$$

Morever there exists an operator-function $Q^{-1}(\lambda)$, $\lambda \in \mathbb{R}_\mu$ such that for some $\delta \in \mathbb{R}$ and any $j \in \mathbb{Z}_+ = \{m \in \mathbb{Z} : m \geq 0\}$

$$\|Q^{-1}(\lambda)Q^{(j)}(\lambda+\eta)\|_{\mathbb{B}(\mathbb{H},\mathbb{H})} \leq c_j < \eta >^{\delta}\ , \quad \lambda \in \mathbb{R}_\mu, \quad \eta \in \mathbb{R},$$

$$\|Q^{(j)}(\lambda+\eta)Q^{-1}(\lambda)\|_{\mathbb{B}(\mathbb{H}',\mathbb{H}')} \leq c_j < \eta >^{\delta}\ , \quad \lambda \in \mathbb{R}_\mu, \quad \eta \in \mathbb{R}.$$

Definition 2.2. Let $Q_j \in \Lambda_\mu(\mathbb{H}_j,\mathbb{H}'_j)$, $j = 1{,}2$, we denote $\mathcal{E}_\mu(Q_1,Q_2)$ a class of operator-functions with values in $\mathbb{B}(\mathbb{H}_1,\mathbb{H}_2)$ defined on $\mathbb{R}_+ \times \mathbb{R}_\mu$ and such that

$$\|Q_2(\lambda)(t\partial_t)^\beta \partial_\lambda^\alpha a(t,\lambda) Q_1^{-1}(\lambda)\|_{\mathbb{B}(\mathbb{H}'_1,\mathbb{H}'_2)} \leq C_{\alpha\beta}(a)$$

for all $\alpha,\beta \in \mathbb{Z}_+$.

The operator defined by the formula

$$(Au)(t) = a(t,\mathcal{D}_t) = (2\pi)^{-1} \int_{\mathbb{R}_\mu} a(t,\lambda)(Mu)(\lambda) t^{i\lambda} dx, \tag{1}$$

is called a Mellin pseudodifferential operator.

In the formula (1) $u \in C_0^\infty(\mathbb{R}_+,\mathbb{H}_1)$,

$$(Mu)(x) = \int_0^\infty u(t) t^{-i\lambda - 1} dt$$

is the Mellin transform of u.

The class of such operators is denoted by $OP\mathcal{E}_\mu(Q_1,Q_2)$.

We can see that $Q(\lambda) \in \mathcal{E}_\mu(Q,I_{\mathbb{H}})$ or $Q(\lambda) \in \mathcal{E}_\mu(I_{\mathbb{H}'},Q^{-1})$ if $Q(\lambda) \in \Lambda_\mu(\mathbb{H},\mathbb{H}')$.

Definition 2.3. We denote by $\mathcal{H}_\delta(\mathbb{R}_+,Q)$ the closure of $C_0^\infty(\mathbb{R}_+,\mathbb{H})$ in the norm

$$\|u\|_{\mathcal{H}_\delta(\mathbb{R}_+,Q)} = \left(\int_{\mathbb{R}_{\delta+1/2}} \|Q(\lambda)(Mu)(\lambda)\|^2_{\mathbb{H}'} d\lambda\right)^{1/2},$$

where $Q \in \Lambda_{\delta+1/2}(\mathbb{H},\mathbb{H}')$.

All standard formulas of the calculus of pseudodifferential operators on $\mathbb{R}^n$ (formula of composition, formula for formal adjoint and other ones) are transfered in an obvious manner to $M\Psi DO$ evidently.

The next proposition follows from the composition formula and the Calderon-Vaillancourt theorem for ΨDO with bounded operator symbol.

Proposition 2.1. An operator $A = a(t,\mathcal{D}_t) \in OP\mathcal{E}_{\delta+1/2}(Q_1,Q_2)$ is a bounded operator from $\mathcal{H}_\delta(\mathbb{R}_+,Q_1)$ into $\mathcal{H}_\delta(\mathbb{R}_+,Q_2)$ and there are constants $C > 0$ and $N > 0$ which are independent of A such that

$$\|Au\|_{\mathcal{H}_\delta(\mathbb{R}_+,Q_2)} \leq C\Big(\sum_{|\alpha+\beta|\leq N} C_{\alpha\beta}(a)\Big)\|u\|_{\mathcal{H}_\delta(\mathbb{R}_+,Q_1)}.$$

Definition 2.4. A symbol $a(t,\lambda) \in \mathcal{E}_\mu(Q_1,Q_2)$ is called slowly varying at the point 0, if

$$\lim_{t\to+0}\sup_{\lambda\in\mathbb{R}_\mu}\|Q_2(\lambda)(t\partial_t a(t,\lambda))Q_1^{-1}(\lambda)\|_{\mathbb{B}(\mathbb{H}_1',\mathbb{H}_2')} = 0.$$

The class of all slowly varying symbols is denoted by $\widetilde{\mathcal{E}}_\mu(Q_1,Q_2)$ and the corresponding class of $M\Psi DO$ is denoted by $OP\widetilde{\mathcal{E}}_\mu(Q_1,Q_2)$.

We denote $\mathcal{E}^0_\mu(Q_1,Q_2)$ the class of symbols $a(t,\lambda) \in \mathcal{E}_\mu(Q_1,Q_2)$ such that

$$\lim_{t\to+0}\sup_{\lambda\in\mathbb{R}_\mu}\|Q_2(\lambda)a(t,\lambda)Q_1^{-1}(\lambda)\|_{\mathbb{B}(\mathbb{H}_1',\mathbb{H}_2')} = 0$$

and by $OP\mathcal{E}^0_\mu(Q_1,Q_2)$ the corresponding class of $M\Psi DO$.

Proposition 2.2.

(i) Let $A = a(t,\mathcal{D}_t) \in OP\widetilde{\mathcal{E}}_{\delta+1/2}(Q_1,Q_2)$, $B = b(t,\mathcal{D}_t) \in OP\mathcal{E}_{\delta+1/2}(Q_2,Q_3)$ then $BA \in OP\mathcal{E}_{\delta+1/2}(Q_1,Q_3)$ and the symbol $\sigma_{BA}(t,\lambda)$ of the operator BA is given by the formula

$$\sigma_{BA}(t,\lambda) = b(t,\lambda)a(t,\lambda) + r(t,\lambda),$$

where $r(t,\lambda) \in OP\mathcal{E}^0_{\delta+1/2}(Q_1,Q_3)$.

(ii) If $A = a(t,\mathcal{D}_t) \in OP\widetilde{\mathcal{E}}_{\delta+1/2}(Q_1,Q_3)$ and if it is considered as a map from $\mathcal{H}_\delta(\mathbb{R}_+,Q_1)$ into $\mathcal{H}_\delta(\mathbb{R}_+,Q_2)$ then the adjoint operator

$$A^* \in OP\mathcal{E}_{\delta+1/2}(Q_2^{-1*},Q_1^{-1*})$$

and its symbol $\sigma_{A^*}(t,\lambda)$ is given by the formula

$$\sigma_{A^*}(t,\lambda) = a^*(t,\lambda) + r(t,\lambda),$$

where $r(t,\lambda) \in OP\mathcal{E}^0_{\delta+1/2}(Q_2^{-1*},Q_1^{-1*})$.

3. Criteria of local invertibility of $M\Psi DO$.

The main results of the paper are contained in the following theorems.

Theorem 3.1. Let $= a(t,\mathcal{D}_t) \in OP\widetilde{\mathcal{E}}_{\delta+1/2}(Q_1,Q_2)$ then the following statements are equivalent:

(a) The operator $A : \mathcal{H}_\delta(\mathbb{R}_+,Q_1) \to \mathcal{H}_\delta(\mathbb{R}_+,Q_2)$ is locally invertible at the point 0;

(b) There exists $\varepsilon > 0$ such that the operator-function $a(t,\lambda)$ is invertible for all $(t,\lambda) \in (0,\varepsilon) \times \mathbb{R}_{\delta+1/2}$ and

$$\sup_{(0,\varepsilon)\times\mathbb{R}_{\delta+1/2}} \|Q_1(\lambda)a^{-1}(t,\lambda)Q_2^{-1}(\lambda)\|_{\mathbb{B}(\mathbb{H}_2',\mathbb{H}_1')} < \infty; \tag{2}$$

(c) There exists a local inverse operator for A at the point 0 belonging to $OP\widetilde{\mathcal{E}}_{\delta+1/2}(Q_2,Q_1)$.

Remark 3.1. If we introduce in the space $\mathbb{H}_j$ the norm $\|\cdot\|_{\mathbb{H}_j(\lambda)}$ depending on the parameter λ by means of the formula

$$\|u\|_{\mathbb{H}_j(\lambda)} = \|Q_j(\lambda)u\|_{\mathbb{H}_j'}, \quad j = 1,2,$$

then the condition (2) takes the form

$$\sup_{(0,\varepsilon)\times\mathbb{R}_{\delta+1/2}} \|a^{-1}(t,\lambda)\|_{\mathbb{B}(\mathbb{H}_2(\lambda),\mathbb{H}_1(\lambda))} < \infty. \tag{3}$$

The proof of Theorem 3.1 is based on the Proposition 2.1 and Proposition 2.2 (i).

The next theorem gives asymptotic estimates of the solutions of the equation

$$a(t,\mathcal{D}_t)u = f \tag{4}$$

in the neigborhood of the point 0 where $a(t,\mathcal{D}_t) \in OP\widetilde{\mathcal{E}}_{\delta+1/2}(Q_1,Q_2)$ for all $\delta \in (\delta_1,\delta_2)$.

Let $\chi \in C^\infty(\overline{\mathbb{R}}_+)$ and $\chi(t) = 1$ when $t \in [0,1/2]$ and $\chi(t) = 0$ when $t \geq 1$, $\chi_\varepsilon(t) = \chi\left(\frac{t}{\varepsilon}\right)$.

Theorem 3.2. Let condition (2) (or equivalent condition (3)) be fulfilled for all $\delta \in [\alpha_1,\alpha_2] \subset (\delta_1,\delta_2)$, then if $u \in \mathcal{H}_{\alpha_2}(\mathbb{R}_+,Q_1)$ and $\chi_{\varepsilon_0}(t) \in \mathcal{H}_{\alpha_1}(\mathbb{R}_+,Q_2)$ for some $\varepsilon_0 > 0$ then there exists $\varepsilon \in (0,\varepsilon_0)$ such that $\chi_\varepsilon u \in \mathcal{H}_{\alpha_1}(\mathbb{R}_+,Q_1)$.

4. Local invertibility of operators from Boutet-de-Monvel's algebra at conic points.

We will consider operators from Bontet-de-Monvel's algebra when the symbols of the operators incoming to boundary value problem have second kind discontinuties of the second kind at the conic point of boundary.

Application of Theorems 3.1 and 3.2 allows to give criteria of local invertibility and to estimate the behavior of solutions of the boundary value problem at conic points.

Let Ω be a smooth compact manifold with boundary $\partial\Omega$, $\mathbb{K} = \mathbb{R}_+ \times \Omega$ is a cone in coordinates $(r,\omega) \in \mathbb{R}_+ \times \Omega$; E, (F_1,F_2) are vector bundles over Ω $(\partial\Omega)$.

We denote by $OP\mathfrak{S}^{m,d}(\Omega,E,F_1,F_2)$ a class of Boutet-de-Monvel's operators on Ω order of $m \in \mathbb{Z}$ and type of $d \in \mathbb{Z}_+$ (see [11], [12]).

An operator $A \in OP\mathfrak{S}^{m,d}(\Omega,E,F_1,F_2)$ maps $H^s(\Omega,E,F_1)$ into $H^{s-m}(\Omega,E,F_2)$ where $s > d - 1/2$, and $H^s(\Omega,E,F_j) = H^s(\Omega,E) \oplus H^{s+1/2}(\Omega,F_j)$, $j = 1,2$.

$H^s(\Omega,E),H^s(\partial\Omega,F)$ are the Sobolev spaces of cross-sections of the bundle E, F_j respectively.

As usually ([12]) we introduce the Sobolev spaces $H^s_{<\lambda>}(\Omega,E,F_j)$ of cross-sections with norm in the natural way depending of the parameter $< \lambda >$, $\lambda \in \mathbb{C}$.

Definition 4.1. Let $a(r,\lambda)$ be defined on $\mathbb{R}_+ \times \mathbb{R}_\mu$ with values in the space $OP\mathfrak{S}^{m,d}(\Omega,E,F_1,F_2)$.

We say that $a(r,\lambda) \in A^{m,d}_\mu(\mathbb{K},E,F_1,F_2)$ if for all $\alpha,\beta \in \mathbb{Z}_+$ the following estimates are satisfied:

$$\|(r\partial_r)^\beta \partial^\alpha_\lambda a(r,\lambda)\|_{\mathbb{B}(H^s_{<\lambda>}(\Omega,E,F_1),H^s_{<\lambda>}(\Omega,E,F_2))} \le C_{\alpha\beta s} \tag{5}$$

for any $s > d - 1/2$, and we say that $a(r,\lambda) \in \widetilde{\mathfrak{A}}^{m,d}(\mathbb{K}_\Omega,E,F_1,F_2)$ if in addition to the estimates (5)

$$\lim_{r\to+0} \sup_{\lambda\in\mathbb{R}_\mu} \|(r\partial_r)a(r,\lambda)\|_{\mathbb{B}(\mathbb{H}^s_{<\lambda>}(\Omega,E,F_1),H^{s-m}_{<\lambda>}(\Omega,E,F_2))} = 0.$$

Let us assoociate a $M\Psi DO$ $a(r,\mathcal{D}_r)$ to the operator-function $a(r,\lambda)$ by the formula (1) and let us call this operator by operator of boundary value problem in cone.

The set of all such operators will be denoted $OP\mathfrak{A}^{m,d}_\mu(\mathbb{K}, E, F_1, F_2)$.

The notation $OP\widetilde{\mathfrak{A}}^{m,d}(\mathbb{K}_\Omega,E,F_1,F_2)$ has an obvious meaning.

Definition 4.2. We denote $V^s_\beta(\mathbb{K}_\Omega,E,F_j)$, $s,\beta \in \mathbb{R}$, the completion of

$$C^\infty_0(\mathbb{K}_\Omega,E) \oplus C^\infty_0(\partial\mathbb{K}, F_j)$$

in the norm

$$\|v\|_{V^s_\beta(\mathbb{K}_\Omega,E,F_j)} = \left(\int_{\mathbb{R}_{\beta-s+n/2}} \|(Mu)(\lambda)\|^2_{H^s_{<\lambda>}(\Omega,E,F_j)} d\lambda\right)^{\frac{1}{2}}.$$

Proposition 4.1. An operator $A \in OP\mathfrak{A}^{m,d}_{\beta-s+n/2}(\mathbb{K}_\Omega, E, F_1, F_2)$ maps $V^s_\beta(\mathbb{K}_\Omega, E, F_1)$ into $V^{s-m}_\beta(\mathbb{K}_\Omega, E, F_2)$ if $s > d - 1/2$.

As it is shown in [11], [12] there exist reduction order operators

$$Q^s_j(\lambda) : H^s_{<\lambda>}(\Omega, E, F_j) \to L_2(\Omega, E) \oplus L_2(\partial\Omega, F_j) = \mathbb{H}'_j,$$

depending on the parameter $<\lambda>$. It is not so hard to check that $Q^s_j(\lambda)$ satisfies all conditions of Definition 2.1 in which $\mathbb{H}_j = H^s(\Omega, E, F_j)$ and $\mathbb{H}'_j$ have been defined above.

As follows from the estimates (5), if $A \in OP\mathfrak{A}^{m,d}_{\beta-s+n/2}(\mathbb{K}_\Omega, E, F_1, F_2)$ ($A \in OP\widetilde{\mathfrak{A}}^{m,d}_{\beta-s+n/2}(\mathbb{K}_\Omega, E, F_1, F_2)$) then

$$A \in OP\mathcal{E}_{\beta-s+n/2}(Q^s_1, Q^s_2) \quad (A \in OP\widetilde{\mathcal{E}}_{\beta-s+n/2}(Q^s_1, Q^s_2)).$$

Thus the next theorems are consequences of the Theorems 3.1, 3.2.

Theorem 4.1. An operator $A \in OP\widetilde{\mathfrak{A}}^{m,d}_{\beta-s+n/2}(\mathbb{K}_\Omega, E, F_1, F_2)$ acting from $V^s_\beta(\mathbb{K}_\Omega, E, F_1)$ into $V^{s-m}_\beta(\mathbb{K}_\Omega, E, F_2)$, $s > d - 1/2$, is locally invertible at the point 0 if there exists $\varepsilon > 0$ such that operator symbol $a(r,\lambda)$ is invertible when $r \in (0,\varepsilon)$, $\lambda \in \mathbb{R}_{\beta-s+n/2}$ and

$$\sup_{(0,\varepsilon)\times\mathbb{R}_{\beta-s+n/2}} \|a^{-1}(r,\lambda)\|_{\mathbb{B}(H^{s-m}_{<\lambda>}(\Omega,E,F_2), H^s_{<\lambda>}(\Omega,E,F_1))} < \infty.$$

Theorem 4.2. Let $A \in OP\widetilde{\mathfrak{A}}^{m,d}_{\beta-s+n/2}(\mathbb{K}_\Omega, E, F_1, F_2)$ for all $\beta \in (\alpha_1, \alpha_2)$ and assume that the condition of the previous theorem is fulfilled for all $\beta \in [\beta_1, \beta_2] \subset (\alpha_1, \alpha_2)$. Then if $u \in V^s_{\beta_2}(\mathbb{K}_\Omega, E, F_1)$, $Au = f, \chi_{\varepsilon_0}(r) f \in V^{s-m}_{\beta_1}(\mathbb{K}_\Omega, E, F_2)$ for some $\varepsilon_0 > 0$ then there exists $\varepsilon \in (0, \varepsilon_0)$ such that $\chi_\varepsilon(r) u \in V^s_{\beta_1}(\mathbb{K}_\Omega, E, F_1)$.

Let us consider the boundary value problem in $\mathbb{K}_\Omega = \{(r,\omega) \in \mathbb{R}_+ \times \Omega\}$ where Ω is a domain with smooth boundary on the unit sphere S^{n-1}:

$$Au = f, \quad on \quad \mathbb{K}_\Omega, B_j u = f_j, \quad on \quad \partial\mathbb{K}_\Omega, \quad j = 1, \dots, \ell, \tag{6}$$

where A and B_j are differential operators written in the polar coordinate system (r,ω):

$$A = r^{-m} \sum_{k=0}^{m} \sum_{|\alpha| \le m-k} a_{\alpha k}(r,\omega) D^\alpha_\omega (r, D_r)^k = r^{-m} \mathcal{A}(r, rD_r, \omega, D_\omega),$$

$$B_j = r^{-m_j} \sum_{k=0}^{m_j} \sum_{|\alpha| \le m_j-k} b_{\alpha kj}(r,\omega) D^\alpha_\omega (rD_r)^k = r^{-m_j} \mathcal{B}_j(r, rD_r, \omega, D_\omega),$$

where

$$a_{\alpha k}(r,\omega) \in \widetilde{\mathcal{E}}^\infty(\mathbb{K}_\Omega) \otimes \mathbb{B}(\mathbb{C}^N, \mathbb{C}^N),$$

$$b_{\alpha kj}(r,\omega) \in \widetilde{\mathcal{E}}^{\infty}(\mathbb{K}_{\Omega}) \otimes \mathbb{B}(\mathbb{C}^N,\mathbb{C}^{N_j}),$$

$$\begin{aligned}\widetilde{\mathcal{E}}^{\infty}(\mathbb{K}_{\Omega}) &= \{a(r,\omega) \in C^{\infty}(\mathbb{R}_+ \times \Omega) : |(r\partial_r)^j \partial_{\omega}^{\alpha} a(r,\omega)| \le c_{\alpha j} \quad \textit{and} \\ &\lim_{r\to+0} \sup_{\omega\in\overline{\Omega}} (r\partial_r)a(r,\omega) = 0\}.\end{aligned}$$

For example, the function $\exp(i\beta|\ln r|^{\alpha})b(\omega)$, where $b(\omega) \in C^{\infty}(\overline{\Omega})$, $\beta \in \mathbb{R}$, $\alpha \in (0,1)$ belongs to $\widetilde{\mathcal{E}}^{\infty}(\mathbb{K}_{\Omega})$.

Since the boundary value problem is reduced to a problem in the Boutet-de-Monvel's algebra we can formulate a theorem about local invertibility of problem (6). For the formulation of this theorem we need in the following notation

$$\begin{aligned}L(r,\lambda)u &= \{\mathcal{A}(r,\lambda,\omega,D_{\omega})u(\omega), w \in \Omega; \\ &\mathcal{B}_1(r,\lambda,\omega,D_{\omega})u(\omega), \dots, \mathcal{B}_{\ell}(r,\lambda,\omega,D_{\omega})u(\omega), \omega \in \partial\Omega\}.\end{aligned}$$

Theorem 4.3. The operator A of the boundary value problem (6) acting from $V_{\beta}^{s}(\mathbb{K},\mathbb{C}^N)$ into $V_{\beta}^{s-m}(\mathbb{K},\mathbb{C}^N) \oplus_{j=1}^{\ell} V_{\beta}^{s-m_j-1/2}(\partial\mathbb{K},\mathbb{C}^{N_j})$, $s > \max_j\{m_j+1/2\}$ is locally invertible at the point 0 iff the operator-function $L(r,\lambda) : H^s(\Omega,\mathbb{C}^N) \to H^{s-m}(\Omega,\mathbb{C}^N) \oplus_{j=1}^{\ell} H^{s-m_j-1/2}(\Omega,\mathbb{C}^{N_j})$ is invertible on $(0,\varepsilon) \times \mathbb{R}_{\beta-s+n/2}$ for some $\varepsilon > 0$ and

$$\sup_{(0,\varepsilon)\times\mathbb{R}_{\beta-s+n/2}} \|L^{-1}(r,\lambda)\|_{\mathbb{B}(H^{s-m}_{<\lambda>}(\Omega,\mathbb{C}^N)\oplus_{j=1}^{l} H^{s-m_j-1/2}_{<\lambda>}(\partial\Omega,\mathbb{C}^{N_j}),H^{s}_{<\lambda>}(\Omega,\mathbb{C}^N))} < \infty.$$

Similarly we could formulate a theorem about the local improving of behavior of solution of boundary value problem (6) at the point 0.

Let us consider the Dirichlet and Neumann problems for the Laplace operator in the domain $\Omega \in \mathbb{R}^2$ described in polar coordinate system (ϱ,θ) by condition:

$$\Omega = \{(\varrho,\theta) : \varrho \in \mathbb{R}_+, f_1(\varrho) < \theta < f_2(\varrho)\},$$

where $f_j(\varrho) \in \widetilde{\mathcal{E}}^{\infty}(\mathbb{R}_+)$, $0 \le f_1(\varrho) < f_2(\varrho) < 2\pi$.

Let $\varphi(\varrho) = f_2(\varrho) - f_1(\varrho)$ and we suppose that $\inf_{\mathbb{R}_+} \varphi(\varrho) > 0$. Let us introduce the change of variables

$$\Phi = \{\varrho = r, \theta = (1-\omega)f_1(r) + \omega f_2(r), \quad r \in \mathbb{R}_+, \omega \in (0,1)\}$$

transforming the corner

$$\mathbb{K} = \{(r,\omega) \in \mathbb{R}_+ \times (0,1)\}$$

on the domain Ω.

The change of variables transforms the original problems in the domain Ω to Dirichlet and Neumann problems in $\mathbb{K}$ for some differential operator. The operator-function $L(r,\lambda)$ (see Theorem 4.3) is given by formula

$$\begin{aligned} L(r,\lambda) &= \{\lambda^2 + \varphi^{-2}(r)\frac{\partial^2}{\partial\omega^2}, \quad \omega \in (0,1), \\ u(0) &= u(1) = 0 \quad \text{for Dirichlet problem and} \\ u'(0) &= u'(1) = 0 \quad \text{for Neumann problem}\}. \end{aligned}$$

Let

$$\Gamma_{j,\varepsilon} = \{r + i\mu \in \mathbb{C} : \mu = \pi j \varphi^{-1}(r), \quad r \in (0,\varepsilon), j \in \mathbb{Z}\},$$

$V_\beta^s(\Omega)$ $(V_\beta^{s-1/2}(\partial\Omega), s > 1/2)$ is the space of restrictions of distributions from $V_\beta^s(\mathbb{R}^2)$ to Ω $(\partial\Omega)$.

Using the well-known spectral properties of Sturm-Liouville problems for $\frac{d^2}{d\omega^2}$ and Theorems 4.1 and 4.2 we obtain the following result.

Theorem 4.4. Let

$$\underline{\lim}_{\varepsilon\to+0}\text{dist}(\mathbb{R}_{\beta-s-1},\Gamma_{j,\varepsilon}) > 0 \tag{7}$$

for any $j \in \mathbb{Z} \setminus 0$, then the operator A_D of Dirichlet problem is locally invertible at the point zero as an operator acting from $V_\beta^s(\Omega)$ into $V_\beta^{s-2}(\Omega) \oplus V_\beta^{s-1/2}(\partial\Omega)$.

If condition (7) is satisfied for all $\beta \in [\beta_1,\beta_2]$, $u \in V_{\beta_2}^s(\mathcal{D})$, $A_D u = f$, and $\chi_{\varepsilon_0}(r)f \in V_{\beta_1}^{s-2}(\Omega) \oplus V_{\beta_1}^{s-1/2}(\partial\Omega)$ then there exists $\varepsilon \in (0,\varepsilon_0)$ such that $\chi_\varepsilon(r)u \in V_{\beta_1}^s(\mathcal{D})$.

A similar result is valid for the Neumann problem operator but we must replace $\mathbb{Z} \setminus 0$ by $\mathbb{Z}$ and $V_\beta^{s-1/2}$ by $V_\beta^{s-3/2}$ for its formulation.

References

[1] V. A. Kondratyev. Boundary problems for elliptic equations with conic or salient points [in Russian]. Trudy Mosk. Mat. Ob., Vol.16 (1963), 121–292.

[2] B.-W. Schulze. Pseudodifferential operators on manifolds with singularities. Amsterdam: North-Holland, 1991.

[3] B.-W. Schulze. Pseudodifferential operators and analysis on manifolds with corners, part I–IV, VI–IX, Reports of the Karl-Weierstrass-Institute, Berlin, 1989–1991, parts XII, XIII: preprints no.214 and 220, SFB 256, Univ Borm., 1992.

[4] E. Schrohe. Boundary value problems in Boutet de Monvel's algebra for manifolds with conical singularities. Abstract of the Conference "Partial Differential Equations", Potsdam, Sept. 6–10, 1993.

[5] B.-W. Schulze, Pseudodifferential operators on manifolds with eges, Symp. "Part. Diff. Equ." Holzhau, 1988, Teubuer Texte zur Mathematik 112, 259–288, Leipzig, 1989.

[6] B.-W. Schulze. The Mellin pseudodifferential calculus on manifolds with corners. Proc. Symp. "Analysis on manifolds with singularities", Breitenbrunn, 1990, Teubner Texte zur Mathematik 131, 208–290.

[7] V.S. Rabinovich. Criteria of local invertibility of Mellin pseudodifferential operators with operator symbols and its applications [in Russian]. Doklady Ros. Akademii Nauk, Math., 1993, vol. 333, no. 2, 147–150.

[8] V.S. Rabinovich. Criteria of local invertibility of pseudodifferential operators with operator symbols and some its applications. It will be published the journ. "Trudy Sankt–Peterburg Mat. Ob.".

[9] V.V. Grushin. Pseudodifferential operators on $\mathbb{R}^n$ with bounded symbols [in Russian]. Funct. Analis i ego priloz., no. 4, 1970, 202–212.

[10] V.S. Rabinovich. The Noetherian properties of pseudodifferential operators with symbols from class $S^m_{\varrho,\delta}(o \leq \delta = \varrho < 1, \delta < 1)$ [in Russian]. Matem. Zametki, 27, 1980, 226–231.

[11] S. Rempel and B.-W.Schulze. Index Theories of elliptic boundary problems. Akademie-Verlag, Berlin.

[12] G. Grubb. Functional calculus of pseudodifferential boundary value problems. Progress in Math., vol. 65, Boston, Basel, Birkhäuser, 1986.

Vladimir S. Rabinovich
Rostov State University, Department of Mathematics and Mechanics,
Zorge str. 5, 344104 Rostov-na-Donu, Russia.

Operator Theory:
Advances and Applications, Vol. 78

Hypoellipticity of Certain Differential Operators with Degeneration of Infinite Order

M. Reissig, K. Yagdjian

The object of this paper is to study the operator

$$L = D_t^2 + \lambda^2(t) D_x^2 - a(t)\frac{\lambda^2(t)}{\Lambda(t)} D_x \tag{1}$$

on $\mathbb{R}^2$, where $D_t = -i\partial/\partial t, D_x = -i\partial/\partial x$,

$$\Lambda(t) = \exp(i\phi - |t|^{-1}), \quad \phi \in [0,\pi/2), \tag{2}$$

$$a(t) = \begin{cases} a_- \in \mathbb{C} & \text{when} \quad t < 0\,, \\ a_+ \in \mathbb{C} & \text{when} \quad t \geq 0\,, \end{cases}$$

$$\lambda(t) := iD_t\Lambda(t) = (\text{sign } t)t^{-2}\exp(i\phi - |t|^{-1})\,.$$

At the point $t = 0$ the coefficients of operator have a zero of infinite order.

The question to be considered is when the operator L is hypoelliptic or locally solvable.

For the operator

$$P = (D_t - iat^k D_x)(D_t - ibt^k D_x) + ct^{k-1}D_x \tag{3}$$

when k is odd, Gilioli and Treves [3] have proved

Theorem 1. For operator P of form (3) in which k is odd, the following are equivalent:

(i) P is locally solvable at the origin;

(ii) P is hypoelliptic with loss of $2k/(k+1)$ derivatives;

(iii) $\mathrm{Re}\, a \cdot \mathrm{Re}\, b < 0$ and $c/(a-b) \neq n(k+1) + 0$ or 1 for any integer n.

For the case in which k is even, the following theorem has been proved by Menikoff [6]:

Theorem 2. Assume that operator P has form (3) where k is even, then the following are equivalent:

(i) P is locally solvable at the origin;

(ii) P is hypoelliptic (with loss of $2k/(k+1)$ derivatives);

(iii) the estimate

$$\|u\|_{s+2/(k+1)} \le C\left(\|Pu\|_s + \|u\|_s\right), \quad u \in C_0^\infty(K)$$

is valid, where K is any compact subset of $\mathbb{R}^2$;

(iv) either $\operatorname{Re} a \cdot \operatorname{Re} b > 0$, or $\operatorname{Re} a \cdot \operatorname{Re} b < 0$ and $c/(a-b) \neq n(k+1)+1/2$ for $n = 0, \pm 1, \pm 2, \dots$.

Thus in the case $\lambda(t) = t^k$ above mentioned question for the operator L is completely studied. If $a(t) = a \in \mathbb{C}$, then L is hypoelliptic (with loss of $2k/(k+1)$ derivatives) or locally solvable at the origin only in the cases when $a \neq 2n(k+1) - k + 0$ or 2 (k odd) or $a \neq 2n(k+1) - k + 1$ (k even).

The case when $\lambda(t)$ is real and has a zero of infinite order was studied by Hoshiro [4]. The assumption of the main theorem (Theorem 2 [4]) of this paper concerning a sufficient condition for hypoellipticity, applied to (1) is the following:

There exist positive constants δ, R and $\tilde{C}$ such that the inequality

$$\begin{aligned} &\int_{-\delta}^{\delta} \left\{ |v'(t)|^2 + \left(\lambda^2(t)\xi^2 + \left| a(t)\frac{\lambda^2(t)}{\Lambda(t)}\xi \right| \right) |v(t)|^2 \right\} dt \\ &\le \tilde{C} \int_{-\delta}^{\delta} \bar{v}(t) \left\{ D_t^2 + \lambda^2(t)\xi^2 - a(t)\frac{\lambda^2(t)}{\Lambda(t)}\xi \right\} v(t)dt \end{aligned} \tag{4}$$

holds for all $v \in C_0^\infty(-\delta,\delta)$ and all $\xi \in \mathbb{R}, |\xi| \ge R$.

Moreover, there is shown that condition (4) holds under the assumptions, that $\lambda^2(t)$ is strictly monotone in the intervals $(-\delta,0)$ and $(0,\delta)$, and $|a(t)| \le 2C\left|\Lambda(t)\lambda^{-2}(t)D_t\lambda(t)\right|$ on these intervals with $0 < C < 1/2$. If we apply this result to (1) we obtain the condition $\max(|a_+|,|a_-|) \le 2C(1-2|t|)$ with $0 < C < 1/2$ and $t \in (-\delta,\delta)$ for obtaining hypoellipticity.

It will be shown later (see Sec.1) that the operator (1) for example, with $a_- = -2n-1, a_+ = 2\ell+1$ (here n and ℓ are non-negative integer) is hypoelliptic. Otherwise the operator (1) with $a_- = a_+ = 1$ is not hypoelliptic. This means, hypoellipticity cannot be proved for $C = 1/2$, there appear some exceptional values for a_- and a_+.

Our main goal in present paper is to determine all the values of constants a_-,a_+ for which hypoellipticity and local solvability of L holds. In a following paper we shall study the case of operator (1) with more general coefficients.

We prove the following

Theorem 3. Assume that neither $a_- = -2n-1, a_+ = -2\ell-1$ nor $a_- = 2n+1, a_+ = 2\ell+1$, where n and ℓ are non-negative integer. Then the operator L of form (1) with $\phi \in [0,\pi/2)$, is hypoelliptic as well as locally solvable, and the following estimate

$$\|u\|_s \le C_{\phi,K,s}\left(\|Lu\|_{s+\max\{|\operatorname{Re} a_+|,|\operatorname{Re} a_-|\}} + \|u\|_{s-1}\right), \qquad u \in C_0^\infty(K) \tag{5}$$

holds, where K is any compact subset of $\mathbb{R}^2$.

Theorem 4. Assume that either $a_- = -2n-1, a_+ = -2\ell-1$ or $a_- = 2n+1, a_+ = 2\ell+1$, where n and l are non-negative integer. Then:

(i) the operator L with $\phi \in [0,\pi/2)$ is not locally solvable at (0,0);

(ii) the operator L with $\phi \in [0,\pi/2)$ is not hypoelliptic at (0,0).

The last result is closely connected with branching phenomena in the theory of weakly hyperbolic equations [1, 2, 5, 7, 8, 9]. As matter of fact the hyperbolic Cauchy problem for the operator (1)&(2) with $\phi = \pi/2$ has been studied and propagation of singularities has been described by Alexandrian [1].

Then, compare Theorem 4 with Theorem 7 [1] we conclude

Corollary. The weakly hyperbolic homogeneous equation corresponding to operator (1) with $\phi = \pi/2$ has a solution whose WF (wave front set) coincides with the simple ray passes through the origin (0,0) and which changes the direction of motion (is completely reflected by the point (0,0)) if and only if the operator (1) with $\phi = 0$ is not hypoelliptic at the origin (0,0).

Remark. We find it is very interesting that the same phenomena takes place in the case of finite order degeneration, too. Indeed, we propose to be convinced by means of comparison the results of [3], [6] and [8].

References

[1] G.R. Aleksandrian, The parametrix and the propagation of the wave front of solutions of the Cauchy problem for a certain model equation, Soviet J. of Contemporary Math.Anal. **19**(1984) 33–46.

[2] S. Alinhac, Branching of singularities for a class of hyperbolic operators, Indiana Univ.Math.J. **27**(1978) 1027–1037.

[3] A. Gilioli, F. Treves, An example in the solvability theory of linear PDE's, Am. J. Math. **96**(1974) 367–385.

[4] T. Hoshiro, On Levi-type conditions for hypoellipticity of certain differential operators, Commun. in Partial Diff. Equations, **17**(1992), 5&6, 905–922.

[5] V. Ivrii, On wave fronts of the solutions of certain pseudodifferential equations, Funkts. Anal. Prilozhen., **10**(1976), no. 2, 71–72.

[6] A. Menikoff, Some example of hypoelliptic partial differential equations, Math. Ann. **221**(1976) 167–181.

[7] K. Shinkai, Stokes multipliers and a weakly hyperbolic operator, Comm. Part. Diff. Equat. **16**(1991) 667–682.

[8] K. Taniguchi and Y. Tozaki, A hyperbolic equation with double characteristics which has a solution with branching singularities, Math.Japonica **25**(1980) 279–300.

[9] K.H. Yagdjian(=K.A. Yagdzhyan), Pseudodifferential operators with the parameter and the fundamental solution of the Cauchy problem for operators with multiple characteristics, Soviet J. of Contemporary Math. Anal. **21**(1986) 1–29.

M. Reissig
Faculty of Mathematics and Computer Science,
Freiberg University of Mining and Technology
Bernhard-von-Cotta-Str. 2, D-09596 Freiberg, Germany
e-mail:reissig@mathe.ba-freiberg.d400.de

K. Yagdjian
Institute of Mathematics, Armenian Academy of Sciences
Marshal Bagramian Ave. 24B, 375019 Yerevan, Armenia
e-mail:armtsb@adonis.iasnet.com

Operator Theory:
Advances and Applications, Vol. 78

On Approximation of Solutions of Elliptic Boundary Value Problems for Petrovskii Elliptic Systems by Linear Combinations of Fundamental Solutions

I. Roitberg

1. Introduction

Let $G \subset \mathbb{R}^n$ be a bounded domain with the boundary ∂G. In G we consider a Petrovskii properly elliptic system $L = (l_{ij}(x,D))_{i,j=1,\dots,N}$, $ord\ l_{ij} \leq t_j; t_1 \geq \dots \geq t_N \geq 0$, $t_1 + t_2 + \dots + t_N = 2m$.

Suppose that for $L(x,D)$ there exists a global normal fundamental solution $\Phi(x,y)$ $(x,y \in G,\ x \neq y)$ (see section 3, below).

Let $\Omega \subset \overline{\Omega} \subset G$ be a bounded domain with the boundary $\Gamma \in C^\infty$, $G \setminus \overline{\Omega}$ is connected.

In Ω we consider an elliptic boundary value problem

$$Lu|_\Omega = 0; \quad Bu|_\Gamma = (b_{hk}(x,D))_{h=1,\dots,m;\ k=1,\dots,N}\ u = \varphi$$

$$(ord\ b_{hk} \leq \sigma_h + t_k\ ;\ \ \sigma_h < 0\ ; \varphi \in \prod_{h=1}^{m} B^{s-\sigma_h-\frac{1}{p},p}(\Gamma) =: \mathcal{B}^s) \tag{1}$$

with normal boundary conditions.

Let $K \in G \setminus \overline{\Omega}$ be a piece of a smooth $(n-1)$-dimensional surface, and let $(y_k)_{k=1}^\infty \subset G \setminus \overline{\Omega}$ be a sequence of points which is dense in K (i.e. $\overline{(y_k)_{k=1}^\infty} \supset K$).

The following problem was formulated by Uwe Hamann [1]–[3]: Does there exist a sequence of linear combinations

$$u_l(x) = \sum_{k=1}^{l} \sum_{j=1}^{N} \sum_{|\alpha| \leq t_j - 1} c_{k\alpha j}^{(l)}\ D_y^\alpha\ \Phi_j(x,y_k)\ ,l = 1,2,\dots;\ c_{k\alpha j}^{(l)} \in \mathbb{C} \tag{2}$$

such that the relation

$$\|Bu_l - \varphi\|_{\mathcal{B}^s} \longrightarrow 0 \ \ (l \longrightarrow \infty) \tag{3}$$

is true.

The importance of this problem follows from the next fact: relation (3) implies that the sequence (2) converges in the product of the spaces $\prod_{j=1}^{N} \tilde{H}^{t_j+s,p,(t_j)}(\Omega)$ of Sobolev type to the solution of the problem (1) (in the case when the problem has no defect).

This problem was solved by Hamann for the case of constant coefficients ([1]–[3]) and the technique developed in these works relies heavily on the fact that the coefficients are constant.

A different technique based on the Theorems on complete collection of isomorphisms [4]–[6], and on the Theorems on global fundamental solutions ([7], [8]) is developed in the present work. This technique allows us to consider the case of equations with variable coefficients [9], and finally the case of Petrovskii elliptic systems with variable coefficients.

2. Functional spaces.

Let $p,p' \in (1,\infty)$, $\frac{1}{p}+\frac{1}{p'}=1$. For any domain $G \subset \mathbb{R}$ with the smooth boudary ∂G we denote by $H^{s,p}(G)$ $(s \geq 0)$ the space of Liouville classes (Bessel potentials); $H^{-s,p}=(H^{s,p'}(G))^*$; $\|\cdot,G\|_{s,p}$ is a norm in $H^{s,p}(G)$ $(s \in \mathbb{R})$. The space $H^{-s,p'}(G)$ $(s>0)$ is isometrically equivalent to the subspace $H^{-s,p'}_{\overline{G}}(\mathbb{R}^n)=\{u \in H^{-s,p'}(\mathbb{R}^n):\ \operatorname{supp} u \subset \overline{G}\}$ of the space $H^{-s,p'}(\mathbb{R}^n)$. Therefore if $G_o \subset \overline{G_o} \subset G$ then one can say that $H^{-s,p'}(G_o) \subset H^{-s,p'}(G)$ $(s>0)$.

For any $s \geq 0$ we set $\overset{\circ}{H}{}^{s,p}(G) := H^{s,p}_{\overline{G}}(\mathbb{R}^n)=\{u \in H^{s,p}(\mathbb{R}^n) : \operatorname{supp} u \subset \overline{G}\}$. Let

$$H'^{-s,p'}(G)=(\overset{\circ}{H}{}^{s,p}(G))^*;\ \|u,G\|'_{-s,p'}=\sup_{v\in \overset{\circ}{H}{}^{s,p}(G)} |(u,v)_G|(\|v\|_{H^{s,p}(\mathbb{R}^n)})^{-1}$$

is a norm in $H^{s,p'}(G)$. If $s \geq 0$, $s \neq k+1/p$ $(k=0,1,2,\dots)$, then $\overset{\circ}{H}{}^{s,p}(G)$ is isomorphic to the closure of $C_0^\infty(G)$ in $H^{s,p}(G)$. Finally we note that if $u \in H'^{-s,p'}(G)$ $(s>0)$ and $\operatorname{supp} u \subset \overline{G_\circ}$ then $u \in H^{-s,p'}(G_\circ)$ and $\|u\|_{H^{-s,p'}(G_\circ)} = \|u\|_{H'^{-s,p'}(G)}$.

For any $s \in \mathbb{R}$ let $B^{s,p}(\partial G)$ be a Besov space, $B^{s,p}(\partial G)=(B^{-s,p'}(\partial G))^*$; $\langle\langle u,\partial G\rangle\rangle_{s,p}$ is a norm in $B^{s,p}(\partial G)$.

For fixed $r \in \mathbb{N}$ and any $s \in \mathbb{R}$, $s \neq k+1/p$ $(k=0,1,2,\dots,r-1)$ by $\tilde{H}^{s,p,(r)}(G)$ [6] we denote the completion of $C^\infty(\overline{G})$ with respect to the norm

$$|||u,G|||_{s,p,(r)}=(\|u,G\|^p_{s,p}+\sum\nolimits_{j=1}^{r}\langle\langle D_\nu^{j-1}u_j,\,\partial G\rangle\rangle^p_{s-j+1-\frac{1}{p},p})^{\frac{1}{p}} \tag{4}$$

here $D_\nu = i\partial/\partial\nu$, $\partial/\partial\nu$ is the normal derivative with respect to ∂G. For $s=k+1/p$ $(k=0,1,\dots,r-1)$ we define the space $\tilde{H}^{s,p,(r)}(G)$ and the norm (4) with help of complex interpolation. It is clear that the closure $S=S_r$ of the mapping $u \mapsto (u|_{\overline{G}},\ u|_{\partial G},\dots,D_\nu^{r-1}\ u|_{\partial G})$ $(u \in C^\infty(\overline{G}))$ sets an isometry between $\tilde{H}^{s,p,(r)}(G)$ and a subspace of the direct product $H^{s,p}(G)\times\prod_{j=1}^{r}B^{s-j-1-1/p,p}(\partial G)=K^{s,p}$. In addition if $s<1/p$ then $S\ \tilde{H}^{s,p,(r)}(G)=K^{s,p}$. Therefore we can identify $u \in \tilde{H}^{s,p,(r)}(G)$ with the element $Su=(u_0,u_1,\dots,u_r) \in K^{s,p}$; we shall write $u=(u_0,u_1,\dots,u_r) \in \tilde{H}^{s,p,(r)}(G)$, $u|_{\overline{G}}=u_0$ for any $u \in \tilde{H}^{s,p,(r)}(G)$. Let $\tilde{H}^{s,p,(0)}(G) := H^{s,p}(G)$. The space $\tilde{H}^{s,p,(r)}(G)$ was studied in [6].

3. Fundamental solutions of elliptic operators.

Let $\mathcal{L}(x,D) = (\mathcal{L}_{ij}(x,D))_{i,j=1,\ldots}$ be a Douglis-Nirenberg properly elliptic system of order $(T,S) = (t_1,t_2,\ldots,t_n,s_1,s_2,\ldots,s_n)$; $t_1 \geq \ldots \geq t_n \geq 0 = s_1 \geq \cdots \geq s_n$, $ord\ \mathcal{L}_{ij} \leq s_i + t_j$.

Definition. A matrix $\Phi(x,y) = (\Phi_{rk}(x,y))_{r,k=1,\ldots,N}$; $(x,y \in G,\ x \neq y)$ is called a global fundamental solution (G.F.S.) in G of the operator $\mathcal{L}$ if the operator

$$f \mapsto u(x) = \int_G \Phi(x,y)f(y)dy \quad (\forall f \in \prod_{j=1}^{N} H^{-s_j,p}(G)) \tag{5}$$

acts continuously from $\prod_{j=1}^{N} H^{-s_j,p}(G)$ to $\prod_{j=1}^{N} H^{t_j,p}(G)$ if $s_1 = \cdots = s_N = 0$(i.e. $\mathcal{L} = \mathcal{L}$ is a Petrovskii elliptic system) and to $\prod_{j=1}^{N} H^{t_j,p}(G_0)$ for any $G_0 \subset \overline{G_0} \subset G$ if $s_1 + \cdots + s_N \neq 0$; and $Lu = f$ in G. A G.F.S. $\Phi(x,y)$ is called a normal G.F.S. of $\mathcal{L}$ if the conjugate matrix $\Phi^*(x,y) := (\overline{\Phi_{kr}(y,x)})_{k,r=1,\ldots,N}$ is a GFS of the formally adjoint operator $\mathcal{L}^+$. We note that for a Petrovskii elliptic system L the formally adjoint expression $L^+(x,D) = (l^+_{ji}(x,D))_{j,i=1,\ldots,N}$ is a Douglis Nirenberg elliptic system of order $(T',S') = (t'_1,\ldots,t'_N,s'_1,\ldots,s'_N)$, $t'_j = t_1$; $s'_j = t_j - t_1$ $(j = 1,\ldots,N)$ $ord\ l^+_{ji}(x,D) \leq t'_j + s'_j$.

We note [7] that the necessary and sufficient condition for the existence of the normal GFS of $\mathcal{L}$ is the uniqueness of the Cauchy problem both for the system $\mathcal{L}\sqcap = \prime$ and for the system $\mathcal{L}^+ v = 0$. Properties of regularity of the GFS $\Phi(x,y)$ are studied in [7], [8]. In particular the closure of the mapping

$$f \mapsto u_j = (\textstyle\int_G \Phi(x,y)f(y)dy)_j \quad (f \in (L_p(G))^N,\ j = 1,\ldots,N)$$

acts continuously in the pair of spaces

$$(H^{s,p}(G))^N \rightarrow \begin{cases} H^{t_j+s,p}(G) & (s > 0,\ t_j + s \geq 0) \\ H'^{\ t_j+s,p}(G) & (s < 0,\ t_j + s < 0) \end{cases};$$

a closure of the mapping $g \mapsto v_j = (\int_G \Phi^*(x,y)g(y)dy)_j$ $(g \in \prod_{j=1}^{N} H^{-s'_i,p'}(G),\ j = 1,\ldots,N)$ acts continuously in the spaces

$$\prod_{j=1}^{N} H^{-s'_j+s,p'}(G) \longrightarrow \begin{cases} H^{t_1+s,p'}(G) & (s < s'_N = t_N - t_1,\ t_1 + s \geq 0) \\ H'^{t_1+s,p'}(G) & (s \leq s'_N,\ t_1 + s < 0) \\ H^{t_N,p'}(G) & (s'_N < s < 0). \end{cases}$$

4. Main results.

Theorem 1. Let (i) $s \in \mathbb{R}$, $p \in (1,+\infty)$; (ii) in G there exists a normal GFS $\Phi(x,y)$ of $L(x,D)$ (iii) ϕ such that the elliptic problem (1) is solvable. Then there exists a sequence (2) satisfying the relation (3).

Theorem 2. Under the conditions of Theorem 1 let K be the smooth boundary of a domain U $(U \subset \overline{U} \subset G\backslash\overline{\Omega})$ and let the problem

$$L^{+}v|_U = 0;\ D_\nu^k v_j|_K = 0\ \ (k = 0,\dots,h_j - 1; j = 1,\dots,N; h_1 + \dots + h_N = m)\ (6)$$

have only a trivial solution. Then there exists a sequence (2) satisfying the relation (3) with $|\alpha| \le h_j - 1$ in place of $|\alpha| \le t_j - 1$ in (2).

Theorem 3. Under the conditions of Theorem 1 let the sequence $(y_k)_{k=1}^{\infty}$ be dense in an open set U $(U \subset \overline{U} \subset G\backslash\overline{\Omega})$. Then there exists a sequence (2) satisfying the relation (3) with $|\alpha| = 0$ in place of $|\alpha| \le t_j - 1$ in (2).

5. Proof of the Theorem 1.

Let the matrix $C(x,D)$ complement the matrix of the boundary expressions $B(x,D)$ to the Dirichlet system. Then the Green formula [10]

$$(Lu,v)_\Omega + \langle Bu,C'v\rangle_\Gamma = (u,L^{+}v)_\Omega + \langle Cu,B'v\rangle_\Gamma \tag{7}$$

$(u,v \in (C^\infty(\overline{\Omega}))^N)$
Here and below $(\cdot,\cdot)_\Omega$ and $\langle\cdot,\cdot\rangle_\Gamma$ mean scalar products in $(L_2(\Omega))^N$ and $(L_2(\Gamma))^N$ or their extensions, respectively. The problem

$$L^{+}v = g\ \ (in\ \Omega),\ B'v|_\Gamma = \psi \tag{8}$$

is called a formal adjoint to the problem (1). It is an elliptic problem also. Then the kernel $N^* = \{v \in (C^\infty(\overline{\Omega}))^N : L^{+}v = 0 \text{ in } \Omega\ \ B'v = 0 \text{ on } \Gamma\}$ is finite dimensional and the problem (1) is solvable in $\prod_{j=1}^{N} \tilde{H}^{t_j+s,p}$ if and only if the relation

$$\langle\phi, C'v\rangle_\Gamma = 0\ \ (\forall v \in N^*) \tag{9}$$

is valid.

To prove Theorem 1 it is sufficient to prove that if $\eta \in \mathcal{B}^{-s} := (\mathcal{B}^s)^*$ satisfies the relation

$$\langle B(x,D)D_y^\alpha\Phi_j(x,y_k),\ \eta\rangle_\Gamma = 0, \tag{10}$$

$$\langle\eta,\ C'v\rangle_\Gamma = 0 \tag{11}$$

then $\eta = 0$.

Since $(B'(x,D),\ C'(x,D))$ is a Dirichlet system then by a Theorem on extension [6] there exists an element $w \in (\tilde{H}^{-s,p',(t_1)}(\Omega))^N$ such that

$$B'w|_\Gamma = 0,\ \ C'w|_\Gamma = \eta;\ |||w,\Omega|||_{-s,p',(t_1)} \le C\langle\langle\eta\rangle\rangle_{\mathcal{B}^{-s}} \tag{12}$$

From Green's formula (7) with $u = D_y^\alpha\Phi_j(x,y),\ \ (x \in \overline{\Omega},\ y \in G\backslash\overline{\Omega})$ and $v = w$ we obtain

$$\langle B(x,D_x)\ D_y^\alpha\Phi_j(x,y),\ \eta\rangle_\Gamma = (D_y^\alpha\Phi(x,y),\ g(x))_\Omega; \tag{13}$$

$$g = L^+ w|_{\overline{\Omega}} \in \prod_{j=1}^{N} H^{-s-t_j,p'}(\Omega). \tag{14}$$

If $-s - t_j < 0$ then we can say that $g_j \in H^{-s-t_j}$ supp $g_j \subset \overline{\Omega}$ (see above). Then $\omega_{0j} := ((\Phi(\cdot,y), g)_\Omega)_j \in \{H^{-s,p'}(G)$ if $-s \geq 0$, $H'^{-s,p'}(G)$ if $-s < 0\}$. In addition $L^+\omega_0 = g$ in Ω and

$$L^+\omega_0(x) = 0 \ \ (x \in G\backslash\overline{\Omega}) \ , \ \omega_0 \in (C^\infty(G\backslash\overline{\Omega}))^N. \tag{15}$$

Formulas (13) and (10) imply that

$$(D^\alpha \omega_{0j})(y_k) = 0 \ \ (|\alpha| \leq t_1 - 1, \ k = 1,2,\dots) \tag{16}$$

and therefore

$$D^\alpha \omega_{0j}|_K = 0 \ \ (|\alpha| \leq t_1 - 1). \tag{17}$$

(15), (16) and uniqueness of the Cauchy problem imply that $\omega_{0j}(y) = 0$ $(\forall y \in G\backslash\overline{\Omega})$. Therefore $\omega_{0j} \in H^{-s,p'}(G)$ and supp $\omega_{0j} \subset \overline{\Omega}$.Then an element $\omega_j = (\omega_{0j},0,\dots,0) \in \tilde{H}^{-s,p',(t_j)}(\Omega)$ satisfies the equalities (for $-s - t_N < 0$)

$$L^+\omega|_{\overline{\Omega}} = g; \ B'\omega|_\Gamma = 0; \ C'\omega|_\Gamma = 0. \tag{18}$$

Then (12), (14) and (18) imply that

$$L^+z|_{\overline{\Omega}} = 0, \ B'z|_\Gamma = 0, \ C'z|_\Gamma = \eta \tag{19}$$

where $z = w - \omega_0$. The equalities (19) imply that $z \in N^*$ and therefore the equality $\eta = 0$ follows from (19) and (11). And the proof of Theorem 1 is completed in the case $-s - t_N < 0$. The other cases may be considered analogously.

6. To prove Theorem 2 we also construct the function $\omega_0(y)$. Now it satisfies the formulas (16), (17) with $|\alpha| \leq h_j - 1$. Then ω_0 is a solution of the problem (6) in U and $\omega_{0j} \equiv 0$ in U. The uniqueness of the Cauchy problem implies that *supp* $\omega_{0j} \subset \overline{\Omega}$ and the proof is completed as above. The proof of Theorem 3 is similar.

References

[1] U. Hamann. Approximation mittels Linearkombinationen von Fundamentallösungen elliptischer Differentialoperatoren. Math. Nachr. **154**, (1991) 265–284.

[2] U. Hamann. Approximation by Linear Combinations of Fundamental Solutions of the Lamé operator. Math. Nachr. **164**, (1993) 271–282.

[3] U. Hamann. Approximation by Linear Combinations of Fundamental Solutions of Elliptic Systems of Partial Differential Operators. Journal for Analysis and its Applications **13(1)**, (1994) 49–71.

[4] Yu.M. Beresanskii. Expansions in eigenfunctions of selfadjoint operators. "Naukova Dumka", Kiev, (1965).

[5] J.L. Lions; E. Magenes. Non-homogeneous Boundary Value Problems and Applications. Springer-Verlag, Berlin, New York (1972), (Originally published in French by Dunod, Paris 1968).

[6] Ya.A. Roitberg. Elliptic Boundary Value Problems in Spaces of Distributions. Kluwer Acad. Publ., Netherland, (1994).

[7] I. Roitberg. On Fundamental Solutions of Elliptic Operators. Dokl. Akad. Nauk., Ukraine **65**, (1991) 14–18.

[8] I. Roitberg. Global Fundametal Solutions of Elliptic Operators. Kandidat thesis, Ukraina Mathematical Institute, Kiev, (1993).

[9] I. Roitberg; Ya. Roitberg. On Approximation of the Solutions of Elliptic Boundary Value Problems by Linear Combinations of Fundamental Solutions. Dokl. Akad. Nauk., Ukraine **12**, (1992) 15–20.

[10] Ya. Roitberg; Z. Sheftel'. Theorem on a Homeomorphism for Elliptic System and its Applications. Mat. Sbornik 78, **(120):3**, (1969) 446–472.

Inna Roitberg
Chernigov State Pedagogical Institute
Sverdlova str. 53
250038 Chernigov, Ukraine

Operator Theory:
Advances and Applications, Vol. 78

The Reduced Wave Operator with two Unbounded Media

Yoshimi Saitō

1. Introduction.

Consider the differential expression

$$h = -\mu(x)^{-1}\Delta. \tag{1.1}$$

Here Δ is the Laplacian in $\mathbb{R}^N$ with $N \geq 2$, and $\mu(x)$ is a positive function on $\mathbb{R}^N$ given by

$$\mu(x) = \begin{cases} \mu_1 & (x \in \Omega_1), \\ \mu_2 & (x \in \Omega_2), \end{cases} \tag{1.2}$$

where $\mu_1, \mu_2 > 0$, $\mu_1 \neq \mu_2$, and Ω_ℓ, $\ell = 1, 2$, are open sets of $\mathbb{R}^N$ such that

$$\begin{cases} \Omega_1 \cap \Omega_2 = \emptyset, \\ \overline{\Omega_1} \cup \Omega_2 = \Omega_1 \cup \overline{\Omega_2} = \mathbb{R}^N, \end{cases} \tag{1.3}$$

$\overline{\Omega_\ell}$, $\ell = 1, 2$, being the closure of Ω_ℓ. It is easy to see that a selfadjoint realization H of h is given by

$$\begin{cases} D(H) = H^2(\mathbb{R}^N), \\ Hu = hu, \end{cases} \tag{1.4}$$

in the Hilbert space

$$X = L_2(\mathbb{R}^N; \mu(x)dx), \tag{1.5}$$

where $D(H)$ denotes the domain of H, $H^2(\mathbb{R}^N)$ denotes the second-order Sobolev space over $\mathbb{R}^N$, and hu should be taken in the sense of distributions. We denote the resolvent of H by $R(z)$, that is,

$$R(z) = (H - z)^{-1}. \tag{1.6}$$

The separating surface S is defined by

$$S = \overline{\Omega_1} \cap \overline{\Omega_2} = \partial\Omega_1 = \partial\Omega_2, \tag{1.7}$$

$\partial\Omega_\ell$, $\ell = 1, 2$, being the boundary of Ω_ℓ. Let

$$n^{(\ell)}(x) = (n_1^{(\ell)}(x), n_2^{(\ell)}(x), \cdots, n_N^{(\ell)}(x)) \qquad (\ell = 1, 2) \tag{1.8}$$

be the unit outward normal of $\partial\Omega_\ell$ at $x \in \partial\Omega_\ell = S$. Obviously we have $n^{(1)}(x) + n^{(2)}(x) = 0$ for $x \in S$.

This article is going to review work concerning the operator H under certain conditions imposed on the separating surface S. In a paper published in 1986,

D. Eidus started the study of H under the condition that the separating surface S has a *cone-like* shape (Eidus [2]). One of his results is the limiting absorption principle for H, i.e., the existence of the extended resolvent

$$\lim_{\eta \downarrow 0} R(\lambda \pm i\eta) = R^{\pm}(\lambda) \tag{1.9}$$

for $\lambda > 0$ in a topology which will be specified in the following section. After some improvement by Saitō [7], G. Roach and B. Zhang studied the problem from a new angle ([4], [5], [6]). They, among others, proved that the limiting function $u_{\pm} = R^{\pm}(\lambda) f$ satisfies the (modified) *radiation condition*

$$\left(\frac{\partial}{\partial |x|} \mp ik\right) u_{\pm}(x) = \text{small at } x = \infty \tag{1.10}$$

for appropriate functions f, where $k = k(x) = \sqrt{\mu(x)\lambda}$. We shall discuss these work in § 2 in more details.

In a recent work by W. Jäger and Y. Saitō [8] the operator H was studied under another assumption on the separating surface S which can be applied where S is a cylinder or a plane. Another feature of this work is that the limiting absorption principle has been proved only through a priori estimates of the radiation condition term. The results and sketches of the proofs will be given in § 3. The detailed demonstration will be published elsewhere. § 3 will also contain some concluding remarks.

2. Cone-Shaped Separating Surfaces.

Throughout this section we assume that $N \geq 3$. Eidus [2] considered the operator H under the following

Assumption 2.1. The separating surface S is the union of a finite number of C^1 surfaces. There exist positive constants c_1 and c_2 such that

$$\left| n_N^{(1)}(x) \right| \geq c_1 \tag{2.1}$$

and

$$|x \cdot n^{(1)}(x)| \leq c_2 \tag{2.2}$$

for any $x \in S$ at which S is smooth. Here $x \cdot n^{(1)}(x)$ is the $\mathbb{R}^N$-inner product of x and the unit outward normal $n^{(1)}(x)$ of $\partial\Omega_1$ at $x \in \partial\Omega_1 = S$.

Note that a cone having its vertex at the origin and the positive x_N-axis as its axis satisfies (1.8) and (1.9).

For $t \in \mathbb{R}$ let the weighted L_2 space $L_{2,t}(\mathbb{R}^N)$ be defined by

$$L_{2,t}(\mathbb{R}^N) = \{f \ : \ (1 + |x|)^t f(x) \in L_2(\mathbb{R}^N)\}, \tag{2.3}$$

with its inner product of norm

$$\begin{cases} (u,v)_t = \int_{\mathbb{R}^N} (1+|x|)^{2t} u(x)\overline{v(x)}\,dx, \\ \|u\|_t = \left[(u,u)_t\right]^{1/2}. \end{cases} \tag{2.4}$$

For $t \in \mathbb{R}$ and a nonnegative integer k the weighted Sobolev space $H_t^k(\mathbb{R}^N)$ is defined by

$$H_t^k(\mathbb{R}^N) = \{f \ : \ (1+|x|)^t D^\alpha f(x) \in L_2(\mathbb{R}^N), |\alpha| \le k\} \tag{2.5}$$

where

$$\begin{cases} D^\alpha = \partial_1^{\alpha_1}\partial_2^{\alpha_2}\cdots\partial_N^{\alpha_N}, \\ \partial_j = \partial/\partial x_j \qquad (j = 1,2,\cdots,N), \\ |\alpha| = \alpha_1+\alpha_2+\cdots+\alpha_N. \end{cases} \tag{2.6}$$

The inner product $(\ ,\ )_{t,k}$ and norm $\|\ \|_{t,k}$ are given by

$$\begin{cases} (u,v)_{t,k} = \int_{\mathbb{R}^N} (1+|x|)^{2t} \sum_{|\alpha|\le k} D^\alpha u(x)\overline{D^\alpha v(x)}\,dx, \\ \|u\|_{t,k} = \left[(u,u)_{t,k}\right]^{1/2}. \end{cases} \tag{2.7}$$

Let M be a positive number and let $R(z)$ be the resolvent of the operator H as in § 1. Then, under the above assumption, Eidus [2] showed the a priori estimate

$$\|R(z)f\|_{-\beta,1} \le c\,\|f\|_\beta \qquad (z \in \mathbb{R} - \mathbb{C}, \quad \operatorname{Re} z \le M), \tag{2.8}$$

where $f \in L_{2,\beta}(\mathbb{R}^N)$ and β is a constant such that $\beta = 3$ if $N > 3$, and $\beta > 3$ if $N = 3$. The constant c does not depend on $f \in L_{2,\beta}(\mathbb{R}^N)$ and $z \in \mathbb{C} - \mathbb{R}$ with $\operatorname{Re} z \le M$. Starting with this a priori estimate, Eidus [2], among others, obtained the following limiting absorption principle:

Theorem 2.2 ([2]). Suppose that Assumption 2.1 holds. Then, for each $\lambda > 0$, there exist the limits

$$\lim_{\epsilon\downarrow 0} R(\lambda \pm i\epsilon) = R^\pm(\lambda) \tag{2.9}$$

in $\mathbf{B}(L_{2,1}(\mathbb{R}^N), L_{2,-1}(\mathbb{R}^N))$.

Eidus [2] also discussed the continuity of the operators $R^\pm(\lambda)$ and the limiting amplitude principle for H. However, since the radiation condition for the operator H is not defined, the characterization of the extended resolvent $R^\pm(\lambda)$ is not given in [2].

The original goal of Saitō [7] was to find the radiation condition for H as well as the improvement of the topology in which the convergence (2.9) takes place. Let $u = R(z)f$. Although the radiation condition of the type (1.10) seems to be natural since the function $\mu(x)$ is constant on each Ω_j for $j = 1,2$, we have to make sure that the radiation condition guarantees the uniqueness of the solution of the equation

$$(-\mu(x)^{-1}\Delta - \lambda)u = f. \tag{2.10}$$

Saitō [7] tried to apply the radiation condition method which was successful in the study of the 2-body Schrödinger operator. Let us introduce the following notation:

(1) $k = k(x) = k(x,z) = [z\mu(x)]^{1/2}$, where the branch is taken so that $\operatorname{Im} k(x,z) \geq 0$;

(2) $b = b(x) = b(x,z) = \operatorname{Im} k(x,z)$;

(3) $\mathcal{D}u = \nabla u + \{(N-1)/(2r)\}\widetilde{x}u - ik(x)\widetilde{x}u$, where $\widetilde{x} = x/|x|$;

(4) $\mathcal{D}_r u = \mathcal{D}u \cdot \widetilde{x} = \partial u/\partial r + \{(N-1)/(2r)\}u - ik(x)u$;

Multiplying both sides of (2.10) by $\varphi(x)\overline{\mathcal{D}_r u}$ with a weight function $\varphi(x)$, integrating on $B_{rR} = \{x \in \mathbb{R}^N : r < |x| < R\}$, and taking the real parts of both sides, we have

$$\begin{aligned}
&\int_{B_{rR}} \left(b\varphi + \frac{1}{2}\frac{\partial \varphi}{\partial r}\right)|\mathcal{D}u|^2\,dx + \sum_{\ell=1}^{2}\int_{\partial\Omega_\ell \cap B_{rR}} \varphi \operatorname{Im}\left\{\overline{k}\frac{\partial u}{\partial n}\overline{u}\right\} dS \\
&\qquad + \int_{B_{rR}} \left(\frac{\varphi}{r} - \frac{\partial \varphi}{\partial r}\right)(|\mathcal{D}u|^2 - |\mathcal{D}_r u|^2)\,dx \\
&\qquad + c_N \int_{B_{rR}} r^{-2}\left(\frac{\varphi}{r} - 2^{-1}\frac{\partial \varphi}{\partial r} + b\varphi\right)|u|^2\,dx \\
&= \operatorname{Re}\int_{B_{rR}} \varphi\mu(x) f\overline{\mathcal{D}_r u}\,dx \\
&\qquad + 2^{-1}\sum_{\ell=1}^{2}\int_{\partial\Omega_\ell \cap B_{rR}} \varphi\left\{\frac{(N-1)b}{r} + |k|^2\right\}(\widetilde{x}\cdot n)|u|^2\,dS \\
&\qquad + 2^{-1}\int_{S_R} \varphi(2|\mathcal{D}_r u|^2 - |\mathcal{D}u|^2 - c_N r^{-2}|u|^2)\,dS, \\
&\qquad - 2^{-1}\int_{S_r} \varphi(2|\mathcal{D}_r u|^2 - |\mathcal{D}u|^2 - c_N r^{-2}|u|^2)\,dS
\end{aligned} \tag{2.11}$$

(cf. [3], Lemma 3.1 and [7], Lemma 3.1). Here S_t is the $N-1$-sphere of $\mathbb{R}^N$ with center at the origin and radius t, and the weight function φ has the form

$$\varphi(x) = \xi(|x|)\alpha(x) \geq 0, \tag{2.12}$$

where $\alpha(x)$ is constant on each Ω_j, $j = 1, 2$. In order to evaluate the radiation condition term $\mathcal{D}u$, it is crucial to get a good evaluation for the surface integral term (the second term of the left-hand side of (2.11)). In [7] the surface integral was not handled satisfactorily, so [7] could only show that the limits (2.9) in Theorem 2.2 converge in a weaker topology which has been used in the case of Schrödinger operators.

Theorem 2.3 ([7]). Let $\delta > 1/2$. Suppose that Assumption 2.1 holds. Then, for each $\lambda > 0$, the limit (2.9) exists in $\mathbf{B}(L_{2,\delta}(\mathbb{R}^N), L_{2,-\delta}(\mathbb{R}^N))$.

Then Roach and Zhang [4] showed that the radiation condition does guarantee the uniqueness of the solution of the equation (2.9). In fact they proved the following:

Theorem 2.4 ([4]). Let $u \in H^2(\mathbb{R}^N)_{\mathrm{loc}}$ be a solution of the homogeneous equation

$$-\mu^{-1}\Delta u - \lambda u = 0 \tag{2.13}$$

with $\lambda > 0$ such that

$$\lim_{R\to\infty} R^{-1} \int_{|x|<R} |\nabla u - i\sqrt{\lambda\mu(x)}u|^2\, dx = 0. \tag{2.14}$$

Then u is identically 0.

Their proof is based on an identity for solutions of the equation (2.13)-(2.14). This identity is similar to (2.11) but not the same. Having established the uniqueness through the radiation condition, they showed the limiting absorption principle for H. The pair of Besov spaces $B(\mathbb{R}^N)$ and $B^*(\mathbb{R}^N)$ (see Agmon and Hörmander [1]) are used in [4] instead of $L_{2,\delta}(\mathbb{R}^N)$ and $L_{2,-\delta}(\mathbb{R}^N)$, and their proof of the convergence of (2.9) is also new in the sense that they utilized the convergence of the Green kernel of $-\Delta$ after expressing $R(z)$ by the Green kernel. Thus the extended resolvent $R^{\pm}(\lambda)$ was uniquely determined.

Theorem 2.5 ([4]). Under Assumption 2.1, the limit (2.9) exists for each $\lambda > 0$ in $\mathbf{B}(B(\mathbb{R}^N), B^*(\mathbb{R}^N))$. Let $u = R^+(\lambda)f$ where $f \in B(\mathbb{R}^N)$. Then u is a unique solution of the equation (2.10)-(2.14).

Roach and Zhang [5], [6] discuss the limiting amplitude principle and scattering theory.

3. Cylinder-Shaped Separating Surfaces.

In this section we assume that $N \geq 2$.

Assumption 3.1. We assume that

$$(\mu_2 - \mu_1)(x \cdot n^{(1)}) = (\mu_1 - \mu_2)(x \cdot n^{(2)}) \geq 0 \tag{3.1}$$

for almost all $x \in S$.

The condition (3.1) requires that the inner products $x \cdot n^{(1)}$ and $x \cdot n^{(2)}$ do not change their signs almost always on S. Note that the above assumption is satisfied if Ω_1 is a cylindrical domain, $\mu_1 < \mu_2$, and the origin is put in Ω_1. Also the above assumption can be applied when the separating surface S is a plane.

Jäger and Saitō [8] studied the operator H under the above assumption. The method is almost entirely based on the identity (2.11). As for the uniqueness, starting with (2.11), we have

Theorem 3.2 ([8]). Let $\delta > 1/2$ and suppose that Assumption 3.1 holds. Let $u \in H^2(\mathbb{R}^N)_{\mathrm{loc}}$ be a solution of the homogeneous equation (2.13) with $\lambda > 0$ such that

$$\liminf_{R\to\infty} \int_{S_R} \left(\left| \frac{\partial u}{\partial r} \right|^2 + |u|^2 \right) dS = 0. \tag{3.2}$$

Then u is identically zero. The condition (3.2) can be replaced by either

$$\liminf_{R\to\infty} \int_{S_R} \left| \frac{\partial u}{\partial r} - iku \right|^2 dS = 0 \tag{3.3}$$

or

$$\int_{|x|>1} \frac{1}{r} \left| \frac{\partial u}{\partial r} - iku \right|^2 dx < \infty. \tag{3.4}$$

The identity (2.11) can be used again to have an a priori estimate for the radiation condition term $\mathcal{D}u$ by choosing the weight function $\varphi(x) = \xi(|x|)/\sqrt{\mu(x)}$. Note that the surface integral term (the second term of the left-hand side) of (2.11) becomes zero by this choice of φ. Then we have

Theorem 3.3 ([8]). Suppose that Assumption 3.1 holds. Let $1/2 < \delta \le 1$. Let $u = R(z)f$, where $f \in L_{2,\delta}(\mathbb{R}^N)$ and $z \in \mathbb{C} - \mathbb{R}$ with $\mathrm{Re}\, z > 0$. Then there exists a positive constant $C = C(\delta)$ such that

$$\|\mathcal{D}u\|_{\delta-1} \le C\|f\|_\delta, \tag{3.5}$$

where the constant $C(\delta)$ is independent of f and z.

Starting with these two estimates, we can prove the limiting absorption principle for H with certain estimates for the extended resolvent $R^{\pm}(\lambda)$.

Theorem 3.4 ([8]). Suppose that Assumption 3.1 holds. Let $1/2 < \delta \le 1$. Then there exist the limits (2.9) in $\mathbf{B}(L_{2,\delta}(\mathbb{R}^N), H^2_{-\delta}(\mathbb{R}^N))$ for each $\lambda > 0$. Further we have

$$\|R^{\pm}(\lambda)\|_{(0,\delta)}^{(0,-\delta)} \le \frac{C_1}{\sqrt{\lambda}} \qquad (\lambda > 0), \tag{3.6}$$

where $\|R^{\pm}(\lambda)\|_{(0,\delta)}^{(0,-\delta)}$ is the operator norm in $\mathbf{B}(L_{2,\delta}(\mathbb{R}^N), L_{2,-\delta}(\mathbb{R}^N))$ and $C_1 = C_1(\delta)$ is a positive constant depending only on δ, and

$$\|R(\lambda)\|_{(0,\delta)}^{(2,-\delta)} \le C_2 \qquad (c \le \lambda \le d) \tag{3.7}$$

for $0 < c < d < \infty$, where $\|R^{\pm}(\lambda)\|_{(0,\delta)}^{(2,-\delta)}$ is the operator norm in $\mathbf{B}(L_{2,\delta}(\mathbb{R}^N), H^2_{-\delta}(\mathbb{R}^N))$ and $C_2 = C_2(\delta, c, d)$ is a positive constant depending only on δ, c, d.

In [8] we obtained the whole proof through the evaluations of the radiation condition term $\mathcal{D}u$, and the identity (2.11) turned out to contain almost all necessary information for our proof. In this sense the operator H is one of the operators to which the radiation condition method can be applied effectively. Also we remark that this method can be used with slight modification to give another proof of the limiting absorption principle when the separating surface has a cone-like shape. Thus we are now able to treat both cases in a uniform way.

References

[1] S. Agmon and L. Hörmander, Asymptotic properties of solutions of differential equations with simple characteristics, J. Analyse Math. **30** (1976), 1–37.

[2] D. Eidus, The limiting absorption and amplitude problems for the diffraction problem with two unbounded media, Comm. Math. Phys. **107** (1986), 29–38.

[3] T. Ikebe and Y. Saitō, Limiting absorption method and absolute continuity for the Schrödinger operator, J. Math. Kyoto Univ. **12** (1972), 513–612.

[4] G. Roach and B. Zhang, On Sommerfeld radiation conditions for the diffraction problem with two unbounded media, Proc. Royal Soc. Edinburgh **121A** (1992), 149–161.

[5] G. Roach and B. Zhang, The limiting amplitude principle for the wave propagation problem with two unbounded media, Math. Proc. Cambridge Philos. Soc. **112** (1992), 207–223.

[6] G. Roach and B. Zhang, Spectral representation and scattering theory for the wave equation with two unbounded media, Math. Proc. Cambridge Philos. Soc. **113** (1993), 423–447.

[7] Y. Saitō, A remark on the limiting absorption principle for the reduced wave equation with two unbounded media, Pacific J. Math. **136** (1989), 183–208.

[8] W. Jäger and Y. Saitō, The Limiting absorption principle for the reduced wave operator with cylindrical discontinuity. Preprint. 1994.

Yoshimi Saitō
Department of Mathematics
University of Alabama at Birmingham
Birmingham, Alabama 35294
United States of America

Operator Theory:
Advances and Applications, Vol. 78

Mellin Quantization in the Cone Calculus for Boutet de Monvel's Algebra

E. Schrohe, B.-W. Schulze

Abstract

We present a pseudodifferential calculus for boundary value problems on manifolds with conical singularities. We then show how to associate to each totally characteristic (Fuchs type) pseudodifferential symbol with values in Boutet de Monvel's algebra an operator-valued Mellin symbol is such a way that the difference between the two corresponding operators is smoothing in the interior. This allows us to extend the action of the operators to weighted Mellin-Sobolev spaces.

AMS Subject Classification: 35 S 15, 58 G 20, 46 E 35, 46 H 35.

Key Words: Manifolds with conical singularities, boundary value problems, Boutet de Monvel's calculus, Mellin calculus

Introduction

Following the work of Kondrat'ev, Plamenevskij, and Schulze it is now a widely accepted idea that the analysis on manifolds with conical singularities should be based on (i) a pseudodifferential calculus using totally characteristic (or Fuchs type) symbols near the singularities and (ii) weighted Sobolev spaces.

A basic motivation is the interest in an index theory within a pseudodifferential algebra that contains what one considers the typical differential operators: Near a conical singularity we identify the manifold with the cylinder $X \times \mathbb{R}_+$ where X is a smooth compact manifold with boundary. Then the typical differential operators are those of the form $A = t^{-\mu} \sum_{j=0}^{\mu} A_j(t)(t\partial_t)^j$ with families $A_j(\cdot)$ of differential operators of order $\mu - j$ on X which are smooth up to $t = 0$. One reason for this point of view is that any Laplace-Beltrami operator associated with the Riemannian metric $t^2 h_X(t) + dt^2$ of a warped cone provides an example for an operator of this kind whenever $h_X(\cdot)$ is a family of Riemannian metrics on X which is smooth up to $t = 0$. Another justification stems from the observation that one can introduce an 'artificial' conical point on a smooth manifold by introducing polar coordinates and that under this operation any differential operator assumes this form; for details see [14]. In particular, suppose we are given a differential operator and a conical domain D in Euclidean space; then the operator will have the above type in the natural polar coordinates for D.

Operators of this kind are called totally characteristic or Fuchs type operators. Correspondingly, a totally characteristic pseudodifferential operator of order μ on the cylinder $X \times \mathbb{R}_+$ is one whose symbol, up to the weight factor

$t^{-\mu}$, is of the form $a(x,t,\xi,\tau) = b(x,t,\xi,t\tau)$, where b is a usual pseudodifferential symbol which is smooth up to $t = 0$. The weighted Mellin-Sobolev space $\mathcal{H}^{s,\gamma}$ are best described for $s \in \mathbb{N}$, when they consist of all functions u such that $t^{n/2-\gamma}(t\partial_t)^k D_x^\alpha u \in L^2(X \times \mathbb{R}_+)$ for all $k+|\alpha| \le s$; for general s one can use duality and interpolation. Here, x and t are the variables on X and $\mathbb{R}_+$, respectively, ξ and τ are the corresponding covariables; n is the dimension of X.

Outside a neighborhood of the singularities, Boutet de Monvel's calculus in its standard form is the natural choice.

The principal analytical problem then is to define a quantization that associates with a totally characteristic pseudodifferential symbol a continuous operator on the weighted Sobolev spaces. This plays a crucial role for example in the construction of parametrices to totally characteristic differential operators. The Leibniz inversion of the symbol yields a totally characteristic pseudodifferential symbol. A priori it is by no means clear how to define from this symbol a continuous action on the weighted Sobolev spaces.

This is the question we address in this note. Our answer is what we call 'Mellin quantization'. We show that, for every weight γ and each totally characteristic pseudodifferential symbol a with values in Boutet de Monvel's algebra on X which is smooth up to $t = 0$, we find a Mellin symbol f such that the difference $\operatorname{op} a - \operatorname{op}_M^\gamma f$ is a regularizing operator in Boutet de Monvel's calculus on the nonsingular part of the manifold. Here, $\operatorname{op}_M^\gamma f$ is the weighted Mellin pseudodifferential operator associated with the operator-valued Mellin symbol f. It naturally acts on the spaces $\mathcal{H}^{s,\gamma}$ and therefore solves the problem.

There are two more satisfying aspects to this solution. For one thing, the Mellin calculus allows an intrinsic description of the totally characteristic pseudodifferential operators in the sense that it respects the natural $\mathbb{R}_+$ structure of the space near the singularities. Secondly, it turns out that there is a notion of ellipticity that allows us to characterize the Fredholm property between the spaces $\mathcal{H}^{s,\gamma}$ while the degeneracy of the symbols at $t = 0$ in general prevents these operators from being Fredholm on the usual Sobolev spaces.

The case of manifolds *without* boundary is automatically included in the analysis, since the operators on the boundary also belong to the calculus. The methods to treat the latter case have been developed by Schulze [16], [18].

It should be mentioned that the analysis of [14], [15] can be viewed as part of a more general concept pointed out by Schulze [16]: Given a parameter-dependent pseudodifferential calculus on a space X it should be possible to construct a pseudodifferential calculus also for the 'cone' $X \times \mathbb{R}_+$. In the present case the space X is a manifold with boundary, and the calculus is a parameter-dependent version of Boutet de Monvel's calculus on X. Since Boutet de Monvel's calculus in its standard form is already rather complex, the parameter-dependent version has been established in [14] in a new efficient way; the central idea is to use the concept of parameter-dependent pseudodifferential operators based on group actions and wedge Sobolev spaces.

Mellin quantization is also relevant for pseudodifferential boundary value problems without the transmission property, such as those considered by Vishik&Eskin [19] and Eskin [5]. The interior normal direction then plays the role of the cylinder axis $\mathbb{R}_+$, and the asymptotics of solutions can be described in terms of Mellin-Sobolev spaces; the asymptotics for problems with the transmission property then correspond to Taylor asymptotics near $t = 0$, while much more general asymptotics are possible, cf. [18].

1 Parameter-Dependent and Fuchs Type Operators in Boutet de Monvel's Calculus

1.1 Manifolds with Conical Singularities

An n-dimensional manifold with boundary is a topological (second countable) Hausdorff space M such that each point $m \in M$ has a neighborhood which is diffeomorphic to either $\mathbb{R}^n$ or the closed half-space $\overline{\mathbb{R}}^n_+$. The former points are called the interior points of M, the latter the boundary points. We will use the standard notation $\operatorname{int} M$ and ∂M.

1.1 Definition. A *manifold with boundary and conical singularities D of dimension $n+1$* is a topological (second countable) Hausdorff space with a finite subset $\Sigma \subset D$ ('singularities') such that $D \backslash \Sigma$ is an $n+1$-dimensional manifold with boundary, and for every $v \in \Sigma$ there is an open neighborhood U of v, a compact manifold with boundary X of dimension n, and a system $\mathcal{F} \neq \emptyset$ of mappings with the following properties

(1) For all $\phi \in \mathcal{F}$, $\phi : U \to X \times [0,1)/X \times \{0\}$ is a homeomorphism with $\phi(v) = X \times \{0\}/X \times \{0\}$.

(2) Given $\phi_1, \phi_2 \in \mathcal{F}$, the restriction $\phi_1 \phi_2^{-1} : X \times (0,1) \to X \times (0,1)$ extends to a *diffeomorphism* $X \times (-1,1) \to X \times (-1,1)$.

(3) The charts $\phi \in \mathcal{F}$ are compatible with the charts for the manifold for $D \backslash \Sigma$: The restriction $\phi : U \backslash \{v\} \to X \times (0,1)$ is a diffeomorphism.

We can and will assume that for each singularity $v \in \Sigma$, the system $\mathcal{F}$ is maximal with respect to the properties (1), (2), and (3).

1.2 Definition and Remark. By assumption, $D \backslash \Sigma$ is a manifold with boundary. Properties 1.1(1) and (2) imply that any neighborhood of a point $v \in \Sigma$ contains points of the topological boundary of $D \backslash \Sigma$, namely of $\partial X \times (0,1)$.

A point $x \in D$ is an *interior point of D* if there is an open neighborhood of x which is homeomorphic to an open ball in $\mathbb{R}^{n+1}$, and $\operatorname{int} D$ is the collection of all interior points; $\partial D = D \backslash \operatorname{int} D$ is the boundary of D. We always have $\Sigma \subset \partial D$.

1.3 Remark. Let D be a manifold with boundary and conical singularities. Then the topological boundary ∂D of D is a (boundaryless) manifold with conical singularities in the sense of [18, Definition 1.1.15].

1.4 Notation and Assumptions. In a neighborhood of one of the singularities, X will denote the cross-section as in 1.1; it is a manifold with boundary of dimension n, in particular, X contains its boundary. For practical purposes, this is often inconvenient. We shall therefore agree to denote by X the open interior, and by $\overline{X}$ the manifold including the boundary. We let $X^\wedge = X \times \mathbb{R}_+; \overline{X}^\wedge = \overline{X} \times \mathbb{R}_+$, $Y = \partial X$ is the topological boundary of $X; Y$ is a closed manifold of dimension $n-1$. We let $Y^\wedge = Y \times \mathbb{R}_+$.

It is on the cylinder $X^\wedge$ that the analysis in this paper is performed. We assume that X is endowed with a Riemannian metric and embedded in a closed Riemannian manifold Ω and that $X^\wedge$ carries the canonical (cylindrical) metric.

1.2 Parameter-Dependent Symbols and Sobolev Spaces

In a collar neighborhood of the boundary Y of X we introduce normal coordinates. A point there can be written $x = (y,r)$ with $y \in Y, r \geq 0$. If U is an open subset of $\mathbb{R}^{n-1}$, then coordinates in $U \times \mathbb{R}$ will also be written in the form $x = (x',r)$ or likewise $x = (x',x_n)$, with $x' \in U$ and $r, x_n \in \mathbb{R}$.

1.5 Sobolev Spaces on $\mathbb{R}^n$ and $\mathbb{R}^n_+$. Let U be an open subset of $\mathbb{R}^{n-1}$. For a function or distribution u on $U \times \mathbb{R}$ let r^+u denote its restriction to $U \times \mathbb{R}_+$. We shall also use the operator r^+ to indicate the restriction of functions or distributions on Ω to X.

$H^s(\mathbb{R}^n), s \in \mathbb{R}$, is the usual Sobolev space over $\mathbb{R}^n$. We let $H^s(\mathbb{R}^n_+) = \mathrm{r}^+H^s(\mathbb{R}^n)$ and $H^s_0(\mathbb{R}^n_+) = \{u \in H^s(\mathbb{R}^n) : \operatorname{supp} u \subseteq \overline{\mathbb{R}}^n_+\}$. Equivalently, $H^s_0(\mathbb{R}^n_+)$ is the closure of $C^\infty_0(\mathbb{R}^n_+)$ in the topology of $H^s(\mathbb{R}^n)$.

The notation extends to the case of compact manifolds via a partition of unity. This yields the spaces $H^s(\Omega), H^s(X)$, and $H^s_0(X)$. We shall also employ the notation $H^s(X \times \mathbb{R}_+), H^s_0(X \times \mathbb{R}_+)$, etc, understanding that we use the canonical choice of $L^2(X \times \mathbb{R}_+)$.

For functions on $U \times \mathbb{R}_+$ or distributions in $H^s(\mathbb{R}^n_+), s > -\frac{1}{2}$, we let e^+ denote the operator of extension (by zero) to $U \times \mathbb{R}$ and $H^s(\mathbb{R}^n)$. Again this carries over to the manifold case and yields a bounded map $\mathrm{e}^+ : H^s(X) \to H^s(\Omega), -\frac{1}{2} < s < \frac{1}{2}$.

$\mathcal{S}(\mathbb{R}^n)$ denotes the space of all rapidly decreasing functions on $\mathbb{R}^n$, and $\mathcal{S}(\mathbb{R}^n_+)$ is the space of all restrictions of functions in $\mathcal{S}(\mathbb{R}^n)$ to $\mathbb{R}^n_+$. $\mathcal{S}'(\mathbb{R}^n_+)$ is the dual space of $\mathcal{S}(\mathbb{R}^n_+)$. Note that $\mathcal{S}'(\mathbb{R}^n_+)$ contains distributions with support in $\partial\mathbb{R}^n_+$. If we define the weighted Sobolev space $H^{s,\sigma}(\mathbb{R}^n_+)$ and $H^{s,\sigma}_0(\mathbb{R}^n_+)$ as the sets of all $\langle x\rangle^{-\sigma} u$, where u is an element of $H^s(\mathbb{R}^n_+)$ and $H^s_0(\mathbb{R}^n_+)$, respectively, then $\mathcal{S}(\mathbb{R}^n_+) = \mathrm{proj} - \lim_{s,\sigma\to\infty} H^{s,\sigma}(\mathbb{R}^n_+)$ and $\mathcal{S}'(\mathbb{R}^n_+) = \mathrm{ind} - \lim_{s,\sigma\to-\infty} H^{s,\sigma}_0(\mathbb{R}^n_+)$.

It was an important point in [14] to develop a new approach to Boutet de Monvel's calculus based on group actions and operator valued symbols. Since this connection is going to play a role also in this paper we repeat the essential notions.

1.6 Group Actions and Operator-Valued Symbols. Let E,F be Banach spaces with strongly continuous group actions $\{\kappa_\lambda : \lambda \in \mathbb{R}_+\}$ and $\{\tilde{\kappa}_\lambda : \lambda \in \mathbb{R}_+\}$. By definition this means that

(i) $\lambda \mapsto \kappa_\lambda \in C(\mathbb{R}_+,\mathcal{L}_\sigma(E))$, $\lambda \mapsto \tilde{\kappa}_\lambda \in C(\mathbb{R}_+,\mathcal{L}_\sigma(F))$ (strong continuity of κ and $\tilde{\kappa}$), and

(ii) $\kappa_\lambda\kappa_\mu = \kappa_{\lambda\mu}, \tilde{\kappa}_\lambda\tilde{\kappa}_\mu = \tilde{\kappa}_{\lambda\mu}$.

Here $\mathcal{L}_\sigma(\cdot)$ refers to the space $\mathcal{L}(\cdot)$ endowed with the strong topology.

Let $U \subseteq \mathbb{R}^k$ and $p \in C^\infty(U \times \mathbb{R}^n,\mathcal{L}(E,F)), \mu \in \mathbb{R}$. We will write $p \in S^\mu(U,\mathbb{R}^n;E,F)$ provided that for every $K \subset\subset U$ and all multi-indices α,β, there is a constant $C = C(K,\alpha,\beta)$ with

$$\|\tilde{\kappa}_{\langle\eta\rangle^{-1}}\{D_\eta^\alpha D_y^\beta p(y,\eta)\}\kappa_{\langle\eta\rangle}\|_{\mathcal{L}(E,F)} \le C\,\langle\eta\rangle^{\mu-|\alpha|}, \quad y \in K, \tag{1}$$

cf. [16, 3.2.1, Definition 1]. The space $S^\mu(U,\mathbb{R}^n;E,F)$ is a Fréchet space topologized by the choice of the best constants C.

For the usual or weighted Sobolev spaces on $\mathbb{R}_+$, we will always employ the group action

$$[\kappa_\lambda f](r) = \lambda^{\frac{1}{2}} f(\lambda r). \tag{2}$$

On $E = \mathbb{C}$ we use the trivial group action $\kappa_\lambda \equiv id$. For $E = F = \mathbb{C}$ we shall write $S^\mu(U,\mathbb{R}^n)$ instead of $S^\mu(U,\mathbb{R}^n;\mathbb{C},\mathbb{C})$. The above definition then coincides with the standard symbol class notation.

If $F_1 \hookleftarrow F_2 \hookleftarrow \ldots$ is a sequence of Banach spaces with the same group action, and F is the Fréchet space given as the projective limit of the F_k, then let

$$S^\mu(U,\mathbb{R}^n;E,F) = \text{proj} - \lim_k S^\mu(U,\mathbb{R}^n;E,F_k). \tag{3}$$

Vice versa, if E is the inductive limit of the Banach spaces $E_1 \hookrightarrow E_2 \hookrightarrow \ldots$ with the same group action, then

$$S^\mu(U,\mathbb{R}^n;E,F) = \text{ind} - \lim_k S^\mu(U,\mathbb{R}^n;E_k,F), \tag{4}$$

Finally, a symbol p belongs to $S^\mu(U,\mathbb{R}^n;E,F), E = \text{ind} - \lim E_k, F = \text{proj} - \lim F_l$, if the group actions coincide on the E_k and F_l, respectively, and $p \in S^\mu(U,\mathbb{R}^n;E_k,F_l)$ for all k and l. We give it the topology induced by all the topologies of the spaces $S^\mu(U,\mathbb{R}^n;E_k,F_l)$.

We will, in particular, deal with the spaces $S^\mu(U,\mathbb{R}^n;\mathcal{S}'(\mathbb{R}_+),\mathcal{S}(\mathbb{R}_+))$. For the inductive and projective limit constructions we shall then use the representation of $\mathcal{S}'(\mathbb{R}_+)$ and $\mathcal{S}(\mathbb{R}_+)$, respectively, as limits of weighted Sobolev spaces over $\mathbb{R}_+$.

In view of the nuclearity of $C^\infty(U)$ we have

$$S^\mu(U,\mathbb{R}^n;E,F) = C^\infty(U)\hat{\otimes}_\pi S^\mu(\mathbb{R}^0,\mathbb{R}^n;E,F), \tag{5}$$

the functions in the last space on the right hand side being independent of y.

1.7 Definition. Let $U = U_1 \times U_2 \subseteq \mathbb{R}^n \times \mathbb{R}^n$ be open and $p \in S^\mu(U,\mathbb{R}^n;E,F)$ an operator-valued symbol. Then the pseudodifferential operator $\operatorname{op} p$ is defined by

$$[\operatorname{op} p(f)](y) = (2\pi)^{-n} \int\int_{U_2} e^{i(y-y')\eta} p(y,y',\eta) f(y')dy'd\eta \tag{1}$$

for $f \in C_0^\infty(U_2,E), y \in U_1$. This reduces to

$$[\operatorname{op} p(f)](y) = (2\pi)^{-\frac{n}{2}} \int e^{iy\eta} p(y,\eta)\hat{f}(\eta)d\eta, \tag{2}$$

for 'simple' symbols, i.e. those that are independent of y'. Here, $\hat{f}(\eta) = (2\pi)^{-\frac{n}{2}} \cdot \int e^{-iy\eta} f(y)dy$ is the vector-valued Fourier transform of f.

We may also consider the case where a part of the covariables serves as parameters: For $U \subseteq \mathbb{R}^n$ open, $p \in S^\mu(U_y, \mathbb{R}^n_\eta \times \mathbb{R}^l_\lambda; E,F)$ then defines a parameter-dependent operator $\operatorname{op} p(\lambda)$ by

$$[\operatorname{op} p(\lambda)f](y) = (2\pi)^{-n/2} \int e^{iy\eta} p(y,\eta,\lambda)\hat{f}(\eta)d\eta, \tag{3}$$

$f \in C_0^\infty(U,E)$, similarly for 'double' symbols $p(y,y',\eta,\lambda)$.

1.8 The Manifold Case. Let Ω be a smooth manifold and E,F Banach spaces with strongly continuous group actions. Moreover, let $P : C_0^\infty(\Omega,E) \to C^\infty(\Omega,F)$ be a continuous operator. We shall say that $P \in \operatorname{op}S^\mu(\Omega,\mathbb{R}^n;E,F)$ if the following holds:

(i) For all C_0^∞ functions ϕ,ψ, supported in the same coordinate neighborhood, the operator $(\phi P\psi)_* : C_0^\infty(U,E) \to C^\infty(U,F)$ induced on $U \subseteq \mathbb{R}^n$ by $\phi P\psi$ and the coordinate maps has the form $(\phi P\psi)_* = \operatorname{op} p$ for some $p \in S^\mu(U,\mathbb{R}^n;E,F)$.

(ii) For all C_0^∞ functions ϕ,ψ, with disjoint supports, the operator $\phi P\psi$ is given as an integral operator with a kernel in $C^\infty(\Omega \times \Omega,\mathcal{L}(E,F))$ (more precisely a kernel section, see [3, Section 23.4]).

If P depends on a parameter $\lambda \in \mathbb{R}^l$, then (i) carries over, while in (ii) we ask that the integral kernel belongs to $\mathcal{S}(\mathbb{R}^l,C^\infty(\Omega \times \Omega,\mathcal{L}(E,F)))$.

Suppose we are given a locally finite covering of the manifold by relatively compact coordinate neighborhoods $\{\Omega_j\}$ with associated coordinate maps $\chi_j : \Omega_j \to U_j$. Then we may find $p_j \in S^\mu(U,\mathbb{R}^n;E,F)$ and an integral operator K_j with C^∞ kernel such that $P(f \circ \chi_j)(\chi_j^{-1}(x)) = \operatorname{op} p_j(f)(x) + K_j f(x)$ for all $f \in C_0^\infty(U_j,E)$. We shall call the tuple $\{p_j\}$ the symbol of P.

Let now $\Omega_j \cap \Omega_k \neq \emptyset$, and suppose that both ϕ and ψ are supported in the intersection. Denote by P_j and P_k the operators on $C_0^\infty(\Omega,E)$ induced by $(\phi \circ \chi_j^{-1}) \operatorname{op} p_j\, (\psi \circ \chi_j^{-1})$ and $(\phi \circ \chi_k^{-1}) \operatorname{op} p_k\, (\psi \circ \chi_k^{-1})$. Then $P_j - P_k$ is an integral operator with a kernel in $C^\infty(\Omega \times \Omega, \mathcal{L}(E,F))$. Vice versa, given a tuple $\{p_j\}$ with this property, we can define an operator $P : C_0^\infty(\Omega,E) \to C_0^\infty(\Omega,F)$ whose symbol is $\{p_j\}$. Hence the notion $S^\mu(\Omega,\mathbb{R}^n;E,F)$ makes sense.

Given a parameter-dependent operator $P(\lambda) : C_0^\infty(\Omega,E) \to C^\infty(\Omega,F)$, we define the operator $P_+(\lambda) : C_0^\infty(\overline{X},E) \to C^\infty(X,F)$ by

$$P_+(\lambda) = \mathrm{r}^+ P(\lambda) \mathrm{e}^+. \tag{1}$$

Just like before, the distributions have to be sufficiently smooth in order to allow an extension by zero.

1.3 Boutet de Monvel's Calculus

1.9 Definition. Let $\mu \in \mathbb{R}, d \in \mathbb{N}$ and $U \subseteq \mathbb{R}^{n-1}$ open. In the following definition the parameter-dependence will always refer to the parameter $\lambda \in \mathbb{R}^l$.

(a) A regularizing parameter-dependent singular Green operator (s.G.o.) on $U \times \mathbb{R}_+$ of type 0 is a family of integral operators

$$G_0(\lambda) : C_0^\infty(U,\mathcal{S}(\mathbb{R}_+)) \to C^\infty(U,C^\infty(\overline{\mathbb{R}}_+))$$

given by a kernel in $\mathcal{S}(\mathbb{R}^l, C^\infty(U \times \overline{\mathbb{R}}_+ \times U \times \overline{\mathbb{R}}_+))$. Here we identify $C_0^\infty(U,\mathcal{S}(\mathbb{R}_+))$ and $C^\infty(U,C^\infty(\overline{\mathbb{R}}_+))$ with subsets of $C^\infty(U \times \mathbb{R}_+)$. A regularizing s.G.o. G_0 of type d is a parameter-dependent operator of the form $G_0(\lambda) = \sum_{j=0}^d G_{0j}(\lambda)\partial_r^j$ with regularizing parameter-dependent s.G.o's G_{0j} of type zero and the derivative ∂_r on $\mathbb{R}_+$.

A parameter-dependent s.G.o. of order μ and type d on U is an operator

$$G(\lambda) : C_0^\infty(U,\mathcal{S}(\mathbb{R}_+)) \to C^\infty(U,\mathcal{S}(\mathbb{R}_+))$$

that can be written $G = \sum_{j=0}^d \operatorname{op} g_j \partial_r^j + G_0$, where each g_j is a (parameter-dependent and operator-valued) symbol g_j in $S^{\mu-j}(U,\mathbb{R}^{n-1} \times \mathbb{R}^l; \mathcal{S}'(\mathbb{R}_+),\mathcal{S}(\mathbb{R}_+))$ and G_0 is a regularizing parameter-dependent s.G.o..

(b) A regularizing parameter-dependent trace operator of type 0 is an operator

$$T_0(\lambda) : C_0^\infty(U,\mathcal{S}(\mathbb{R}_+)) \to C^\infty(U)$$

with an integral kernel in $\mathcal{S}(\mathbb{R}^l, C^\infty(U \times U \times \overline{\mathbb{R}}_+))$. A regularizing trace operator T_0 of type d is a sum $T_0(\lambda) = \sum_{j=0}^d T_{0j}\partial_r^j$; each T_{0j} being regularizing of type 0.

A parameter-dependent trace operator T of order μ and type d on U is an operator that can be written $T = \sum_{j=0}^d \operatorname{op} t_j \partial_r^j + T_0$, with t_j in $S^{\mu-j}(U,\mathbb{R}^{n-1} \times \mathbb{R}^l; \mathcal{S}'(\mathbb{R}_+),\mathbb{C})$ and a regularizing trace operator T_0 of type d.

(c) A regularizing parameter-dependent potential operator on U is an operator

$$K_0(\lambda) : C_0^\infty(U) \to C^\infty(U,C^\infty(\overline{\mathbb{R}}_+))$$

given by an integral kernel in $\mathcal{S}(\mathbb{R}^l, C^\infty(U \times \overline{\mathbb{R}}_+ \times U))$; a parameter-dependent potential operator K of order μ is a sum $K = \operatorname{op} k + K_0$ with a pseudodifferential symbol k in $S^\mu(U,\mathbb{R}^{n-1} \times \mathbb{R}^l; \mathbb{C},\mathcal{S}(\mathbb{R}_+))$ and a regularizing parameter-dependent potential operator K_0.

(d) All these spaces of operators carry Fréchet topologies in a natural way: We use the topology of non-direct sums of Fréchet spaces in connection with the natural topologies on the symbol spaces and on the spaces $\mathcal{S}(\mathbb{R}^l, \ldots)$ for the integral kernels.

1.10 Remark. (Non-direct sums of Fréchet spaces) Let E,F be Fréchet spaces and suppose both are continuously embedded in the same Hausdorff vector space.

The exterior direct sum $E \oplus F$ is Fréchet and has the closed subspace $\Delta = \{(a, -a) : a \in E \cap F\}$. The non-direct sum of E and F then is the Fréchet space $E + F := E \oplus F/\Delta$.

1.11 Parameter-Dependent Operators in Boutet de Monvel's Calculus. Let $U \subseteq \mathbb{R}^{n-1}$ be open. A *parameter-dependent operator of order* $\mu \in \mathbb{R}$ *and type* $d \in \mathbb{N}$ *in Boutet de Monvel's calculus* on U is a family $\{A(\lambda) : \lambda \in \mathbb{R}^l\}$ of operators

$$A(\lambda) = \begin{bmatrix} P_+(\lambda) + G(\lambda) & K(\lambda) \\ T(\lambda) & S(\lambda) \end{bmatrix} : \begin{matrix} C_0^\infty(U \times \overline{\mathbb{R}}_+) \\ \oplus \\ C_0^\infty(U) \end{matrix} \quad \longrightarrow \quad \begin{matrix} C^\infty(U \times \overline{\mathbb{R}}_+) \\ \oplus \\ C^\infty(U) \end{matrix} , \qquad (1)$$

where

$P(\cdot)$ $= \operatorname{op} p(\cdot)$ with $p \in S^\mu_{tr}(U \times \overline{\mathbb{R}}_+ \times U \times \overline{\mathbb{R}}_+,\mathbb{R}^n; \mathbb{R}^l)$, $P_+ = \mathrm{r}^+ P \mathrm{e}^+$,
$G(\cdot)$ is a parameter-dependent singular Green operator of order μ, type d,
$K(\cdot)$ is a parameter-dependent potential operator of order μ,
$T(\cdot)$ is a parameter-dependent trace operator of order μ and type d,
$S(\cdot)$ is a parameter-dependent pseudodifferential operator of order μ on U.

The subscript 'tr' indicates that the symbol p satisfies the transmission condition (see [13, Section 2.2.2.1]) at the boundary $U \times \{0\}$. Note that the decomposition $P_+ + G$ is not unique; the regularizing pseudodifferential operators provide examples for operators that belong to both classes. We shall write $A \in \mathcal{B}^{\mu,d}(U \times \mathbb{R}_+; \mathbb{R}^l)$. The topology on this space is that of a non-direct sum of Fréchet spaces induced by (1) and the topologies on the spaces of pseudodifferential, singular Green, trace, and potential operators.

A parameter-dependent *regularizing* operator A of type d in Boutet de Monvel's calculus on U is one that can be written in the form (1) with all entries being regularizing operators. We shall write $A \in \mathcal{B}^{-\infty,d}(U \times \mathbb{R}_+; \mathbb{R}^l)$, and give this space the obvious Fréchet topology.

It is a consequence of 1.9 that the operators in (1) indeed have the desired mapping properties.

Given an operator $A \in \mathcal{B}^{\mu,d}(U \times \mathbb{R}_+; \mathbb{R}^l)$ we have a symbol a for A, namely the quintuple $a = \{p,g,k,t,s\}$ of the symbols for the operators P,G,K,T, and S, respectively. As pointed out before, there is a certain ambiguity in the choice of the symbols; we understand them as equivalence classes of tuples inducing the same operator modulo $\mathcal{B}^{-\infty,d}(U \times \mathbb{R}_+; \mathbb{R}^l)$.

1.12 Boutet de Monvel's Algebra on a Manifold. Let X be an n-dimensional C^∞ manifold with boundary Y, embedded in an n-dimensional manifold Ω without boundary, all not necessarily compact. Let V_1, V_2 be vector bundles over Ω and W_1, W_2 be vector bundles over Y.

Let $\{\Omega_j\}$ denote a locally finite open covering of Ω and suppose that the coordinate charts map $X \cap \Omega_j$ to $U_j \times \mathbb{R}_+ \subset \mathbb{R}^n_+$ and $Y \cap \Omega_j$ to $U_j \times \{0\}$ for a suitable open set $U_j \subseteq \mathbb{R}^{n-1}$, unless $\Omega_j \cap Y = \emptyset$.

For a smooth function ϕ on Ω write M_ϕ for the multiplication operator with the diagonal matrix $\operatorname{diag}\{\phi, \phi|_Y\}$. We will say that $A \in \mathcal{B}^{\mu,d}(X; \mathbb{R}^l)$, if

$$A(\lambda): \begin{array}{c} C_0^\infty(\overline{X}, V_1) \\ \oplus \\ C_0^\infty(Y, W_1) \end{array} \to \begin{array}{c} C^\infty(\overline{X}, V_2) \\ \oplus \\ C^\infty(Y, W_2) \end{array}, \tag{1}$$

is an operator with the following properties:

(i) For all C_0^∞ functions ϕ, ψ, supported in one and the same coordinate neighborhood Ω_j intersecting the boundary, the operator

$$(M_\phi A(\lambda) M_\psi)_*: \begin{array}{c} C_0^\infty(U_j \times \overline{\mathbb{R}}_+, V_1) \\ \oplus \\ C_0^\infty(U_j, W_1) \end{array} \to \begin{array}{c} C^\infty(U_j \times \overline{\mathbb{R}}_+, V_2) \\ \oplus \\ C^\infty(U_j, W_2) \end{array},$$

induced on $U_j \times \mathbb{R}_+$ by $M_\phi A(\lambda) M_\psi$ and the coordinate maps, is an operator in the class $\mathcal{B}^{\mu,d}(U_j \times \mathbb{R}_+; \mathbb{R}^l)$ of Boutet de Monvel's calculus on $\mathbb{R}^n_+$ in the sense of 1.11.

(ii) If ϕ, ψ are as before, but the coordinate chart does not intersect the boundary, then all entries in the matrix $(M_\phi A(\lambda) M_\psi)_*$ – except for the pseudodifferential part – are regularizing.

(ii) If the supports of the functions $\phi, \psi \in C_0^\infty(\Omega)$ are disjoint, then $(M_\phi A(\lambda) M_\psi)_*$ is an integral operator whose kernel density is C^∞ and a rapidly decreasing function of λ in all semi-norms defining the Fréchet topology of the smooth densities.

In each coordinate patch Ω_j intersecting the boundary we may associate a symbol tuple with A by asking that the operator A_j which is locally induced by A and the coordinate maps has a symbol tuple $a_j = \{p_j, g_j, k_j, t_j, s_j\}$ as in 1.8 and 1.11. In an interior chart, only the pseudodifferential part in the matrix for A is non-regularizing; it has a symbol p_j in the sense of an equivalence class of symbols. We shall call the tuple $\{a_j\}$ a symbol for A.

1.4 Sobolev Spaces Based on the Mellin Transform

1.13 Parameter-Dependent Order Reductions on Ω. For $\mu \in \mathbb{R}$ there is a parameter-elliptic pseudodifferential operator $\Lambda^\mu \in \operatorname{op} S^\mu(\Omega,\mathbb{R}^n;\mathbb{R})$, depending on the parameter $\tau \in \mathbb{R}$ such that

$$\Lambda^\mu(\tau) : H^s(\Omega,V) \to H^{s-\mu}(\Omega,V)$$

is an isomorphism for all τ. *Parameter-ellipticity* simply means that there is a symbol $q \in S^{-\mu}(\Omega,\mathbb{R}^n;\mathbb{R})$ such that $\lambda^\mu q - 1$ and $q\lambda^\mu - 1$ both are elements of $S^{-1}(\Omega,\mathbb{R}^n;\mathbb{R})$.

In order to construct such an operator one can e.g. start with symbols of the form $\langle \xi,(\tau,C)\rangle^\mu \in S^\mu(\mathbb{R}^n,\mathbb{R}^n_\xi;\mathbb{R}_\tau)$ with a large constant $C > 0$ and patch them together to an operator on the manifold Ω with the help of a partition of unity and cut-off functions.

Alternatively, one can choose a Hermitean connection on V and consider the operator $(C + |\tau|^2 - \Delta)^{\frac{\mu}{2}}$, where Δ denotes the connection Laplacian and C is a large positive constant.

1.14 Weighted Mellin-Sobolev Spaces. (a) Let $\{\Lambda^\mu : \mu \in \mathbb{R}\}$ be a family of parameter-elliptic pseudodifferential operators as in 1.13. For $s,\gamma \in \mathbb{R}$, the space $\mathcal{H}^{s,\gamma}(\Omega^\wedge)$ is the closure of $C_0^\infty(\Omega^\wedge)$ in the norm

$$\|u\|_{\mathcal{H}^{s,\gamma}(\Omega^\wedge)} = \left\{ \int_{\Gamma_{\frac{n+1}{2}-\gamma}} \|\Lambda^s(\operatorname{Im} z) Mu(z)\|^2_{L^2(\Omega)}\, |dz| \right\}^{\frac{1}{2}}. \tag{1}$$

Here n is the dimension of X and Ω , $\Gamma_\beta = \{z \in \mathbb{C} : \operatorname{Re} z = \beta\}$, and M is the Mellin transform, cf. 2.1.

(b) We let $\mathcal{H}^{s,\gamma}(X^\wedge) = \{\mathrm{r}^+ f : f \in \mathcal{H}^{s,\gamma}(\Omega^\wedge)\}$. The space $\mathcal{H}^{s,\gamma}(X^\wedge)$ carries the quotient norm:

$$\|u\|_{\mathcal{H}^{s,\gamma}(X^\wedge)} = \inf\{\|f\|_{\mathcal{H}^{s,\gamma}(\Omega^\wedge)} : f \in \mathcal{H}^{s,\gamma}(\Omega^\wedge), \mathrm{r}^+ f = u\}.$$

(c) For $s = l \in \mathbb{N}$ we obtain the alternative description

$$u \in \mathcal{H}^{l,\gamma}(\Omega^\wedge) \quad \text{iff} \quad t^{\frac{n}{2}-\gamma}(t\partial_t)^k Du(x,t) \in L^2(\Omega^\wedge)$$

for all $k \le l$ and all differential operators D of order $\le l - k$ on Ω, cf. [16, Section 2.1.1, Proposition 2].

(d) The space $\mathcal{H}^{s,\gamma}(X^\wedge)$ is independent of the particular choice of the order-reducing family.

(e) $\mathcal{H}^{s,\gamma}(X^\wedge) \subseteq H^s_{loc}(X^\wedge)$; $\mathcal{H}^{s,\gamma}(X^\wedge) = t^\gamma \mathcal{H}^{s,0}(X^\wedge)$;
$\mathcal{H}^{0,0}(X^\wedge) = t^{-\frac{n}{2}} L^2(X^\wedge)$.

(f) If ϕ is the restriction to $X^\wedge$ of a function in $C_0^\infty(\Omega \times \mathbb{R})$, then the operator M_ϕ of multiplication by ϕ

$$M_\phi : \mathcal{H}^{s,\gamma}(X^\wedge) \to \mathcal{H}^{s,\gamma}(X^\wedge)$$

is bounded for all $s,\gamma \in \mathbb{R}$, and the mapping $\phi \mapsto M_\phi$ is continuous in the corresponding topology.

Notice that (d) is a simple consequence of the fact that if $\{\Lambda^\mu : \mu \in \mathbb{R}\}$ and $\{\tilde{\Lambda}^\mu : \mu \in \mathbb{R}\}$ are two order-reducing families, then for each μ, the operator $\Lambda^\mu \tilde{\Lambda}^{-\mu}$ is parameter-elliptic of order zero. (f) is immediate from (c) and interpolation.

2 Mellin Quantization

2.1 Mellin Symbols

In Section 4 of [14] we considered Mellin symbols with asymptotics; they are meromorphic functions on $\mathbb{C}$ with values in Boutet de Monvel's algebra. For the definition of the Mellin operator $\mathrm{op}_M^\gamma a$ associated with the Mellin symbol a, we only need to know a on the line $\Gamma_{\frac{1}{2}-\gamma}$, and we certainly do not need analyticity. We can extend the calculus to larger classes of Mellin symbols by considering the case where the symbols additionally depend on the space variables t and t' – comparable to studying pseudodifferential 'double' symbols $p(x,y,\xi)$ after having treated Fourier multipliers $p(\xi)$.

2.1 Mellin Transforms. For $\beta \in \mathbb{R}$, Γ_β denotes the vertical line $\{z \in \mathbb{C} : \mathrm{Re}\, z = \beta\}$. We recall that the classical Mellin transform Mu of a complex-valued $C_0^\infty(\mathbb{R}_+)$-function u is given by

$$(Mu)(z) = \int_0^\infty t^{z-1} u(t)\, dt. \tag{1}$$

M extends to an isomorphism $M : L^2(\mathbb{R}_+) \to L^2(\Gamma_{\frac{1}{2}})$. Of course, (1) also makes sense for functions with values in a Fréchet space E. The fact that $Mu|_{\Gamma_{\frac{1}{2}-\gamma}}(z) = M_{t\to z}(t^{-\gamma}u)(z+\gamma)$ motivates the following definition of the *weighted Mellin transform* M_γ:

$$M_\gamma u(z) = M_{t\to z}(t^{-\gamma}u)(z+\gamma), \quad u \in C_0^\infty(\mathbb{R}_+,E).$$

2.2 Notation. In the following let $\mu \in \mathbb{Z}$ and $d \in \mathbb{N}$ be fixed. Given $f \in C^\infty(\mathbb{R}_+ \times \mathbb{R}_+, \mathcal{B}^{\mu,d}(X;\Gamma_{\frac{1}{2}-\gamma}))$ we shall write $f = f(t,t',z)$, where z indicates the variable in $\Gamma_{\frac{1}{2}-\gamma}$. For t,t',z fixed, $f(t,t',z)$ is a boundary value problem in Boutet de Monvel's calculus, so it acts on sections of vector bundles over $\overline{X}$ and Y. In order to fix the notation, assume that

$$f(t,t',z) : \begin{array}{c} C^\infty(\overline{X},V_1) \\ \oplus \\ C^\infty(Y,W_1) \end{array} \rightarrow \begin{array}{c} C^\infty(\overline{X},V_2) \\ \oplus \\ C^\infty(Y,W_2) \end{array} \tag{1}$$

with smooth vector bundles V_1, V_2, over $\overline{X}$ and W_1, W_2, over Y.

2.3 Definition. Let $f \in C^\infty(\mathbb{R}_+ \times \mathbb{R}_+, \mathcal{B}^{\mu,d}(X;\Gamma_{\frac{1}{2}-\gamma}))$. For $u \in C_0^\infty(\overline{X}^\wedge, V_1) \oplus C_0^\infty(Y^\wedge, W_1) = C_0^\infty(\mathbb{R}_+, C^\infty(\overline{X}, V_1) \oplus C^\infty(Y, W_1))$ let

$$[\mathrm{op}_M^\gamma(f)u](t) = \frac{1}{2\pi i} \int_{\Gamma_{\frac{1}{2}-\gamma}} \int_0^\infty (t/t')^{-z} f(t,t',z) u(t') \frac{dt'}{t'} dz \quad . \tag{1}$$

The right hand side of (1) is to be understood as an iterated integral. If f is independent of t', or equivalently $f \in C^\infty(\mathbb{R}_+, \mathcal{B}^{\mu,d}(X;\Gamma_{\frac{1}{2}-\gamma}))$, then (1) reduces to

$$[\mathrm{op}_M^\gamma(f)u](t) = \frac{1}{2\pi i} \int_{\Gamma_{\frac{1}{2}-\gamma}} t^{-z} f(t,z)[M_\gamma u](z) dz. \tag{2}$$

We did not specify the variable x in (1) or (2), understanding that for fixed $t', u(t') = u(\cdot, t')$ is in $C^\infty(\overline{X}, V_1) \oplus C^\infty(Y, W_1)$ and that $f(t,t',z)$ acts as an operator in Boutet de Monvel's calculus with respect to the x-variables.

Like pseudodifferential double symbols, Mellin double symbols are not uniquely determined. It is immediate from integration by parts in (1) that

$$\mathrm{op}_M^{\frac{1}{2}}[\ln^k(t/t') f(t,t',z)] = \mathrm{op}_M^{\frac{1}{2}}[\partial_z^k f(t,t',z)]. \tag{3}$$

For $f \in C^\infty(\mathbb{R}_+ \times \mathbb{R}_+, \mathcal{B}^{\mu,d}(X;\Gamma_{\frac{1}{2}-\gamma}))$ or $f \in C^\infty(\mathbb{R}_+, \mathcal{B}^{\mu,d}(X;\Gamma_{\frac{1}{2}-\gamma}))$ we will have a continuous map

$$\mathrm{op}_M^\gamma f : \begin{array}{c} C_0^\infty(\overline{X}^\wedge, V_1) \\ \oplus \\ C_0^\infty(Y^\wedge, W_1) \end{array} \to \begin{array}{c} C^\infty(\overline{X}^\wedge, V_2) \\ \oplus \\ C^\infty(Y^\wedge, W_2) \end{array} . \tag{4}$$

Smoothness of f up to zero yields continuity of $\mathrm{op}_M^\gamma f$ on the weighted Mellin-Sobolev spaces, cf. Theorem 2.4; the preceding relation (3), however, shows that smoothness is not necessary.

2.4 Theorem. Let $f \in C^\infty(\overline{\mathbb{R}}_+ \times \overline{\mathbb{R}}_+, \mathcal{B}^{\mu,d}(X;\Gamma_{\frac{1}{2}-\gamma})), s > d - \frac{1}{2}$. Given $\omega_1, \omega_2 \in C_0^\infty(\overline{\mathbb{R}}_+)$, there is a bounded extension

$$\omega_1[\mathrm{op}_M^\gamma f]\omega_2 : \begin{array}{c} \mathcal{H}^{s,\gamma+\frac{n}{2}}(X^\wedge, V_1) \\ \oplus \\ \mathcal{H}^{s,\gamma+\frac{n-1}{2}}(Y^\wedge, W_1) \end{array} \to \begin{array}{c} \mathcal{H}^{s-\mu,\gamma+\frac{n}{2}}(X^\wedge, V_2) \\ \oplus \\ \mathcal{H}^{s-\mu,\gamma+\frac{n-1}{2}}(Y^\wedge, W_2) \end{array} . \tag{1}$$

A proof can be found in [15]. We will also need the following results. They show that, just as in the case of pseudodifferential operators, one has asymptotic summation of symbols and the possibility of switching from operators with 'double' symbols to those with 'simple' ones.

2.5 Asymptotic Summation. Let $d \in \mathbb{N}$ be fixed, $\mu_1,\mu_2,\ldots$ a sequence in $\mathbb{Z}$ tending to $-\infty$, $f_j \in C^\infty(\overline{\mathbb{R}}_+ \times \overline{\mathbb{R}}_+,\mathcal{B}^{\mu_j,d}(X;\Gamma_{\frac{1}{2}-\gamma}))$, and $\mu = \max \mu_j$. Then there is an

$$f \in C^\infty(\overline{\mathbb{R}}_+ \times \overline{\mathbb{R}}_+,\mathcal{B}^{\mu,d}(X;\Gamma_{\frac{1}{2}-\gamma}))$$

such that for any $N \in \mathbb{N}$ there is a J with

$$f - \sum_{j=1}^{J} f_j \in C^\infty(\overline{\mathbb{R}}_+ \times \overline{\mathbb{R}}_+,\mathcal{B}^{\mu-N,d}(X;\Gamma_{\frac{1}{2}-\gamma})). \tag{1}$$

This f is unique modulo $C^\infty(\overline{\mathbb{R}}_+ \times \overline{\mathbb{R}}_+,\mathcal{B}^{-\infty,d}(X;\Gamma_{\frac{1}{2}-\gamma}))$. We shall write $f \sim \sum_{j=1}^\infty f_j$. The same result is true with $\overline{\mathbb{R}}_+ \times \overline{\mathbb{R}}_+$ replaced by $\overline{\mathbb{R}}_+, \mathbb{R}_+ \times \mathbb{R}_+$, or $\mathbb{R}_+$.

2.6 Theorem. For $f \in C^\infty(\mathbb{R}_+ \times \mathbb{R}_+,\mathcal{B}^{\mu,d}(X;\Gamma_0))$ there is a $g \in C^\infty(\mathbb{R}_+,\mathcal{B}^{\mu,d}(X;\Gamma_0))$ such that

$$\mathrm{op}_M^{\frac{1}{2}} f(t,t',z) - \mathrm{op}_M^{\frac{1}{2}} g(t,z) \in \mathcal{B}^{-\infty,d}(X^\wedge)\ . \tag{1}$$

In particular, relation (1) holds for any symbol g with the asymptotic expansion

$$g(t,z) \sim \sum_{j=0}^{\infty} \frac{1}{j!}(-t'\partial_{t'})^j \partial_z^j f(t,t',z)|_{t'=t}\ . \tag{2}$$

2.2 Pseudodifferential Action and Mellin Quantization

2.7 Pushforward of Pseudodifferential Operators. Let U,V be open sets in $\mathbb{R}^n$, $\chi : U \to V$ a diffeomorphism. Moreover, let E,F be Banach spaces with group action. Given an operator

$$P : C_0^\infty(U,E) \to C^\infty(U,F)\ ,$$

the pushforward $\chi_* P : C_0^\infty(V,E) \to C^\infty(V,F)$ is defined by

$$(\chi_* P)f(x) = [P(f \circ \chi)][\chi^{-1}(x)]\ .$$

If $P = \mathrm{op}\, p$ for some $p \in S^\mu(U,\mathbb{R}^n; E,F)$ then there is a symbol $q \in S^\mu(V,\mathbb{R}^n; E,F)$ with $\mathrm{op}\, q = \chi_* P$ modulo regularizing operators, and q is unique up to symbols in $S^{-\infty}(V,\mathbb{R}^n; E,F)$. In this sense χ defines a pushforward also on the symbol level:

$$\chi_* : S^\mu(U,\mathbb{R}^n; E,F)/S^{-\infty}(U,\mathbb{R}^n; E,F) \to S^\mu(V,\mathbb{R}^n; E,F)/S^{-\infty}(V,\mathbb{R}^n; E,F)\ .$$

The mapping is an isomorphism; the inverse is induced by the pushforward via χ^{-1}. The same statements are true for symbols with the transmission property.

One way of proving this is to first convert the symbol p to a 'double' symbol $p_1(y,y',\eta)$ by multiplying p with a cut-off function $\phi = \phi(y,y')$ near the diagonal

$\{y = y'\}$; $\operatorname{op} p$ and $\operatorname{op} p_1$ only differ by a regularizing operator. Then one can compute a 'double' symbol $q_1 \in S^\mu(V \times V, \mathbb{R}^n; E, F)$ with $\chi_* \operatorname{op} p_1 = \operatorname{op} q_1$ and finally switch to a y'-independent symbol q with $\operatorname{op} q_1 \equiv \operatorname{op} q$ modulo regularizing operators.

In what follows it will often be possible to find a 'double' symbol q_1 with $\chi_* \operatorname{op} p = \operatorname{op} q_1$ by a straightforward substitution in oscillatory integrals. We will then also write $q_1 = \chi_* p$.

2.8 Corollary. Let $\chi : U \to V$ be a diffeomorphism of open sets in $\mathbb{R}$, and let $a \in C^\infty(U, \mathcal{B}^{\mu,d}(X; \mathbb{R}))$ induce a pseudodifferential action by

$$\operatorname{op} a(u)(y) = \frac{1}{2\pi} \int \int_U e^{i(y-y')\eta} a(y,\eta) u(y') dy' d\eta \tag{1}$$

for $u \in C_0^\infty(U, C^\infty(\overline{X}, V_1) \oplus C^\infty(Y, W_1))$. For the pushforward $\chi_* \operatorname{op} a$ we then have

$$\chi_* \operatorname{op} a = \operatorname{op} b + G \ , \tag{2}$$

where

(i) the symbol b belongs to $C^\infty(V, \mathcal{B}^{\mu,d}(X; \mathbb{R}))$. It is determined via the symbol pushforward of the various local symbols for a. In this sense we shall use the notation $b = \chi_* a$.

(ii) The operator G belongs to $\mathcal{B}^{-\infty,d}(X^\wedge)$. In other words, we can write

$$G = \sum_{j=0}^{\infty} G_j \begin{bmatrix} \partial_r^j & 0 \\ 0 & I \end{bmatrix} ; \tag{3}$$

here ∂_r is the normal derivative on X, and each G_j is a matrix of integral operators with kernel functions which are smooth up to the boundary of X.

Proof. We have $C^\infty(U, \mathcal{B}^{\mu,d}(X; \mathbb{R})) = C^\infty(U) \hat{\otimes}_\pi \mathcal{B}^{\mu,d}(X; \mathbb{R})$. Since convergence of the symbols implies convergence of the associated operators, it is sufficient to assume that $a(y,\eta) = \psi(y) A(\eta)$ with $\psi \in C^\infty(U)$ and $A \in \mathcal{B}^{\mu,d}(X; \mathbb{R})$. The assertion is certainly true for regularizing A: In this case, op a already has the form (3); hence the pushforward is of the same type and (2) holds with $b = 0$, for ∂_r is not affected. We can therefore localize with respect to a coordinate neigborhood Ω_j for Ω and assume that A is given locally by a quintuple of parameter-dependent symbols in Boutet de Monvel's calculus, $(p_j, g_j, k_j, t_j, s_j)$, where $p_j = p_j(x, \xi, \eta) \in S^\mu_{tr}(X_j, \mathbb{R}^n_\xi \times \mathbb{R}_\eta)$, $X_j = \Omega_j \cap \overline{X}$, is a pseudodifferential symbol with the transmission property, g is a parameter–dependent singular Green symbol, etc., cf. 1.12. We then have to show that their pushforward is preserved.

In order to see this, let us focus on p_j; the arguments for the other symbols are similar. We have

$$S^\mu_{tr}(U \times X_j, \mathbb{R}^n \times \mathbb{R}) = C^\infty(U) \hat{\otimes}_\pi S^\mu_{tr}(X_j, \mathbb{R}^n \times \mathbb{R}); \tag{4}$$

thus $\psi(y)p(x,\xi,\eta) \in S^{\mu}_{tr}(U \times X_j,\mathbb{R}^n \times \mathbb{R})$. We know that $S^{\mu}_{tr}(U \times X_j,\mathbb{R}^n \times \mathbb{R})$ is invariant under coordinate transforms, therefore the pushforward $\chi_*[\psi(y)p(x,\xi,\eta)]$ belongs to $S^{\mu}_{tr}(V \times X_j,\mathbb{R}^n \times \mathbb{R})$ modulo $S^{-\infty}(V \times X_j,\mathbb{R}^n \times \mathbb{R})$. Employing now (4) with U replaced by V plus the fact that $C^\infty(V,\mathcal{F}) = C^\infty(V)\hat{\otimes}_\pi\mathcal{F}$ for every Fréchet space $\mathcal{F}$, we see that $\chi_*[\psi(y)p(x,\xi,\eta)] \in C^\infty(V,S^{\mu}_{tr}(X_j,\mathbb{R}^n \times \mathbb{R}))$ may be considered the pseudodifferential part (with transmission property) of a parameter-dependent symbol tuple in $C^\infty(V,\mathcal{B}^{\mu,d}(X_j;\mathbb{R}))$. Applying the same argument for the four other components g_j,k_j,t_j, and s_j we obtain the symbol $b \in C^\infty(V,\mathcal{B}^{\mu,d}(X;\mathbb{R}))$. $\triangleleft$

2.9 Pseudodifferential and Mellin Symbols.
Given $f \in C^\infty(\mathbb{R}_+,\mathcal{B}^{\mu,d}(X;\Gamma_0))$ let

$$b(y,\eta) = f(e^y, -i\eta) \quad , \quad y,\eta \in \mathbb{R} \ . \tag{1}$$

Denoting by exp the diffeomorphism $y \mapsto e^y$ from $\mathbb{R}$ to $\mathbb{R}_+$ we have

$$\mathrm{op}_M^{\frac{1}{2}} f = \exp_* \mathrm{op}\, b \ . \tag{2}$$

In more detail: For $u \in C_0^\infty(\mathbb{R}_+,C^\infty(\overline{X},V) \oplus C^\infty(Y,V_2))$ let $u^*(y) = u(e^y)$; then $[\mathrm{op}_M^{\frac{1}{2}} f(u)](e^y) = [\mathrm{op}\, b(u^*)](y)$. This is a simple consequence of the identity

$$\frac{1}{2\pi i}\int_{\Gamma_0}\int_0^\infty \left(\frac{e^y}{t'}\right)^{-z} f(e^y,z)u(t')\frac{dt'}{t'}dz = \frac{1}{2\pi}\int_{-\infty}^{\infty}\int_{-\infty}^{\infty} e^{i(y-y')\eta} f(e^y,-i\eta)u^*(y')dy'd\eta \ .$$

Equation (1) implies that $b \in C^\infty(\mathbb{R},\mathcal{B}^{\mu,d}(X;\mathbb{R}))$. According to Corollary 2.8, we will have $\exp_*\mathrm{op}\, b \equiv \mathrm{op}\, a$ modulo $\mathcal{B}^{-\infty,d}(X^\wedge)$. Hence,

$$\mathrm{op}_M^{\frac{1}{2}} f \equiv \mathrm{op}\, a \quad \text{modulo } \mathcal{B}^{-\infty,d}(X^\wedge) \ .$$

We shall now analyze the relationship between f and a.

2.10 Definition and Remark. For $\mu \in \mathbb{Z}$ and $d \in \mathbb{N}$ let

$$\mathring{M}\mathcal{B}^{\mu,d}(X^\wedge) = \{\mathrm{op}_M^{\frac{1}{2}} f + G \ : \ f \in C^\infty(\mathbb{R}_+,\mathcal{B}^{\mu,d}(X;\Gamma_0)),\ G \in \mathcal{B}^{-\infty,d}(X^\wedge)\}.$$

For $f \in C^\infty(\mathbb{R}_+,\mathcal{B}^{-\infty,0}(X;\Gamma_0))$, $\mathrm{op}_M^{\frac{1}{2}} f$ is an integral operator with smooth kernel on $\overline{X}^\wedge$. Hence $\mathring{M}\mathcal{B}^{-\infty,d}(X^\wedge) := \bigcap_\mu \mathring{M}\mathcal{B}^{\mu,d}(X^\wedge) = \mathcal{B}^{-\infty,d}(X^\wedge)$.

The following lemma may be considered a 'coarse' quantization result. It shows that pseudodifferential and Mellin symbols induce the same operators modulo $\mathcal{B}^{-\infty,d}(X^\wedge)$ as long as we consider symbol classes with arbitrary behavior near $t = 0$.

2.11 Lemma.

$$\mathring{M}\mathcal{B}^{\mu,d}(X^\wedge)/\mathring{M}\mathcal{B}^{-\infty,d}(X^\wedge) \cong \mathcal{B}^{\mu,d}(X^\wedge)/\mathcal{B}^{-\infty,d}(X^\wedge) \ .$$

The isomorphism is given by $f \mapsto \exp_* b$ with $b(y,\eta) = f(e^y, -i\eta)$; the inverse by $a \mapsto f$ with $f(t,z) = [\ln_* a](\ln t, iz)$.

Proof. By 2.8 and 2.9 the mapping $f \mapsto \exp_* b$, where $b = f(e^y, -i\eta)$, maps the left hand side to the right hand side injectively. A direct computation then yields the above inverse. ◁

For what follows it will be interesting to know more precisely what the pushforward by exp looks like. We start with a formal calculation.

2.12 Lemma. Let $p \in S^\mu(\mathbb{R},\mathbb{R})$. Then $\exp_* \mathrm{op}\, p$ is the pseudodifferential operator with the 'double' symbol

$$(\exp_* p)(t,t',\tau) = p(\ln t, M(t,t')^{-1}\tau)\frac{1}{t'}M(t,t')^{-1} \ . \tag{1}$$

Here $M(t,t') = \frac{\ln t - \ln t'}{t-t'}$ is C^∞ and strictly positive on $\mathbb{R}_+ \times \mathbb{R}_+$.

Proof. For $u \in C_0^\infty(\mathbb{R}_+), t' = e^{y'}$, we have

$$\begin{aligned}
[\mathrm{op}\, p(u \circ \exp)](\ln t) &= \frac{1}{2\pi}\int\!\!\int e^{i(\ln t - y')\eta} p(\ln t,\eta) u(e^{y'}) dy' d\eta \\
&= \frac{1}{2\pi}\int\!\!\int_0^\infty e^{i(\ln t - \ln t')\eta} p(\ln t,\eta) u(t')\frac{dt'}{t'} d\eta \\
&= \frac{1}{2\pi}\int\!\!\int_0^\infty e^{i(t-t')M(t,t')\eta} p(\ln t,\eta) u(t')\frac{dt'}{t'} d\eta \\
&= \frac{1}{2\pi}\int\!\!\int_0^\infty e^{i(t-t')\tau} p(\ln t, M(t,t')^{-1}\tau) \\
&\qquad u(t')\frac{1}{t'}M(t,t')^{-1} dt' d\tau.
\end{aligned}$$

This gives (1). The function $M(t,t')$ is smooth and ≥ 0, for ln is monotonely increasing. Moreover, M has no zero since, for $t = t'$, we have $M(t,t) = \frac{1}{t} > 0$. ◁

2.13 Lemma. (a) $\partial_{t'}^k M(t,t') \,|_{t'=t} = c_k t^{-k-1}$ for suitable $c_k \in \mathbb{R}$, $k = 0,1,\dots$. In particular, $(t'\partial_{t'})^k [t' M(t,t')]|_{t'=t}$ is smooth up to $t = 0$.
(b) $t^{k-1}\partial_{t'}^k [M(t,t')^{-1}]|_{t'=t}$ is smooth up to $t = 0, k = 0,1,\dots$.

Proof. (a) Let $u,v \in \mathbb{R}_+$. We have for $1 + x = \frac{u}{v}$, $|x| < 1$

$$\ln u - \ln v = \ln(\frac{u}{v}) = \ln(1+x) = \sum_{j=1}^\infty \frac{(-1)^{j+1}}{j} x^j = \sum_{j=1}^\infty \frac{(-1)^{j+1}}{j}\frac{(u-v)^j}{v^j} \ ,$$

hence

$$M(u,v) = \frac{\ln u - \ln v}{u - v} = \sum_{k=0}^{\infty} \frac{(-1)^k}{k+1} \frac{(u-v)^k}{v^{k+1}}.$$

Therefore

$$\partial_u^k M(u,v) \mid_{u=v} = k! \frac{(-1)^k}{k+1} v^{-k-1} \quad .$$

This proves the first statement. Now it is easily checked that, for $k \geq 1$, $(t\partial_t)^k$ is a linear combination of terms $t^j \partial_t^j$, $j = 1,...,k$, so we obtain the second statement, too.

(b) By induction, $\partial_{t'}^k[M(t,t')^{-1}]$ is a linear combination of terms of the form

$$M(t,t')^{-r-1} \prod_{l=1}^{r} \partial_{t'}^{j_l} M(t,t'),$$

where $r \leq k$ and $\sum_{l=1}^r j_l = k$. This implies that $\partial_{t'}^k[M(t,t')^{-1}] \mid_{t'=t}$ is a linear combination of terms $t^{r+1} t^{-r-k}$, $0 \leq r \leq k$. ◁

2.14 Definition. Let $\mu \in \mathbb{Z}$, $d \in \mathbb{N}$. By $C^\infty(\overline{\mathbb{R}}_+, \widetilde{\mathcal{B}}^{\mu,d}(X;\mathbb{R}))$ we denote the set of all $a \in C^\infty(\mathbb{R}_+, \mathcal{B}^{\mu,d}(X;\mathbb{R}))$ for which there is a $b \in C^\infty(\overline{\mathbb{R}}_+, \mathcal{B}^{\mu,d}(X;\mathbb{R}))$ such that

$$a(t,\tau) = b(t,t\tau) \quad .$$

2.15 Theorem. For $f \in C^\infty(\overline{\mathbb{R}}_+ \mathcal{B}^{\mu,d}(X;\Gamma_0))$ there is an $a \in C^\infty(\mathbb{R}_+ \widetilde{\mathcal{B}}^{\mu,d}(X;\mathbb{R}))$ with

$$\operatorname{op} a \equiv \operatorname{op}_M^{\frac{1}{2}} f \quad \text{modulo } \mathcal{B}^{-\infty,d}(X^\wedge) \quad . \tag{1}$$

Proof. We know from 2.9 that $\operatorname{op}_M^{\frac{1}{2}} f \equiv \operatorname{op}(\exp_* p)$ where $p(y,\eta) = f(e^y, -i\eta)$, and, by Lemma 2.12, $c_1(t,t',\tau) = [\exp_* p](t,t',\tau) = p(\ln t, M(t,t')^{-1}\tau)\frac{1}{t'}M(t,t')^{-1} = f(t, -iM(t,t')^{-1}\tau)\frac{1}{t'}M(t,t')^{-1}$ with the notation of 2.7. Let us convert the 'double' symbol c_1 to a symbol $c \in C^\infty(\mathbb{R}_+, \mathcal{B}^{\mu,d}(X;\mathbb{R}))$ independent of t':

$$c(t,\tau) \sim \sum_{k=0}^{\infty} \frac{1}{k!} \partial_{t'}^k D_\tau^k c_1(t,t',\tau) \mid_{t'=t} \quad . \tag{2}$$

Now

$$\partial_{t'}^k D_\tau^k c_1(t,t',\tau) = \partial_{t'}^k \{(-i)^k (\partial_z^k f)(t, -iM(t,t')^{-1}\tau) \frac{1}{t'} M(t,t')^{-k-1}\} \quad . \tag{3}$$

By induction this is a linear combination of terms of the form

$$(\partial_z^{k+j} f)(t, -iM(t,t')^{-1}\tau)\, \tau^j\, g_{kj}(t,t') \quad , \quad j = 0, \ldots, k, \tag{4}$$

where $g_{kj}(t,t')$ is a linear combination of terms

$$(t')^{-1-l_0} \prod_{i=1}^{r} \partial_{t'}^{l_i} \{M(t,t')^{-1}\}.$$

Here $r = k+1+j$, and $l_0 + \sum_{i=1}^{r} l_i = k$. Using Lemma 2.13 we conclude that $t^{-j} g_{kj}(t,t)$ is smooth up to $t = 0$.

Combining (3) and (4) we see that $\partial_{t'}^k D_\tau^k c_1(t,t',\tau)|_{t'=t}$ is a linear combination of terms of the form $(\partial_z^{k+j} f)(t, -it\tau)\,(t\tau)^j s_{kj}(t)$, where s_{kj} is a smooth function on $\overline{\mathbb{R}}_+$. Since $(\partial_z^{k+j} f)(t, -it\tau) \in C^\infty(\overline{\mathbb{R}}_+, \widetilde{\mathcal{B}}^{\mu-k-j,d}(X;\mathbb{R}))$, we obtain the symbol a by asymptotic summation in $C^\infty(\overline{\mathbb{R}}_+, \widetilde{\mathcal{B}}^{\mu,d}(X;\mathbb{R}))$. Note that there is asymptotic summation for symbols in this class: Given a sequence $\{a_j\}$ with $a_j \in C^\infty(\overline{\mathbb{R}}_+, \widetilde{\mathcal{B}}^{\mu-j,d}(X;\mathbb{R}))$ and $a_j(t,\tau) = b_j(t,t\tau)$ for $b_j \in C^\infty(\overline{\mathbb{R}}_+, \mathcal{B}^{\mu-j,d}(X;\mathbb{R}))$ choose $b \sim \sum b_j$ and let $a(t,\tau) = b(t,t\tau)$. We will then have $a - \sum_{j=0}^N a_j \in C^\infty(\overline{\mathbb{R}}_+, \widetilde{\mathcal{B}}^{\mu-N,d}(X;\mathbb{R})) \subseteq C^\infty(\mathbb{R}_+, \mathcal{B}^{\mu-N,d}(X;\mathbb{R}))$; hence $\operatorname{op} c - \operatorname{op} a \in \mathcal{B}^{-\infty,d}(X^\wedge)$.
$\triangleleft$

2.16 Theorem. (Mellin Quantization) Let $a \in C^\infty(\overline{\mathbb{R}}_+, \widetilde{\mathcal{B}}^{\mu,d}(X;\mathbb{R}))$. Then there is an $f \in C^\infty(\overline{\mathbb{R}}_+, \mathcal{B}^{\mu,d}(X;\Gamma_0))$ such that

$$\operatorname{op}_M^{\frac{1}{2}} f \equiv \operatorname{op} a \quad \text{modulo} \quad \mathcal{B}^{-\infty,d}(X^\wedge) \quad . \tag{1}$$

Proof. We know from Lemma 2.11 that $\operatorname{op} a \equiv \operatorname{op}_M^{\frac{1}{2}} g$ with

$$g(t,t',z) = [\ln_* a](\ln t, \ln t', iz) \ ; \tag{2}$$

here, we use the 'double' symbol of $[\ln_* a]$ one obtains by straightforward substitution in the oscillatory integral. Given a symbol $q \in S^\mu(\mathbb{R}_+,\mathbb{R})$ a computation similar to that in 2.12 shows that

$$(\ln_* q)(y,y',\eta) = q(e^y, M(e^y,e^{y'})\eta) e^{y'} M(e^y,e^{y'})$$

with the function $M(t,t') = \frac{\ln t - \ln t'}{t-t'}$ introduced in 2.12. Hence, in our case,

$$g(t,t',i\tau) = a(t, -M(t,t')\tau)\, t' M(t,t') \quad . \tag{3}$$

Now we apply Theorem 2.6. We have $\operatorname{op}_M^{\frac{1}{2}} g \equiv \operatorname{op}_M^{\frac{1}{2}} f$ modulo $\mathcal{B}^{-\infty,d}(X^\wedge)$ whenever $f \in C^\infty(\mathbb{R}_+, \mathcal{B}^{\mu,d}(X;\Gamma_0))$ has the asymptotic expansion

$$f(t,z) \sim \sum_{k=0}^{\infty} \frac{1}{k!} (-t'\partial_{t'})^k \partial_z^k g(t,t',z)\, |_{t'=t} \quad . \tag{4}$$

By assumption, the symbol $b(t,\tau) = a(t,t^{-1}\tau)$ is an element of $C^\infty(\overline{\mathbb{R}}_+, \mathcal{B}^{\mu,d}(X;\mathbb{R}))$. Thus $t^{-k}(\partial_\tau^k a)(t,t^{-1}\tau) = \partial_\tau^k b(t,\tau) \in C^\infty(\overline{\mathbb{R}}_+, \mathcal{B}^{\mu,d}(X;\mathbb{R}))$. The function $(t'\partial_{t'})^j (t' M(t,t'))\, |_{t'=t}$ is smooth up to $t = 0$ for $j = 0,1,\ldots$, by Lemma 2.13. So all the terms on the right hand side of (4) are smooth up to $t = 0$ and the asymptotic summation can be carried out in $C^\infty(\overline{\mathbb{R}}_+, \mathcal{B}^{\mu,d}(X;\Gamma_0))$. $\triangleleft$

2.3 Mellin Quantization for Arbitrary Weights

In the previous section we studied the question how to associate to a totally characteristic pseudodifferential symbol $a \in C^\infty(\overline{\mathbb{R}}_+,\widetilde{\mathcal{B}}^{\mu,d}(X;\mathbb{R}))$ a Mellin symbol $f_{1/2} \in C^\infty(\overline{\mathbb{R}}_+,\mathcal{B}^{\mu,d}(X;\Gamma_0))$ with $\operatorname{op} a = \operatorname{op}_M^{\frac{1}{2}} f_{1/2}$ modulo $\mathcal{B}^{-\infty,d}(X^\wedge)$. Given an arbitrary weight $\gamma \in \mathbb{R}$ this result allows us to easily find a Mellin symbol $f_\gamma \in C^\infty(\overline{\mathbb{R}}_+,\mathcal{B}^{\mu,d}(X;\Gamma_{\frac{1}{2}-\gamma}))$ such that $\operatorname{op} a = \operatorname{op}_M^\gamma f_\gamma$ modulo $\mathcal{B}^{-\infty,d}(X^\wedge)$:

2.17 Theorem. For every $a \in C^\infty(\overline{\mathbb{R}}_+,\widetilde{\mathcal{B}}^{\mu,d}(X;\mathbb{R}))$ and every $\gamma \in \mathbb{R}$ there is an $f_\gamma \in C^\infty(\overline{\mathbb{R}}_+,\mathcal{B}^{\mu,d}(X;\Gamma_{\frac{1}{2}-\gamma}))$ such that

$$\operatorname{op}_M^\gamma f_\gamma \equiv \operatorname{op} a \quad \text{modulo } \mathcal{B}^{-\infty,d}(X^\wedge) \ . \tag{1}$$

Proof. The Mellin symbol f_γ can be computed in terms of the function $f = f_{1/2}$ in Theorem 2.16. The definition of $\operatorname{op}_M^\gamma$ shows that

$$\operatorname{op} a \equiv \operatorname{op}_M^{\frac{1}{2}} f_{1/2} = \operatorname{op}_M^\gamma g_\gamma,$$

where $g_\gamma(t,t',z) = (t/t')^{1/2-\gamma} f_{1/2}(t,z - \frac{1}{2} + \gamma)$. We convert g_γ to a t'-independent symbol f_γ with

$$\begin{aligned} f_\gamma(t,z) \quad &\sim \quad \sum_{k=0}^\infty \frac{1}{k!}(-t'\partial_{t'})^k \partial_z^k g_\gamma(t,t',z)|_{t'=t} \\ &\sim \quad \sum_{k=0}^\infty \frac{1}{k!}(-t'\partial_{t'})^k (\frac{t}{t'})^{\frac{1}{2}-\gamma}|_{t'=t}\ \partial_z^k f_{1/2}(t,z - \frac{1}{2} + \gamma) \\ &\sim \quad \sum_{k=0}^\infty \frac{1}{k!}(\frac{1}{2} - \gamma)^k \partial_z^k f_{1/2}(t,z - \frac{1}{2} + \gamma), \end{aligned} \tag{2}$$

where we used that $(-t'\partial_{t'})^k (t/t')^{1/2-\gamma}|_{t'=t} = (x\partial_x)^k x^{1/2-\gamma}|_{x=1} = (1/2-\gamma)^k$. Since $f_{1/2}$ is smooth up to $t = 0$, the asymptotic summation can be carried out in $C^\infty(\overline{\mathbb{R}}_+,\mathcal{B}^{\mu,d}(X;\Gamma_{\frac{1}{2}-\gamma}))$, and we obtain the assertion. ◁

2.18 Remark. In [14] we defined spaces $M_P^{\mu,d}(X)$ of (t,t')-independent Mellin symbols of order μ and type d. They are meromorphic functions on $\mathbb{C}$, their only singularities are poles described in terms of the 'asymptotic type' P. We can then consider the classes $C^\infty(\overline{\mathbb{R}}_+,M_P^{\mu,d}(X))$ and the associated Mellin operators according to 2.3, cf. [15]. If the singularity set P is empty we shall write $h \in C^\infty(\overline{\mathbb{R}}_+,M_O^{\mu,d}(X))$. Then $h(t,\cdot)$ is an entire function, and Cauchy's theorem implies that $\operatorname{op}_M^\gamma h = \operatorname{op}_M^{\gamma'} h$ for all $\gamma,\gamma' \in \mathbb{R}$. We now let $f_{1/2} = h|_{\Gamma_0}$. According to Theorem 2.17 $\operatorname{op}_M^\gamma f_\gamma \equiv \operatorname{op}_M^\gamma (h|_{\Gamma_{\frac{1}{2}-\gamma}})$ modulo $\mathcal{B}^{-\infty,d}(X^\wedge)$. Therefore, $f_\gamma - h|_{\Gamma_{\frac{1}{2}-\gamma}} \in C^\infty(\overline{\mathbb{R}}_+,\mathcal{B}^{-\infty,d}(X;\Gamma_{\frac{1}{2}-\gamma}))$. This can be viewed as a slightly different convergence result for the Taylor series on the right hand side of 2.17(2).

2.19 Remark. The present Mellin quantization which ensures a control of the operators in Boutet de Monvel's algebra up to the conical singularity will play a crucial role in [15] as part of a cone algebra without asymptotics. This will then be an important step towards the construction of the corresponding cone algebra *with* asymptotics that will also be established in [15].

References

[1] Agranovič, M.S., and Višik, M.I.: Elliptic problems with a parameter and parabolic problems of general type (Russ.), *Usp. Mat. Nauk*, **19** (1963), 53–161 (Engl transl. *Russ. Math. Surveys,* **19** (1963), 53–159).

[2] Boutet de Monvel, L.: Boundary problems for pseudo-differential operators, *Acta Math.* **126** (1971), 11–51.

[3] Dieudonné, J.: *Grundzüge der modernen Analysis*, vol. 7. VEB Deutscher Verlag der Wissenschaften, Berlin 1982.

[4] Dorschfeldt, Ch., and Schulze, B.-W.: Pseudo-differential operators with operator-valued symbols in the Mellin-edge-approach, preprint, SFB 288, Berlin 1993, *Ann. Global Anal. and Geom.* **12** (1994), 135–171.

[5] Eskin, G.I.: *Boundary Value Problems for Elliptic Pseudodifferential Equations* (Russ.), Moscow 1973 (Engl. transl. Amer. Math. Soc. Translations of Math. Monographs **52**, Providence, R.I. 1981).

[6] Grubb, G.: *Functional Calculus of Pseudo-Differential Boundary Problems,* Progress in Mathematics 65, Birkhäuser, Boston, Basel 1986.

[7] Hörmander, L.: *The Analysis of Linear Partial Differential Operators,* vols. I–IV, Springer, Berlin, New York, Tokyo 1983–1985.

[8] Kondrat'ev, V.A.: Boundary value problems in domains with conical or angular points, Transactions Moscow Math. Soc. **16** (1967) 227–313.

[9] Lewis, J.E., and Parenti, C.: Pseudodifferential operators of Mellin type, *Comm. in Partial Diff. Equ.* **8** (1983), 477–544.

[10] Melrose, R.: Transformation of boundary problems, *Acta Math.* **147** (1981), 149–236.

[11] Melrose, R.: *The Atiyah-Patodi-Singer Index Theorem,* A K Peters, Wellesley, MA 1993.

[12] Plamenevskij, B. A.: *Algebras of Pseudodifferential Operators* (Russ.), Nauka, Moscow 1986.

[13] Rempel, S., and Schulze, B.-W.: *Index Theory of Elliptic Boundary Problems,* Akademie-Verlag, Berlin 1982.

[14] Schrohe, E., and Schulze, B.-W.: Boundary value problems in Boutet de Monvel's algebra for manifolds with conical singularities I. In: *Advances in Partial Differential Equations* **1**: *Pseudodifferential Operators and Mathematical Physics.* Akademie Verlag, Berlin, 1994, 97–209.

[15] Schrohe, E., and Schulze, B.-W.: Boundary value problems in Boutet de Monvel's algebra for manifolds with conical singularities II. in preparation

[16] Schulze, B.-W.: *Pseudo-Differential Operators on Manifolds with Singularities,* North-Holland, Amsterdam 1991.

[17] Schulze, B.-W.: Mellin representations of pseudo-differential operators on manifolds with corners. *Ann. Global Anal. and Geometry* **8** (1990), 261–297.

[18] Schulze, B.-W.:*Pseudo-Differential Boundary Value Problems, Conical Singularities and Asymptotics*, Akademie Verlag, Berlin 1994.

[19] Višik, M.I., and Eskin, G.I.: Convolution equations in bounded domains in spaces with weighted norms, *Math. USSR Sb.* **69** (1966), 65–110.

Elmar Schrohe and B.-W. Schulze
Max-Planck-Arbeitsgruppe
"Partielle Differentialgleichungen und Komplexe Analysis"
Universität Potsdam, 14415 Potsdam, Germany

Operator Theory:
Advances and Applications, Vol. 78

Transmission Algebras on Singular Spaces with Components of Different Dimensions

B.-W. Schulze

Abstract

The pseudo-differential calculus on a manifold with conical singularities extends in a natural way to spaces of hedgehog shape, e.g., to manifolds that are transversally intersected by intervals. For treating the higher-dimensional analogues for spaces with components of different dimensions it is nessecary to establish a corresponding theory in the framework of an algebra which is the range of operator-valued symbols in the Fourier-edge approach. The present paper constructs such an algebra on hedgehog configurations, based on Mellin pseudo-differential operators, acting in weighted Sobolev spaces with (and without) asymptotics. We introduce ellipticity, obtain Fredholm operators and parametrix constructions within the algebras. The elliptic regularity of solutions will be formulated with (and without) asymptotics in the weighted Mellin Sobolev spaces.

AMS Subject Classification: 35 S 05, 35 J 70, 58 G 15.

Key Words: Pseudo-differential calculus on singular spaces, conical singularities, Mellin symbols, different-dimensional components.

0. Introduction

This paper is part of the program to develop a pseudo-differential calculus for solving elliptic equations on spaces with components of different dimensions. The corresponding analysis on manifolds with a piece-wise smooth geometry (conical points, edges, corners, ...) in the sense of Schulze [8], [9], [10] will be regarded as a special case when the dimension is constant. An example for our new situation is $M \cup I$ with a C^∞ manifold M and an interval I, both being embedded in $\mathbb{R}^N$, with transversal intersection. More general examples can be obtained by replacing I by a cone with vertex in M or by a wedge $\cong \{\text{model cone}\} \times \{\text{edge}\}$ with $\{\text{edge}\} \subset M$. We are interested here in an algebra of operators that contains a certain class of typical differential operators. The operators in the algebra have a symbolic structure that allows the concept of ellipticity and a parametrix construction within the algebra. This paper will study the case $B \cup I$, with a manifold B with (finitely many) conical singularities $B_0 := \{v_1, \dots, v_k\}$ and a C^∞ curve I with end points b,c and $\{b\} \subseteq B_0$ or $\{b,c\} \subseteq B_0$. Recall from [9] that the structure of B entails that $B \setminus B_0$ is a C^∞ manifold and that to every $v \in B_0$ there is an open neighbourhood

V of v in B and a closed compact C^∞ manifold X (dependent on v) and a system of homeomorphisms

$$\varphi : V \longrightarrow X^\Delta := (\overline{\mathbb{R}}_+ \times X)/(\{0\} \times X)$$

such that $\varphi_0 := \varphi|_{V\setminus\{v\}} : V \setminus \{v\} \longrightarrow X^\wedge := \mathbb{R}_+ \times X$ is a diffeomorphism. Another condition is that for any second homeomorphism ψ from that system the composition $\varphi_0\psi_0^{-1} : X^\wedge \longrightarrow X^\wedge$ is the restriction of some diffeomorphism $\mathbb{R} \times X \longrightarrow \mathbb{R} \times X$ to $\mathbb{R}_+ \times X$.

Assume, in particular, that X is embedded in the unit sphere S^{N-2} of $\mathbb{R}^{N-1}$, $n = \dim X$. Then $X^\Delta = \{x' \in \mathbb{R}^{N-1} : x' = 0 \text{ or } x'/|x'| \in X\}$. Let $\tilde{x} = (x', x_N)$ denote the coordinates in $\mathbb{R}^N$. Then a useful model situation for $B \cup I$ near a conical point intersecting I is

$$\begin{aligned} X^\Delta \cup \mathbb{R}_+ \quad = \quad & \{\tilde{x} \in \mathbb{R}^N : \tilde{x} = 0 \text{ or } \tilde{x} = (x',0) \text{ with } x'/|x'| \in X\} \\ & \cup \{\tilde{x} \in \mathbb{R}^N : x' = 0, x_N \geq 0\}\ . \end{aligned}$$

Incidentally the intersections between the components of different dimensions will be called the contact points. In $X^\Delta \cup \mathbb{R}_+$ we have one contact point $\tilde{x} = 0$. We shall study the case of the infinite cone together with a half axis separately. Then we can pass to $B \cup I$ by a localization argument. On the other hand $X^\Delta \cup \mathbb{R}_+$ is also of independent interest for future applications to the case of a generalized wedge $(X^\Delta \cup \mathbb{R}_+) \times \mathbb{R}^q$ that is the local model of the corresponding configuration mentioned in the beginning. The calculus on the infinite "cone" $X^\Delta \cup \mathbb{R}_+$ for cone axis variable t tending to ∞ employs the analysis of pseudo-differential operators on a non-compact manifold in the sense of Cordes [3], Schrohe [5]. We shall use here the "classical" variant formulated in Schulze [8], Hirschmann [16], cf. also Egorov, Schulze [9]. We will not describe here the technical details once again but tacitly use the results and refer to that theory as the pseudo-differential calculus with exit conditions.

In view of the fact that we want to obtain pseudo-local operators it will be the main point of the present paper to establish the appropriate class of smoothing transmission operators that define a connection between the different-dimensional components over the contact points. They will be singular near the contact points and in general non-compact as operators in the weighted Sobolev spaces. That means that the non-smoothing operators in our calculus are modulo the transmission operators the elements from the cone algebras on the different manifolds with conical singularities, cf. Section 1. The cone algebras contain, in particular, the typical differential operators which are of Fuchs type, e.g., in the case $X^\Delta \times \mathbb{R}_+$, those over $X^\wedge = \mathbb{R}_+ \times X$ and $\mathbb{R}_+$, respectively.

In Section 2 we shall formulate an operator algebra together with the transmission operators in the context of the discrete asymptotics. It would be reasonable also to do that for the continuous asymptotics, cf. [8], [9], for being able to treat the "edgified" theory in the framework of the variable discrete asymptotics. However this would exceed the allowed size of this article. Section 3 will present the concept of ellipticity and the construction of parametrices, the Fredholm property, and the elliptic regularity of solutions, including the discrete asymptotics.

1. Elements of the cone calculus

Our transmission algebra will generalize the cone algebras from [8], [9]. We will recall here only some elements from that calculus. By definition a manifold B with conical singularities B_0 has in a neighbourhood of any $v \in B_0$ the structure X^Δ. In calculations on X^Δ we will always pass to the open stretched cone $X^\wedge = \mathbb{R}_+ \times X$. Now $B \setminus \{v\}$ can be regarded near v as a manifold with a cylindrical end of the form $(0,1) \times X$. To B we associate the stretched manifold $\mathbb{B}$ obtained from $B \setminus B_0$ by replacing the ends by $[0,1) \times X$ for every $v \in B_0$ (remember that X may depend on v). Then $\mathbb{B}$ is a C^∞ manifold with boundary $\partial\mathbb{B}$ which is the disjoint union over all occurring cone bases. Moreover we have $\mathbb{B} \setminus \partial\mathbb{B} \cong B \setminus B_0$.

Points in $X^\wedge$ will often be denoted by (t,x). On $\mathbb{R}_+ \ni t$ we have the classical Mellin transform

$$(Mu)(z) = \int_0^\infty t^{z-1} u(t)dt ,$$

first being defined on $u \in C_0^\infty(\mathbb{R}_+)$, with $z \in \mathbb{C}$, and then extended to various distribution spaces, also E-valued ones for suitable locally convex vector spaces. If $G \subseteq \mathbb{C}$ is an open set then $\mathcal{A}(G)$ will denote the space of all holomorphic functions in G. Analogously we write $\mathcal{A}(G,E)$ for the space of holomorphic E-valued functions in G. We have $(Mu)(z) \in \mathcal{A}(\mathbb{C})$ for $u \in C_0^\infty(\mathbb{R}_+)$ though we often look at the restriction of the Mellin transform to a weight line

$$\Gamma_\beta = \{z \in \mathbb{C} : \text{Re } z = \beta\}$$

for suitable real β. In particular, $M : C_0^\infty(\mathbb{R}_+) \longrightarrow \mathcal{A}(\mathbb{C})\big|_{\Gamma_{\frac{1}{2}}}$ extends by continuity to an isomorphism $M : L^2(\mathbb{R}_+) \longrightarrow L^2(\Gamma_{\frac{1}{2}})$ with the inverse $(M^{-1}g)(t) = (2\pi i)^{-1} \int t^{-z} g(z)dz$ with the integral being taken over $\Gamma_{\frac{1}{2}}$. Here and in the sequel we also use function or distribution spaces on the weight lines, analogously to those over $\mathbb{R}$. In particular, we have the Schwartz space $\mathcal{S}(\Gamma_\beta)$ or the Sobolev spaces $H^s(\Gamma_\beta)$ of smoothness $s \in \mathbb{R}$. We will also use the weighted Mellin transform $(M_\gamma u)(z) = M(t^{-\gamma}u)(z+\gamma)$ and the weighted Mellin pseudo-differential operators

$$(\text{op}_M^\gamma(h)u)(t) = t^\gamma M^{-1}(T^{-\gamma}h(z)M(t^{-\gamma}u)(z)) \tag{1.1}$$

for $M := M_0$ and $T^{-\gamma}h(z) := h(z-\gamma)$. Here $h(z)$, given on $\Gamma_{\frac{1}{2}-\gamma}$, is a symbol in Hörmander's symbol class of order $\mu \in \mathbb{R}$ (with constant coefficients) which is on the real axis denoted by $S^\mu(\mathbb{R})$. Classical symbols will be indicated by subscript cl. Then, according to our convention for notations, we have the symbol classes $S^\mu(\Gamma_\beta)$ and $S^\mu_{\text{cl}}(\Gamma_\beta)$, respectively. The operator

$$\text{op}_M^\gamma(h) : C_0^\infty(\mathbb{R}_+) \longrightarrow C^\infty(\mathbb{R}_+)$$

will be extended to distribution spaces in the usual way, and the notation will also be used in various vector-valued analogues. If we write $\mathrm{op}_M(\cdot) = \mathrm{op}_M^0(\cdot)$, then

$$\mathrm{op}_M^\gamma(h) = t^\gamma \mathrm{op}_M(T^{-\gamma}h)t^{-\gamma} . \tag{1.2}$$

On $X^\wedge$ we have the weighted Sobolev spaces

$$\mathcal{H}^{s,\gamma}(X^\wedge) \quad \text{for} \quad s,\gamma \in \mathbb{R} .$$

Let us briefly recall the meaning. If $\chi : U \longrightarrow X$ is a chart on X then an element $u \in \mathcal{H}^{s,\gamma}(X^\wedge)$ that has compact support in U with respect to the variable on the base X is characterized in the local coordinates $(t,x) \in \mathbb{R}_+ \times \mathbb{R}^n$ by

$$\left\{ \frac{1}{2\pi i} \int\limits_{\Gamma_{\frac{n+1}{2}-\gamma}} \int\limits_{\mathbb{R}^n} (1+|z|^2+|\xi|^2)^s \left|(M_{\gamma-\frac{n}{2},t\to z}F_{x\to\xi}u)(z,\xi)\right|^2 dz\, d\xi \right\}^{\frac{1}{2}} < \infty .$$

Here $F = F_{x\to\xi}$ is the Fourier transform in $\mathbb{R}^n$. We will not repeat here the functional analytic properties of those spaces. Let us only mention that $\mathcal{H}^{s,\gamma}(X^\wedge) \subset H^s_{\mathrm{loc}}(X^\wedge)$ for all $s,\gamma \in \mathbb{R}$. A cut-off function on $\mathbb{R}_+$ in this paper will be any $\omega(t) \in C_0^\infty(\overline{\mathbb{R}}_+)$ that equals 1 in a neighbourhood of $t = 0$. Then we can form the spaces

$$\mathcal{K}^{s,\gamma}(X^\wedge) = \omega\mathcal{H}^{s,\gamma}(X^\wedge) + (1-\omega)H^s(X^\wedge) . \tag{1.3}$$

Here an element $u \in H^s(X^\wedge)$ with $u = 0$ near $t = 0$ which is supported in U with respect to x may be characterized as follows. Let $\kappa : U \longrightarrow U_1$ be a diffeomorphism to an open set $U_1 \subset S^n$ (= unit sphere in $\mathbb{R}^{n+1} \ni \tilde{x}$) and let $\pi : \mathbb{R}^{n+1} \setminus \{0\} \longrightarrow \mathbb{R}_+ \times S^n \ni (t,\varphi)$ be polar coordinates in $\mathbb{R}^{n+1} \setminus \{0\}$. Then $\tilde{u}(\tilde{x}) := \pi^* u_1$ with $u_1(t,\varphi) := u(t,\kappa^{-1}(\varphi))$ satisfies

$$\left\{ \int\limits_{\mathbb{R}^{n+1}} (1+|\tilde{\xi}|^2)^s |(F_{\tilde{x}\to\tilde{\xi}}\tilde{u})(\tilde{\xi})|^2 d\tilde{\xi} \right\}^{\frac{1}{2}} < \infty .$$

The space $\mathcal{K}^{s,\gamma}(X^\wedge)$ is independent of the choice of ω. Note, in particular, that our weight convention is connected with $n = \dim X$. For $n = 0$ we have the spaces $\mathcal{K}^{s,\gamma}(\mathbb{R}_+)$ that coincide outside a neighbourhood of $t = 0$ with $H^s(\mathbb{R}_+) = H^s(\mathbb{R})\big|_{\mathbb{R}_+}$. Finally if B is a manifold with conical singularities which is compact as a topological space we denote by $\mathcal{H}^{s,\gamma}(\mathbb{B})$ the subspace of all $u \in H^s_{\mathrm{loc}}(\mathrm{int}\,\mathbb{B})$ such that for an $\tilde{\omega} \in C^\infty(\mathbb{B})$ that equals 1 in a collar neighbourhood of $\partial\mathbb{B}$ and vanishes outside another such neighbourhood we have $\psi_*(\tilde{\omega}u) \in \mathcal{K}^{s,\gamma}(X^\wedge)$ under a diffeomorphism ψ of the collar neighbourhood to $\mathbb{R}_+ \times X$. This definition is independent of the concrete choice of $\tilde{\omega}$ and ψ. It will be necessary also to look at the asymptotics of elements in $\mathcal{K}^{s,\gamma}(X^\wedge) \ni u(t,x)$ for $t \to 0$. Analogous asymptotics then make sense over $\mathbb{B}$ near $\partial\mathbb{B}$. Let $\Theta = (\vartheta,0]$, $\vartheta < 0$, and set

$$\mathcal{K}^{s,\gamma}_\Theta(X^\wedge) = \bigcap_{\varepsilon>0} \mathcal{K}^{s,\gamma-\vartheta-\varepsilon}(X^\wedge) ,$$

endowed with the corresponding inductive limit topology. The condition $u(t,x) \in \mathcal{K}^{s,\gamma}_{\Theta}(X^\wedge)$ will be interpreted as flatness of u of order $-\vartheta - 0$ for $t \to 0$ with respect to the reference weight γ. Now let

$$P = \{(p_j,m_j,L_j)\}_{j=0,\dots,N}\ , \quad N = N(P) \tag{1.4}$$

be a sequence with $p_j \in \mathbb{C}$, $\frac{n+1}{2} - \gamma + \vartheta < \operatorname{Re} p_j < \frac{n+1}{2} - \gamma$, $m_j \in \mathbb{N} = \{0,1,2,\dots\}$ and $L_j \subset C^\infty(X)$ being a vector subspace of finite dimension. Then we define a subspace $\mathcal{K}^{s,\gamma}_P(X^\wedge)$ of $\mathcal{K}^{s,\gamma}(X^\wedge) \ni u$ by the property

$$\omega\Big\{u(t,x) - \sum_{j=0}^{N}\sum_{k=0}^{m_j} c_{jk}(x) t^{-p_j} \log^k t\Big\} \in \mathcal{K}^{s,\gamma}_{\Theta}(X^\wedge)$$

with certain u-dependent (unique) coefficients $c_{jk} \in L_j$, $0 \le k \le m_j$ and a cut-off function ω. We will say that (1.4) is an asymptotic type and denote the set of those P by $\operatorname{As}(X,g)$ with the weight data $\underline{g} = (\gamma,\Theta)$. The spaces $\mathcal{K}^{s,\gamma}_{\Theta}(X^\wedge)$, $\mathcal{K}^{s,\gamma}_P(X^\wedge)$ are Fréchet in a natural way. For the infinite weight interval Θ we will employ analogous notations. In this case $\operatorname{As}(X,g)$ will consist of infinite sequences

$$P = \{(p_j,m_j,L_j)\}_{j\in\mathbb{N}}$$

with $\operatorname{Re} p_j < \frac{n+1}{2} - \gamma$ for all j and $\operatorname{Re} p_j \to -\infty$ as $j \to \infty$. If we set $P_k = \{(p,m,L) \in P : \operatorname{Re} p > \frac{n+1}{2} - \gamma - k\}, k \in \mathbb{N} \setminus \{0\}$,we have $\mathcal{K}^{s,\gamma}_{P_{k+1}}(X^\wedge) \subset \mathcal{K}^{s,\gamma}_{P_k}(X^\wedge)$ for all k, and we define

$$\mathcal{K}^{s,\gamma}_P(X^\wedge) = \bigcap_{k\in\mathbb{N}\setminus\{0\}} \mathcal{K}^{s,\gamma}_{P_k}(X^\wedge)\ ,$$

endowed with the topology of the projective limit. For $\dim X = 0$ the spaces L_j in the asymptotic types disappear and we write $\operatorname{As}(\underline{g})$ for the corresponding set of asymptotic types. These notions extend in an obvious manner to $\mathbb{B}$, where we fix a collar neighbourhood V of $\partial\mathbb{B}$ and a diffeomorphism $V \longrightarrow \overline{\mathbb{R}}_+ \times X$. Thus we obtain the spaces $\mathcal{H}^{s,\gamma}_{\Theta}(\mathbb{B})$, $\mathcal{H}^{s,\gamma}_P(\mathbb{B})$. The case with subscript Θ may be regarded as the empty asymptotic type, i.e. where

$$\pi_{\mathbb{C}} P = \{p_j\}_{j\in\mathbb{N}} \quad (\text{or} \quad \pi_{\mathbb{C}} P = \{p_j\}_{j=1,\dots,N})$$

is empty. Let us set

$$\mathcal{S}^{\gamma}_P(X^\wedge) = \omega\mathcal{K}^{\infty,\gamma}_P(X^\wedge) + (1-\omega)\mathcal{S}(X^\wedge)$$

for $P \in \operatorname{As}(X,(\gamma,\Theta))$, with any cut-off function ω, and $\mathcal{S}(X^\wedge) = \mathcal{S}(\mathbb{R},C^\infty(X))\big|_{\mathbb{R}_+}$. The space $\mathcal{S}^{\gamma}_P(X^\wedge)$ is independent of the choice of the concrete ω and it is Fréchet in a natural way.

Definition 1. Let $\gamma,\delta \in \mathbb{R}$, $\Theta = (\vartheta,0]$ be given and fix $P \in \mathrm{As}(X,(\delta,\Theta))$, $Q \in \mathrm{As}(X,(-\gamma,\Theta))$. Then a

$$G \in \bigcap_{s\in\mathbb{R}} \mathcal{L}(\mathcal{K}^{s,\gamma}(X^\wedge),\mathcal{K}^{\infty,\delta}(X^\wedge))$$

is said to be a Green operator of the class

$$C_G(X^\wedge,\underline{g})_{P,Q} \quad \text{with} \quad \underline{g} = (\gamma,\delta,\Theta)$$

if G induces continuous operators

$$\begin{array}{rcl} G \ : \ \mathcal{K}^{s,\gamma}(X^\wedge) & \longrightarrow & \mathcal{S}^{\delta}_{P}(X^\wedge)\ , \\ G^* : \ \mathcal{K}^{s,-\delta}(X^\wedge) & \longrightarrow & \mathcal{S}^{-\gamma}_{Q}(X^\wedge) \end{array}$$

for all $s \in \mathbb{R}$, G^* being the formal adjoint of G with respect to the $\mathcal{K}^{0,0}(X^\wedge)$-scalar product. Analogously a

$$G \in \bigcap_{s\in\mathbb{R}} \mathcal{L}(\mathcal{H}^{s,\gamma}(\mathbb{B}),\mathcal{H}^{\infty,\delta}(\mathbb{B}))$$

is called a Green operator of the class

$$C_G(\mathbb{B},\underline{g})_{P,Q} \quad \text{with} \quad \underline{g} = (\gamma,\delta,\Theta)$$

if G induces continuous operators

$$\begin{array}{rcl} G \ : \ \mathcal{H}^{s,\gamma}(\mathbb{B}) & \longrightarrow & \mathcal{H}^{\infty,\delta}_{P}(\mathbb{B})\ , \\ G^* : \ \mathcal{H}^{s,-\delta}(\mathbb{B}) & \longrightarrow & \mathcal{H}^{\infty,-\gamma}_{Q}(\mathbb{B}) \end{array}$$

for all $s \in \mathbb{R}$, G^* being the formal adjoint of G with respect to the $\mathcal{H}^{0,0}(\mathbb{B})$-scalar product. The union of the Green operator spaces over all P,Q will be denoted by $C_G(X^\wedge,\underline{g})$ and $C_G(\mathbb{B},\underline{g})$, respectively.

From now on we shall freely use the calculus of pseudo-differential operators on a (paracompact) C^∞ manifold Ω. If $U \subseteq \mathbb{R}^m$ is open then $S^\mu(U\times\mathbb{R}^n) \ni a(x,\xi)$ is Hörmander's symbol class, defined by the estimates

$$|D^\alpha_x D^\beta_\xi a(x,\xi)| \leq c(1+|\xi|)^{\mu-|\beta|}$$

for all $\alpha \in \mathbb{N}^m$, $\beta \in \mathbb{N}^n$, with constants $c = c(\alpha,\beta,K)$, for all $x \in K$, $K \Subset U$, $\xi \in \mathbb{R}^n$. Subscript *cl* will indicate the subclasses of classical elements. If $\Omega \subseteq \mathbb{R}^n$ is an open set and $U = \Omega\times\Omega$, then we will also write (x,x') for the points in U, and we set for $a(x,x',\xi) \in S^\mu(\Omega\times\Omega\times\mathbb{R}^n)$

$$\mathrm{Op}(a)u(x) = \iint e^{i(x-x')\xi} a(x,x',\xi)u(x')dx'\, đ\xi\ ,$$

$đ\xi = (2\pi)^{-n} d\xi$. Let $L^\mu(\Omega)$ be the set of all $\mathrm{Op}(a)$ with $a(x,x',\xi) \in S^\mu(\Omega\times\Omega\times\mathbb{R}^n)$ and $L^\mu_{\mathrm{cl}}(\Omega)$ the subspace of all classical pseudo-differential operators in Ω, defined by $\mathrm{Op}(a)$, $a(x,x',\xi) \in S^\mu_{\mathrm{cl}}(\Omega \times \Omega \times \mathbb{R}^n)$. Moreover, let $L^{-\infty}(\Omega)$ be the space of all operators with kernels in $C^\infty(\Omega \times \Omega)$. In general, if Ω is a C^∞ manifold, we use the same notations, now in the meaning that the elements in $L^\mu(\Omega)$ $(L^\mu_{\mathrm{cl}}(\Omega))$ are locally of the indicated structure, modulo an operator in $L^{-\infty}(\Omega)$ which is the space of operators with kernels in $C^\infty(\Omega \times \Omega)$ (we suppose once and for all that on the manifold in consideration there is a fixed Riemannian metric and then an associated measure dx). The symbol spaces as well as the operator spaces have natural Fréchet topologies. The same is true of the parameter-dependent analogues $L^\mu(\Omega;\mathbb{R}^l)$ $(L^\mu_{\mathrm{cl}}(\Omega;\mathbb{R}^l))$, locally being described by symbols $a(x,x',\xi,\lambda) \in S^\mu(\Omega \times \Omega \times \mathbb{R}^{n+l}_{\xi,\lambda})$ $(\in S^\mu_{\mathrm{cl}}(\Omega \times \Omega \times \mathbb{R}^{n+l}_{\xi,\lambda}))$, modulo $L^{-\infty}(\Omega;\mathbb{R}^l) = \mathcal{S}(\mathbb{R}^l,L^{-\infty}(\Omega))$, $\mathcal{S}(\ldots)$ denoting the Schwartz space of $L^{-\infty}(\Omega)$-valued functions on $\mathbb{R}^l$. We are mainly interested here in the case $l = 1$ and $X = \Omega$ with the base X of the cone that is compact and closed. Let

$$R = \{(r_j,n_j,N_j)\}_{j\in\mathbb{Z}}$$

be a sequence with $r_j \in \mathbb{C}$, $|\mathrm{Re}\, r_j| \to \infty$ as $j \to \infty$, $n_j \in \mathbb{N}$, and N_j be a finite-dimensional subspace of finite-dimensional operators in $L^{-\infty}(X)$. Any such R will be called an asymptotic type of a Mellin symbol. Let $\underline{\mathrm{As}}(X)$ denote the set of those R. For $\dim X = 0$ the spaces N_j disappear and we write $\underline{\mathrm{As}}$. Set $\pi_{\mathbb{C}}R = \{r_j\}_{j\in\mathbb{Z}}$. If $A \subset \mathbb{C}$ is a set we call a $\chi(z) \in C^\infty(\mathbb{C})$ an A-excision function if $|\chi(z)| \le 1$, $\chi(z) = 0$ for $\mathrm{dist}(z,\overline{A}) < \varepsilon_0$, $\chi(z) = 1$ for $\mathrm{dist}(z,\overline{A}) > \varepsilon_1$ with certain constants $0 < \varepsilon_0 < \varepsilon_1 < \infty$.

Definition 2. Let $\mu \in \mathbb{R}$, $R \in \underline{\mathrm{As}}(X)$. Then $M^\mu_R(X)$ denotes the space of all

$$h(z) \in \mathcal{A}(\mathbb{C} \setminus \pi_{\mathbb{C}}R, L^\mu_{\mathrm{cl}}(X)) \tag{1.5}$$

with the following properties

(i) $h(z)$ is meromorphic with poles at the points r_j of multiplicities $n_j + 1$, and Laurent coefficients of $(z - r_j)^{-(k+1)}$ belonging to N_j, $0 \le k \le n_j$,

(ii) if $\chi(z)$ is any $\pi_{\mathbb{C}}R$-excision function then

$$\chi(z)h(z)\big|_{\Gamma_\beta} \in L^\mu_{\mathrm{cl}}(X;\Gamma_\beta) \tag{1.6}$$

for every $\beta \in \mathbb{R}$ and uniformly in $c \le \beta \le c'$ for every $c < c'$.

In the case $n = 0$ we have to write

$$h(z) \in \mathcal{A}(\mathbb{C} \setminus \pi_{\mathbb{C}}R) \quad \text{and} \quad \chi(z)h(z)\big|_{\Gamma_\beta} \in S^\mu_{\mathrm{cl}}(\Gamma_\beta)\ ,$$

instead of (1.5) and (1.6), respectively.

In the latter case we write M_R^μ for the resulting space. If $\pi_{\mathbb{C}} R = \varnothing$ we obtain by definition the spaces $M_O^\mu(X)$ and M_O^μ, respectively. All occurring spaces are Fréchet. In particular, it makes sense to talk about

$$C^\infty(\overline{\mathbb{R}}_+,M_R^\mu(X)) , \qquad C^\infty(\overline{\mathbb{R}}_+,M_R^\mu) .$$

The functions $h(t,z) \in C^\infty(\overline{\mathbb{R}}_+,M_R^\mu(X))$ $(C^\infty(\overline{\mathbb{R}}_+,M_R^\mu))$ will be regarded as operator-valued Mellin symbols with the covariable $z \in \Gamma_\beta$ for a certain β such that $\Gamma_\beta \cap \pi_{\mathbb{C}} R = \varnothing$. Let us discuss from now on mainly the case with non-trivial X. The case $\dim X = 0$ is an obvious modification.

Let $P \in \mathrm{As}(X,g)$ with $\underline{g} = \left(\gamma + \frac{n}{2},\Theta\right)$ and set

$$\mathcal{H}_P^{s,\gamma+\frac{n}{2}}(X^\wedge) := \omega\mathcal{K}_P^{s,\gamma+\frac{n}{2}}(X^\wedge) + (1-\omega)\mathcal{H}^{s,\gamma+\frac{n}{2}}(X^\wedge) ,$$

endowed with the corresponding Fréchet structure.

Theorem 3. Let $h(t,z) \in C^\infty(\overline{\mathbb{R}}_+,M_R^\mu(X))$ be independent of t for $t >$ const. Then the operator

$$\mathrm{op}_M^\gamma(h) : C_0^\infty(X^\wedge) \longrightarrow C^\infty(X^\wedge)$$

(that is defined by (1.2) including the operator-valued action on X) has continuous extensions

$$\mathrm{op}_M^\gamma(h) : \mathcal{H}^{s,\gamma+\frac{n}{2}}(X^\wedge) \longrightarrow \mathcal{H}^{s-\mu,\gamma+\frac{n}{2}}(X^\wedge) \tag{1.7}$$

for all $s \in \mathbb{R}$ and all $\gamma \in \mathbb{R}$ with $\pi_{\mathbb{C}} R \cap \Gamma_{\frac{1}{2}-\gamma} = \varnothing$. Moreover, for every $P \in \mathrm{As}(X,\underline{g})$, $\underline{g} = (\gamma + \frac{n}{2},\Theta)$, there is a $Q \in \mathrm{As}(X,\underline{g})$ such that (1.7) restricts to a continuous operator

$$\mathrm{op}_M^\gamma(h) : \mathcal{H}_P^{s,\gamma+\frac{n}{2}}(X^\wedge) \longrightarrow \mathcal{H}_Q^{s-\mu,\gamma+\frac{n}{2}}(X^\wedge)$$

for every $s \in \mathbb{R}$.

A proof of this theorem as well as of the other results of this section may be found in [8].

Let us now return to the differential operators on $X^\wedge$ that are regarded as the typical ones for the conical singularity. These operators have the form

$$A = t^{-\mu}\sum_{j=0}^{\mu} a_j(t)\left(-t\frac{\partial}{\partial t}\right)^j \tag{1.8}$$

with coefficients $a_j(t) \in C^\infty(\overline{\mathbb{R}}_+,\mathrm{Diff}^{\mu-j}(X))$. Here $\mathrm{Diff}^\nu(X)$ is the space of all differential operators on X of order ν in its natural Fréchet topology. Then $h(t,z) = \sum_{j=0}^{\mu} a_j(t)z^j$ belongs to $C^\infty(\overline{\mathbb{R}}_+,M_O^\mu(X))$, and we can write $A = t^{-\mu}\mathrm{op}_M^\gamma(h)$ for every $\gamma \in \mathbb{R}$.

The cone operator algebra to be described here will contain the differential operators of the form (1.8) near $t = 0$ (or near $\partial\mathbb{B}$). In order to obtain the pseudo-differential operators we fix a finite open covering $\mathcal{U} = \{U_1,\ldots,U_N\}$ of X by coordinate neighbourhoods and a system of charts $\chi_j : U_j \longrightarrow \mathbb{R}^n$, furthermore functions $\varphi_j,\psi_j \in C_0^\infty(U_j)$ with $\varphi_j\psi_j = \varphi_j$ for all j, where $\{\varphi_1,\dots,\varphi_N\}$ is a subordinate partition of unity. Let us fix a system of symbols

$$p_{U,0}(t,x,t\tau,\xi) \quad \text{with} \quad p_{U,0}(t,x,\tilde{\tau},\xi) \in S^\mu_{\mathrm{cl}}(\overline{\mathbb{R}}_+ \times \mathbb{R}^n \times \mathbb{R}^{1+n}_{\tilde{\tau},\xi}) ,$$

$U \in \mathcal{U}$, x being local coordinates in U, and set

$$P_0 u = t^{-\mu} \sum_{j=1}^{N} \varphi_j (1 \times \chi_j)^* \mathrm{Op}(p_{U_j,0})(1 \times \chi_j)_* \psi_j u . \tag{1.9}$$

Here $1 \times \chi_j : \mathbb{R}_+ \times U_j \longrightarrow \mathbb{R}_+ \times \mathbb{R}^n$ is the diffeomorphism associated with χ_j and $(1 \times \chi_j)^*$ is the function pull-back, $(1 \times \chi_j)_* = ((1 \times \chi_j)^*)^{-1}$. We then have $P_0 \in L^\mu_{\mathrm{cl}}(X^\wedge)$. In order to obtain continuous operators between spaces of $\mathcal{K}^{s,\gamma}(X^\wedge)$-type we have to modify P_0 near $t = 0$ by means of a so-called Mellin operator convention. In addition we have to impose an adequate behaviour of symbols for $t \to \infty$. The latter precautions will not be formulated here explicitly once again. We refer to Cordes [3], Schrohe [5] and to the material in [8] for this point. In this context we have the exit property of symbols for $t \to \infty$. It is required to be classical in (t,τ,ξ) of order 0 in t and of order μ in (τ,ξ), cf. also Egorov, Schulze [9], Hirschmann [16]. If $p_{U,\infty}(t,x,\tau,\xi)$ are such symbols for $U \in \mathcal{U}$ and if we set

$$p_U(t,x,\tau,\xi) = t^{-\mu}\omega(t) p_{U,0}(t,x,t\tau,\xi) + (1 - \omega(t)) p_{U,\infty}(t,x,\tau,\xi)$$

for some cut-off function $\omega(t)$, then

$$Pu = \sum_{j=1}^{N} \varphi_j (1 \times \chi_j)^* \mathrm{Op}(p_{U_j})(1 \times \chi_j)_* \psi_j u \tag{1.10}$$

belongs to $L^\mu_{\mathrm{cl}}(X^\wedge)$, again, and equals P_0 near $t = 0$.

The Mellin operator convention is the following result, cf. [7], [8].

Theorem 4. There exists an $h(t,z) \in C^\infty(\overline{\mathbb{R}}_+, M^\mu_{\mathcal{O}}(X))$ such that

$$\sum_{j=1}^{N} \varphi_j (1 \times \chi_j)^* \mathrm{Op}(p_{U_j,0})(1 \times \chi_j)_* \psi_j = \mathrm{op}^\delta_M(h) \quad \mathrm{mod} \quad L^{-\infty}(X^\wedge)$$

for every $\delta \in \mathbb{R}$.

Now, if ω,ω_0,ω_1 are arbitrary cut-off functions, the operator

$$A_\psi := t^{-\mu}\omega(t)\mathrm{op}_M^{\gamma-\frac{n}{2}}(h)\omega_0(t) + (1 - \omega(t))P(1 - \omega_1(t))$$

belongs to $L^\mu_{\mathrm{cl}}(X^\wedge)$ and induces continuous operators

$$A_\psi : \mathcal{K}^{s,\gamma}(X^\wedge) \longrightarrow \mathcal{K}^{s-\mu,\gamma-\mu}(X^\wedge)$$

for all $s \in \mathbb{R}$.

Definition 5. Let $\mu,\nu \in \mathbb{R}$, $\mu - \nu \in \mathbb{N}$, $\gamma \in \mathbb{R}$, and $\Theta = (-k,0]$, $k \in \mathbb{N} \setminus \{0\}$. Then $C^\nu(X^\wedge,\underline{g})$ for $\underline{g} = (\gamma,\gamma-\mu,\Theta)$ is the space of all operators of the form

$$A = t^{-\mu}\omega(t)\mathrm{op}_M^{\gamma-\frac{n}{2}}(h)\omega_0(t) + (1-\omega(t))P(1-\omega_1(t)) + W + G\,. \tag{1.11}$$

Here ω,ω_0,ω_0 are cut-off functions with $\omega\omega_0 = \omega$, $\omega\omega_1 = \omega_1$, and $P \in L^\nu_{\mathrm{cl}}(X^\wedge)$ has the form (1.10), here for the order ν, $h(t,z) \in C^\infty(\overline{\mathbb{R}}_+,M^\nu_{\mathcal{O}}(X))$ is associated with P via Theorem 4 (for the order ν), moreover $G \in C_G(X^\wedge,\underline{g})$, and W is a finite linear combination of operators of the form

$$\omega_2(t)t^{-\nu+j}\mathrm{op}_M^{\gamma_j-\frac{n}{2}}(h_j)\omega_3(t)\ , \quad j \in \mathbb{N}$$

with $h_j(z) \in M^{-\infty}_{R_j}(X)$, $\Gamma_{\frac{n+1}{2}-\gamma_j} \cap \pi_{\mathbb{C}}R_j = \emptyset$, $\gamma_j \le \gamma \le \gamma_j + j + \mu - \nu$ for all j, and cut-off functions ω_2,ω_3. Analogously $C^\nu(\mathbb{B},\underline{g})$ is the subspace of all pseudo-differential operators $A \in L^\nu_{\mathrm{cl}}(\mathrm{int}\,\mathbb{B})$ with $\omega A\omega_0 \in C^\nu(X^\wedge,\underline{g})$ in a collar neighbourhood of $\partial\mathbb{B}$ in the coordinates (t,x), ω,ω_0 being cut-off functions.

Theorem 6. Every $A \in C^\nu(X^\wedge,\underline{g})$ induces continuous operators

$$A : \mathcal{K}^{s,\gamma}(X^\wedge) \longrightarrow \mathcal{K}^{s-\nu,\gamma-\mu}(X^\wedge) \tag{1.12}$$

for all $s \in \mathbb{R}$. Moreover for every $P \in \mathrm{As}(X,\underline{a})$, $\underline{a} = (\gamma,\Theta)$ there is a $Q \in \mathrm{As}(X,\underline{b})$, $\underline{b} = (\gamma-\mu,\Theta)$, such that (1.12) restricts to a continuous operator

$$A : \mathcal{K}^{s,\gamma}_P(X^\wedge) \longrightarrow \mathcal{K}^{s-\nu,\gamma-\mu}_Q(X^\wedge)$$

for every $s \in \mathbb{R}$. Moreover every $A \in C^\nu(\mathbb{B},\underline{g})$ induces continuous operators

$$A : \mathcal{H}^{s,\gamma}(\mathbb{B}) \longrightarrow \mathcal{H}^{s-\nu,\gamma-\mu}(\mathbb{B}) \tag{1.13}$$

for all $s \in \mathbb{R}$, and for every $P \in \mathrm{As}(X,\underline{a})$ there exists a $Q \in \mathrm{As}(X,\underline{b})$ such that (1.13) restricts to a continuous operator

$$A : \mathcal{H}^{s,\gamma}_P(\mathbb{B}) \longrightarrow \mathcal{H}^{s-\nu,\gamma-\mu}_Q(\mathbb{B})$$

for every $s \in \mathbb{R}$.

Note that every $A \in C^\nu(\mathbb{B},g)$, given by (1.13), has the property

$$A - P \in L^{-\infty}(X^\wedge)\ .$$

Hence the homogeneous principal symbol $\sigma^\nu_\psi(A)$ of A coincides with that of P and it has in local coordinates $(t,x) \in \mathbb{R}_+ \times \mathbb{R}^n$ near $t=0$ the form

$$t^{-\nu}p_{(\nu)}(t,x,t\tau,\xi)$$

where $p_{(\nu)}(t,x,\tilde{\tau},\xi) \in C^\infty(\overline{\mathbb{R}}_+ \times \mathbb{R}^n \times (\mathbb{R}^{1+n}_{(\tilde{\tau},\xi)} \setminus \{0\}))$ satisfies $p_{(\nu)}(t,x,\lambda\tilde{\tau},\lambda\xi) = \lambda^\nu p_{(\nu)}(t,x,\tilde{\tau},\xi)$ for all $\lambda > 0$ and all $t,x,\tilde{\tau},\xi$ with $(\tilde{\tau},\xi) \neq 0$. Furthermore, the operator function

$$\sigma_M^\nu(A)(z) := h(0,z) + h_0(z) : H^s(X) \longrightarrow H^{s-\nu}(X) ,$$

parametrized by $z \in \Gamma_{\frac{n+1}{2}-\gamma}$, is called the conormal symbol of A of order ν.

Analogously every $A \in C^\nu(\mathbb{B},g)$ has a pair of principal symbols $(\sigma_\psi^\nu(A),\sigma_M^\nu(A))$, where $\sigma_\psi^\nu(A)$ is the homogeneous principal symbol of A of order ν and $\sigma_M^\nu(A)$ the conormal symbol of order ν that is defined near $\partial\mathbb{B}$ in the same way as before.

Theorem 7. $A \in C^\nu(X^\wedge,\underline{g})$ with $\underline{g} = (\gamma,\gamma-\mu,\Theta)$, $\tilde{A} \in C^{\tilde{\nu}}(X^\wedge,\underline{\tilde{g}})$ with $\underline{\tilde{g}} = (\gamma-\mu,\gamma-\mu-\tilde{\mu},\Theta)$ implies $\tilde{A}A \in C^{\tilde{\nu}+\nu}(X^\wedge,\underline{h})$ with $\underline{h} = (\gamma,\gamma-\mu-\tilde{\mu},\Theta)$ and

$$\sigma_\psi^{\tilde{\nu}+\nu}(\tilde{A}A) = \sigma_\psi^{\tilde{\nu}}(\tilde{A})\sigma_\psi^\nu(A) , \qquad \sigma_M^{\nu+\tilde{\nu}}(\tilde{A}A) = (T^\nu\sigma_M^{\tilde{\nu}}(\tilde{A}))\sigma_M^\nu(A) .$$

An analogous result holds over $\mathbb{B}$.

Definition 8. An $A \in C^\mu(X^\wedge,\underline{g})$ with $\underline{g} = (\gamma,\gamma-\mu,\Theta)$, is called elliptic if

(i) $\sigma_\psi^\mu(A) \neq 0$ as a function on $T^*X^\wedge \setminus 0$, and moreover (in local coordinates near $t=0$) $p_{(\mu)}(t,x,\tilde{\tau},\xi) \neq 0$ for $(\tilde{\tau},\xi) \neq 0$, up to $t=0$,

(ii) $\sigma_M^\mu(A)(z) : H^s(X) \longrightarrow H^{s-\mu}(X)$ is an isomorphism for all $z \in \Gamma_{\frac{n+1}{2}-\gamma}$ and any fixed $s \in \mathbb{R}$,

(iii) A is exit-elliptic for $t \to \infty$, of order 0 in t.

The ellipticity of an $A \in C^\mu(\mathbb{B},g)$ is defined by an obvious analogue of (i), (ii).

Theorem 9. If $A \in C^\mu(X^\wedge,\underline{g})$, $\underline{g} = (\gamma,\gamma-\mu,\Theta)$, is elliptic there exists a parametrix $P \in C^{-\mu}(X^\wedge,\underline{g}^{-1})$, $\underline{g}^{-1} = (\gamma-\mu,\gamma,\Theta)$, in the sense

$$PA - I \in C_G(X^\wedge,\underline{g}_l) , \qquad AP - I \in C_G(X^\wedge,\underline{g}_r)$$

with $\underline{g}_l = (\gamma,\gamma,\Theta)$, $\underline{g}_r = (\gamma-\mu,\gamma-\mu,\Theta)$. An analogous result holds over $\mathbb{B}$.

2. The singular transmission algebra

The singular transmission algebra on $X^\wedge \cup \mathbb{R}_+$ with $n = \dim X$ will consist of block matrices of operators

$$\mathcal{A} = \begin{pmatrix} A & C_- \\ C_+ & A' \end{pmatrix} : \begin{matrix} \mathcal{K}^{s,\gamma+\frac{n}{2}}(X^\wedge) \\ \oplus \\ \mathcal{K}^{s,\gamma}(\mathbb{R}_+,\mathbb{C}^{N_-}) \end{matrix} \longrightarrow \begin{matrix} \mathcal{K}^{s-\nu,\gamma-\mu+\frac{n}{2}}(X^\wedge) \\ \oplus \\ \mathcal{K}^{s-\nu,\gamma-\mu}(\mathbb{R}_+,\mathbb{C}^{N_+}) \end{matrix}$$

that are continuous for all $s \in \mathbb{R}$, with certain $\gamma \in \mathbb{R}$, interior orders $\nu \in \mathbb{R}$ and weight shift orders $\mu \in \mathbb{R}$, where we assume as in the cone algebra that $\mu - \nu \in \mathbb{N}$. The dimensions N_-, N_+ will depend on $\mathcal{A}$, and compositions of operators will be possible when the corresponding dimensions fit together. The diagonal elements are assumed as follows:

$$A \in C^\nu(X^\wedge, \underline{g})$$

with $\underline{g} = (\gamma + \frac{n}{2}, \gamma - \mu + \frac{n}{2}, \Theta)$,

$$A' \in C^\nu(\mathbb{R}_+, \underline{h}) \otimes \mathbb{C}^{N_+} \otimes \mathbb{C}^{N_-}$$

with $\underline{h} = (\gamma, \gamma - \mu, \Theta)$. Here $\Theta = (-k, 0]$, $k \in \mathbb{N} \cup \{+\infty\}$. It remains to define the operators C_+, C_-, called the transmission operators. They will be of similar structure as the smoothing Mellin+Green operators of the cone classes. Our definitions will be formulated for C_- and C_+ at the same time. It will be convenient to introduce at once the classes of operator block matrices

$$C_G(X^\wedge \cup \mathbb{R}_+, \underline{g}; N_-, N_+)\ , \quad C_{M+G}(X^\wedge \cup \mathbb{R}_+, \underline{g}; N_-, N_+)$$

where we allow once again also non-vanishing elements in the diagonals, belonging to the C_G and C_{M+G} classes over $X^\wedge$ and $\mathbb{R}_+$, respectively (up to some weight shift on $X^\wedge$).

Definition 1. Let $\gamma, \delta \in \mathbb{R}$ and $\underline{g} = (\gamma, \delta, \Theta)$ with $\Theta = (\vartheta, 0]$, $-\infty \leq \vartheta < 0$. A

$$\mathcal{G} \in \bigcap_{s \in \mathbb{R}} \mathcal{L}(\mathcal{K}^{s,\gamma+\frac{n}{2}}(X^\wedge) \oplus \mathcal{K}^{s,\gamma}(\mathbb{R}_+, \mathbb{C}^{N_-}), \mathcal{K}^{\infty,\delta+\frac{n}{2}}(X^\wedge) \oplus \mathcal{K}^{\infty,\delta}(\mathbb{R}_+, \mathbb{C}^{N_+}))$$

is called a Green operator of the class

$$C_G(X^\wedge \cup \mathbb{R}_+, \underline{g}; N_-, N_+)$$

if it induces continuous operators

$$\begin{array}{lcll} \mathcal{G} & : & \mathcal{K}^{s,\gamma+\frac{n}{2}}(X^\wedge) \oplus \mathcal{K}^{s,\gamma}(\mathbb{R}_+, \mathbb{C}^{N_-}) & \longrightarrow & \mathcal{S}_P^{\delta+\frac{n}{2}}(X^\wedge) \oplus \mathcal{S}_{P'}^{\delta}(\mathbb{R}_+, \mathbb{C}^{N_+})\ , \\ \mathcal{G}^* & : & \mathcal{K}^{s,-\delta+\frac{n}{2}}(X^\wedge) \oplus \mathcal{K}^{s,-\delta}(\mathbb{R}_+, \mathbb{C}^{N_+}) & \longrightarrow & \mathcal{S}_Q^{-\gamma+\frac{n}{2}}(X^\wedge) \oplus \mathcal{S}_{Q'}^{-\gamma}(\mathbb{R}_+, \mathbb{C}^{N_-}) \end{array}$$

for all $s \in \mathbb{R}$, with asymptotic types $P \in \mathrm{As}\left(X, (\delta + \frac{n}{2}, \Theta)\right)$, $P' \in \mathrm{As}(\delta, \Theta)$, $Q \in \mathrm{As}\left(X, (-\gamma + \frac{n}{2}, \Theta)\right)$, $Q' \in \mathrm{As}(-\gamma, \Theta)$. Here $\mathcal{G}^*$ is the formal adjoint of $\mathcal{G}$ in the sense

$$(u, \mathcal{G}^* v)_{\mathcal{K}^{0,\frac{n}{2}}(X^\wedge) \oplus \mathcal{K}^{0,0}(\mathbb{R}_+, \mathbb{C}^{N_-})} = (\mathcal{G}u, v)_{\mathcal{K}^{0,\frac{n}{2}}(X^\wedge) \oplus \mathcal{K}^{0,0}(\mathbb{R}_+, \mathbb{C}^{N_+})}$$

for all $u \in C_0^\infty(X^\wedge) \oplus C_0^\infty(\mathbb{R}_+, \mathbb{C}^{N_-})$, $v \in C_0^\infty(X^\wedge) \oplus C_0^\infty(\mathbb{R}_+, \mathbb{C}^{N_+})$.

Note that in contrast to the formal adjoints of Green operators of the previous section we have here used the non-degenerate sesqui-linear pairings

$$\mathcal{K}^{s,\gamma+\frac{n}{2}}(X^\wedge) \times \mathcal{K}^{-s,-\gamma+\frac{n}{2}}(X^\wedge) \longrightarrow \mathbb{C}$$

induced by the scalar product in $\mathcal{K}^{0,\frac{n}{2}}(X^\wedge) = L^2(X^\wedge)$ with the measure $dt\,dx$. This was only made for convenience.

Remark 2. Every $\mathcal{G} \in C_G(X^\wedge \cup \mathbb{R}_+,\underline{g}; N_-,N_+)$ induces a compact operator

$$\mathcal{G} : \mathcal{K}^{s,\gamma+\frac{n}{2}}(X^\wedge) \oplus \mathcal{K}^{s,\gamma}(\mathbb{R}_+,\mathbb{C}^{N_-}) \longrightarrow \mathcal{K}^{s,\delta+\frac{n}{2}}(X^\wedge) \oplus \mathcal{K}^{s,\delta}(\mathbb{R}_+,\mathbb{C}^{N_+})$$

for every $s \in \mathbb{R}$.

Proposition 3. Let $\mathcal{G} \in C_G(X^\wedge \cup \mathbb{R}_+,\underline{g}; N,N)$ for $\underline{g} = (\gamma,\gamma,\Theta)$, and let

$$1 + \mathcal{G} : \mathcal{K}^{s,\gamma+\frac{n}{2}}(X^\wedge) \oplus \mathcal{K}^{s,\gamma}(\mathbb{R}_+,\mathbb{C}^{N}) \longrightarrow \mathcal{K}^{s,\gamma+\frac{n}{2}}(X^\wedge) \oplus \mathcal{K}^{s,\gamma}(\mathbb{R}_+,\mathbb{C}^{N})$$

be an invertible operator for one fixed $s \in \mathbb{R}$. Then, this operator is invertible for all $s \in \mathbb{R}$, and there is a $\mathcal{G}_1 \in C_G(X^\wedge \cup \mathbb{R}_+,\underline{g}; N,N)$ with

$$(1+\mathcal{G})^{-1} = 1 + \mathcal{G}_1 \ .$$

An asymptotic type of a Mellin transmission symbol is a sequence

$$R = \{(r_j,n_j,N_j)\}_{j\in\mathbb{Z}}$$

with $r_j \in \mathbb{C}$, $|\mathrm{Re}\ r_j| \to \infty$ as $|j| \to \infty$, $n_j \in \mathbb{N}$, and

$$N_j \subset \begin{pmatrix} C^\infty(X \times X) & C^\infty(X) \otimes \mathbb{C}^{N_-} \\ \mathbb{C}^{N_+} \otimes C^\infty(X) & \mathbb{C}^{N_+} \otimes \mathbb{C}^{N_-} \end{pmatrix}$$

being a finite-dimensional subspace of finite-dimensional operators for every $j \in \mathbb{Z}$. Here an element $d(x,x') \in C^\infty(X \times X)$ is interpreted as an operator $D : f \longrightarrow \int d(x,x')f(x')dx'$, $D : H^s(X) \longrightarrow H^\infty(X)$. Moreover, a $d_-(x) = (d_{-,i}(x))_{i=1,\dots,N_-} \in C^\infty(X) \otimes \mathbb{C}^{N_-}$ corresponds to an operator

$$D_-c = \sum_{i=1}^{N_-} d_{-,i}(x)c_i \quad \text{for} \quad c = (c_1,\dots,c_{N_-}) \in \mathbb{C}^{N_-} \ , \quad D_- : \mathbb{C}^{N_-} \longrightarrow H^\infty(X) \ ,$$

and an element $d_+(x) = (d_{+,k}(x))_{k=1,\dots,N_+}$ corresponds to a

$$D_+ : H^s(X) \longrightarrow \mathbb{C}^{N_+} \ , \quad (D_+f)_k = \int d_{+,k}(x)f(x)dx \ , \quad k = 1,\dots,N_+ \ .$$

Set

$$\pi_{\mathbb{C}}R = \{r_j\}_{j\in\mathbb{Z}} \ .$$

$\underline{\mathrm{As}}(X; N_-,N_+)$ will denote the set of all those asymptotic types R.

Definition 4. $M_R^{-\infty}(X; N_-,N_+)$ for $R \in \underline{\mathrm{As}}(X; N_-,N_+)$ is the space of all functions

$$h(z) \in \mathcal{A}\left(\mathbb{C} \setminus \pi_{\mathbb{C}}R, \bigcap_{s\in\mathbb{R}} \mathcal{L}\left(\begin{array}{c} H^s(X) \\ \oplus \\ \mathbb{C}^{N_-} \end{array}, \begin{array}{c} H^\infty(X) \\ \oplus \\ \mathbb{C}^{N_+} \end{array}\right)\right)$$

with the following properties

(i) $h(z)$ is meromorphic with poles at r_j of multiplicities n_j+1, with Laurent coefficients at $(z-r_j)^{-(k+1)}$ belonging to N_j, $0 \leq k \leq n_j$, $j \in \mathbb{Z}$,

(ii) if $\chi(z)$ is a $\pi_{\mathbb{C}}R$-excision function, then,

$$\chi(z)h(z)\big|_{\Gamma_\beta} \in \mathcal{S}\left(\Gamma_\beta, \bigcap_{s\in\mathbb{R}} \mathcal{L}\left(\begin{array}{c} H^s(X) \\ \oplus \\ \mathbb{C}^{N_-} \end{array}, \begin{array}{c} H^\infty(X) \\ \oplus \\ \mathbb{C}^{N_+} \end{array}\right)\right)$$

uniformly in $c \leq \beta \leq c'$ for every $c < c'$.

The definition gives rise to a natural Fréchet topology in $M_R^{-\infty}(X;N_-,N_+)$. We will denote by $M_O^{-\infty}(X;N_-,N_+)$ the subspace with $\pi_{\mathbb{C}}R = \varnothing$. Let $h(z) \in M_R^{-\infty}(X;N_-,N_+)$, $\pi_{\mathbb{C}}R \cap \Gamma_{\frac{1}{2}-\gamma} = \varnothing$. Then we can form the operators

$$\operatorname{op}_M^\gamma(h): \begin{array}{c} \mathcal{H}^{s,\gamma+\frac{n}{2}}(X^\wedge) \\ \oplus \\ \mathcal{H}^{s,\gamma}(\mathbb{R}_+,\mathbb{C}^{N_-}) \end{array} \longrightarrow \begin{array}{c} \mathcal{H}^{\infty,\gamma+\frac{n}{2}}(X^\wedge) \\ \oplus \\ \mathcal{H}^{\infty,\gamma}(\mathbb{R}_+,\mathbb{C}^{N_+}) \end{array}$$

which are continuous for all $s \in \mathbb{R}$. If $\omega(t),\tilde{\omega}(t)$ are arbitrary cut-off functions then

$$\omega\operatorname{op}_M^\gamma(h)\tilde{\omega}: \begin{array}{c} \mathcal{K}^{s,\gamma+\frac{n}{2}}(X^\wedge) \\ \oplus \\ \mathcal{K}^{s,\gamma}(\mathbb{R}_+,\mathbb{C}^{N_-}) \end{array} \longrightarrow \begin{array}{c} \mathcal{K}^{\infty,\gamma+\frac{n}{2}}(X^\wedge) \\ \oplus \\ \mathcal{K}^{\infty,\gamma}(\mathbb{R}_+,\mathbb{C}^{N_+}) \end{array}$$

are continuous for all $s \in \mathbb{R}$.

Proposition 5. Let $h(z) \in M_R^{-\infty}(X;N_-,N_+)$ and $\pi_{\mathbb{C}}R \cap \Gamma_{\frac{1}{2}-\gamma} = \varnothing$. Then to every pair of asymptotic types (P,P') with $P \in \operatorname{As}\left(X, (\gamma+\frac{n}{2},\Theta)\right)$, $P' \in \operatorname{As}(\gamma,\Theta)$, there is a (Q,Q') with $Q \in \operatorname{As}\left(X, (\gamma+\frac{n}{2},\Theta)\right)$, $Q' \in \operatorname{As}(\gamma,\Theta)$ such that

$$\omega\operatorname{op}_M^\gamma(h)\tilde{\omega}: \begin{array}{c} \mathcal{K}_P^{s,\gamma+\frac{n}{2}}(X^\wedge) \\ \oplus \\ \mathcal{K}_{P'}^{s,\gamma}(\mathbb{R}_+,\mathbb{C}^{N_-}) \end{array} \longrightarrow \begin{array}{c} \mathcal{K}_Q^{\infty,\gamma+\frac{n}{2}}(X^\wedge) \\ \oplus \\ \mathcal{K}_{Q'}^{\infty,\gamma}(\mathbb{R}_+,\mathbb{C}^{N_+}) \end{array}$$

is continuous for all $s \in \mathbb{R}$.

Proposition 6. Let $h(z) \in M_R^{-\infty}(X;N_0,N_+)$, $f(z) \in M_S^{-\infty}(X;N_-,N_0)$ and $\omega,\omega_0,\tilde{\omega}$ be cut-off functions. Then, if $\pi_{\mathbb{C}}R \cap \Gamma_{\frac{1}{2}-\gamma} = \pi_{\mathbb{C}}S \cap \Gamma_{\frac{1}{2}-\gamma} = \varnothing$ we have

$$\omega\operatorname{op}_M^\gamma(h)(1-\omega_0)\operatorname{op}_M^\gamma(f)\tilde{\omega} \in C_G(X^\wedge \cup \mathbb{R}_+, \underline{g}; N_-,N_+)$$

with $\underline{g} = \left(\gamma+\frac{n}{2}, \gamma+\frac{n}{2}, \Theta\right)$ for every Θ.

Definition 7. $C_{M+G}(X^\wedge \cup \mathbb{R}_+, \underline{g}; N_-, N_+)$ with $\underline{g} = (\gamma, \gamma - \mu, \Theta)$ for $\Theta = (-k, 0]$, $k \in \mathbb{N} \setminus \{0\}$ is the space of all operators $\mathcal{M} + \mathcal{G}$ with $\mathcal{G} \in C_G(X^\wedge \cup \mathbb{R}_+, \underline{g}; N_-, N_+)$ and $\mathcal{M}$ being a finite sum of operators of the form

$$\omega(t) t^{-\mu+j} \mathrm{op}_M^{\gamma_j}(h_j) \tilde{\omega}(t) \ , \quad j \in \mathbb{N} \ ,$$

with arbitrary $h_j \in M_{R_j}^{-\infty}(X; N_-, N_+)$ and $\gamma_j \in \mathbb{R}$ satisfying $\pi_{\mathbb{C}} R_j \cap \Gamma_{\frac{1}{2}-\gamma_j} = \varnothing$, and

$$\gamma_j \leq \gamma \leq \gamma_j + j \quad \text{for all} \quad j \ .$$

If $\Theta = (-\infty, 0]$ we define the C_{M+G}-class as the intersection of the corresponding ones over all $k \in \mathbb{N} \setminus \{0\}$. Let us set

$$\sigma_M^{\mu-j}(\mathcal{A})(z) = h_j(z) \quad \text{for} \quad \mathcal{A} \in C_{M+G}(X^\wedge \cup \mathbb{R}_+, \underline{g}; N_-, N_+) \ .$$

If $\Theta = (-k, 0]$ is finite, the so-called conormal symbols $\sigma_M^{\mu-j}(\mathcal{A})$ of orders $\mu - j$ are of interest only for $j = 0, \ldots, k-1$, since

$$\omega(t) t^{-\mu+j} \mathrm{op}_M^{\gamma_j}(h_j) \tilde{\omega}(t) \in C_G(X^\wedge \cup \mathbb{R}_+, \underline{g}; N_-, N_+)$$

for all $j \geq k$.

Proposition 8. $\mathcal{A} \in C_{M+G}(X^\wedge \cup \mathbb{R}_+, \underline{g}; N_-, N_+)$ and $\sigma_M^{\mu-j}(\mathcal{A}) = 0$ for all j implies $\mathcal{A} \in C_G(X^\wedge \cup \mathbb{R}_+, \underline{g}; N_-, N_+)$.

Theorem 9. $\mathcal{A} \in C_{M+G}(X^\wedge \cup \mathbb{R}_+, \underline{g}; N_-, N_0)$, $\underline{g} = (\gamma, \gamma - \mu, \Theta)$, $\tilde{\mathcal{A}} \in C_{M+G}(X^\wedge \cup \mathbb{R}_+, \underline{\tilde{g}}; N_0, N_+)$, $\underline{\tilde{g}} = (\gamma - \mu, \gamma - \mu - \tilde{\mu}, \Theta)$ implies $\tilde{\mathcal{A}}\mathcal{A} \in C_{M+G}(X^\wedge \cup \mathbb{R}_+, \underline{h}; N_-, N_+)$, $\underline{h} = (\gamma, \gamma - \mu - \tilde{\mu}, \Theta)$, and

$$\sigma_M^{\tilde{\mu}+\mu-j}(\tilde{\mathcal{A}}\mathcal{A})(z) = \sum_{p+q=j} (T^{\mu-q} \sigma_M^{\tilde{\mu}-p}(\tilde{\mathcal{A}}))(z) \sigma_M^{\mu-q}(\mathcal{A})(z)$$

for all j ($j = 0, \ldots, k-1$ for finite k). $\mathcal{A} \in C_G(\ldots)$ or $\tilde{\mathcal{A}} \in C_G$ implies $\tilde{\mathcal{A}}\mathcal{A} \in C_G(\ldots)$.

Definition 10. Let $\mu, \nu \in \mathbb{R}$, $\mu - \nu \in \mathbb{N}$, $\Theta = (-k, 0]$, $k \in (\mathbb{N} \setminus \{0\}) \cup \{\infty\}$. Then $C^\nu(X^\wedge \cup \mathbb{R}_+, \underline{g}; N_-, N_+)$ for $\underline{g} = (\gamma, \gamma - \mu, \Theta)$ is the space of all operators

$$\mathcal{A} = \begin{pmatrix} A & C_- \\ C_+ & A' \end{pmatrix} \tag{2.1}$$

with $A \in C^\nu\left(X^\wedge, (\gamma + \frac{n}{2}, \gamma - \mu + \frac{n}{2}, \Theta)\right)$, $A' \in C^\nu(\mathbb{R}_+, (\gamma, \gamma - \mu, \Theta)) \otimes \mathbb{C}^{N_+} \otimes \mathbb{C}^{N_-}$, and

$$\mathcal{C} := \begin{pmatrix} 0 & C_- \\ C_+ & 0 \end{pmatrix} \in C_{M+G}(X^\wedge \cup \mathbb{R}_+, \underline{g}; N_-, N_+)$$

with $\sigma_M^{\mu-l}(\mathcal{C}) = 0$ for all $l \in \mathbb{N}$ with $\mu - l > \nu$.

Note that every $\mathcal{A} \in C^\nu(X^\wedge \cup \mathbb{R}_+, g; N_-, N_+)$ could alternatively be written

$$\mathcal{A} = \begin{pmatrix} B & 0 \\ 0 & B' \end{pmatrix} + \mathcal{C}$$

with arbitrary $\mathcal{C} \in C_{M+G}(X^\wedge \cup \mathbb{R}_+, g; N_-, N_+)$ satisfying $\sigma_M^{\mu-l}(\mathcal{C}) = 0$ for all $\mu - l > \nu$, and

$$B = t^{-\nu}\omega \operatorname{op}_M^\gamma(h)\omega_0 + (1-\omega)P(1-\omega_1) ,$$

$h(t,z) \in C^\infty(\overline{\mathbb{R}}_+, M_O^\nu(X))$, $P \in L_{\mathrm{cl}}^\nu(X^\wedge)$ like in Definition 10, and analogously B', here related to $\mathbb{R}_+$,

$$B' = t^{-\nu}\omega \operatorname{op}_M^\gamma(h')\omega_0 + (1-\omega)P'(1-\omega_1)$$

with the corresponding $N_+ \times N_-$-matrix valued $h'(t,z)$ and P', respectively. Every $\mathcal{A} \in C^\nu(X^\wedge \cup \mathbb{R}_+, g; N_-, N_+)$ has a pair of homogeneous principal symbols of order ν namely

$$\sigma_\psi^\nu(\mathcal{A}) = (\sigma_\psi^\nu(A), \sigma_\psi^\nu(A')) ,$$

the second component being an $N_+ \times N_-$-matrix, and a principal conormal symbol of order ν

$$\sigma_M^\nu(\mathcal{A})(z) := \begin{pmatrix} h(0,z) & 0 \\ 0 & h'(0,z) \end{pmatrix} + \sigma_M^\nu(\mathcal{C})(z) . \tag{2.2}$$

In addition there are the principal exit symbols for $t \to \infty$, where t is regarded both as the cone axis variable of $X^\wedge$ and as the variable on $\mathbb{R}_+$.

Similarly as above for C_{M+G} we might define also the lower order conormal symbols of operators in $C^\nu(X^\wedge \cup \mathbb{R}_+, g; N_-, N_+)$, but we shall omit this here.

Proposition 11. $\mathcal{A} \in C^\nu(X^\wedge \cup \mathbb{R}_+, \underline{g}; N_-, N_+)$ for $\Theta = (-k, 0]$, $k \geq 2$, $\sigma_\psi^\nu(\mathcal{A}) = 0$, $\sigma_M^\nu(\mathcal{A}) = 0$ implies $\mathcal{A} \in C^{\nu-1}(X^\wedge \cup \mathbb{R}_+, \underline{g}; N_-, N_+)$.

Theorem 12. Every $\mathcal{A} \in C^\nu(X^\wedge \cup \mathbb{R}_+, \underline{g}; N_-, N_+)$ for $\underline{g} = (\gamma, \gamma - \mu, \Theta)$ induces continuous operators

$$\mathcal{A} : \begin{matrix} \mathcal{K}^{s,\gamma+\frac{n}{2}}(X^\wedge) \\ \oplus \\ \mathcal{K}^{s,\gamma}(\mathbb{R}_+, \mathbb{C}^{N_-}) \end{matrix} \longrightarrow \begin{matrix} \mathcal{K}^{s-\nu,\gamma-\mu+\frac{n}{2}}(X^\wedge) \\ \oplus \\ \mathcal{K}^{s-\nu,\gamma-\mu}(\mathbb{R}_+, \mathbb{C}^{N_+}) \end{matrix} \tag{2.3}$$

for all $s \in \mathbb{R}$. Moreover, for every $P \in \mathrm{As}\left(X, \left(\gamma + \frac{n}{2}, \Theta\right)\right)$, $P' \in \mathrm{As}(\gamma, \Theta)$ there are $Q \in \mathrm{As}\left(X, \left(\gamma - \mu + \frac{n}{2}, \Theta\right)\right)$, $Q' \in \mathrm{As}(\gamma - \mu, \Theta)$ such that (2.3) restricts to a continuous operator

$$\mathcal{A} : \begin{matrix} \mathcal{K}_P^{s,\gamma+\frac{n}{2}}(X^\wedge) \\ \oplus \\ \mathcal{K}_{P'}^{s,\gamma}(\mathbb{R}_+, \mathbb{C}^{N_-}) \end{matrix} \longrightarrow \begin{matrix} \mathcal{K}_Q^{s-\nu,\gamma-\mu+\frac{n}{2}}(X^\wedge) \\ \oplus \\ \mathcal{K}_{Q'}^{s-\nu,\gamma-\mu}(\mathbb{R}_+, \mathbb{C}^{N_+}) \end{matrix}$$

for every $s \in \mathbb{R}$.

Remark 13. If $\mathcal{A} \in C^\nu(X^\wedge \cup \mathbb{R}_+,\underline{g}; N_-,N_+)$ satisfies $\sigma_\psi^\nu(\mathcal{A}) = 0$, $\sigma_M^\nu(\mathcal{A}) = 0$ (and if also the pairs of principal exit symbols with respect to $X^\wedge$ and $\mathbb{R}_+$ vanish) then (2.3) is a compact operator for every $s \in \mathbb{R}$.

Theorem 14. $\mathcal{A} \in C^\nu(X^\wedge \cup \mathbb{R}_+,\underline{g}; N_-,N_0)$, $\underline{g} = (\gamma,\gamma-\mu,\Theta)$, $\tilde{\mathcal{A}} \in C^{\tilde{\nu}}(X^\wedge \cup \mathbb{R}_+,\underline{\tilde{g}}; N_0,N_+)$, $\underline{\tilde{g}} = (\gamma-\mu,\gamma-\mu-\tilde{\mu},\Theta)$ implies $\tilde{\mathcal{A}}\mathcal{A} \in C^{\tilde{\nu}+\nu}(X^\wedge \cup \mathbb{R}_+,\underline{h}; N_-,N_+)$, $\underline{h} = (\gamma,\gamma-\mu-\tilde{\mu},\Theta)$, and

$$\sigma_\psi^{\tilde{\nu}+\nu}(\tilde{\mathcal{A}}\mathcal{A}) = \sigma_\psi^{\tilde{\nu}}(\tilde{\mathcal{A}})\sigma_\psi^\nu(\mathcal{A})$$

(with obvious component-wise multiplication),

$$\sigma_M^{\tilde{\nu}+\nu}(\tilde{\mathcal{A}}\mathcal{A})(z) = (T^\nu \sigma_M^{\tilde{\nu}}(\tilde{\mathcal{A}}))(z)\sigma_M^\nu(\mathcal{A})(z)\ .$$

If $\tilde{\mathcal{A}}$ or $\mathcal{A}$ belongs to C_{M+G} (C_G) then $\tilde{\mathcal{A}}\mathcal{A}$ belongs to C_{M+G} (C_G).

Let us now consider the case $\mathbb{B} \cup \mathbb{R}_+$. From this we can easily derive the formulations for $\mathbb{B} \cup I$ with a finite interval I by interpreting the second end point as the conical singularity, cf. also the remarks at the end of the following Section 4.

Definition 15. Let $\gamma,\delta \in \mathbb{R}$ and $\underline{g} = (\gamma,\delta,\Theta)$ with $\Theta = (\vartheta,0]$, $-\infty \leq \vartheta < 0$. A

$$\mathcal{G} \in \bigcap_{s\in\mathbb{R}} \mathcal{L}(\mathcal{H}^{s,\gamma+\frac{n}{2}}(\mathbb{B}) \oplus \mathcal{K}^{s,\gamma}(\mathbb{R}_+,\mathbb{C}^{N_-}), \mathcal{H}^{\infty,\delta+\frac{n}{2}}(\mathbb{B}) \oplus \mathcal{K}^{\infty,\delta}(\mathbb{R}_+,\mathbb{C}^{N_+}))$$

is called a Green operator of the class

$$C_G(\mathbb{B} \cup \mathbb{R}_+,\underline{g}; N_-,N_+)$$

if it induces continuous operators

$$\begin{array}{rlcl} \mathcal{G} & : \ \mathcal{H}^{s,\gamma+\frac{n}{2}}(\mathbb{B}) \oplus \mathcal{K}^{s,\gamma}(\mathbb{R}_+,\mathbb{C}^{N_-}) & \longrightarrow & \mathcal{H}_P^{\infty,\delta+\frac{n}{2}}(\mathbb{B}) \oplus \mathcal{S}_{P'}^{\delta}(\mathbb{R}_+,\mathbb{C}^{N_+})\ , \\ \mathcal{G}^* & : \ \mathcal{H}^{s,-\delta+\frac{n}{2}}(\mathbb{B}) \oplus \mathcal{K}^{s,-\delta}(\mathbb{R}_+,\mathbb{C}^{N_+}) & \longrightarrow & \mathcal{H}_Q^{\infty,-\gamma+\frac{n}{2}}(\mathbb{B}) \oplus \mathcal{S}_{Q'}^{-\gamma}(\mathbb{R}_+,\mathbb{C}^{N_-}) \end{array}$$

for all $s \in \mathbb{R}$, with asymptotic types P,P',Q,Q' as in the above Definition 1. $\mathcal{G}^*$ is here the formal adjoint of $\mathcal{G}$ in the sense

$$(u,\mathcal{G}^*v)_{\mathcal{H}^{0,\frac{n}{2}}(\mathbb{B})\oplus\mathcal{K}^{0,0}(\mathbb{R}_+,\mathbb{C}^{N_-})} = (\mathcal{G}u,v)_{\mathcal{H}^{0,\frac{n}{2}}(\mathbb{B})\oplus\mathcal{K}^{0,0}(\mathbb{R}_+,\mathbb{C}^{N_+})}$$

for all $u \in C_0^\infty(\text{int}\ \mathbb{B}) \oplus C_0^\infty(\mathbb{R}_+,\mathbb{C}^{N_-})$, $v \in C_0^\infty(\text{int}\ \mathbb{B}) \oplus C_0^\infty(\mathbb{R}_+,\mathbb{C}^{N_+})$.

Note that the above Proposition 3 has an immediate analogue for the case $\mathcal{G} \in C_G(\mathbb{B} \cup \mathbb{R}_+,\underline{g}; N,N)$.

Definition 16. $C_{M+G}(\mathbb{B} \cup \mathbb{R}_+,\underline{g};N_-,N_+)$ with $\underline{g} = (\gamma,\gamma-\mu,\Theta)$ for $\Theta = (-k,0]$, $k \in \mathbb{N} \setminus \{0\}$, is the space of all operators $\mathcal{M} + \mathcal{G}$ with $\mathcal{G} \in (\mathbb{B} \cup \mathbb{R}_+,\underline{g};N_-,N_+)$ and $\omega\mathcal{M}\omega_0$ belonging to $C_{M+G}(X^\wedge \cup \mathbb{R}_+,\underline{g};N_-,N_+)$ for any choice of cut-off functions $\omega(t)$, $\omega_0(t)$ with t being the variable on both "sides" of $X^\wedge \cup \mathbb{R}_+$ near the "stretched contact point", and $\omega\mathcal{M}\omega_0$ being interpreted as the operator in the corresponding (t,x)-coordinates from a collar neighbourhood of $\partial\mathbb{B}$ in $\mathbb{B}$. The conormal symbols $\sigma_M^{\mu-j}(\mathcal{A})$ can be defined for $\mathcal{A} \in C_{M+G}(\mathbb{B} \cup \mathbb{R}_+,\underline{g};N_-,N_-)$ in the same way as above for the C_{M+G}-classes over $X^\wedge \cup \mathbb{R}_+$, since they only refer to the contact point. There is then an obvious analogue of Proposition 8 for the C_{M+G}-class over $\mathbb{B} \cup \mathbb{R}_+$ as well as of Theorem 9.

Theorem 17. Every $\mathcal{A} \in C_{M+G}(\mathbb{B} \cup \mathbb{R}_+,\underline{g};N_-,N_+)$ induces continuous operators

$$\mathcal{A}: \begin{array}{c} \mathcal{H}^{s,\gamma+\frac{n}{2}}(\mathbb{B}) \\ \oplus \\ \mathcal{K}^{s,\gamma}(\mathbb{R}_+,\mathbb{C}^{N_-}) \end{array} \longrightarrow \begin{array}{c} \mathcal{H}^{\infty,\gamma-\mu+\frac{n}{2}}(\mathbb{B}) \\ \oplus \\ \mathcal{K}^{\infty,\gamma-\mu}(\mathbb{R}_+,\mathbb{C}^{N_+}) \end{array} \tag{2.4}$$

for every $s \in \mathbb{R}$. Moreover for every $P \in \mathrm{As}\left(X,\left(\gamma+\frac{n}{2},\Theta\right)\right)$, $P' \in \mathrm{As}(\gamma,\Theta)$ there exist $Q \in \mathrm{As}\left(X,\left(\gamma-\mu+\frac{n}{2},\Theta\right)\right)$, $Q' \in \mathrm{As}(X,(\gamma-\mu,\Theta))$ such that (2.4) restricts to continuous operators

$$\mathcal{A}: \begin{array}{c} \mathcal{H}_P^{s,\gamma+\frac{n}{2}}(\mathbb{B}) \\ \oplus \\ \mathcal{K}_{P'}^{s,\gamma}(\mathbb{R}_+,\mathbb{C}^{N_-}) \end{array} \longrightarrow \begin{array}{c} \mathcal{H}_Q^{\infty,\gamma-\mu+\frac{n}{2}}(\mathbb{B}) \\ \oplus \\ \mathcal{K}_{Q'}^{\infty,\gamma-\mu}(\mathbb{R}_+,\mathbb{C}^{N_+}) \end{array}$$

for every $s \in \mathbb{R}$. An analogous result holds for the C_{M+G}-operators over $X^\wedge \cup \mathbb{R}_+$.

Remark 18. An $\mathcal{A} \in C_{M+G}(\mathbb{B} \cup \mathbb{R}_+,\underline{g};N_-,N_+)$ with $\underline{g} = (\gamma,\gamma-\mu,\Theta)$, regarded as a continuous operator

$$\mathcal{A}: \begin{array}{c} \mathcal{H}^{s,\gamma+\frac{n}{2}}(\mathbb{B}) \\ \oplus \\ \mathcal{K}^{s,\gamma}(\mathbb{R}_+,\mathbb{C}^{N_-}) \end{array} \longrightarrow \begin{array}{c} \mathcal{H}^{s,\gamma-\mu+\frac{n}{2}}(\mathbb{B}) \\ \oplus \\ \mathcal{K}^{s,\gamma-\mu}(\mathbb{R}_+,\mathbb{C}^{N_+}) \end{array}$$

for any fixed $s \in \mathbb{R}$ is compact if and only if $\sigma_M^{\mu}(\mathcal{A}) = 0$.

An analogous remark is true, of course, also for the C_{M+G}-class over $X^\wedge \cup \mathbb{R}_+$.

Definition 19. Let $\mu,\nu \in \mathbb{R}$, $\mu - \nu \in \mathbb{N}$, $\Theta = (-k,0]$, $k \in (\mathbb{N} \setminus \{0\}) \cup \{\infty\}$. Then $C^\nu(\mathbb{B} \cup \mathbb{R}_+,\underline{g};N_-,N_+)$ for $\underline{g} = (\gamma,\gamma-\mu,\Theta)$ is the space of all operators $\mathcal{A}$ that have the form of a block matrix like (2.1) with $A' \in C^\nu\left(\mathbb{B},\left(\gamma+\frac{n}{2},\gamma-\mu+\frac{n}{2},\Theta\right)\right)$, $A' \in C^\nu(\mathbb{R}_+,(\gamma,\gamma-\mu,\Theta)) \otimes \mathbb{C}^{N_+} \otimes \mathbb{C}^{N_-}$, and

$$\mathcal{C} := \begin{pmatrix} 0 & C_- \\ C_+ & 0 \end{pmatrix} \in C_{M+G}(\mathbb{B} \cup \mathbb{R}_+,\underline{g};N_-,N_+)$$

with $\sigma_M^{\mu-l}(\mathcal{C}) = 0$ for all $l \in \mathbb{N}$ with $\mu - l > \nu$.

On the space $C^\nu(\mathbb{B}\cup\mathbb{R}_+,g;N_-,N_+)$ we have the principal symbols $(\sigma_\psi^\nu(A), \sigma_\psi^\nu(A'))$ of order ν, where the first component corresponds to the homogeneous principal symbol of A regarded as an element of $L_{\mathrm{cl}}^\nu(\mathrm{int}\,\mathbb{B})$. Moreover we have $\sigma_M^\nu(\mathcal{A})$ that follows immediately from the definition for the operators on $X^\wedge\cup\mathbb{R}_+$. The elements of the calculus on $\mathbb{B}\cup\mathbb{R}_+$ are now formally anlogous to those on $X^\wedge\cup\mathbb{R}_+$. This concerns, in particular, Proposition 11, Theorem 12, Remark 13 and Theorem 14. We shall omit formulating the obvious details.

Remark 20. If $\mathcal{A}\in C^\nu(\mathbb{B}\cup\mathbb{R}_+,(\gamma,\gamma-\mu,\Theta);N_-,N_+)$ $(\in C^\nu(X^\wedge\cup\mathbb{R}_+,(\gamma,\gamma-\mu,\Theta);N_-,N_+))$ then we have for the formal adjoint $\mathcal{A}^*\in C^\nu(\mathbb{B}\cup\mathbb{R}_+,(-\gamma+\mu,-\gamma,\Theta);N_+,N_-)$ $(\in C^\nu(X^\wedge\cup\mathbb{R}_+,(-\gamma+\mu,-\gamma,\Theta);N_+,N_-))$. We can easily express the symbols of $\mathcal{A}^*$ by those of $\mathcal{A}$ (cf. [8], [9] for the cone calculus). Also the C_{M+G}-classes remain preserved under formal adjoints, up to the indicated modification of weight data and interchanged N_-,N_+.

3. Ellipticity and parametrices

Definition 1. An operator $\mathcal{A}\in C^\mu(X^\wedge\cup\mathbb{R}_+,\underline{g};N,N)$ with $\underline{g}=(\gamma,\gamma-\mu,\Theta)$ is called elliptic if

(i) the diagonal elements are elliptic with respect to the principal interior symbolic levels σ_ψ^μ in the sense of the classes $C^\mu\left(X^\wedge,\left(\gamma+\frac{n}{2},\gamma-\mu+\frac{n}{2},\Theta\right)\right)$ and $C^\mu(\mathbb{R}_+,(\gamma,\gamma-\mu,\Theta))\otimes\mathbb{C}^N\otimes\mathbb{C}^N$, respectively,

(ii)

$$\sigma_M^\mu(\mathcal{A})(z):\begin{array}{c}H^s(X)\\ \oplus\\ \mathbb{C}^N\end{array}\longrightarrow\begin{array}{c}H^{s-\mu}(X)\\ \oplus\\ \mathbb{C}^N\end{array}$$

is an isomorphism for every $z\in\Gamma_{\frac{1}{2}-\gamma}$ and some fixed $s\in\mathbb{R}$,

(iii) the diagonal elements of $\mathcal{A}$ are exit elliptic for $t\to\infty$, where t is the cone axis variable for $X^\wedge$ and $\mathbb{R}_+$, respectively (cf. also Section 1, Definition 8).

Analogously an $\mathcal{A}\in C^\mu(X^\wedge\cup\mathbb{R}_+,g;N,N)$ with $\underline{g}=(\gamma,\gamma-\mu,\Theta)$ is called elliptic if the diagonal elements are elliptic in the sense of the classes $C^\mu\left(X^\wedge,\left(\gamma+\frac{n}{2},\gamma-\mu+\frac{n}{2},\Theta\right)\right)$ and $C^\mu(\mathbb{R}_+,(\gamma,\gamma-\mu,\Theta))\otimes\mathbb{C}^N\otimes\mathbb{C}^N$, respectively, if (ii) holds, and if furthermore the $N\times N$-diagonal element that refers to $\mathbb{R}_+$ is exit-elliptic for $t\to\infty$.

Remark 2. For an adequate notion of ellipticity in $C^\mu(\ldots,\underline{g};N_-,N_+)$ it is clear that the diagonal elements have to be elliptic in the sense of the σ_ψ^μ-components of the principal symbols. Since the part on $\mathbb{R}_+$ is an operator matrix, this condition concerns the determinant and hence we have necessarily $N_-=N_+$.

Theorem 3. Let $\mathcal{A} \in C^{\mu}(X^{\wedge} \cup \mathbb{R}_+, \underline{g}; N,N)$ with $\underline{g} = (\gamma,\gamma-\mu,\Theta)$ be elliptic. Then, there exists a parametrix $\mathcal{P} \in C^{-\mu}(X^{\wedge} \cup \mathbb{R}_+, \underline{g}^{-1}; N,N)$, $\underline{g}^{-1} = (\gamma-\mu,\gamma,\Theta)$, in the sense

$$\mathcal{P}\mathcal{A} - 1 \in C_G(X^{\wedge} \cup \mathbb{R}_+, \underline{g}_l) \, , \quad \mathcal{A}\mathcal{P} - 1 \in C_G(X^{\wedge} \cup \mathbb{R}_+, \underline{g}_r)$$

with $\underline{g}_l = (\gamma,\gamma,\Theta)$, $\underline{g}_r = (\gamma-\mu,\gamma-\mu,\Theta)$. An analogous result holds over $\mathbb{B} \cup \mathbb{R}_+$.

Theorem 4. Let $\mathcal{A} \in C^{\mu}(X^{\wedge} \cup \mathbb{R}_+, \underline{g}; N,N)$ with $\underline{g} = (\gamma,\gamma-\mu,\Theta)$ be elliptic. Then

$$\mathcal{A}u \in \mathcal{K}^{s-\mu,\gamma-\mu+\frac{n}{2}}(X^{\wedge}) \oplus \mathcal{K}^{s-\mu,\gamma-\mu}(\mathbb{R}_+,\mathbb{C}^N)$$

and

$$u \in \mathcal{K}^{-\infty,\gamma+\frac{n}{2}}(X^{\wedge}) \oplus \mathcal{K}^{-\infty,\gamma}(\mathbb{R}_+,\mathbb{C}^N) \tag{3.1}$$

imply

$$u \in \mathcal{K}^{s,\gamma+\frac{n}{2}}(X^{\wedge}) \oplus \mathcal{K}^{s,\gamma}(\mathbb{R}_+,\mathbb{C}^N)$$

for every $s \in \mathbb{R}$. Furthermore

$$\mathcal{A}u \in \mathcal{K}_Q^{s-\mu,\gamma-\mu+\frac{n}{2}}(X^{\wedge}) \oplus \mathcal{K}_{Q'}^{s-\mu,\gamma-\mu}(\mathbb{R}_+,\mathbb{C}^N)$$

for certain $Q \in \mathrm{As}\big(X, (\gamma-\mu+\frac{n}{2},\Theta)\big)$, $Q' \in \mathrm{As}(\gamma,\Theta)$ and the condition (3.1) imply the existence of asymptotic types $P \in \mathrm{As}\big(X, (\gamma+\frac{n}{2},\Theta)\big)$, $P' \in \mathrm{As}(\gamma,\Theta)$ such that

$$u \in \mathcal{K}_P^{s,\gamma+\frac{n}{2}}(X^{\wedge}) \oplus \mathcal{K}_{P'}^{s,\gamma}(\mathbb{R}_+,\mathbb{C}^N) \, .$$

An analogous result holds over $\mathbb{B} \cup \mathbb{R}_+$.

In the beginning we have mentioned the case $\mathbb{B} \cup I$ with a finite interval I. This can easily be included by considering the second endpoint as a separate conical singularity. In a neighbourhood of that point the ordinary elements of the cone calculus can be applied, and operators as well as spaces over $\mathbb{B} \cup I$ are then obtained by glueing together the corresponding objects from $\mathbb{B} \cup \mathbb{R}_+$ near $t = 0$ and from the endpoint of I, using a corresponding partition of unity. The details are easy and will not be formulated explicitly here.

4. Remarks

As noted in the beginning the present calculus can be generalized to more complicated spaces with components of different dimension. The most immediate generalization concerns the case when a cone base X is a finite disjoint union of closed compact C^{∞} manifolds of various dimensions and if we look at X^{Δ}. This will be subject of a forthcoming paper. It is also possible to look at compact cone bases X with C^{∞} boundary in the sense of Schrohe, Schulze [6], where from the vertex there starts an interval or another cone of different dimension. Finally these resulting generalized cone operator algebras ("transmission algebras") in the version of infinite cones should be "edgified" analogously to the edge pseudo-differential calculus of [8], [9], [11]. The geometric meaning of such configurations is obvious. The interest in the corresponding "elliptic theories" is also motivated by the problem of analyzing heat conduction on such singular bodies where the analogous parabolic theories would be necessary. The concept of parabolicity on spaces with "usual" conical or edge singularities is in preparation in a series of joint papers of T. Buchholz, B.-W. Schulze, where it is obvious how the operator algebra aspects and the constructions from the ellipticity are involved.

References

[1] H. Cordes, A global parametrix for pseudo-differential operators over $\mathbb{R}^n$, with applications, Preprint, SFB 72, Universität Bonn, 1976.

[2] Ch. Dorschfeldt and B.-W. Schulze, Pseudo-differential operators with operator-valued symbols in the Mellin-edge approach, Ann. Global Analysis Geom., **12**, 135–171, (1994).

[3] Yu. V. Egorov and B.-W. Schulze, Pseudo-differential operators, singularities, applications, Birkhäuser Verlag, Basel, (to appear).

[4] T. Hirschmann, Pseudo-differential operators on manifolds with corners Va, Reports Karl-Weierstraß-Institut für Mathematik, 04, Berlin, 1991.

[5] E. Schrohe, Spaces of weighted symbols and weighted Sobolev spaces on manifolds, In Pseudo-Differential Operators, Springer Lecture Notes in Math., **1256**, 360–377, (1987).

[6] E. Schrohe and B.-W. Schulze, Boundary value problems in Boutet de Monvel's algebra for manifolds with conical singularities I, In Advances in Partial Differential Equations: (Pseudo-Differential Calculus and Mathematical Physics), 97–209, Akademie Verlag, Berlin, 1994.

[7] B.-W. Schulze, Mellin representations of pseudo-differential operators on manifolds with corners, Ann. Global Analysis Geom., **8**, 261–297.

[8] B.-W. Schulze, Pseudo-differential operators on manifolds with singularities, North-Holland, Amsterdam , 1991

[9] B.-W. Schulze, Pseudo-differential boundary value problems, conical singularities, and asymptotics, Akademie Verlag, Berlin, 1994.

[10] B.-W. Schulze, The Mellin pseudo-differential calculus on manifolds with corners, In Proc. Int. Symp. "Analysis in Domains and on Manifolds with Singularities", Breitenbrunn 1990, Teubner-Texte zur Math., **131**, Leipzig, 271–289, 1992.

[11] B.-W. Schulze, Pseudo-differential operators, ellipticity, and asymptotics on manifolds with edges, Math. Research, Vol. **82**, "Partial Differential Equations; Models in Physics and Biology", pages 291–330. Akademie Verlag Berlin 1994.

Bert-Wolfgang Schulze
Max-Planck-Arbeitsgruppe "Partielle Differentialgleichungen und Komplexe Analysis"
Universität Potsdam, 14415 Potsdam, Germany

Operator Theory:
Advances and Applications, Vol. 78

On Approximation by Solutions of non-local Elliptic Problems

Z. G. Sheftel[1]

1. Introduction.

In this paper we study the question whether an arbitrary function on a manifold may be approximated by solutions of a non-local boundary value problem for an elliptic equation of order $2m$ with discontinuous coefficients (the precise statement of the question is given in Section 3). Such questions for usual elliptic boundary value problems were studied by many authors since 1960 (see [1], [2], [3], [10], [11] and their references). The mentioned non-local elliptic problems are studied since 1970 when they were introduced by Ya. Roitberg and the author in [4]; similar problems arise for instance in magnetohydrodynamics. The definition of such problems is given in Section 2.

2. Non-local elliptic problems.

Let $G \subset \mathbb{R}^n$ be a bounded domain with the boundary $\Gamma \subset C^\infty$. G_1 is a subdomain of G with the boundary $\gamma \subset C^\infty$, $\Gamma \cap \gamma = \emptyset$ $G_2 = G \setminus \bar{G}_1$. Let us assume that there exists a diffeomorphism $\hat{\alpha} : \Gamma \to \gamma$. Since Γ and γ are infinitely smooth we can find such $\varepsilon > 0$ that for $|t| < \varepsilon$ the mapping $\alpha : x + \nu_\Gamma t \longrightarrow \hat{\alpha}x + \nu_\gamma t$ (ν_Γ is the unit vector of the inward normal to Γ at the point x and ν_γ is the unit vector of the inward (with respect to G_1) normal to γ at the point $\hat{\alpha}x$) is a diffeomorphism of some neighbourhood $U(\Gamma)$ in $\mathbb{R}^n$ of Γ onto a neighbourhood $V(\gamma)$ in $\mathbb{R}^n$ of the surface γ. These neighbourhoods are assumed to be sufficiently small to satisfy the conditon $U(\Gamma) \cap V(\gamma) = \emptyset$. For each function $u(y)$ $(y \in V(\gamma))$ we put $(Ju)(x) = u(\alpha x)$ $(x \in U(\Gamma))$. Let in G_i $(i = 1,2)$ be given linear differential expressions

$$L_i(x, D) := \sum_{|\beta| \leq 2m_i} a_{i\beta}(x) D^\beta$$

with complex-valued coefficients; we put $\ell := m_1 + 2m_2$. On γ we prescribe 2ℓ linear differential expressions $B_{ji}(x, D)$ $(x \in \gamma;\ j = 1, \dots, \ell;\ i = 1,2)$ and on Γ we prescribe ℓ expressions $B_{j3}(x, D)$ $(x \in \Gamma;\ j = 1, \dots, \ell)$:

$$B_{ji}(x,D) := \sum_{|\beta| \leq m_{ij}} b_{ij\beta}(x) D^\beta \quad (j = 1, \dots, \ell;\ i = 1,2,3).$$

[1]The research was partially supported by the foundation of fundamental researches under grant of DKNT of Ukraine.

The coefficients of all expressions in consideration are for simplicity assumed to be infinitely smooth. We consider the non-local problem

$$L_i u_i(x) = f_i(x) \quad (x \in G_i;\ i = 1{,}2), \tag{1}$$

$$\begin{gathered} B_j u := J(B_{j1}u_1(y) + B_{j2}u_2(y))(x) + B_{j3}u_3(x) = \varphi_j\ (x) \\ (x \in \Gamma;\ y = \alpha x \in \gamma;\quad j = 1,\dots,\ell;\ \ell = m_1 + 2m_2), \end{gathered} \tag{2}$$

or more shortly

$$Lu = f, \quad Bu|_\Gamma = \varphi. \tag{3}$$

This class of problems includes in particular the transmission problems (the problems with discontinuous coefficients). In the sequel we shall assume that the problem (3) is elliptic (see [4], [6]); this means that the expressions L_i are properly elliptic in G_i $(i = 1{,}2)$ and the boundary expressions B_{ji} $(j = 1,\dots,\ell;\ i = 1{,}2{,}3)$ satisfy certain algebraic condition of Lopatinski type. In addition we assume the system of boundary expressions to be 2μ-normal (see [5], [7]), where $\mu = (m_1, m_2)$ is the multi-index of the system. The 2μ-normality of the system $\{B_{ji}\}$ means that the $(\ell \times 3)$-matrix

$$B(x, D) := (B_{ji}(x, D) :\quad j = 1,\dots,\ell;\ i = 1{,}2{,}3)$$

can be completed with new rows of the same type

$$C(x, D) := (C_{ji}(x, D) :\quad j = 1,\dots,\ell;\ i = 1{,}2{,}3)$$

to a Dirichlet matrix of order 2μ. The precise definition of the Dirichlet matrix is given in [5], [7]. Here we are given only a typical example of Dirichlet matrix — the matrix of Cauchy data. For the equation (1) it is the matrix of following form:

$$\left(\begin{array}{ccc} \begin{pmatrix} D_{\nu_\gamma}^{2m_1-1} & 0 & 0 \\ 0 & D_{\nu_\gamma}^{2m_2-1} & 0 \\ 0 & 0 & D_{\nu_\Gamma}^{2m_2-1} \end{pmatrix} \\ \cdots \quad \cdots \quad \cdots \\ \begin{pmatrix} D_{\nu_\gamma}^{2m_1-2m_2} & 0 & 0 \\ 0 & 1 & 0 \\ 0 & 0 & 1 \end{pmatrix} \\ \begin{matrix} D_{\nu_\gamma}^{2m_1-2m_2-1} & 0 & 0 \end{matrix} \\ \cdots \quad \cdots \quad \cdots \\ \begin{matrix} 1 & 0 & 0 \end{matrix} \end{array} \right) \quad (m_1 > m_2)$$

$$\begin{pmatrix}
\begin{pmatrix} D^{2m_1-1}_{\nu_\gamma} & 0 & 0 \\ 0 & D^{2m_2-1}_{\nu_\gamma} & 0 \\ 0 & 0 & D^{2m_2-1}_{\nu_\Gamma} \end{pmatrix} \\
\ldots \quad \ldots \quad \ldots \\
\begin{pmatrix} 1 & 0 & 0 \\ 0 & D^{2m_2-2m_1}_{\nu_\gamma} & 0 \\ 0 & 0 & D^{2m_2-2m_1}_{\nu_\Gamma} \end{pmatrix} \\
\ldots \quad \ldots \quad \ldots \\
\begin{pmatrix} 0 & D^{2m_2-2m_1-1}_{\nu_\gamma} & 0 \\ 0 & 0 & D^{2m_2-2m_1-1}_{\nu_\Gamma} \end{pmatrix} \\
\ldots \quad \ldots \quad \ldots \\
\begin{pmatrix} 0 & 1 & 0 \\ 0 & 0 & 1 \end{pmatrix}
\end{pmatrix} \quad (m_1 < m_2)$$

In the case $m_1 = m_2$ all the blocks of this matrix consist of 3 rows. Under prescribed assumptions the Green's formula is valid [5], [7]:

$$\begin{gathered}(Lu,v) + < Bu, C'v > = (u, L^+v) + < Cu, B'v > \\ (u = (u_1, u_2), \;\; v = (v_1, v_2); \quad u,v \in C^\infty(G_1) \times C^\infty(G_2)).\end{gathered} \tag{4}$$

Here $L^+ = (L_1^+, L_2^+)$ is the expression formally adjoint to $L = (L_1, L_2)$; $(\cdot,\cdot)$ is the inner product in $L_2(G_1) \times L_2(G_2)$; $< \cdot,\cdot >$ is the inner product in $(L_2(\Gamma))^\ell$; $C = C(x, D)$ is the matrix of boundary expressions which completes the matrix $B(x, D)$ to a Dirichlet matrix of order 2μ; B' and C' are also 2μ-normal matrices. The problem

$$L^+v = g, \quad B'v|_\Gamma = \psi \tag{5}$$

is called adjoint to the problem (3) with respect to the Green's formula (4). The ellipticity of the problem (3) implies the ellipticity of the adjoint problem (5).

3. Formulation of the problem.

Let Λ_1 be a smooth $(n-1)$-dimensional manifold without border, Λ is an open subset of Λ_1 having smooth boundary and situated inside G. For arbitrary sufficiently smooth in $G_1 \cup G_2$ solution $u = (u_1, u_2)$ of the problem (3) we put

$$\nu_r u := (u|_\Lambda, \ldots, D_\nu^{r-1} u|_\Lambda), \tag{6}$$

where $D_\nu = i\frac{\partial}{\partial\nu}$, ν is the normal to Λ. Let G_0 be a domain of arbitrarily small diameter, situated inside G, Γ_0 an arbitrarily small open subset of Γ. Let us change in any way the function $f = (f_1, f_2)$ in G_0 (or change φ_j on Γ_0). The question is studied whether it is possible to approximate by the obtained vectors $\nu_r u$ any defined on Λ vector-function $\Psi = (\Psi_1, \dots, \Psi_r)$. This problem is studied in Sections 5–7.

4. Function spaces.

For any bounded domain $\Omega \subset \mathbb{R}^n$ with the boundary $\partial\Omega \in C^\infty$ we denote $H^{s,p}(\Omega)$ $(s \geq o,\ 1 < p < \infty)$ the space of Bessel potentials, $H^{-s,p}(\Omega)$ is the dual space to $H^{s,p'}(\Omega)$ $(\frac{1}{p} + \frac{1}{p'} = 1)$ with respect to the extension $(\cdot,\cdot)$ of the inner product in $L_2(\Omega)$; $\|\cdot\|_{s,p}$ is the norm in $H^{s,p}(\Omega)$ $(s \in \mathbb{R})$: The Besov spaces are denoted $B^{s,p}(\partial\Omega)$ $(s \in \mathbb{R},\ 1 < p < \infty)$; the spaces $B^{-s,p'}(\partial\Omega)$ and $B^{s,p}(\partial\Omega)$ are mutually dual with respect to the extension $< \cdot,\cdot >$ of the inner product in $L_2(\partial\Omega)$; $\ll \cdot \gg_{s,p}$ is the norm in $B^{s,p}(\partial\Omega)$. For the domain G under consideration divided into two parts G_1 and G_2 and for the multi-index $\mu = (m_1,m_2)$ we put $H^{s,p}(G_1,G_2) := H^{s,p}(G_1) \times H^{s,p}(G_2)$; $H^{2\mu+s}(G_1,G_2) :=$ $:= H^{2m_1+s,p}(G_1) \times H^{2m_2+s,p}(G_2)$; the norm in these spaces will be denoted $\|\cdot\|_{s,p}$ and $\|\cdot\|_{2\mu+s,p}$ respectively; the inner product in $H^{0,2}(G_1,G_2) = L_2(G_1) \times L_2(G_2)$ and $L_2(G_2)$ will be denoted $(\cdot,\cdot)$. It will be clear from the context in what sense we use the notations $\|\cdot\|_{s,p}$ and $(\cdot,\cdot)$. We put also $C^\infty(G_1,G_2) := C^\infty(G_1) \times C^\infty(G_2)$.

5. The case of connected complement of $\bar{\Lambda}$.

Let as in Section 3 G_0 be a domain of arbitrarily small diameter situated inside G. We put

$$M(G_0) := \{u \in C^\infty(G_1, G_2) : \ \text{supp}\ Lu \subset G_0,\ Bu|_\Gamma = 0\},$$
$$\nu_r M(G_0) := \{\nu_r u : u \in M(G_0)\},$$

where $\nu_r u$ is defined by equality (6).

Theorem 1. Let be $G_0 \cap G_1 \neq \emptyset$, $\Lambda \subset G_1$, $G_1 \setminus \bar{\Lambda}$ connected and let for the expression L_1^+ in G_1 the property of uniqueness for Cauchy problem be valid: if $L_1^+ v = 0$ in $G' \subset G_1$ and $v = 0$ in $G'' \subset G'$ then $v = 0$ in G'. Then the set $\nu_{2m_1} M(G_0)$ is dense in $\prod_{j=1}^{2m_1} B^{s_j,p}(\Lambda)$ for any $s_j \geq 0$ $(j = 1,\dots,2m_1)$ and $1 < p < \infty$.

Proof. It's enough to show that for any $s \geq 0$, $1 < p < \infty$ the set $\nu_{2m_1} M(G_0)$ is dense in $\prod_{r=1}^{2m_1} B^{2m_1+s-r+1-\frac{1}{p},p}(\Lambda)$.
We must verify that if $t_r \in B^{-(2m_1+s-r+1-\frac{1}{p}),p'}(\Lambda)$ $(r = 1,\dots,2m_1)$ and

$$\sum_{r=1}^{2m_1} < t_r, D_\nu^{r-1} u >= 0 \quad (\forall u \in M(G_0)), \tag{7}$$

then $t_1 = \cdots = t_{2m_1} = 0$. We can rewrite the equality (7) in such form:

$$\left(\sum_{r=1}^{2m_1} D_\nu^{r-1}(t_r \times \delta_\Lambda), u\right) = 0 \quad (\forall u \in M(G_0)) \tag{8}$$

where δ_Λ is the Dirac measure concentrated in Λ. It is easy to show by estimating that

$$w := \sum_{r=1}^{2m_1} D_\nu^{r-1}(t_r \times \delta_\Lambda) \in H^{-2\mu-s,p'}(G_1, G_2).$$

Let us denote by $\mathcal{M}$ the kernel of the problem (3). To the ellipticity $\mathcal{M} \subset C^\infty(G_1, G_2)$ we obtain that $\mathcal{M} \subset M(G_0)$ and it follows from (8) that

$$(w,u) = 0 \quad (\forall u \in \mathcal{M}). \tag{9}$$

But then [5], [7], [8] there exists a solution $v = (v_1, v_2) \in H^{-s,p'}(G_1, G_2)$ of the problem with homogeneous boundary conditions

$$L^+ v = w, \quad B'v|_\Gamma = 0, \tag{10}$$

which is adjoint to the problem (3) with respect to the Green's formula (4). According to the local increasing of smoothness theorem [8] $v = (v_1, v_2) \in C^\infty(\bar{G}_1 \setminus \bar{\Lambda}, \bar{G}_2)$ and satisfies the homogeneous boundary conditions (10) in usual sense. Furthermore due to (4),(10) and (9) we have

$$\begin{gathered}(v,Lu) = (L^+v,u) = (w,u) = 0 \quad (\forall u \in M(G_0)), i.e. \\ (v,Lu) = 0 \quad (\forall u \in M(G_0)).\end{gathered} \tag{11}$$

Let us assume for simplicity that the problem (3) with $\varphi = 0$ is solvable for any f. Then it follows from (11) that $v = 0$ in G_0; in general case it is not difficult to show that there exists a solution of the problem (10) vanishing in G_0 (cf [10]). Since $\bar{G}_1 \setminus \bar{\Lambda}$ is connected it follows from the uniqueness for the Cauchy problem for L_1^+ that $v = (v_1, v_2) = 0$ in $\bar{G}_1 \setminus \bar{\Lambda}$. Thus $v_1 \in H^{-s,p'}(G_1)$ and supp $v_1 \subset \bar{\Lambda}$, therefore (see [9], lemma)

$$v_1 = \sum_{j=0}^{k} D_\nu^j(\tau_j \times \delta_\Lambda),$$

where $\tau_j \in B^{-(s-j-\frac{1}{p}),p'}(\Lambda)$ $(j = 0, \ldots, k)$, k is the greateset integer less than $s - \frac{1}{p}$. Now we have from (10)

$$L_1^+\left(\sum_{j=0}^{k} D_\nu^j(\tau_j \times \delta_\Lambda)\right) = \sum_{r=1}^{2m_1} D_\nu^{r-1}(t_r \times \delta_\Lambda). \tag{12}$$

Taking into account the ellipticity of L_1^+ and linear independence of the Dirac measure and its derivatives we obtain successively $\tau_k = \cdots = \tau_0 = 0$. Then it follows from (12) that $t_r = 0 \quad (r = 1, \ldots, 2m_1)$. q.e.d.

Similar result holds in the case when $G_0 \cap G_2 \neq \emptyset \;\; \Lambda \subset G_2$. Now we shall consider the case when G_0 and Λ are situated in different parts of G.

Theorem 2. Let be

a) $G_0 \subset G_2, \quad \Lambda \subset G_1, \quad G_1 \setminus \bar{\Lambda}$ is connected;

b) the expression L_2^+ has in G_2 the property of uniqueness for Cauchy problem;

c) the 'Cauchy problem' $L_1^+ v_1 = 0$ in $G_1 \setminus \bar{\Lambda}, \quad D_\nu^{j-1} v_1 = 0$ on $\gamma \;\; (j = 1, \ldots, 2m_1)$ has only zero solution.

Then the set $\nu_{2m_1} M(G_0)$ is dense in $\prod_{j=1}^{2m_1} B^{s_j,p}(\Lambda)$ for any $s_j \geq 0$ and $1 < p < \infty$.

Proof. As in the proof of the theorem 1 we obtain a vanishing in G_0 solution of the problem (10). Hence due to the condition b) $v = 0$ in G_2 and the problem (10) is reduced to

$$L_1^+ v_1 = w \;\; in \;\; G_1, \;\; B'_{j1} v_1|_\gamma = 0 \quad (j = 1, \ldots, \ell). \tag{13}$$

Since supp $w \subset \bar{\Lambda}$ we obtain $L_1^+ v_1 = 0$ in $G_1 \setminus \bar{\Lambda}$; now taking into consideration the properties of the 2μ-normal systems and Dirichlet matrices [7] we can rewrite the problem (13) in such from:

$$L_1^+ v_1 = 0 \text{ in } G_1 \setminus \bar{\Lambda}, \;\; D_\nu^{j-1} v_1 = 0 \text{ on } \gamma \quad (j = 1, \ldots, 2m_1).$$

Due to the assumption c) we obtain $v_1 = 0$ in $G_1 \setminus \bar{\Lambda}$, i.e. supp $v_1 \subset \bar{\Lambda}$. Now we complete the proof as in the theorem 1. q.e.d.

Similar result holds also in the case when $G_0 \subset G_1$, $\Lambda \subset G_2$, $G_2 \setminus \bar{\Lambda}$ is connected.

6. The case of disconnected complement of $\bar{\Lambda}$.

Now we consider the case when $\bar{\Lambda} = \Lambda_1$ is the boundary of an n-dimensional domain i.e. the complement of the set $\bar{\Lambda}$ is disconnected. Here we restrict ourselves to two of possible cases; in other cases similar statements hold.

Theorem 3. Let be

a) $G_0 \cap G_1 \neq \emptyset$, $\Lambda \subset G_1$, $\bar{\Lambda}$ is the boundary of a subdomain $G' \subset G_1, \bar{G}' \cap \bar{G}_0 = \emptyset$;

b) the expression L_1^+ has in G_1 the property of uniqueness for the Cauchy problem;

c) the Dirichlet problem for the equation $L_1^+ v_1 = 0$ in G' has no more than one solution.

Then $\nu_{m_1} M(G_0)$ is dense in $\prod_{r=1}^{m_1} B^{m_1 - r + 1 - \frac{1}{p}, p}(\Lambda)$.

Proof. It's enough to show that if the elements

$$t_r \in B^{-(m_1-r+1-\frac{1}{p}),p'}(\Lambda) \quad (r = 1,\dots,m_1)$$

satisfy the condition

$$\sum_{r=1}^{m_1} < t_r, D_\nu^{r-1} u >= 0 \quad (\forall u \in M(G_0)) \tag{14}$$

then $t_1 = \cdots = t_{m_1} = 0$. We can rewrite the equality (14) in such form:

$$\left(\sum_{r=1}^{m_1} D_\nu^{r-1}(t_r \times \delta_\Lambda), u \right) = 0 \quad (\forall u \in M(G_0)).$$

Now the estimate

$$|(D_\nu^{r-1}(t_r \times \delta_\Lambda), u)| \leq \ll t_r \gg_{(m_1-r+1-\frac{1}{p}),p'} \|u\|_{\mu,p}$$

implies that

$$w := \sum_{r=1}^{m_1} D_\nu^{r-1}(t_r \times \delta_\Lambda) \in H^{-\mu,p'}(G) \tag{15}$$

and $(w,u) = 0 \quad (\forall u \in M(G_0))$. Therefore the problem (10) has a solution $v = (v_1, v_2) \in H^{\mu,p'}(G_1, G_2)$; due to the local increasing of smoothness theorem $v \in C^\infty(\bar{G}_1 \setminus \bar{\Lambda}, \bar{G}_2)$. As in the proof of the theorem 1 we can find a solution v that vanishes in $\bar{G}_1 \setminus \bar{G}'$. For this solution we have $v_1 \in H^{m_1,p'}(G_1)$ and $v_1 = 0$ in $\bar{G}_1 \setminus \bar{G}'$, therefore $D_\nu^{j-1} v_1|_\Lambda = 0 \quad (j = 1,\dots,m_1)$, hence $v_1 = 0$ in G', i.e. $v_1 = 0$ in G_1. But now it follows from (10) that $w = 0$ and therefore $t_1 = \cdots = t_{m_1} = 0$. q.e.d.

A similar result holds in the case when $G_0 \cap G_2 \neq \emptyset, \quad \Lambda \subset G_2$. Now we consider another case.

Theorem 4. Let be

a) $G_0 \subset G_2, \ \Lambda \subset G_1, \ \bar{\Lambda}$ is the boundary of a subdomain $G' \subset G_1$;

b) the expression L_2^+ has in G_2 the property of uniqueness for the Cauchy problem;

c) the Dirichlet problem for the equation $L_1^+ v_1 = 0$ in G' has no more than one solution;

d) the 'Cauchy problem' $L_1^+ v_1 = 0$ in $G_1 \backslash \bar{G}', \ D_\nu^{j-1} v_1 = 0$ on γ $(j = 1,\dots,2m_1)$ has only zero solution.

Then $\nu_{m_1} M(G_0)$ is dense in $\prod_{r=1}^{m_1} B^{m_1-r+1-\frac{1}{p},p}(\Lambda)$.

Proof. By the same reasoning as in the proof of the theorem 3 we can find an element w defined by (15) and satisfying the condition $(w,u) = 0 \quad (\forall u \in M(G_0))$. The equation (10) with this w has a solution $v = (v_1,v_2) \in H^{\mu,p'}(G_1,G_2)$ and due to the local increasing of smoothness theorem $v \in C^\infty(\bar{G}_1 \setminus \bar{\Lambda},\bar{G}_2)$. In addition the solution v can be taken vanishing in G_0. Then according to the assumption b) $v_2 = 0$ in $\bar{G}_2$ and the problem (10) is reduced to

$$L_1^+ v_1 = w \ \ in \ \ G_1, \ \ B'_{j1} v_1 = 0 \ \ on \ \ \gamma \ \ (j = 1,\dots,l). \tag{16}$$

Since supp $w \subset \bar{\Lambda}$ we obtain $L_1^+ v_1 = 0$ in $G_1 \setminus \bar{G}'$; now taking into consideration the properties of the 2μ-normal systems and Dirichlet matrices [7] we can rewrite the problem (16) in such form:

$$L_1^+ v_1 = 0 \ \ in \ \ G_1 \setminus \bar{G}', \ \ D_\nu^{j-1} v_1 = 0 \ \ on \ \ \gamma \quad (j = 1,\dots,2m_1).$$

Due to the assumption d) we obtain $v_1 = 0$ in $\bar{G}_1 \setminus \bar{G}'$. Since $v_1 \in H^{m_1,p'}(G_1)$ it implies $D_\nu^{j-1} v_1|_\Lambda = 0 \quad (j = 1,\dots,m_1)$. Thus v_1 is a solution of the Dirichlet problem $L_1^+ v_1 = 0$ in G', $D_\nu^{j-1} v_1 = 0$ on $\Lambda \quad (j = 1,\dots,m_1)$ and $v_1 = 0$ in G' due to the assumption c), i.e. $v_1 = 0$ in G_1. Now it follows from (10) that $w = 0$ and therefore $t_1 = \dots = t_{m_1} = 0$. q.e.d.

7. Some other results.

Let Γ_0 be a given open subset of Γ; the diameter of Γ_0 may be arbitrarily small. We set

$$M(\Gamma_0) := \{u \in C^\infty(G_1,G_2) : \ \ Lu = 0, \ \ \text{supp} \ \ Bu|_\Gamma \subset \Gamma_0\},$$

$$\nu_r M(\Gamma_0) := \{\nu_r u : \ \ u \in M(\Gamma_0)\},$$

where $\nu_r u$ is defined by the equality (6).

Theorem 5. Let be $\Lambda \subset G_2$, $G_2 \setminus \bar{\Lambda}$ is connected and let the expression L_2^+ has in G_2 the property of uniqueness for Cauchy problem. Then the set $\nu_{2m_2} M(\Gamma_0)$ is dense in $\prod_{j=1}^{2m_2} B^{s_j,p}(\Lambda)$ for any $s_j \geq 0$ and $1 < p < \infty$.

Proof. Let us form a containing G domain G'' with smooth boundary $\Gamma' \supset \supset \Gamma \setminus \Gamma_0$. In addition we take G'' so close to G that $G'' \setminus G \subset U(\Gamma)$ (see the beginning of Section 2) and that the coefficients of the differenial expressions L_2, L_1, B_{j3}, B_{ji} $(i = 1,2)$ could be extended into $G'' \setminus G$, $\alpha(G'' \setminus G)$ and onto Γ', $\alpha\Gamma'$ respectively with preservation of smoothness and of the ellipticity of the problem. Let now G_0 be an open subset of $G'' \setminus G$. Then according to the theorem 1 (in which G_1 is replaced by G_2) for the domain G'' we obtain that

$$\overline{\nu_{2m_2} M(G_0)} = \prod_{j=1}^{2m_2} B^{s_j,p}(\Lambda). \tag{17}$$

But the functions from $M(G_0)$ satisfy on Γ zero boundary conditions, hence if $u \in M(G_0)$ then supp $Bu|_\Gamma \subset \Gamma_0$. Therefore $M(G_0) \subset M(\Gamma_0)$ and the required density follows from (17). q.e.d.

Theorem 6. Let be

a) $\Lambda \subset G_1$, $G_1 \setminus \bar{\Lambda}$ is connected;

b) the expression L_2^+ has in G_2 the property of uniqueness for the Cauchy problem;

c) the 'Cauchy problem' $L_1^+ v_1 = 0$ in $G_1 \setminus \bar{\Lambda}$, $D_\nu^{j-1} v_1 = 0 \quad (j = 1, \dots, 2m_1)$ has only zero solution.

Then $\nu_{2m_1} M(\Gamma_0)$ is dense in $\prod_{j=1}^{2m_1} B^{s_j,p}(\Lambda)$ for any $s_j \geq 0$ and $1 < p < \infty$.

Proof. We form the domain G'' as in the proof of the theorem 5 and apply the theorem 2 to a domain $G_0 \subset G'' \setminus G$. q.e.d.

The next two assertions follow from the analogy of the theorem 3 and from the theorem 4, respectively.

Theorem 7. Let be

a) $\Lambda \subset G_2$, $\bar{\Lambda}$ is the boundary of a subdomain $G' \subset G_2$;

b) the expression L_2^+ has in G_2 the property of uniqueness for the Cauchy problem;

c) the Dirichlet problem for the equation $L_2^+ v_2 = 0$ in G' has no more than one solution.

Then $\nu_{m_2} M(\Gamma_0)$ is dense in $\prod_{r=1}^{m_2} B^{m_2-r+1-\frac{1}{p},p}(\Lambda)$.

Theorem 8. Let be

a) $\Lambda \subset G_1$, $\bar{\Lambda}$ is the boundary of a subdomain $G' \subset G_1$;

b) the expression L_2^+ in G_2 has the property of uniqueness for the Cauchy problem;

c) the Dirichlet problem for the equation $L_1^+ v_1 = 0$ in the domain G' has no more than one solution;

d) the 'Cauchy problem' $L_1^+ v_1 = 0$ in $G_1 \setminus \bar{G'}$, $D_\nu^{j-1} v_1 = 0$ on γ $(j = 1, \dots, 2m_1)$ has only zero solution.

Then $\nu_{m_1} M(\Gamma_0)$ is dense in $\prod_{r=1}^{m_1} B^{m_1-r+1-\frac{1}{p},p}(\Lambda)$.

The author expresses his gratitude to Yu. Berezanski, A. Prilepko, Ya. Roitberg and A. Skubachevski for useful discussions.

References

[1] F.E. Browder, Approximation by solutions of partial differential equations, Amer. Journal of Math., **84**, 134–160, (1962).

[2] F.E. Browder, Functional analysis and partial differential equations II., Mathematische Annalen, **145**, 81–226, (1962).

[3] U. Hamann, Approximation durch Normalableitungen von Loesungen elliptischer Randwertprobleme in beliebigen Sobolev-Räumen, Math. Nachrichten, **128**, 199–214, (1986).

[4] Ya.A. Roitberg; Z.G. Sheftel, On a class of general non-local elliptic problems (Russian), Dokl. Akad Nauk SSSR, **192**(3), 511–513, (1970), Engl. transl.: Soviet Math. Dokl., **11**(3), 681–684, (1970).

[5] Ya.A. Roitberg; Z.G. Sheftel, Green's formula and homeomorphism theorems for non-local elliptic boundary value problems (Russian), Dokl. Akad Nauk SSSR, **201**(5), 1059–1062, (1971), Engl. transl.: Soviet Math. Dokl., **12**(6), 1797–1800, (1971).

[6] Ya.A. Roitberg; Z.G. Sheftel, Nonlocal problems for elliptic equations and systems (Russian), Sibirski Mat. Journal, **13**(1), 165–181 (1972).

[7] Ya.A. Roitberg; Z.G. Sheftel, Green's formula and solvability conditions for non-local elliptic boundary value problems (Russian), Ukrainski Math. Journal, **25**(4), 479–491, (1973).

[8] Ya.A. Roitberg; Z.G. Sheftel, The isomorphism theorems for non-local elliptic boundary value problems and their applications, Ukrainski Math. Journal, **25**(6), 761–771, (1973).

[9] Ya.A. Roitberg; Z.G. Sheftel, On approximation of functions on manifolds by solutions of elliptic boundary value problems, Non-linear boundary value problems, **1**, Kiev, 86–90, (1989).

[10] Ya.A. Roitberg; Z.G. Sheftel, On the density of solutions of elliptic problems with localized right sides in function spaces on manifolds (Russian), Dokl. Akad Nauk SSSR, **305**, 1317–1320, (1989), Engl. transl.: Soviet Math. Dokl., **39**(2),411–414, (1989)

[11] Ya.A. Roitberg; Z.G. Sheftel, On density of the solutions of boundary value problems for ellipticity in the sence of Petrovski systems in function spaces on manifolds (Russian), Dokl. Akad Nauk SSSR, **9(Ser. A)**, 17–20, (1990).

Zinovi G. Sheftel
Department of Mathematics
Educational Institute
Sverdlova str. 53
250038 Chernigov, Ukraine

Operator Theory:
Advances and Applications, Vol. 78

A Stability Set in the Cauchy Problem for Elliptic Systems

A. Shlapunov, N. Tarkhanov[1]

Let P be an elliptic differential operator of order p with real analytic coefficients on an open set $X \subset \mathbb{R}^n$.

Given a domain $\mathcal{D} \subset\subset X$ with smooth boundary $\partial\mathcal{D}$, fix an open connected part S of $\partial\mathcal{D}$.

Suppose $B_j, j = 0, 1, \ldots, p-1$, is a Dirichlet system of order $p-1$ on S. In the sequel, b_j stands for the order of the boundary operator B_j, so that $0 \leq b_j \leq p-1$ and $b_i \neq b_j$ for $i \neq j$. We will consider only those Dirichlet systems on S which can be extended to a Dirichlet system on the whole boundary.

For any function f in Sobolev space $H^s(\mathcal{D})$, the traces of $B_j f$ on S are well-defined as elements of $H^{s-bj-1/2}(S)$, provided that s is an integer $\geq p$. If, in addition, $Pf = 0$ in $\mathcal{D}$, then this remains valid for *any* integer s.

Consider the following Cauchy problem.

Problem 1. Given functions $f_j \in H^{s-bj-1/2}(S)$, $0 \leq j \leq p-1$, on S, it is required to find a function $f \in H^s(\mathcal{D})$ such that

$$\begin{cases} Pf = 0 & \text{in} \quad \mathcal{D}, \\ B_j f = f_j & \text{on} \quad S \quad (0 \leq j \leq p-1). \end{cases}$$

This problem was studied in our paper [1]. It is ill-posed, and we are aimed here in finding a *stability set* in the problem. That means a set Σ of solutions $f \in H^s(\mathcal{D})$ to $Pf = 0$ with the following property. Given any sequence $\{f^{(\mu)}\} \subset \Sigma$, if $B_j f^{(\mu)} \to 0$ on S $(0 \leq j \leq p-1)$, then $f^{(\mu)} \to 0$ in $\mathcal{D}$.

Let us fix a domain $\mathcal{O} \subset\subset X$ such that $\mathcal{D} \subset \mathcal{O}$ and $\partial\mathcal{D} \cap \mathcal{O} = S$, and a domain $\sigma \subset\subset \mathcal{O}\backslash\overline{\mathcal{D}}$ with smooth boundary such that $\mathcal{O}\backslash\sigma$ has no compact connected components.

Denote by $\text{Sol}(\mathcal{O})$ the subspace of $H^s(\mathcal{O})$ formed by weak solutions of the equation $Pf = 0$ in $\mathcal{O}$, and analogously for $\text{Sol}(\sigma)$. These spaces, if equipped with the induced Hermitian structures, are separable Hilbert spaces. According to [1], there is an orthonormal basis $\{e_\nu\}$ in $\text{Sol}(\mathcal{O})$ such that the restrictions of e_ν to σ ensure an orthogonal basis of $\text{Sol}(\sigma)$.

Consider the Green integral

$$G(\oplus f_j)(x) = \int_S \sum_{j=0}^{p-1} C_j \Phi(x,.) f_j ds \qquad (x \notin S),$$

[1]This research was supported by the Alexander-von-Humboldt Foundation.

$\{C_j\}_{j=0}^{p-1}$ being the Dirichlet system on S adjoint to $\{B_j\}_{j=0}^{p-1}$ with respect to Green's formula, and Φ being a fundamental solution of P on X.

This integral, when restricted to σ, is in $\mathrm{Sol}(\sigma)$. Denote by $k_\nu(G(\oplus f_j))$ its Fourier coefficients with respect to the orthogonal system $\{e_\nu|_\sigma\}$. Exactly,

$$k_\nu(G(\oplus f_j)) = \int_S \sum_{j=0}^{p-1} C_j k_\nu(\Phi(.,y)) f_j ds \qquad (\nu \in \mathbf{N}),$$

where $k_\nu(\Phi(.,y))$ are Fourier coefficients of the fundamental solution $\Phi(x,y)|_{x\in\sigma}$, y being on S.

The main result of [1] is that the Cauchy problem is solvable if and only if

$$\sum_{\nu=1}^{\infty} |k_\nu(G(\oplus f_j))|^2 < \infty.$$

We complete this result in the following way.

Theorem 1. Given a sequence $\{f^{(\mu)}\} \subset \mathrm{Sol}(\mathcal{D})$, if

$$\sum_{\nu=1}^{\infty} |k_\nu(G(\oplus B_j f^{(\mu)}))|^2 \leq 1$$

and $B_j f^{(\mu)} \to 0$ in the norm of $H^{s-b_j-1/2}(S)$ for all $0 \leq j \leq p-1$, then $f^{(\mu)} \to 0$ in the topology of $H^s_{\mathrm{loc}}(\mathcal{D} \cup S)$.

Of course, the adequate conclusion here should be that $f^{(\mu)} \to 0$ in the norm of $H^s(\mathcal{D})$, but we are not able to prove this.

Proof. Fix a sequence $\{f^{(\mu)}\}$ satisfying the condition above.

It was proved in [1] that

$$f^{(\mu)} = -G(\oplus B_j f^{(\mu)})(x) + F^{(\mu)}(x) \qquad (x \in \mathcal{D}), \tag{1}$$

where $F^{(\mu)} \in \mathrm{Sol}(\mathcal{O})$ is given by the Fourier series

$$F^{(\mu)} = \sum_{\nu=1}^{\infty} k_\nu(G(\oplus B_j f^{(\mu)}))\, e_\nu.$$

We see at once that the restrictions of $F^{(\mu)}$ and $G(\oplus B_j f^{(\mu)})$ to σ coincide.

Since $B_j f^{(\mu)} \to 0$ in the norm of $H^{s-b_j-1/2}(S)$ for all $0 \leq j \leq p-1$, the first term in the right hand side of (1) tends to zero in the topology of $H^s_{loc}(\mathcal{D} \cup S)$.

We claim that $F^{(\mu)} \to 0$ in the topology of $C^\infty_{\mathrm{loc}}(\mathcal{O})$. To prove this it suffices to show that each subsequence of $\{f^{(\mu)}\}$ has a subsequence which converges to zero in $C^\infty_{\mathrm{loc}}(\mathcal{O})$.

Indeed, assume that this is true whereas $F^{(\mu)}$ does not converge to zero. As

$$\|F^{(\mu)}\|^2_{H^s(\mathcal{O})} = \sum_{\nu=1}^{\infty} |k_\nu(G(\oplus B_j f^{(\mu)}))|^2 \le 1$$

and the embedding

$$\mathrm{Sol}(\mathcal{O}) \hookrightarrow \{F \in C^\infty_{\mathrm{loc}}(\mathcal{O}) : \ PF = 0 \ \text{ in } \ \mathcal{O}\}$$

is compact, each subsequence of $\{F^{(\mu)}\}$ contains a subsequence convergent in the topology of $C^\infty_{\mathrm{loc}}(\mathcal{O})$. Therefore, $\{F^{(\mu)}\}$ has a subsequence which converges to a nonzero element of $C^\infty_{\mathrm{loc}}(\mathcal{O})$. This contradicts our assumption.

We now turn to proving the relation $F^{(\mu)} \to 0$ in $C^\infty_{\mathrm{loc}}(\mathcal{O})$. To this end, pick a subsequence of $\{F^{(\mu)}\}$. By abuse of notation, we continue to write $\{F^{(\mu)}\}$ for this subsequence.

From the discussion above it follows that $\{F^{(\mu)}\}$ has a subsequence $\{F^{(\mu_i)}\}$ which converges in the topology of $C^\infty_{\mathrm{loc}}(\mathcal{O})$ to a function $F \in C^\infty_{\mathrm{loc}}(\mathcal{O})$ satisfying $PF = 0$ in $\mathcal{O}$. In particular, the sequence $\{G(\oplus B_j f^{(\mu_i)})|_\sigma\}$ converges to $F|_\sigma$.

On the other hand, $\{G(\oplus B_j f^{(\mu)})|_\sigma\}$ converges to zero on σ because the Cauchy data of $f^{(\mu)}$ on S tends to zero. Thus, $F = 0$ on σ, and so $F \equiv 0$ in the domain $\mathcal{O}$.

This completes the proof. □

References

[1] A.A. Shlapunov and N.N. Tarkhanov, Bases with double orthogonality in the Cauchy problem for systems with injective symbol, Dokl. Ross. AN (1) **326** (1992), 45–49 (to appear in Proc. London Math. Soc. **69** (1994)).

A. Shlapunov
Scuola Normale Superiore, Piazza dei Cavalieri 7
56126 Pisa, Italy

N. Tarkhanov
Max-Planck-Arbeitsgruppe "Partielle Differentialgleichungen und Komplexe Analysis", FB Mathematik, Universität Potsdam, Postfach 60 15 53
14415 Potsdam, Germany

Operator Theory:
Advances and Applications, Vol. 78

Discrete spectrum asymptotics for the Schrödinger operator in a moderate magnetic field

A. V. Sobolev [1]

1. Results and discussion.

Let $\mathbf{a} \in \left(L^\infty(\mathbb{R}^d)\right)^d$, $d \geq 2$ be a real vector-valued function. The Schrödinger operator $H_0 = H_0(h,\mu)$ with the magnetic potential $\mu\mathbf{a}$ is defined as an operator associated with the quadratic form

$$\|(-ih\nabla - \mu\mathbf{a})u\|^2, \; u \in C_0^\infty(\mathbb{R}^d),$$

$h \in (0, h_0]$ and $\mu \geq 0$ being the Planck constant and the intensity of the magnetic field respectively. We study spectral properties of the perturbed operator

$$H_\mathfrak{a} = H_0 + V, \quad \mathfrak{a} = \{\mathbf{a}, V\},$$

with a real-valued function V (electric potential). Precisely, we analyse the asymptotics as[2] $h \to 0$, $\mu \geq 0$, $\mu h \leq C$ of traces of the form

$$\mathcal{M}_s(h,\mu) = \mathcal{M}_s(h,\mu;\psi,\mathfrak{a}) = \operatorname{tr}\{\psi g_s(H_\mathfrak{a})\}. \tag{1.1}$$

Here $\psi \in C_0^\infty(\mathbb{R}^d)$ and $g_s, s \geq 0$, is defined as follows:

$$\left.\begin{array}{rcl} g_s(\lambda) & = & |\lambda|^s, \lambda < 0, \\ g_s(\lambda) & = & 0, \lambda \geq 0. \end{array}\right\}$$

The condition $\mu h \leq C$ ensures that the leading terms in the asymptotic formulae which we obtain, do not depend on the magnetic field. In this sense the magnetic field under consideration is moderate, though the inequality $\mu h \leq C$ allows μ to grow as $h \to 0$. Further discussion of this point is postponed until the end of this section.

The quantity (1.1) can be viewed as a "local version" of the sum

$$\mathcal{M}_s(h,\mu;1,\mathfrak{a}) = \sum_{k\geq 1} |\lambda_k|^s,$$

where $\lambda_k = \lambda_k(h,\mu;\mathfrak{a})$, $k \geq 1$, are negative eigenvalues of $H_\mathfrak{a}$ enumerated in the non-decreasing order. In particular, $\mathcal{M}_0$ is a local counterpart of the number of all

[1] EPSRC fellow

[2] Here and in what follows we denote by C and c (with or without indices) various positive constants whose precise value is of no importance.

negative eigenvalues. Sometimes one calls $\mathcal{M}_s(h,\mu;1)$ ($\mathcal{M}_s(h,\mu;\psi)$) Riesz mean (local Riesz mean) of order s for the operator $H_{\mathfrak{a}}$. Note that due to the truncation ψ the trace (1.1) can be finite even if negative spectrum of $H_{\mathfrak{a}}$ is not discrete. In case $\mu = 0$ we write $H_V(h)$ (or H_V) and $\mathcal{N}_s(h;\psi,V)$ instead of $H_{\mathbf{a},V}(h,\mu)$ and $\mathcal{M}_s(h,\mu;\psi,\mathfrak{a})$ respectively.

The asymptotics of $\mathcal{M}_s(h,\mu)$ has been analysed in [10] in the case $\mathbf{a}, V \in C_0^\infty(\mathbb{R}^d)$. It was shown there that for any $s \in [0,1]$, $h \to 0, 0 \le \mu \le Ch^{-1}$, the trace (1.1) obeys

$$\mathcal{M}_s(h,\mu;\psi,V) = \mathfrak{W}_s(h;\psi,V) + \langle\mu\rangle^{s+1}O(h^{-(d-s-1)}),\ \langle\mu\rangle = (1+\mu^2)^{\frac{1}{2}}, \tag{1.2}$$

with the standart Weylian leading term

$$\mathfrak{W}_s(h) = \mathfrak{W}_s(h;\psi,V) = \Xi_s h^{-d}\int \psi(x)\big(V_-(x)\big)^{s+\frac{d}{2}}dx, \tag{1.3}$$

$$\Xi_s = \frac{|\mathbb{S}^{d-1}|}{(2\pi)^d}\int_0^1 t^{d-1}(1-t^2)^s dt,$$

where $|\mathbb{S}^{d-1}|$ stands for the surface area of the $(d-1)$-dimensional unit sphere. Corresponding result for the case $\mu = 0$ was obtained in [5], [6].

In the present paper the smoothness assumption is removed. Instead, we assume that

$$\left.\begin{aligned} V, \mathbf{a} \in C^\infty(\mathbb{R}^d\setminus\{0\}),\ \mathbf{a}\in C(\mathbb{R}^d);\\ |\partial^m V(x)| \le C_m|x|^{-\beta-|m|},\ \ 0\le\beta<2,\ |m|\ge 0;\\ |\partial^m \mathbf{a}(x)| \le C_m|x|^{1-|m|},\ \ |m|\ge 1.\end{aligned}\right\} \tag{1.4}$$

Moreover, V is asymptotically homogeneous at $x = 0$:

$$V(x) = |x|^{-\beta}\big(\Phi(\hat{x}) + U(x)\big), \tag{1.5}$$

with a function $\Phi \in C^\infty(\mathbb{S}^{d-1})$ and some $U \in L^\infty(\mathbb{R}^d)$ such that $U(x) \to 0$, $|x| \to 0$. As a rule we use the notation

$$W(x) = \Phi(\hat{x})|x|^{-\beta}. \tag{1.6}$$

Notice that (1.4) contains no estimates on the function $\mathbf{a}$ itself, but only on its derivatives. This fact is quite natural, since the constant component of $\mathbf{a}$ can be chosen arbitrarily or eventually eliminated by a simple gauge transformation.

Since $\beta < 2$ in (1.4), the operator V is H_0-form bounded with a relative bound $\epsilon < 1$. This is a consequence of the $H_{0,0}$-form boundedness of V and the diamagnetic inequality (see [2]). Hence $H_{\mathfrak{a}}$ is well defined as a form sum.

To study the local trace (1.1) it will be sufficient to assume that the operator under consideration is of the form $H_{\mathfrak{a}}$ only in a neighbourhood of $\operatorname{supp}\psi$. Its behaviour outside is irrelevant. To distinguish such an operator from the "true"

Schrödinger operator $H_{\mathfrak{a}}$ we shall use the notation $A_{\mathfrak{a}} = A_{\mathfrak{a}}(h,\mu)$ (or simply A). Let us specify the assumptions on $A_{\mathfrak{a}}$. Below $B(r)$ denotes the closed ball of radius $r > 0$ centered at the origin. We suppose that $\operatorname{supp}\psi \subset B(E/2)$ with some $E > 0$.

Assumption 1.1. The operator $A_{\mathfrak{a}}$ is selfadjoint in $L^2(\mathbb{R}^d)$ and satisfies the following conditions:

(1) It is semibounded from below;

(2) There exist real-valued functions $V, \mathbf{a}$ which obey (1.4), such that for any $\chi \in C_0^\infty(B(4E))$ and $u \in D(H_{\mathfrak{a}})$, $\mathfrak{a} = (\mathbf{a}, V)$, the function χu belongs to $D(A_{\mathfrak{a}})$ and

$$A_{\mathfrak{a}}(\chi u) = H_{\mathfrak{a}}(\chi u).$$

For $\operatorname{tr}\{\psi g_s(A_{\mathfrak{a}})\}$ we keep the notation $\mathcal{M}_s(h,\mu;\psi,\mathfrak{a})$. This will not cause any confusion in what follows.

Under conditions (1.4), (1.5) the formula (1.2) is not necessarily true. The answer depends on the relation between β, s and d. A natural parameter that determines the form of the asymptotics in this case is

$$\omega = \omega(\beta, s) = \frac{2\beta s}{2-\beta}. \tag{1.7}$$

The values d and $d-s-1$ (which are the orders of h in the leading and remainder term in (1.2) respectively) serve as "threshlods" — when ω crosses either of them, the asymptotics changes its form. Note in particular, that the condition $\omega < d$ is necessary and sufficient for $\mathfrak{W}_s$ to be finite. We discuss separately all possible cases in Theorems 1.2–1.6 below. It will be always assumed that the operator $A_{\mathfrak{a}}$ satisfies Assumption 1.1, that $0 < h \le C$, $0 \le \mu \le Ch^{-1}$, and, unless otherwise stated, $0 \le s \le 1$. Emphasize that no quantitative information on $A_{\mathfrak{a}}$ outside $B(4E)$ is involved. The asymptotic formulae will be uniform in the functions $V, \mathbf{a}, \psi$ in the sense that the remainders will depend only on the constants C_m in the bounds (1.4) and $|\partial^m \psi(x)| \le C_m$.

1. $\omega > d$. The leading term of $\mathcal{M}_s$ is completely determined by the asymptotic potential W defined in (1.6):

Theorem 1.2 Let $\omega > d$. Then

$$\mathcal{M}_s(h,\mu;\psi,\mathfrak{a}) = h^{-\omega}\Big(\psi(0)\mathcal{N}_s(1;1,W) + o(1)\Big). \tag{1.8}$$

Finiteness of the trace $\mathcal{N}_s(1;1,W)$ in (1.8) for $d \ge 3$ can be easily obtained from the classical Rosenblum-Lieb-Cwickel estimate:

$$\mathcal{N}_0(1;1;W+\kappa) \le C \int (W(x)+\kappa)_-^{\frac{d}{2}} dx \le C\kappa^{\frac{d}{2}-\frac{d}{\beta}},\ C = C(\beta,\Phi). \tag{1.9}$$

Indeed, using the simple relation

$$\mathcal{N}_s(1;1,W) = -\int_0^\infty \kappa^s d\mathcal{N}_0(1;1,W+\kappa) = s\int_0^\infty \kappa^{s-1}\mathcal{N}_0(1;1,W+\kappa)d\kappa,$$

one can conclude from (1.9) that

$$\mathcal{N}_s(1;1,W) < \infty, \ \omega(\beta,s) > d.$$

For $d = 2$ one uses another Cwickel-type bound (see [3]).

2. $\omega = d$. The asymptotics is still determined by W. However, in contrast to (1.8) the leading term can be calculated effectively. Denote

$$\mathfrak{B}_s(\Phi) = \frac{2\Xi_s}{2-\beta}\int_{\mathbb{S}^{d-1}} \Phi_-(\vartheta)^{\frac{d}{2}+s} d\vartheta.$$

Theorem 1.3. Let $\omega = d$. Then

$$\mathcal{M}_s(h,\mu;\psi,\mathfrak{a}) = h^{-d}|\ln h|\psi(0)\mathfrak{B}_s(\Phi) + (|\ln h|+1)o(h^{-d}). \tag{1.10}$$

We emphasize that the remainder estimates in (1.8), (1.10) are bounded uniformly in $\mu \le Ch^{-1}$.

3. $d-s-1 < \omega < d$. The leading order is given by the classical Weyl coefficient (1.3). The singularity of the potential gives rise to a second term, that occupies an intermediate position between the main term and the remainder in (1.2):

Theorem 1.4. Let $d-s-1 < \omega < d$. Then

$$\begin{aligned}\mathcal{M}_s(h,\mu;\psi,\mathfrak{a}) \ &= \ \mathfrak{W}_s(h;\psi,V) + h^{-\omega}\psi(0)\Theta_s \\ &\quad + o(h^{-\omega}) + O(\langle\mu\rangle^{s+1}h^{s+1-d}),\end{aligned} \tag{1.11}$$

where

$$\Theta_s = \Theta_s(\Phi,\beta) = \lim_{\kappa\to 0}\big(\mathcal{N}_s(1;1,W+\kappa) - \mathfrak{W}_s(1;1,W+\kappa)\big). \tag{1.12}$$

The definition (1.12) can be viewed as a regularization of the formula

$$\Theta_s = \mathcal{N}_s(1;1,W) - \mathfrak{W}_s(1;1,W).$$

Both terms in the r.h.s. are infinite because of the inequality $\omega < d$. The finiteness of the limit (1.12) is guaranteed by the condition $d-s-1 < \omega$. One can check that the following regularized version gives the same value of Θ_s:

$$\left.\begin{aligned}\Theta_s = \Theta_s(\Phi,\beta) = \ &\lim_{\rho\to\infty}\big(\mathcal{N}_s(1;\phi_\rho,W) - \mathfrak{W}_s(1;\phi_\rho,W)\big), \\ &\phi\in C_0^\infty(\mathbb{R}^d),\ \phi(0)=1, \quad \phi_\rho(x) = \phi(x\rho^{-1}).\end{aligned}\right\} \tag{1.13}$$

Presumably, in general, Θ_s does not admit any explicit representation in terms of W. We can only say that it is homogeneous in Φ:

$$\Theta_s(q\Phi,\beta) = q^{\frac{2s}{2-\beta}}\Theta_s(\Phi,\beta), \ \forall q \geq 0. \tag{1.14}$$

This follows from the homogeneity of the potential W and the function g_s. However, in the particular case of a purely Coulomb potential, the coefficient Θ_1 can be calculated explicitly. Let $\beta = 1, \Phi(\theta) = -q, q \geq 0$. Then

$$\Theta_1(-q,1) = -\frac{q^2}{8}. \tag{1.15}$$

Note that this formula is consistent with (1.14). The expression for Θ_1 has been known for a long time in connection with the so-called Scott correction to the ground state energy of a large atom. We refer to [5] for details and further references. A proof of (1.15) based on the relation (1.12) and a precise formula for the eigenvalues of H_W, can be found in [5]. It is worth mentioning that the choice $\beta = 1, s = 1$ in combination with the condition $d - s - 1 < \omega < d$ implies that $d = 3$. This means that the Scott correction term is meaningful only in the three-dimensional case.

4. $\omega = d - s - 1$. As in the previous border-line case ($\omega = d$), the asymptotics acquires a logarithmic factor — this time in the remainder:

Theorem 1.5. Let $\omega = d - s - 1$. Then

$$\mathcal{M}_s(h,\mu;\psi,\mathfrak{a}) = \mathfrak{W}_s(h;\psi,V) + O(|\ln h| h^{s+1-d}) + O(\langle\mu\rangle^{s+1}h^{s+1-d}). \tag{1.16}$$

5. $\omega < d - s - 1$. The asymptotics looks like (1.2) — the contribution from the singularity gets "absorbed" by the remainder:

Theorem 1.6. Let $\omega < d - s - 1$. Then

$$\mathcal{M}_s(h,\mu;\psi,\mathfrak{a}) = \mathfrak{W}_s(h;\psi,V) + O(\langle\mu\rangle^{s+1}h^{s+1-d}). \tag{1.17}$$

In particular, the local counting function $\mathcal{M}_0$ satisfies (1.17).

We point out that in all five cases above the leading term of $\mathcal{M}_s$ does not depend on μ. In cases 3–5 the asymptotics gets more precise if $\mu h \to 0$. On the contrary, if $\mu h = const$, the formulae (1.11), (1.16), (1.17) loose their asymptotic character since the remainders have the same order as $\mathfrak{W}_s$ in this situation. This observation is consistent with the well known fact that the Weyl term no more describes the behaviour of $\mathcal{M}_s(h,\mu)$ if $\mu h \geq c$. We refer to [9](see also [11]), where the asymptotics of $\mathcal{M}_1(h,\mu;1,\mathfrak{a})$ was studied for $d = 3$ and a homogeneous magnetic field for any $\mu \geq 0$. It was shown that for $\mu h \geq c$ the leading term is to be replaced by another coefficient that takes into account the magnetic field.

We stress again that no information on $A_{\mathfrak{a}}$ outside $B(4E)$ is involved in Theorems 1.2–1.6. In particular, the lower bound of $A_{\mathfrak{a}}$ is irrelevant. This "locality" property can be used so as to extend the obtained results to the case when the potential V has several singularities. To that end it suffices to use an appropriate partition of unity and a translation transformation. The partition of unity argument permits also to replace $B(4E)$ in Assumption 1.1 by any open bounded set $\mathcal{D}$.

2. Multiscale analysis

The centerpiece of our method is a very useful though elementary technical tool, suggested by V. Ivrii in [6]–[8] (see also the joint work with I.M. Sigal [5]), which allows one to deduce from the "local" result (Proposition 2.2 below) an asymptotics of $\mathcal{M}_s$ in a much more general setting. Here we give a schematic description of the approach, referring to [10] for the detailed presentation.

In this section we shall be dealing with smooth functions V and $\mathbf{a}$. We use the notation $B(z,\ell) = \{x \in \mathbb{R}^d : |x-z| \le \ell\}$, $B(\ell) = B(0,\ell)$ with $z \in \mathbb{R}^d$, $\ell > 0$. Corresponding open balls are denoted by $\mathring{B}(z,\ell)$, $\mathring{B}(\ell)$ respectively.

Local asymptotics. Our starting point is the formula (1.2) obtained in [10]. To state the needed result in precise terms we need the following

Assumption 2.1. The operator $A_{\mathfrak{a}}$ is self-adjoint and satisfies the following conditions:

(1) It is semibounded from below.

(2) For some open set $\mathcal{D} \subset \mathbb{R}^d$ and some functions $V, \mathbf{a} \in C_0^\infty(\mathbb{R}^d)$, one has $C_0^\infty(\mathcal{D}) \subset D(A_{\mathfrak{a}})$ and

$$A_{\mathfrak{a}} u = H_{\mathfrak{a}} u, \ \forall u \in C_0^\infty(\mathcal{D}).$$

Next Proposition describes the asymptotics of $\mathcal{M}_s(h,\mu;\psi,\mathfrak{a})$ for $\mathcal{D} = \mathring{B}(4E)$ and $\psi \in C_0^\infty(B(E/2))$, $E > 0$.

Proposition 2.2.[10] Let operator $A_{\mathfrak{a}}$ obey Assumption 2.1 with $\mathcal{D} = \mathring{B}(4E)$. Then for any $\psi \in C_0^\infty(B(E/2))$ and $\mu \le Ch^{-1}$

$$\big|\mathcal{M}_s(h,\mu;\psi,\mathfrak{a}) - \mathfrak{W}_s(h;\psi,V)\big| \le Cr(h,\mu). \tag{2.1}$$

Here $r(h,\mu) = (1+\mu^{s+1})h^{s+1-d}$. The constant C in (2.1) depends only on E and the constants in the estimates

$$\left.\begin{array}{r} |\partial_x^m \mathbf{a}(x)| \le C_m, \ |m| \ge 1; \\ |\partial_x^m V(x)| \le C_m, \ |\partial_x^m \psi(x)| \le C_m, \ |m| \ge 0, \end{array}\right\} x \in B(4E). \tag{2.2}$$

Let us point out that information on $A_{\mathfrak{a}}$ outside $B(4E)$ is irrelevant.

Multiscale analysis. On the basis of Proposition 2.2 we shall find the asymptotics of $\mathcal{M}_s$ in the following general situation. Let $\mathcal{D} \in \mathbb{R}^d$ be an arbitrary open set and let $f \in C(\bar{\mathcal{D}}), \ell \in C^1(\bar{\mathcal{D}})$ be two functions such that

$$f(x) > 0, \ \ell(x) > 0, \ x \in \bar{\mathcal{D}};$$

$$|\partial_x \ell(x)| \le \varrho < 1, \ x \in \mathcal{D}; \tag{2.3}$$

$$cf(y) \le f(x) \le Cf(y), \ \forall x \in \mathcal{D} \cap B(y, \ell(y)), \ y \in \mathcal{D}; \tag{2.4}$$

Our objective is the asymptotics of $\mathcal{M}_s(h, \mu; \psi, \mathfrak{a})$ for an operator $A_{\mathfrak{a}}$ satisfying Assumption 2.1 with the domain $\mathcal{D}$ and some functions $V, \mathbf{a} \in C_0^\infty(\mathbb{R}^d)$, $\psi \in C_0^\infty(\mathcal{D})$ which obey the bounds

$$\left.\begin{array}{l} |\partial_x^m \mathbf{a}(x)| \le C_m \ell(x)^{1-|m|}, \ \ |m| \ge 1; \\ |\partial_x^m V(x)| \le C_m f(x)^2 \ell(x)^{-|m|}, \ |\partial_x^m \psi(x)| \le C_m \ell(x)^{-|m|}, |m| \ge 0, \end{array}\right\} x \in \mathcal{D}. \tag{2.5}$$

One can think of $f(x)^2$ as a measure of the size of $V(x)$, while $\ell(x)$ characterizes the behaviour of $V(x), \mathbf{a}(x)$ and $\psi(x)$ under differentiation. Emphasize that the functions $f(x), \ell(x)$ are allowed to depend on h, μ. We require only that

$$f(x)\ell(x) \ge ch; \ \ f(x)^2 \ge c\mu h, \ \ x \in \mathcal{D}. \tag{2.6}$$

We also need the following condition on $\operatorname{supp} \psi$:

$$\bigcup B(x, 8\ell(x)) \subset \mathcal{D}, \tag{2.7}$$

where the union is taken over those $x \in \mathcal{D}$, for which $B(x, \ell(x)) \cap \operatorname{supp} \psi \ne \emptyset$. Let $r(h, \mu)$ be the function from (2.1) and let

$$R(h, \mu) = \int_{\mathcal{D}} f(x)^{2s} r\left(\frac{h}{\ell(x) f(x)}, \frac{\mu \ell(x)}{f(x)} \right) \ell(x)^{-d} dx.$$

Our aim is to prove

Theorem 2.3. Let the operator A obey Assumption 2.1 for an open set $\mathcal{D}$ with the functions $V, \mathbf{a}$ and ψ satisfying conditions (2.3)–(2.7) with $\varrho < 1/8$. Then

$$\left| \mathcal{M}_s(h, \mu; \psi, \mathfrak{a}) - \mathfrak{W}_s(h; \psi, V) \right| \le CR(h, \mu), \tag{2.8}$$

where the constant C is uniform in the functions $\mathbf{a}, V, f, \ell, \psi$ satisfying (2.3)–(2.7).

Using the explicit form of the function $r(h, \mu)$, one can rewrite $R(h, \mu)$:

$$\left.\begin{array}{rcl} R(h, \mu) & = & h^{s+1-d} I_1(h, \mu) + \mu^{s+1} h^{s+1-d} I_2(h, \mu), \\ I_1(h, \mu) & = & \int_{\mathcal{D}} f(x)^{d+s-1} \ell(x)^{-s-1} dx, \ \ I_2(h, \mu) = \int_{\mathcal{D}} f(x)^{d-2} dx. \end{array}\right\} \tag{2.9}$$

The integral I_1 makes sense since $\ell(x)$ is positive in $\bar{\mathcal{D}}$.

Proof of Theorem 2.3. We first construct

(1) a covering of $\mathcal{D}$ by the balls $\mathring{B}(z_j, \ell_j)$ centered at the points $z_j \in \mathcal{D}, j = 1, 2, \dots$ of radii $\ell_j = \ell(z_j)$, such that for some fixed integer $N = N(\varrho)$ the intersection of more than N balls is empty;

(2) a partition of unity $\chi_j \in C_0^\infty(B(z_j, \ell_j))$ subordinate to this covering such that $|\partial_x^m \chi_j(x)| \le C_m \ell_j^{-|m|}$ for all j.

The existence of such a covering and partition of unity follows from [4, Sect. 1.4]. Denote $f_j = f(z_j)$. In view of (2.5) we have

$$|\partial_x^m V(x)| \le C_m f_j^2 \ell_j^{-|m|}, \quad \forall x \in B(z_j, \ell_j).$$

Now, to find the asymptotics of $\mathcal{M}_s(h, \mu; \psi)$ it suffices to calculate the asymptotics of each $\mathcal{M}_s(h, \mu; \psi\chi_j)$ and add up the answers. If one had used Proposition 2.2 for $\mathcal{M}_s(h, \mu; \psi\chi_j)$ directly, one would not have been able to control the dependence of the remainder $r(h,\mu)$ on j. The remedy is to reduce the problem in each $B(z_j, \ell_j)$ to that in the ball $B(1)$ with an auxiliary potentials satisfying (2.2) *uniformly in* j. Precisely, let

$$V_j(x) = f_j^{-2} V(z_j + \ell_j x), \quad \mathbf{a}_j(x) = \ell_j^{-1} \mathbf{a}(z_j + \ell_j x),$$

$$\phi_j(x) = (\chi_j \psi)(z_j + \ell_j x), \quad x \in B(1).$$

Clearly, these functions satisfy (2.2) uniformly in j. One can easily prove that the operator $H_{\mathfrak{a}}(h, \mu)$ is unitarily equivalent to $f_j^2 H_{\mathfrak{a}_j}(\alpha, \nu)$ with the new Planck constant $\alpha = h(f_j \ell_j)^{-1}$, magnetic field intensity $\nu = \mu \ell_j f_j^{-1}$ and $\mathfrak{a}_j = \{\mathbf{a}_j, V_j\}$. Direct calculation shows that

$$\left.\begin{aligned} \mathcal{M}_s(h, \mu; \psi\chi_j, \mathfrak{a}) &= f_j^{2s} \mathcal{M}_s(\alpha, \nu; \phi_j, \mathfrak{a}_j), \\ \mathfrak{W}_s(h; \psi\chi_j, V) &= f_j^{2s} \mathfrak{W}_s(\alpha; \phi_j, V_j). \end{aligned}\right\} \tag{2.10}$$

Due to (2.6) we have $\alpha \le C, \nu\alpha \le C$. Besides, the condition (2.2) is satisfied uniformly in h, μ, j for $E = 2$. Therefore one can use for $\mathcal{M}_s(\alpha, \nu; \phi_j, \mathfrak{a}_j)$ the asymptotics (2.1). Combining it with (2.10) and taking into account (2.3), (2.4), we obtain

$$\begin{aligned} &\left|\mathcal{M}_s(h, \mu; \psi\chi_j, \mathfrak{a}) - \mathfrak{W}_s(h; \psi\chi_j, V)\right| \le C f_j^{2s} r\left(h f_j^{-1} \ell_j^{-1},\ \mu \ell_j f_j^{-1}\right) \\ &\le C \int_{B(z_j, \ell_j)} f(x)^{2s} \ell(x)^{-d} r\left(h f(x)^{-1} \ell(x)^{-1},\ \mu \ell(x) f(x)^{-1}\right) dx \end{aligned}$$

with a constant C uniform in j. It remains to sum these inequalities over j, taking into account that the intersection of more than N balls $B(z_j, \ell_j)$ is empty. This entails (2.8). □

In the next section we apply Theorem 2.3 to prove the results listed in Section 1.

3. Sketch of the proof.

Throughout this section we assume that the operator $A_{\mathfrak{a}}$ obeys Assumption 1.1, $h \in (0, h_0], 0 \leq \mu \leq Ch^{-1}$ and $s \in [0, 1]$. We break up the trace $\mathcal{M}_s(h,\mu;\psi,\mathfrak{a})$ into two parts as follows. Define the parameter

$$r = r_\theta = \left(\frac{h}{\theta}\right)^{\frac{2}{2-\beta}}, \; 0 < \theta \leq 1,$$

and the function

$$\phi(x) = \chi^2(x), \quad \begin{cases} \chi \in C_0^\infty(B(1)); \\ \chi(x) = 1, \; |x| \leq 1/2. \end{cases}$$

Denote

$$\psi_1(x) = \psi(x)\phi_r(x), \quad \psi_2(x) = \psi(x)\big(1 - \phi_r(x)\big), \quad \phi_r(x) = \phi(x/r).$$

By definition

$$|\partial_x^m \psi_k(x)| \leq C_m |x|^{-|m|}, \; x \neq 0, \;\; k = 1, 2. \tag{3.1}$$

Then, obviously,

$$\mathcal{M}_s(h,\mu) = \sum_{k=1}^{2} \mathcal{M}_s(h, \mu; \psi_k).$$

Let us study the traces $\mathcal{M}_s(h, \mu; \psi_k), k = 1, 2$ individually. To be definite, below we outline the proof of Theorem 1.4 only. Proofs of the other Theorems are similar.

Region around the origin. This region contributes to the leading term and entirely determines the second term in the asymptotics (1.11).

Theorem 3.1. Let $\mathbf{a}, V$ obey (1.4) and $d - s - 1 < \omega < d$. Then

$$\mathcal{M}_s(h, \mu; \psi_1, \mathfrak{a}) = \mathfrak{W}_s(h; \psi_1, V) + \psi(0)\Theta_s h^{-\omega} + o(1)h^{-\omega} \tag{3.2}$$

as $h \to 0, \theta \to 0$.

To establish (3.2) one proves first that i) one can neglect the effect of the magnetic field and ii) replace the potential V with its asymptotics W as $x \to 0$. This reduces the problem to the study of $\mathcal{N}_s(h; \psi_1, W)$. To justify this reduction we first compare the resolvents of the operators $H_{\mathfrak{a}}(h, \mu)$ and $H_W(h)$ by means of the resolvent identity. To pass from resolvents to functions of $H_{\mathfrak{a}}(h, \mu), H_W(h)$ we use a representation for functions of selfadjoint operators in terms of their resolvents found in [1]. Using homogeneity of W one can "scale out" the parameter h, after which the definition (1.13) yields almost automatically:

$$\mathcal{M}_s(h,\mu; \psi_1, \mathfrak{a}) \sim \mathcal{N}_s(h; \psi_1, W) = \mathfrak{W}_s(h; \psi_1, W) + \psi(0)h^{-\omega}\Theta_s + o(1)h^{-\omega}.$$

The change W back to V affects only the error.

Outside the origin. An important fact is that the asymptotics of $\mathcal{M}_s(h,\mu;\psi_2)$ does not contain a term of the order $O(h^{-\omega})$ between the Weyl term and remainder:

Theorem 3.2. Under conditions of Theorem 3.1

$$\mathcal{M}_s(h,\mu;\psi_2,\mathfrak{a}) = \mathfrak{W}_s(h;\psi_2,V) + o(1)h^{-\omega} + \langle\mu\rangle^{s+1}O(h^{s+1-d}), \tag{3.3}$$

as $h\to 0, \theta\to 0$.

Proof. We rely upon Theorem 2.3. Let

$$\mathcal{D} = \{x\in\mathbb{R}^d : r_\theta/4 < |x| < 4E\},$$

$$\ell(x) = \varrho|x|,\ f(x) = |x|^{-\frac{\beta}{2}},\ \varrho < 1.$$

It is easy to see that the conditions (2.3), (2.4) are fulfilled. Since $\beta<2$, we have for $x\in\mathcal{D}$

$$f(x)\ell(x) = \varrho|x|^{\frac{2-\beta}{2}} \ge \varrho h\theta^{-1} \ge \varrho h,\ \forall\theta\in(0,1].$$

This yields (2.6). The condition (2.7) is obviously satisfied for sufficiently small ϱ. Furthermore, by (1.4) and (3.1) the conditions (2.5) for V, $\mathbf{a}$ and ψ_2 are also fulfilled. Next, due to Assumption 1.1, operator $A_{\mathfrak{a}}$ obeys Assumption 2.1 with the domain $\mathcal{D}$ defined above. Therefore the conditions of Theorem 2.3 are satisfied. It follows from (2.8), (2.9) that

$$|\mathcal{M}_s(h,\mu;\psi_2,\mathfrak{a}) - \mathfrak{W}_s(h;\psi_2,V)| \le\ C_1h^{s+1-d}I_1(h,\mu) + C_2\mu^{s+1}h^{s+1-d}I_2(h,\mu). \tag{3.4}$$

Let us calculate $I_1 = I_1(h,\mu), I_2 = I_2(h,\mu)$:

$$I_1 = \varrho^{-s-1}\int_{\mathcal{D}}|x|^{-\frac{\beta}{2}(d+s-1)}|x|^{-s-1}dx,\quad I_2 = \int_{\mathcal{D}}|x|^{-\frac{\beta}{2}(d-2)}dx.$$

Since $\beta<2$, the integral I_2 is bounded uniformly in h,μ. One checks directly that

$$I_1(h,\mu) \le C\int_{r/4}^{4E} t^{\sigma-1}dt,\ \sigma = \frac{2-\beta}{2}(d-1-s-\omega).$$

In view of the condition $\omega > d-s-1$, we have

$$I_1 \le Cr^{\sigma} = C(h\theta^{-1})^{d-s-1-\omega}.$$

In combination with (3.4) this yields (3.3). □

Now, to complete the proof of Theorem 1.4 it remains to add up (3.2) and (3.3).

References

[1] W.O. Amrein, A.-M. Boutet de Monvel-Berthier, V. Georgescu, Notes on The N-Body Problem, Part II, University of Geneve, preprint, Geneve, 1991.

[2] J. Avron, I. Herbst, B. Simon, Schrödinger operators with magnetic fields, I., Duke Math. J., **45(4)**, 1978, 847–883.

[3] M. Sh. Birman, M.Z. Solomyak, Estimates for number of negative eigenvalues of the Schrödinger operator and its generalizations, Advances in Soviet Mathematics, (M. Sh. Birman ed.), **7**, AMS, 1991, 1–55.

[4] L. Hörmander, The Analysis of Linear Partial Differential Operators, I, Springer, Berlin, 1983.

[5] V. Ivrii, I.M. Sigal, Asymptotics of the ground state energies of large Coulomb systems, Ann. of Math., **138**, 1993, 243–335.

[6] V. Ivrii, Semiclassical Microlocal Analysis and Precise Spectral Asymptotics, Ecole Polytechnique, Preprints, Palaiseau, 1991–1992.

[7] V. Ivrii, Estimates for the number of negative eigenvalues of the Schrödinger operator with a strong magnetic field, Soviet Math. Dokl., **36(3)**, 1988, 561–564.

[8] V. Ivrii, Estimates for the number of negative eigenvalues of the Schrödinger operator with singular potentials, Proc. Int. Congr. Math. Berkeley, 1986, 1084–1093.

[9] E. H. Lieb, J.P. Solovej, J. Yngvason, Asymptotics of heavy atoms in high magnetic fields: II., Semiclassical regions, Comm. Math. Phys., **161**, 1994, 77–124.

[10] A. Sobolev, Quasi-classical asymptotics of local Riesz means for the Schrödinger operator in a moderate magnetic field, (to appear in Annales de l'I. H. P.)

[11] A. Sobolev, The quasi-classical asymptotics of local Riesz means for the Schrödinger operator in a strong homogeneous magnetic field, Duke Math. J., **74(2)**, 1994, 319–429.

Alexander Sobolev
MAPS, Mathematics and Statistics Subject Group,
University of Sussex, Falmer, Brighton, BN1 9QH, UK

Operator Theory:
Advances and Applications, Vol. 78

Convergence of Schrödinger Operators on Varying Domains

P. Stollman

1. Introduction.

Let $H = -\Delta + V$ be a Schrödinger operator on $\mathbb{R}^d$, where $V = V^+ - V^-$ is a potential with negative part in the Kato class and positive part in L^1_{loc}. For any open set G we write H_G for the corresponding selfadjoint operator in $L_2(G)$ which is defined in the following way: denote by $\mathfrak{h}$ the form associated with H and by $\mathfrak{h}_G$ the closure of $\mathfrak{h}|C_c^\infty(G)$; the associated selfadjoint operator is H_G. If $V \in L^2_{\text{loc}}$ then H_G is just the Friedrichs extension of $H|C_c^\infty(G)$.

In our talk we discuss the following question: Which kind of convergence $G_n \to G$ of domains implies the convergence $H_{G_n} \to H_G$?

In the following section we present a quite satisfactory answer in terms of strong resolvent convergence. The third section is devoted to two Theorems concerning norm resolvent convergence. In connection with Theorem 3 we use a result which we call the "local test theorem". It proved to be useful also in different situations and seems worth noting. While the results of Section 2 as well as Theorem 2 are taken from [14], Theorem 3 is new. A detailed proof will appear in [15].

We end this introductory section by recalling some potential theoretic notions which we need in the sequel: the *capacity* (more precisely the (1,2)-capacity, see [7]) of an open set is given by

$$\text{cap}(U) = \inf \left\{ \int (|\nabla f|^2 + |f|^2) dx; f \in W^{1,2}, f \geq \mathbb{1}_U \right\},$$

and for arbitrary A,

$$\text{cap}(A) = \inf_{A \subset U, U \text{open}} \text{cap}(U).$$

The importance of this set-function lies in the fact that elements in the Sobolev space $W^{1,2}$ have versions which are defined and continuous up to sets of zero capacity (as cap is larger than the measure of a set, this means additional information on elements of Sobolev space which, a priori, are only defined almost everywhere). More precisely: for $f \in W^{1,2}$ the limit

$$\tilde{f}(x) := \lim_{\varepsilon \to 0} |B_\varepsilon(x)|^{-1} \int_{B_\varepsilon(x)} f(y) dy$$

exists up to a set of zero capacity and, of course, equals f almost everywhere. Moreover, $\tilde{f}$ is *quasi-continuous*, which means that for any $\varepsilon > 0$ one can discard an open set V of capacity less than ε such that $\tilde{f}|V^c$ is continuous. It is easy to see that all quasi-continuous representatives of f coincide with $\tilde{f}$ quasi-everywhere, i.e. up to sets of capacity zero.

Let us mention here some complementary results about Dirichlet boundary conditions which are derived by similar techniques, namely [9], where approximation of Dirichlet boundary conditions by multiplication operators is investigated, and [6] which deals with the lowest eigenvalue of Dirichlet Laplacians.

2. Strong resolvent convergence.

In this section we present a criterion for convergence in *strong resolvent sense*

$$H_{G_n} \xrightarrow{\text{srs}} H_G.$$

Since the operators involved are acting in different Hilbert spaces we adopt the convention of [11] and extend the resolvent $(H_{G_n} + i)^{-1}$ by zero to all of $L_2(\mathbb{R}^d)$ and use the same symbol for the extended resolvent. Then

$$H_{G_n} \xrightarrow{\text{srs}} H_G :\Leftrightarrow (H_{G_n} + i)^{-1} f \to (H_G + i)^{-1} f \quad (f \in L_2(\mathbb{R}^d)).$$

In order to state and prove the main result of this section let us look more closely at the form $\mathfrak{h}_G$ and assume $V = 0$ for simplicity of notation. Then

$$D(\mathfrak{h}_G) = W_0^{1,2}(G).$$

It is a well-known fact of potential theory (cf [7]) that

$$D(\mathfrak{h}_G) = W_0^{1,2}(G) = \{f \in W^{1,2}; \tilde{f} = 0 \text{ q.e. on } G^c\}.$$

Now the right hand side of this equations is suitable to define $\mathfrak{h}_G$ for arbitrary subsets of $\mathbb{R}^d$ and so we define, for $M \subset \mathbb{R}^d$

$$D(\mathfrak{h}_M) := W_0^{1,2}(M) := \{f \in W^{1,2}; \tilde{f} = 0 \text{ q.e. on } G^c\}$$

and denote by H_M the associated selfadjoint operator in $\overline{W_0^{1,2}(M)}^{L_2}$. (See also [4, 5] for the definition of general $W_0^{1,2}$-spaces.)

Although this is merely a matter of definition, the use of these spaces is the key to the following Theorem and its simple proof. The reason is that even if one is only interested in sequences G_n of open sets, the limit of such a sequence need not be open. It might however be equivalent to an open set in the following sense:

$$M \sim M' :\Leftrightarrow W_0^{1,2}(M) = W_0^{1,2}(M').$$

Note that by the very definition $\text{cap}(M \triangle M') = 0$ implies $M \sim M'$. The converse is not true, e.g. $M = \{(x,y); x > 0\} \subset \mathbb{R}^d$, in which case $M \sim \overline{M}$ but $\text{cap}(\overline{M} \backslash M) = \infty$.

Theorem 1. Let G_n,G be measurable. If $\overline{\lim}\,(G_n) \sim \underline{\lim}\,(G_n) \sim G$, then

$$H_{G_n} \xrightarrow{\text{srs}} H_G.$$

We refer to [14] for a proof and remark that Theorem 1 covers results by Simon [11], Rauch and Taylor [10], and Weidmann [19]. As a consequence of strong resolvent convergence one has convergence of eigenvalues below the essential spectrum; for details see [19].

3. Convergence in norm resolvent sense.

In this section we present two results concerning norm resolvent convergence $H_{G_n} \xrightarrow{\text{nrs}} H_G$, which means convergence in operator norm of the extended resolvents (see section 2). The next theorem is valid in a quite general framework and based on a convergence theorem for measure perturbations (see [14]):

Theorem 2. Let G_n,G be measurable. Assume that $\overline{\lim}\,(G_n) \sim \underline{\lim}\,(G_n) \sim G$ and that there is a $\Sigma \subset \mathbb{R}^d$ such that $G_n \triangle G \subset \Sigma$ for all $n \in \mathbb{N}$. Then

$$H_{G_n} \xrightarrow{\text{nrs}} H_G.$$

Unfortunately, this result requires that the symmetric difference of G_n and G be "small at infinity". Thus it cannot be applied to periodic domains as, for instance $G = \mathbb{R}^d$, $G_n = \mathbb{R}^d \setminus \bigcup_{k\in\mathbb{Z}^d} n^{-1}B(k)$, where $B(k)$ denotes a ball of radius 1 centered at k. For these sets it is easy to guess that $H_{G_n} \xrightarrow{\text{nrs}} H_G$ and Theorem 2 is obviously not applicable. However one can use

Theorem 3. Let G_n,G be measurable and denote by Γ_L the set of all open cubes of sidelength L. If

$$\sup_{C\in\Gamma_L} \operatorname{cap}((G\triangle G_n)\cap C) \longrightarrow 0 \text{ for } n\to\infty,$$

then

$$H_{G_n} \xrightarrow{\text{nrs}} H_G.$$

Before indicating the ideas of the proof of Theorem 3 let us mention two facts:

- contrary to Theorems 1 and 2, this result uses the geometry of $\mathbb{R}^d$. The condition should be thought of as a uniform local convergence of the sequence G_n to G.

- Secondly, Theorem 3 does not contain Theorem 2 as one can see from the following simple example: $G_n = (1-n^{-1})B(0), G = B(0)$. In this case $\inf_{n\in\mathbb{N}} \operatorname{cap}(G\setminus G_n) > 0$ while Theorem 2 implies norm resolvent convergence.

For the proof of Theorem 3 we use two main ingredients:

The first one is the following result, parts of which are certainly known to many specialists (see e.g. [8]). Nevertheless we think it is worthwile to isolate it in form of a Theorem.

Since we want to apply it to perturbations by potentials and Dirichlet boundary conditions we have chosen to present it in a general version using measure perturbations. To this end we recall that M_0 denotes the class of all non-negative measures which do not charge sets of capacity zero (see [2, 18, 12, 17]) while S_K is the set of measures which satisfy a Kato condition (see [1, 17]). For a measure $\mu \in M_0 - S_K$, i.e. $\mu = \mu^+ - \mu^-$,$\mu^+ \in M_0$,$\mu^- \in S_K$ one can define $-\Delta + \mu$ by its quadratic form. Moreover, certain properties of $\exp(-\Delta + \mu)$ only depend on $c_E(\mu)$, which is defined by

$$c_E(\mu) := \sup_{f\in L_1, \|f\|\le 1} \int (-\Delta + E)^{-1} f d\mu.$$

Those readers who are not familiar with measure perturbations should simply read the μ as a V.

Recall that $(-\Delta + \mu)_G$ is the operator on $L_2(G)$ with Dirichlet boundary conditions as defined in Section 2.

Theorem 4. (Local test) Let μ_n,$\mu \in M_0 - S_K$ satisfy $c_E(\mu_n) \le \gamma$ for all $n \in \mathbb{N}$ and some fixed $\gamma < 1/2$,$E > 0$. Denote by Γ_L the set of cubes of sidelength L. Consider the conditions

(i) $-\Delta + \mu_n \xrightarrow{\text{nrs}} -\Delta + \mu$.

(ii) For all $L > 0$: $\sup_{C\in\Gamma_L} \|(-\Delta + 1\!\!1_C\mu_n + i)^{-1} - (-\Delta + 1\!\!1_C\mu + i)^{-1}\| \to 0$ for $n \to \infty$.

(ii′) For all $L > 0$: $\sup_{C\in\Gamma_L} \|(-\Delta + \mu_n + i)_C^{-1} - (-\Delta + \mu + i)_C^{-1}\| \to 0$ for $n \to \infty$.

(iii) For all (some) $L > 0$: $\sup_{C\in\Gamma_L} \|(-\Delta + \mu_n + i)^{-1} - (-\Delta + \mu + i)^{-1} 1\!\!1_C\| \to 0$ for $n \to \infty$.

Then we have the following implications: (ii)$\Longrightarrow$ (iii) $\Longrightarrow$ (i) $\Longrightarrow$ (iii) and (ii′) $\Longrightarrow$ (iii). If, for every $n \in \mathbb{N}$ either $\mu_n \le \mu$ or $\mu_n \ge \mu$ then (i)–(iii) are equivalent.

Let us sketch the main ideas needed for the proof of the local test theorem: (ii) $\Longrightarrow$ (iii): Fix C and let C' be a much bigger cube with the same center. Then $\mu_n \ne 1\!\!1_{C'}\mu_n$ only outside C' so the difference

$$\|(-\Delta + 1\!\!1_{C'}\mu_n + i)^{-1} 1\!\!1_C - (-\Delta + \mu_n + i)^{-1} 1\!\!1_C\| \to 0$$

exponentially in $\text{dist}(C, \mathbb{R}^d \setminus C')$. This can be seen by using techniques from [16]. Moreover the exponential estimate is uniform in n by the assumption on $c_E(\mu_n)$. (ii′) $\Longrightarrow$ (iii) follows by the same arguments.

(iii) $\Longrightarrow$ (i): Consider a partition of $\mathbb{R}^d$ into unit cubes C_k centered at $k \in \mathbb{Z}^d$ and denote $\mathbb{1}_k := \mathbb{1}_{C_k}$. Then

$$(-\Delta+\mu_n+i)^{-1}-(-\Delta+\mu+i)^{-1} = \sum_{l,k\in\mathbb{Z}^d} \mathbb{1}_l[(-\Delta+\mu_n+i)^{-1}-(-\Delta+\mu+i)^{-1}]\mathbb{1}_k.$$

Now the terms of the sum converge to zero uniformly in $k,\ell \in \mathbb{Z}^d$ and moreover we have exponential decay for these terms if k,ℓ are far apart. Using the Cotlar-Stein lemma (see [3]) or direct estimates, the asserted convergence follows (for a similar argument, [8]).
If $\mu_n \le \mu$ or vice versa, one can use monotonicity arguments based on the Feynman-Kac formula to derive (ii) and (ii′) from (i). □

At this point let me acknowledge helpful comments of E. Mourre who gave me a decisive hint concerning a direct proof of (iii) $\Longrightarrow$ (i) and B. Helffer, who referred me to the Cotlar-Stein lemma.

The interesting implication in the above theorem is of course from (ii) or (ii′) to (i). Moreover one is tempted to think that (ii), (ii′) and (iii) should easily follow from (i). For (iii) this is obvious, for (ii′) we don't know it, but for (ii) it is false, as can be seen from the following example:
Take $V_n := (-1)^n n(\mathbb{1}_{(-n^{-1},0)} + \mathbb{1}_{(0,n^{-1})})$ in $\mathbb{R}$. Then one can check that $-\Delta + V_n \xrightarrow{\text{nrs}} -\Delta$. If we take $C = (0,1)$, then $-\Delta + \mathbb{1}_C V_n$ won't converge: as $\mathbb{1}_C V_n = (-1)^n n\mathbb{1}_{(0,n^{-1})} =: W_n$ it is easy to see that $-\Delta + W_{2n+k} \xrightarrow{\text{nrs}} -\Delta + (-1)^k\delta$ for $k = 0,1$.

To deduce Theorem 3 with the help of the local test we recall first, that there is a measure $\infty_{G_n^c}$ such that $-\Delta + V + \infty_{G_n^c} = (-\Delta + V)_{G_n} = H_{G_n}$. Let $\mu_n := V + \infty_{G_n^c}$ and $\mu := \infty_{G^c}$. Then $\mathbb{1}_C\mu_n$ and $\mathbb{1}_C\mu$ agree outside the set $(G\Delta G_n)\cap C$. The estimate on the capacity of this set allows one to check condition (ii) of Theorem 4 .

This is the second main point in the proof of Theorem 3: obstacles of finite capacity produce perturbations of the semigroup which can be controlled (even with respect to the Hilbert-Schmidt norm) by the capacity. This is the theme of [13] and was also used in [14].

References

[1] S. Albeverio and Z. Ma: Perturbation of Dirichlet forms-Lower semiboundedness, Closability, and Form Cores. J. Funct. Anal. **99** (1991), 332–356.

[2] J. Baxter, G. DalMaso and U. Mosco: Stopping times and Γ-convergence. Trans. Amer. Math. Soc. **303** (1987), 1–38.

[3] C. Fefferman: Recent progress in classical Fourier Analysis. Proc. Int. Congr. Math., Vancouver 1974.

[4] D. Feyel: Ensembles singuliers associées aux espaces de Banach réticules. Ann. Inst. Fourier **31**, 1 (1981), 192–223.

[5] D. Feyel and A. de la Pradelle: Espaces de Sobolev sur les ouvert fins. C. R. Acad. Sci. **280**, série A (1975), 1125–1127.

[6] F. Gesztesy and Zh. Zhao: Domain perturbations, Brownian motion, capacities, and ground states of Dirichlet Schrödinger operators. Math. Z. **215** (1994), 143–150.

[7] L. Hedberg: Spectral synthesis and stability in Sobolev spaces. In: Euclidean Harmonic analysis, Proceedings Maryland, LNM **779** (1980), 73–103.

[8] R. Hempel and I. Herbst: Strong magnetic fields, Dirichlet boundaries, and spectral gaps. Preprint 1994

[9] I. Herbst and Zh. Zhao: Sobolev spaces, Kac regularity, and the Feynman-Kac formula. In: Seminar on Stochastic Processes, Birkhäuser, Boston 1987

[10] J. Rauch and M. Taylor: Potential and Scattering Theory on Wildly perturbed domains. J. Funct. Anal. **18** (1975), 27–59.

[11] B. Simon: A canonical decomposition for quadratic forms with applications to monotone convergence theorems. J. Funct. Anal. **28** (1978), 377–385.

[12] P. Stollmann: Smooth perturbations of regular Dirichlet forms. Proc. Amer. Math. Soc. **117** (1992), 747–752.

[13] P. Stollmann: Scattering at obstacles of finite capacity. J. Funct. Anal. **121** (1994), 416–425.

[14] P. Stollmann: A convergence theorem for Dirichlet forms with applications to boundary value problems with varying domains. Math. Z. to appear.

[15] P. Stollmann: Habilitationsschrift, in preparation.

[16] P. Stollmann and G. Stolz: Singular spectrum for multidimensional Schrödinger operators with potential barriers. J. Operator Theory. to appear.

[17] P. Stollmann and J. Voigt: Perturbation of Dirichlet forms by measures. Potential Analysis, to appear.

[18] T. Sturm: Measures charging no polar sets and additive functionals of Brownian motion. Forum Math. **4** (1992), 257–297.

[19] J. Weidmann: Stetige Abhängigkeit der Eigenwerte und Eigenfunktionen elliptischer Differentialoperatoren vom Gebiet. Math. Scand. **54** (1984), 51–69.

Operator Theory:
Advances and Applications, Vol. 78

Localization for the Poisson Model

G. Stolz

1. The result.

In the quantum mechanical one-electron approximation a solid consisting of nuclei producing single site potentials f_i at points $x_i \in \mathbb{R}^d$ is described by the Schrödinger operator $H = -\Delta + \sum_i f_i(x - x_i)$. Random media are modeled by introducing suitable random parameters into H. There are two types of random media, which are of fundamental interest in physics:

(i) *Crystalline alloys.*
They are described mathematically by the *Anderson model*

$$H_\omega^A = -\Delta + \sum_{n\in\mathbb{Z}} \lambda_n(\omega) f(x-n),$$

where the coupling constants λ_n are independent, identically distributed real random variables and f has suitable decay properties.

(ii) *Media with structural disorder.*
Here the positions of the nuclei are random. An extreme case is given by the *Poisson model*

$$H_\omega^P = -\Delta + \sum_i f(x - X_i(\omega)),$$

where the X_i are the points of a Poisson process with given density $\alpha > 0$. This process is characterized by $\mathrm{Prob}(N(I,\omega) = n) = (\alpha|I|)^n e^{-\alpha|I|}/n!$, where $I \subset \mathbb{R}$ is an interval of length $|I|$, $N(I,\omega)$ the number of points $X_i(\omega)$ in I and $N(I_1)$ is independent of $N(I_2)$ if $I_1 \cap I_2 = \emptyset$.

Here we consider the problem of *localization* for the Poisson model in dimension $d = 1$, i.e. our goal is to prove that

$$\sigma_c(H_\omega^P) = \emptyset \quad \text{for a.e. } \omega$$

under suitable conditions on f, where σ_c denotes the continuous spectrum.

Our strategy of proof will be close to the one used by Kotani and Simon in their proof of localization for the one-dimensional Anderson model. Their result is

Theorem 1 [10]: Let $f : \mathbb{R} \to \mathbb{R}$ be non-negative, continuous and compactly supported. In the Anderson model for $d = 1$ let the distribution of λ_n have an absolutely continuous component. Then

$$\sigma_c(H_\omega^A) = \emptyset \quad \text{for a.e. } \omega.$$

In contrast, our result for the Poisson model is

Main Theorem (Localization for the Poisson model): In the Poisson model for $d = 1$ let $f : \mathbb{R} \to \mathbb{R}$ be non-negative, bounded, measurable and compactly supported. Then

$$\sigma_c(H_\omega^P) = \emptyset \quad \text{for a.e. } \omega.$$

Under the conditions of the Main Theorem it is easily seen that $\sigma(H_\omega^P) = [0,\infty)$ for a.e. ω, so we in fact get dense pure point spectrum in $[0,\infty)$ almost surely.

We remark that in the original result of Kotani and Simon continuity of f could be replaced by boundedness and measurability, leading to only slight changes in the proof. So both above results hold under the same conditions on f.

After a short review of former results for the Poisson model in Section 2 we outline a proof of the Main Theorem in Section 3. Details of the proof will be given in a forthcoming publication [16]. We conclude with some remarks on generalizations and open questions in Section 4.

2. Former results for the Poisson model.

The Poisson model and generalized Poisson models were first treated in connection with the integrated density of states in the 1970's. For results and references in the one-dimensional and multi-dimensional case see [11].

Herbst and Howland [5] study resonances for the one-dimensional Poisson model with an external electric field of constant strength, i.e. the operator $H_\omega^P + Fx$ with $F \neq 0$. They pose the question of localization for $F = 0$, but do not prove it.

A result somewhat weaker than localization is given by

$$\sigma_{ac}(H_\omega^P) = \emptyset \quad \text{for a.e. } \omega. \tag{1}$$

If f is compactly supported, then (1) follows from Kotani's general result on non-deterministic ergodic potentials [8]. Kirsch, Kotani and Simon [7] prove (1) for more general f using a *support theorem* for the absolutely continuous spectrum of Kotani [9]. Later the idea of Simon and Spencer [12] to use trace class methods from scattering theory in proving $\sigma_{ac} = \emptyset$ was applied to H_ω^P in [6], [14] and [15]. We combine most of the known results in

Theorem2: [7, 6, 14, 15] Let either

(i) $f \in L^1(\mathbb{R})$, $f \geq 0$, or

(ii) $\sum_{n\in\mathbb{Z}} \sup_{|x+n|<1} |f(x)| < \infty$ and $f(x) \geq \gamma > 0$ in $[a,b]$ for some $a < b$, or

(iii) f be superexponentially decaying at $\pm\infty$,

then (1) holds.

We finally mention a result of Combes and Hislop [3] for a combined Anderson- and Poisson model in arbitrary dimension d. For the family

$$H_{\omega,\omega'}^{A,P} = -\Delta + \sum_i \lambda_i(\omega) f(x - X_i(\omega'))$$

in $L^2(\mathbb{R}^d)$ they prove localization at low energies for almost every pair (ω,ω') under suitable assumptions for λ_i and f. They cannot treat the pure Poisson model.

3. Ideas of the proof.

Since H_ω^P is an ergodic family of operators, we know that almost surely $\sigma_{ac}(H_\omega^P)$ is the essential closure of $\{E \in \mathbb{R} : \gamma(E) > 0\}$, where $\gamma(\cdot)$ denotes the Lyapunov exponent of H_ω^P [11]. Theorem 2 yields that $\gamma(E) > 0$ for a.e. E. (Since in our situation H_ω^P is non-deterministic this would also follow directly from the results in [8].)

By the Ruelle-Osceledec theorem (and Fubini) this implies for almost every ω: (I) For almost every E there exists a solution u of $H_\omega^P u = Eu$, exponentially decaying at $+\infty$, and a solution v of $H_\omega^P v = Ev$, exponentially decaying at $-\infty$.

At this point it needs to be shown that for most of the E (H_ω-spectrally almost every E) the solutions u and v coincide up to a constant, hence giving that generalized eigenfunctions are in fact L^2-eigenfunctions, i.e. the spectrum is pure point.

This will be established by using *spectral averaging*, a method first used in a proof of localization by Simon and Wolff [13]. Here we need a form of spectral averaging adapted to the Poisson model. Whereas in case of the Anderson model the averaging is done over a coupling constant, we will average over the shift parameter a in the following model:

Let $W_1 \in L^2_{\text{loc}}[0,\infty)$ with $W_1(x) \geq -C|x|^2$ near $+\infty$, $W_2 \in L^2_{\text{loc}}(-\infty,0]$ with $W_2(x) \geq -C|x|^2$ near $-\infty$ and

$$H_a = -d^2/dx^2 + W_1(x-a) + W_2(x+a),$$

meaning the selfadjoint closure of the corresponding operator with domain $C_0^\infty(\mathbb{R})$. Finally, let ρ_a be the Weyl-Titchmarsh spectral measure for H_a (see [2]).

Proposition. (Spectral averaging) Let $0 < a_0 < a_1$ be arbitrary and for any Borel set $B \subset \mathbb{R}$ define

$$\mu(B) := \int_{a_0}^{a_1} \rho_a(B)\, da.$$

Then μ defines a measure which is absolutely continuous on $(0,\infty)$.

We next describe how the Proposition can be used to *prove the Main Theorem:*

Let

$$d_i(\omega) := X_i(\omega) - X_{i-1}(\omega).$$

The d_i are i.i.d. random variables with absolutely continuous distribution $\alpha e^{-\alpha x}dx$. We will first work with fixed d_i, $i \neq 0$ and vary d_0. Taking $\delta > 0$ such that $\operatorname{supp} f \subset [-\delta,\delta]$ and $a := \frac{d_0}{2} - \delta$ (absolutely continuous distributed!) we are in the situation of the above Proposition.

For the so found family H_a let A be the set of those $E \in (0,\infty)$ such that $H_a u = Eu$ does not have an L^2-solution at $+\infty$ or does not have an L^2-solution at $-\infty$. This set A does not depend on the parameter a and from (I) we get that $|A| = 0$ ($|\cdot| =$ Lebesgue measure) at least for almost every choice of d_i, $i \neq 0$. The Proposition gives $\rho_a(A) = 0$ for almost every $a \in (a_0,a_1)$, meaning that for those values of a to spectrally almost every $E \in (0,\infty)$ there are L^2-solutions at $+\infty$ as well as at $-\infty$.

Shnol's theorem guarantees that also spectrally almost everywhere there are solutions globally in $L^2_{-s} = \{f : (1+|\cdot|^2)^{-s/2} f \in L^2\}$ for $s > \frac{1}{2}$. Choosing also $s \leq 1$ a Wronskian argument can be used to show that the L^2-solutions at $\pm\infty$ must coincide with the global L^2_{-s}-solution, therefore yielding the existence of global L^2-solutions for spectrally almost every $E \in (0,\infty)$, i.e. pure point spectrum in $(0,\infty)$. This holds for a.e. choice of d_i, $i \neq 0$, and almost every $a \in (a_0,a_1)$, i.e. with positive probability in d_0. The independence of the d_i together with Pastur's theorem on the non-randomness of the spectral type completes the proof of the Main Theorem. (By $f \geq 0$ there is no spectrum in $(-\infty,0)$.)

We finally note the main ideas in the *proof of the Proposition*: The spectral measures ρ_a are the weak limits as $N \to \infty$ of the spectral measures $(\rho_a)^{N,\beta}_{-N,\alpha}$ corresponding to $-d^2/dx^2 + W_1(x-a) + W_2(x+a)$ on $L^2(-N,N)$ with boundary conditions $\phi(-N)\cos\alpha - \phi'(-N)\sin\alpha = 0$ and $\phi(N)\cos\beta - \phi'(N)\sin\beta = 0$. We also have $\rho_a = \text{w}-\lim_{N\to\infty}(\tilde\rho_a)^N_{-N}$, where $(\tilde\rho_a)^N_{-N}$ is the averaged measure defined by

$$(\tilde\rho_a)^N_{-N}(B) = \frac{1}{\pi^2}\int_0^\pi\int_0^\pi (\rho_a)^{N,\beta}_{-N,\alpha}(B)\,d\alpha\,d\beta.$$

It is well known [1, 10] that $(\tilde\rho_a)^N_{-N}$ is absolutely continuous with density $g_{a,N}(E)$ which can be expressed quite explicitely in terms of the Prüfer amplitude of solutions to $H_a u = Eu$. This expression gives enough information to control $\int_{a_0}^{a_1} g_{a,N}(E)\,da$ uniformly in N and locally uniformly in $E \in (0,\infty)$, establishing the Proposition.

4. Concluding remarks.

(i) In addition to the fact that the Poisson process is stationary we only used that $d_i = X_i - X_{i-1}$ has an absolutely continuous distribution. Actually, all what is needed is that d_i has an absolutely continuous component. This shows that our Main Theorem remains true for many other point processes.

(ii) The Main Theorem should be true under less restrictive assumptions on f. In fact, we do not need in the proof that f in non-negative. But without this sign restriction we can only conclude that $\sigma_c(H_\omega^P) \cap (0,\infty) = \emptyset$ almost surely, not knowing the nature of the spectrum at negative energies. This is likely to be a drawback of our method of proof. (We can not show that the measure μ in the Proposition is absolutely continuous on the whole line.)

We also expect the Main Theorem to be true for f, which are not compactly supported, but quickly decaying.

(iii) For the case of a discrete Anderson model it was shown by del Rio, Makarov and Simon [4, Thm.5.5] that for many exceptional values of ω the spectrum is purely singular continuous. We expect a similar result to be true for the Poisson model, i.e. that "almost every ω" in the Main Theorem leaves open considerable room for other types of spectrum.

(iv) For the Anderson model in $d > 1$ localization at low energies has been established under suitable assumptions (e.g. [10, 3]). At the moment this result can not be extended to a multi-dimensional Poisson model. The most crucial step here would be to prove a result on spectral averaging for a model of the type $H_a = -\Delta + V + W(x-a)$ in $L^2(\mathbb{R}^d)$. This model does not seem to have any kind of "monotonicity" in the parameter a, a property used in all proofs of spectral averaging so far.

References

[1] R. Carmona, J. Lacroix: Spectral theory of random Schrödinger operators, Birkhäuser, Basel – Berlin 1990

[2] E. A. Coddington, N. Levinson: Theory of ordinary differential equations, McGraw-Hill, New York 1955

[3] J.-M. Combes, P. D. Hislop: Localization for continuous random Hamiltonians in d-dimensions, J. Funct. Anal. **124** (1994), 149–180

[4] R. del Rio, N. Makarov, B. Simon: Operators with singular continuous spectrum, II. Rank one operators, to appear in Commun. Math. Phys.

[5] I. W. Herbst, J. S. Howland: The stark ladder and other one-dimensional external field problems, Commun. Math. Phys. **80** (1981), 23–42

[6] P. D. Hislop, S. Nakamura: Stark Hamiltonians with unbounded random potentials, Rev. Math. Phys. **2**(1990), 479–494

[7] W. Kirsch, S. Kotani, B. Simon: Absence of absolutely continuous spectrum for some one dimensional random but deterministic potentials, Ann. Inst. Henri Poincaré **42** (1985), 383–406

[8] S. Kotani: Ljapunov indices determine absolute continuous spectra of stationary one dimensional Schrödinger operators, in: Proc. Taneguchi Intern. Symp. on Stochastic Analysis, Katata and Kyoto 1982, 225–247, North Holland 1983

[9] S. Kotani: Support theorems for random Schrödinger operators, Commun. Math. Phys. **97** (1985), 443–452

[10] S. Kotani, B. Simon: Localization in General One-Dimensional Random Systems, Commun. Math. Phys. **112** (1987), 103–119

[11] L. Pastur, A. Figotin: Spectra of Random and Almost-Periodic Operators, Springer, 1991

[12] B. Simon, T. Spencer: Trace class perturbations and the absence of absolutely continuous spectra, Commun. Math. Phys. **125** (1989), 113–125

[13] B. Simon, T. Wolff: Singular continuous spectrum under rank one perturbations and localization for random Hamiltonians, Commun. Pure Appl. Math. **39** (1986), 75–90

[14] G. Stolz: Note to the paper by P. D. Hislop and S. Nakamura: Stark Hamiltonian with unbounded random potentials, Rev. Math. Phys. **5** (1993), 453–456

[15] G. Stolz: Spectral theory of Schrödinger operators with potentials of infinite barriers type, Habilitationsschrift, Universität Frankfurt, 1994

[16] G. Stolz: Localization for random Schrödinger operators with Poisson potential, in preparation

Günter Stolz
Universität Frankfurt, Fachbereich Mathematik
D-60054 Frankfurt am Main, Germany

Address after September 1, 1994:
University of Alabama at Birmingham
Department of Mathematics
CH 452, Birmingham, AL 35294-1170, USA
e-mail: stolz@math.uab.edu

Operator Theory:
Advances and Applications, Vol. 78

Systems of Partial Differential Equations for a Class of Operator Determinants

C. A. Tracy, H. Widom

I. Introduction

We consider one-dimensional integral operators with kernel of the form

$$K(x,y) = \frac{1}{x-y} \sum_{\alpha,\beta=1}^{n} c_{\alpha,\beta}\, \varphi_\beta(x)\, \varphi_\alpha(y)$$

acting on functions on the union of intervals

$$J = \bigcup_{k=1}^{m} [a_{2k-1}, a_{2k}].$$

If the φ_α belong to C^1 and the matrix $(c_{\alpha,\beta})$ is antisymmetric then the kernel is symmetric and belongs to C^1, and the corresponding operator K is trace class. Modestly generalizing the principal result of [4], where n was equal to 2, we show here that if the functions φ_α satisfy a differential identity of a certain form then there is a system of partial differential equations, with the a_i as independent variables, associated with the determinant $\det(I-K)$. More exactly, the solution of these equations determine the logarithmic derivatives of the determinant with respect to the a_k.

Write C for the matrix $(c_{\alpha,\beta})_{\alpha,\beta=1}^n$, write $\varphi(x)$ for the vector function $(\varphi_\alpha(x))_{\alpha=1}^n$ and denote the inner product in $\mathbb{R}^n$ by $\langle\,\cdot,\cdot\,\rangle$. Then the kernel may be written

$$\frac{\langle\, C\varphi(x), \varphi(y)\,\rangle}{x-y}. \tag{1}$$

If $R(x,y)$ is the resolvent kernel for K, the kernel of the operator $K\,(I-K)^{-1}$, then it is an easy fact (see below) that

$$\frac{\partial}{\partial a_k} \log\det(I-K) = (-1)^{k+1}\, R(a_k,a_k). \tag{2}$$

We define the function $Q(x) = (Q_\alpha(x))_{\alpha=1}^n$ (which depends also on $a_1,\ldots,a_{2m}$) by

$$Q := (I-K)^{-1}\varphi$$

(we restrict ourselves to (a_k) such that $I-K$ is invertible) and set

$$q_k := Q(a_k).$$

These vector functions of $a_1,\ldots,a_{2m}$ will be the quantities of principal interest and will be among the dependent variables in our system of equations. We shall see that for any kernel of the form (1) we have the representation

$$R(a_k,a_k) = \langle C\,\frac{\partial q_k}{\partial a_k}, q_k\,\rangle, \tag{3}$$

so the q_k determine $R(a_k,a_k)$, and that

$$\frac{\partial q_j}{\partial a_k} = (-1)^k\,\frac{\langle C\,q_j, q_k\,\rangle}{a_j - a_k}\,q_k, \qquad (j \neq k), \tag{4}$$

which is the first part of our system of equations. To complete it we need formulas for the derivatives $\partial q_k/\partial a_k$, and this requires introduction of more dependent variables and the differential identity for $\varphi(x)$ alluded to above.

Given n-vectors $v = (v_\alpha)$ and $w = (w_\alpha)$ we denote by $v \otimes w$ the $n \times n$ matrix with α,β entry $v_\alpha\, w_\beta$. The extra dependent variables are the matrices

$$U_i := \int_J x^i\, Q(x) \otimes \varphi(x)\, dx.$$

We also define $Q_i := (I-K)^{-1}\,M^i\,\varphi$, where M denotes multiplication by x, and

$$q_{i,k} := Q_i(a_k).$$

These last are not new since, as we shall show,

$$q_{i+1,k} = a_k\, q_{i,k} + U_i^{\,t}\, C\, q_k, \tag{5}$$

(^{t} denotes transpose) and this determines the $q_{i,k}$ in terms of the U_i and $q_{0,k} = q_k$. The differentiation formula

$$\frac{\partial U_i}{\partial a_k} = (-1)^k\, q_k \otimes q_{i,k} \tag{6}$$

will be the second part of our system of equations.

To obtain formulas for $\partial q_k/\partial a_k$ in terms of the q_k and U_i we use the basic assumption on $\varphi(x)$. This assumption is that there exist a scalar polynomial $m(x)$ and a matrix $A(x)$ with polynomial entries connected with the matrix C by

$$A(x)^t\, C + C\, A(x) = 0 \tag{7}$$

such that

$$m(x)\,\varphi'(x) = A(x)\,\varphi(x). \tag{8}$$

Of course this is the same as taking $m(x) = 1$ and $A(x)$ a rational matrix function, but our final equations are expressed most simply in terms of the coefficients of the polynomials $m(x)$ and $A(x)$. If

$$m(x) = \sum_{l\geq 0} m_l\, x^l, \qquad A(x) = \sum_{l \geq 0} A_l\, x^l, \tag{9}$$

then these equations are

$$m(a_k)\,\frac{\partial q_k}{\partial a_k} = A(a_k)\,q_k + \sum_{\substack{i+j=l-1\\ l\geq 1}} [U_j\,C\,(A_l-(j+1)\,m_{l+1}) - A_l\,U_j\,C]\,q_{i,k}$$

$$-\sum_{j\neq k}(-1)^j\,m(a_j)\frac{\langle\,C\,q_j,q_k\,\rangle}{a_j-a_k}\,q_j. \tag{10}$$

Thus, our unknown functions are the q_k and the U_i, and the complete system of equations consists of (4), (6) and (10). We see from the last that we need include as unknowns only the U_i with $i \leq \max(\deg A - 1, \deg m - 2)$. In particular, if A is constant and m is linear there need be no U_i at all.

II. Derivation of the System of Equations

We may assume until the derivation of (10) that the φ_α are C^1 functions defined on all of $\mathbb{R}$ and have compact support. We denote by χ the characteristic function of J and denote now by K the operator on $L_2(\mathbb{R})$ having kernel $K(x,y)\,\chi(y)$. The determinant $\det(I-K)$ is the same with this definition of K. Moreover since $K(x,y)$ is a C^1 function with compact support K is even a trace class operator on the Sobolev space $H_1(\mathbb{R})$. The mapping $(a_1,\ldots,a_{2m}) \mapsto K$ from $\mathbb{R}^{2m}$ to the space of trace class operators on $H_1(\mathbb{R})$ is strongly differentiable and it is immediate that

$$\frac{\partial K}{\partial a_k} \doteq (-1)^k\,K(x,a_k)\,\delta(y-a_k), \tag{11}$$

where "$\doteq$" means "has kernel" and δ is the Dirac distribution. It follows from this that

$$\frac{\partial}{\partial a_k}\log\det(I-K) = -\mathrm{tr}\,(I-K)^{-1}\,\frac{\partial K}{\partial a_k} = (-1)^{k+1}\,R(a_k,a_k),$$

which is (2). (Note that now $R(x,y)$ is defined for all $x,y \in \mathbb{R}$ and is discontinuous in y. The quantity $R(a_k,a_k)$ denotes the limit of $R(a_k,y)$ as $y \to a_k$ from the interior of J. Similarly for other like quantities.)

We have

$$[M,K] \doteq \langle\,C\,\varphi(x),\varphi(y)\,\rangle\,\chi(y),$$

where the brackets denote commutator, from which it follows upon left and right multiplication by $(I-K)^{-1}$, and using the symmetry of $K(x,y)$, that

$$[M,(I-K)^{-1}] \doteq \langle\,C\,Q(x),Q(y)\,\rangle\,\chi(y). \tag{12}$$

($Q(x)$ is defined precisely as before, but now has domain $\mathbb{R}$.) Hence

$$R(x,y) = \frac{\langle\,C\,Q(x),Q(y)\,\rangle}{x-y}\,\chi(y), \qquad (x\neq y).$$

In particular we have

$$R(a_j,a_k) = \frac{\langle C\, q_j, q_k \rangle}{a_j - a_k}, \qquad (j \neq k) \tag{13}$$

and

$$R(a_k,a_k) = \langle C\, Q'(a_k), q_k \rangle. \tag{14}$$

Here the prime denotes differentiation with respect to x. Of course Q is a function of $a_1, \dots, a_{2m}$ as well as x, and $\langle C\, Q(x), q_k \rangle$ vanishes identically when $x = a_k$ since C is antisymmetric. It follows that the last identity is equivalent to

$$R(a_k,a_k) = -\frac{\partial}{\partial a_k} \langle C\, Q(x), q_k \rangle \,|_{x=a_k} \,. \tag{15}$$

We define $\rho(x,y) := R(x,y) + \delta(x-y)$. This is the kernel of $(I-K)^{-1}$. It follows from (11) that

$$\frac{\partial}{\partial a_k}(I-K)^{-1} \doteq (-1)^k\, R(x,a_k)\, \rho(a_k,y),$$

and applying this to φ gives

$$\frac{\partial Q}{\partial a_k} = (-1)^k\, R(x,a_k)\, q_k. \tag{16}$$

Using this to evaluate the right side of (15) gives (3).

We write Q more explicitly as $Q(x, a_1, \dots, a_{2m})$, displaying its dependence on the a_k. Thus $q_j = Q(a_j, a_1, \dots, a_{2m})$. We see immediately that if $j \neq k$ then

$$\frac{\partial q_j}{\partial a_k} = \frac{\partial Q}{\partial a_k} \,|_{x=a_j} \,.$$

Hence (4) follows from (13) and (16). But

$$\frac{\partial q_k}{\partial a_k} = \frac{\partial Q}{\partial a_k} \,|_{x=a_k} + Q'(a_k) = (-1)^k\, R(a_k,a_k)\, q_k + Q'(a_k), \tag{17}$$

where the prime denotes differentiation with respect to x, and so we will have to find a formula for $Q'(x)$.

But first we derive identities (5) and (6). If we use the fact

$$\langle u,v \rangle\, w = (w \otimes v)\, u, \tag{18}$$

then we see that the former follows immediately upon applying both sides of the operator identity (12) to $M^i \varphi$ and setting $x = a_k$. For the latter, define

$T := \chi(I - K)^{-1}$, where χ denotes multiplication by $\chi(x)$. Then the product formula and (16) give

$$\begin{aligned}\frac{\partial T}{\partial a_k} &= (-1)^k\,(\delta(x-a_k)\,\rho(x-y) + \chi(x)\,R(x,a_k)\,\rho(a_k,y)) \\ &= (-1)^k\,(\delta(x-a_k)\,\rho(a_k-y) + \chi(x)\,R(a_k,x)\,\rho(a_k,y)) \\ &= (-1)^k\,\rho(a_k,x)\,\rho(a_k,y),\end{aligned}$$

where we have used the facts that $R(x,y) = R(y,x)$ for $x,y \in J$ and that $R(a_k,x)$ vanishes outside J. Applying φ to both sides, tensoring on the right with $M^i\varphi$ and integrating over $\mathbb{R}$ give (6).

Everything up to now applied to an arbitrary kernel of the form (1). Now we are going to use (8) and think of our operator K once again as acting on J, more precisely on $H_1(J)$. It is an easy fact that for an operator L with kernel $L(x,y)$ one has

$$[m\,D, L] \doteq \left(m(x)\frac{\partial}{\partial x} + m(y)\frac{\partial}{\partial y} + m'(y)\right)\,L(x,y).$$

(In the above, m denotes multiplication by $m(x)$ and $D = d/dx$.) We apply this to our operator K with kernel $K(x,y)\,\chi(y)$. From (8) we find that

$$\left(m(x)\frac{\partial}{\partial x} + m(y)\frac{\partial}{\partial y}\right)\,\langle\, C\,\varphi(x), \varphi(y)\,\rangle$$

$$= \langle\, C\,A(x), \varphi(x), \varphi(y)\,\rangle + \langle\, C\,\varphi(x), A(y)\,\varphi(y)\,\rangle = \langle\, C\,(A(x) - A(y))\,\varphi(x), \varphi(y)\,\rangle,$$

where we have used (7). Hence

$$\frac{1}{x-y}\,\left(m(x)\frac{\partial}{\partial x} + m(y)\frac{\partial}{\partial y}\right)\,\langle\, C\,\varphi(x), \varphi(y)\,\rangle$$

$$= \langle\, C\,\frac{A(x)-A(y)}{x-y}\,\varphi(x), \varphi(y)\,\rangle = \sum_{\substack{i+j=l-1\\ l\geq 1}} \langle\, C\,A_l\,x^i\varphi(x), y^j\,\varphi(y)\,\rangle.$$

(Recall the notation (9)). We also have

$$\left(m(x)\frac{\partial}{\partial x} + m(y)\frac{\partial}{\partial y} + m'(y)\right)\,\frac{1}{x-y}$$

$$= \frac{m(y)-m(x)}{(x-y)^2} + \frac{m'(y)}{x-y} = -\sum_{\substack{i+j=l-1\\ l\geq 1}} (j+1)\,m_{l+1}\,x^i\,y^j$$

and

$$\left(m(x)\frac{\partial}{\partial x} + m(y)\frac{\partial}{\partial y}\right)\,\chi(y) = -\sum_{j=1}^{2m}(-1)^j\,m(a_j)\,\delta(a_j - y).$$

Putting these things together gives

$$[m\,D,K] \doteq \sum_{\substack{i+j=l-1\\ l\geq 1}} \langle\, C\,(A_l-(j+1)m_{l+1}))\,x^i\,\varphi(x),y^j\,\varphi(y)\,\rangle\,\chi(y)$$

$$-\sum_{j=1}^{2m}(-1)^j\,m(a_j)\,K(x,a_j)\,\delta(a_j-y),$$

and it follows from this in the usual way that

$$[m\,D,(I-K)^{-1}] \doteq \sum_{\substack{i+j=l-1\\ l\geq 1}} \langle\, C\,(A_l-(j+1)m_{l+1}))\,Q_i(x),Q_j(y)\,\rangle\,\chi(y)$$

$$-\sum_{j=1}^{2m}(-1)^j\,m(a_j)\,R(x,a_j)\,\rho(a_j-y).$$

Applying this to φ, using (18) and the symmetry of $K(x,y)$, gives the identity

$$m(x)\,Q'(x)-(I-K)^{-1}\,m\,\varphi' = \sum_{\substack{i+j=l-1\\ l\geq 1}} U_j\,C(A_l-(j+1)m_{l+1}))\,Q_i(x)$$

$$-\sum_{j=1}^{2m}(-1)^j\,m(a_j)\,R(x,a_j)\,q_j. \tag{19}$$

Now by (8), $(I-K)^{-1}\,m\,\varphi' = (I-K)^{-1}\,A\,\varphi$, and so we compute commutators with (multiplication by) A:

$$[A,K] \doteq \frac{A(x)-A(y)}{x-y}\langle\, C\,\varphi(x),\varphi(y)\,\rangle\,\chi(y) = \sum_{\substack{i+j=l-1\\ l\geq 1}} A_l\,\langle\, C\,x^i\,\varphi(x),y^j\,\varphi(y)\,\rangle\,\chi(y),$$

whence

$$[A,(I-K)^{-1}] = \sum_{\substack{i+j=l-1\\ l\geq 1}} A_l\,\langle\, C\,Q_i(x),Q_j(y)\,\rangle\,\chi(y),$$

and this gives

$$(I-K)^{-1}m\,\varphi' = A(x)\,Q(x) - \sum_{\substack{i+j=l-1\\ l\geq 1}} A_l\,U_j\,C\,Q_i(x). \tag{20}$$

Combining (19) and (20) we obtain

$$\begin{aligned} m(x)\,Q'(x) &= A(x)\,Q(x) + \sum_{\substack{i+j=l-1\\ l\geq 1}} [U_j\, C(A_l - (j+1)m_{l+1}) - A_l\, U_j\, C]\, Q_i(x) \\ &\quad - \sum_{j=1}^{2m} (-1)^j\, m(a_j)\, R(x,a_j)\, q_j. \end{aligned}$$

Equation (10) follows from this, (17) and (13).

III. Remarks

Formula (14) for general kernel of the form (1) is by no means new with us. See, for example, Sec. II of [1]. In this reference, which treats certain kernels which have some features in common with ours, the matrix we would call U_0 also plays an important role.

In [4], where $n = 2$, our matrix C was $\begin{pmatrix} 0 & -1 \\ 1 & 0 \end{pmatrix}$, our vector $\varphi(x)$ was denoted by $(\tilde{\varphi}(x), \psi(x))$, our matrix A by $\begin{pmatrix} A & B \\ -C & -A \end{pmatrix}$ and our matrix U_i by $\begin{pmatrix} u_i & \tilde{v}_i \\ v_i & w_i \end{pmatrix}$. The derivation of the equations given here may seem at first sight simpler than the derivation in [4]. In fact the derivations are essentially the same, but because of its generality we were naturally led to use matrix notation here, and this led to an apparent simplification (and a real compactness of notation).

The simplest case is that of the "sine kernel"

$$\lambda \frac{\sin{(x-y)}}{x-y},$$

when $\varphi(x) = \sqrt{\lambda}\, e^{ix}$ and $\psi(x) = \sqrt{\lambda}\, e^{-ix}$. The equations in this case were first derived in [2], where they were obtained as the deformation equations for an isomonodromy problem. It was also shown there that if J is a single interval the determinant can be expressed in terms of a Painlevé transcendent. In [4] several other kernels were studied in detail, some of which also led to Painlevé functions. Recently Palmer [3] showed that all the equations considered in [4] were deformation equations for isomonodromy problems. It may very well be that the same is true of the more general class of equations considered here.

References

[1] A. R. Its, A. G. Izergin, V. E. Korepin and N. A.Slavnov, Differential equations for quantum correlation functions, Int. J. Mod. Phys. **B4** (1990) 1003–1037.

[2] M. Jimbo, T. Miwa, Y. Môri and M. Sato, Density matrix of an impenetrable Bose gas and the fifth Painlevé transcendent, Physica **1D** (1980) 80–158.

[3] J. Palmer, Deformation analysis of matrix models, Physica D, to appear.

[4] C. A. Tracy and H. Widom, Fredholm determinants, differential equations and matrix models, Comm. Math. Phys. **163** (1994) 33–72.

Craig A. Tracy
Department of Mathematics and Institute of Theoretical Dynamics
University of California, Davis, CA 95616, USA

Harold Widom
Department of Mathematics
University of California, Santa Cruz, CA 95064, USA

Operator Theory:
Advances and Applications, Vol. 78

Absorption Semigroups, Feller Property, and Kato Class

J. Voigt

Introduction

Let X be a locally compact space, m a measure on the Borel sets of X (satisfying suitable properties), $U = (U(t);\ t \geq 0)$ a *substochastic* (i.e., positive and contractive) strongly continuous semigroup on $L_1(m)$. Assume further that U^*, the adjoint semigroup, satisfies the Feller property, i.e., $C_0(X)$ (continuous functions tending to zero at infinity) is invariant under U^*, and U^* restricted to $C_0(X)$ is strongly continuous.

Let T denote the generator of U. For suitable $V : X \to [-\infty,\infty]$, an "absorption semigroup" U_V can be associated with the formal expression $T - V$. This paper is concerned with topics connected with the question whether the Feller property carries over from U^* to U_V^*.

In Section 1 we recall definitions and results. In particular the setup sketched above is made precise, and the Kato class associated with a substochastic semigroup is defined. The conditions on V (from [6]) guaranteeing that U_V^* satisfies the Feller property always require V to be locally in the Kato class.

It is shown in Section 2 that this latter property is necessary for the Feller property to carry over, if V is of one sign.

In Section 3 we show that the Kato class associated with U is complete. In fact, we first study the larger class of Miyadera perturbations and show that this class is also complete.

In Section 4 we characterize the Kato class for the heat equation on the half-line $(0,\infty)$, with absorption at $x = 0$. It turns out that at infinity one obtains the same conditions as for the usual one-dimensional Kato class, whereas at $x = 0$ the conditions are less restrictive.

1. Notation and known results

The notation and the results described in this section are mainly contained in [6].

Let X denote a locally compact space, m a measure on the Borel σ-algebra $\mathfrak{B}$ of X having the properties that $C_0(X)$ separates the functions of $L_1(m)$, and that $C_0(X) \to L_\infty(m)$ is injective. For convenience we also assume that $L_\infty(m)$ is the dual of $L_1(m)$.

For a one-parameter semigroup $(W(t); t \geq 0)$ on $L_\infty(m)$ we define the following properties. (By $\mathcal{K}$ we shall denote the system of compact subsets of X), and by $C_b(X)$ the bounded continuous functions on X.)

(F) (*Feller property*) $W(t)(C_0(X)) \subset C_0(X)$ $(t \geq 0)$, and the restriction of W to $C_0(X)$ is strongly continuous.

(L) (*localization property*) $\forall K \in \mathcal{K}\quad \exists \alpha > 0\quad \forall \varepsilon > 0\quad \exists K' \in \mathcal{K}:$

$$\left\|1_K W(t) 1_{X\setminus K'}\right\|_{\infty,\infty} \leq \varepsilon \qquad (0 \leq t \leq \alpha).$$

(S) (*smoothing property*) $W(t)(L_\infty(m)) \subset C_b(X)\quad (t > 0)$.

(SF) (*strong Feller property*) W satisfies (S) and (F).

For examples illustrating these notions we refer to [6; Section2].

Next suppose that $U = (U(t); t \geq 0)$ is a substochastic strongly continuous semigroup on $L_1(m)$, with generator T.

The *enlarged Kato class* associated with U is defined by

$$\widehat{\mathbf{K}} := \{V : X \to [-\infty,\infty] \text{ locally measurable; } V(1-T)^{-1} \text{ bounded}\}.$$

Positivity implies that the boundedness of $V(1-T)^{-1}$ is equivalent to

$$\|V\|_U := \sup_{f\in D(T), \|f\|\leq 1} \int_0^1 \|VU(t)f\|_1 \, dt < \infty.$$

For $V \in \widehat{\mathbf{K}}$, $\alpha > 0$ we define

$$\begin{aligned} c'_\alpha(V) &:= \sup\{\int_0^\alpha \|VU(t)f\| \, dt;\ f \in D(T), \|f\| \leq 1\}, \\ c(V) &:= \lim_{\alpha\to 0} c'_\alpha(V). \end{aligned}$$

The *Kato class* associated with U is then defined by

$$\mathbf{K} := \{V \in \widehat{\mathbf{K}};\ c(V) = 0\}.$$

If $V : X \to [-\infty,\infty]$ is locally measurable then there are certain procedures how to associate a C_0-semigroup U_V with the formal expression "$T - V$", under suitable restrictions on V. We refer to [12], [13], [1] for these procedures as well as for the definition of the notion of U-admissibility of V.

We recall from [3], [4], [11] that $V \in \widehat{\mathbf{K}}$, $c(V) < 1$ implies that $T - V$ is a generator.

To finish this section we quote results from [6] indicating conditions on V which guarantee that (F), (SF) carry over from U^* to U_V^*.

1.1. Proposition. ([6; Proposition 3.2]) Assume that U^* satisfies (F) and let $V \in \overline{C_b(X)}^{\mathbf{K}}$. Then U_V^* satisfies (F).

1.2. Theorem. ([6; Theorem 3.3]) Assume that U^* satisfies (F). Let $V = V_+ - V_-$, $V_\pm \geq 0$ be such that ψV is U-admissible and $U^*_{\psi V}$ satisfies (F) for all $\psi \in C_c(X)_+$ (positive continuous functions with compact support). Further let $V_- \in \widehat{\mathbf{K}}$, $c(V_-) < 1$. Then V is U-admissible, and U^*_V satisfies (F).

In order to state the next result we recall the definition of the local Kato class associated with U,

$$\mathbf{K}_{loc} := \{V : X \to [-\infty,\infty];\ \varphi V \in \mathbf{K} \text{ for all } \varphi \in C_c(X)\}.$$

1.3. Theorem. ([6; Theorems 4.5 and 4.6]) Assume that U^* satisfies (SF). Let $V = V_+ - V_-$, $V_\pm \geq 0$, $V_\pm \in \mathbf{K}_{loc}$, $V_- \in \widehat{\mathbf{K}}$, $c(V_-) < 1$. Then U^*_V satisfies (F). If additionally U^*_V satisfies (L), then U^*_V satisfies (SF).

2. Necessary conditions for the Feller property

In [6] it is shown that for the semigroup associated with the heat equation on $\mathbb{R}^d$ the Feller property for the perturbed semigroup implies that the perturbation is in $\mathbf{K}_{loc}$, if it is of one sign. (More generally, the corresponding property is shown for measure perturbations.) In this section we show that this result holds in the general situation.

2.1. Theorem. Assume that U^* satisfies (F). Let $V \geq 0$ or $V \leq 0$ be U-admissible, and assume that U^*_V satisfies (F). Then $V \in \mathbf{K}_{loc}$.

Proof. (i) Assume $V \geq 0$. For $k \in \mathbb{N}$ let $V^{(k)} := V \wedge k$. For $\alpha > 0$ one then obtains - by monotonicity and the Duhamel formula -

$$\begin{aligned}\int_0^\alpha U_V(\alpha - s)V^{(k)}U(s)\,ds \ &\leq\ \int_0^\alpha U_{V^{(k)}}(\alpha - s)V^{(k)}U(s)\,ds \\ &=\ U(\alpha) - U_{V^{(k)}}(\alpha)\ \leq\ U(\alpha) - U_V(\alpha).\end{aligned}$$

This implies

$$\int_0^\alpha U^*(s)V^{(k)}U^*_V(\alpha - s)\,ds\ \leq\ U^*(\alpha) - U^*_V(\alpha).$$

(The integral occuring here as well as the integrals below containing U^* should be understood as w^*-integrals.) Let $\varphi \in C_c(X)_+$ and choose $\psi \in C_0(X)_+$ such that $\psi = 1$ on $\operatorname{supp}\varphi$. Then there exists $\alpha_0 > 0$ such that $U^*_V(s)\psi \geq \frac{1}{2}$ on $\operatorname{supp}\varphi$ for all $0 \leq s \leq \alpha_0$, by (F) for U^*_V. Therefore

$$\begin{aligned}\int_0^\alpha U^*(s)\varphi V^{(k)}\,ds\ &\leq\ 2\int_0^\alpha U^*(s)\varphi V^{(k)}U^*_V(\alpha - s)\psi\,ds \\ &\leq\ 2\,\|\varphi\|_\infty\,(U^*(\alpha)\psi - U^*_V(\alpha)\psi)\end{aligned}$$

for $0 \le \alpha \le \alpha_0$. Hence

$$\left\| \int_0^\alpha \varphi V^{(k)} U(s)\, ds \right\|_{1,1} = \left\| \int_0^\alpha U^*(s) \varphi V^{(k)}\, ds \right\|_{\infty,\infty}$$
$$\le 2 \|\varphi\|_\infty \|U^*(\alpha)\psi - U_V^*(\alpha)\psi\|_\infty .$$

Taking now $k \to \infty$ we obtain

$$\left\| \varphi V \int_0^\alpha U(s)\, ds \right\|_{1,1} \le 2 \|\varphi\|_\infty \|U^*(\alpha)\psi - U_V^*(\alpha)\psi\|_\infty \to 0$$

for $\alpha \to 0$, by (F) for U^* and U_V^*.

(ii) Assume $V \le 0$. The U-admissibility of V implies that $-V$ is U_V-admissible, therefore U-admissible (use [13; Proposition 1.3(b)]), and $U = (U_V)_{-V}$ (use the evident inequality $(U_V)_{-V} \ge U \ge (U_{-V})_V$ together with the equality of the outer terms implied by [13; Theorem 2.6]). Therefore (i) implies that V is in the local Kato class $\mathbf{K}_{U_V,loc}$ associated with U_V. The inequality $U(t) \le U_V(t)$ $(t \ge 0)$ implies $\mathbf{K}_{U_V} \subset \mathbf{K}_U$, $\mathbf{K}_{U_V,loc} \subset \mathbf{K}_{U,loc}$. ∎

2.2. Remark. In the case $V \ge 0$ of Theorem 2.1, the hypothesis that V is U-admissible is not really needed since it follows from (F) for U_V^*. In fact, recall from [1] that for locally measurable $V \ge 0$ the semigroup $U_V(t)$ is defined for $t > 0$ and that

$$P := s - \lim_{t \to 0} U_V(t)$$

exists (and is a band projection). Assuming that U_V^* has (F) we obtain for $f \in L_1(m)$, $\varphi \in C_0(X)$ that

$$\begin{aligned} \int Pf\varphi\, dm &= \lim_{t\to 0} \int U_V(t) f \varphi\, dm \\ &= \lim_{t \to 0} \int f U_V^*(t) \varphi\, dm = \int f\varphi\, dm, \end{aligned}$$

and this implies $Pf = f$ for all $f \in L_1(m)$, $P = I$. (Compare [6; proof of Theorem 3.3].)

3. Notes on the Kato class

The main observation of this section is that the (enlarged) Kato class is always complete. We start with a related result in an even more general situation.

Let U be a C_0-semigroup on a Banach space E, with generator T. We denote the *Miyadera class* of perturbations of T by

$$\widehat{\mathbf{M}} := \{B \text{ a } T\text{-bounded operator in } X;\ \|B\|_U < \infty\}$$

where

$$\|B\|_U := \sup\{\int_0^1 \|BU(t)x\|\,dt;\ x \in D(T),\ \|x\| \le 1\}.$$

For $B \in \widehat{\mathbf{M}}$, $\alpha > 1$ we define

$$\begin{aligned} c'_\alpha(B) &:= \sup_{x\in D(T),\|x\|\le 1} \int_0^\alpha \|BU(t)x\|\,dt, \\ c(B) &:= \lim_{\alpha\to 0} c'_\alpha(B). \end{aligned}$$

Recall from [3], [4], [11] that $B \in \widehat{\mathbf{M}}$, $c(B) < 1$ implies that $T + B$ is a generator.

3.1. Proposition. $(\widehat{\mathbf{M}}, \|\cdot\|_U)$ is a Banach space, and $\mathbf{M} := \{B \in \widehat{\mathbf{M}};\ c(B) = 0\}$ is a closed subspace of $\widehat{\mathbf{M}}$.

Proof. Without restriction we assume U to be bounded.

We note first that the inclusion mapping $\widehat{\mathbf{M}} \subset L(D_T,E)$ (= the T-bounded operators) is continuous. Here D_T denotes $D(T)$, equipped with the norm $\|\cdot\|_T$,

$$\|x\|_T := \|(1-T)x\|$$

(which is equivalent to the graph norm). Then

$$\|B\|_{L(D_T,E)} = \|B(1-T)^{-1}\|.$$

To show the asserted continuity we take $x \in D(T)$ and estimate

$$\begin{aligned} \|B(1-T)^{-1}x\| &= \|\int_0^\infty Be^{-t}U(t)x\,dt\| \le \sum_{k=0}^\infty e^{-k}\int_0^1 \|Be^{-t}U(t)U(k)x\|\,dt \\ &\le \|B\|_U\,(1 + C\sum_{k=1}^\infty e^{-k})\|x\|, \end{aligned}$$

with $C = \sup_{t\ge 0}\|U(t)\|$.

It is easy to show that the closed unit ball of $\widehat{\mathbf{M}}$ is $L(D_T,E)$-closed, and therefore the completeness of $L(D_T,E)$ implies the completeness of $\widehat{\mathbf{M}}$ (cf. [8; 1.6]).

To show that $\mathbf{M}$ is closed we observe that the family $(c'_\alpha;\ 0 < \alpha \le 1)$ of functions on $\widehat{\mathbf{M}}$ is equicontinuous and therefore $c(\cdot)$ is continuous. Hence $\mathbf{M} = c^{-1}(0)$ is closed. ∎

Next, let $(\Omega, \mathcal{A}, \mu)$ be a measure space, U a positive C_0-semigroup on $L_1(\mu)$, and let $\widehat{\mathbf{K}}$ and $\mathbf{K}$ be as in Section 1.

3.2. Proposition. $\widehat{\mathbf{K}}$ and $\mathbf{K}$ are complete under the norm $\|\cdot\|_U$.

Proof. (i) Let (V_n) be a sequence in the closed unit ball of $\widehat{\mathbf{K}}$, $V_n \to V$ μ-locally almost everywhere. Then $V \in \widehat{\mathbf{K}}$, $\|V\|_U \le 1$. This is an easy consequence of Fatou's lemma.

(ii) For $V \in \widehat{\mathbf{K}}$ the set $\{|V| = \infty\}$ is a local null set.
Indeed, define $V' := 1_{\{|V|=\infty\}}$. Then

$$\|V'\|_U \leq \|\gamma|V|\,\|_U = \gamma\,\|V\|_U$$

for all $\gamma > 0$, hence $\|V'\|_U = 0$. For all $f \in L_1(\mu)$ one obtains $V'U(t)f = 0$ a.e. on [0,1], hence $V'f = V'U(0)f = 0$ by the continuity of $V'U(\cdot)f$. This implies the assertion.

(iii) Using (i) and (ii) one can show the completeness of $\widehat{\mathbf{K}}$ in the same way as the completeness of L_p-spaces for $1 \leq p < \infty$.

(iv) The equality $\mathbf{K} = \widehat{\mathbf{K}} \cap \mathbf{M}$ implies that $\mathbf{K}$ is closed. ∎

Of course, it is well established that the usual Kato classes on $\mathbb{R}^d$ are complete (see [12], and also Erratum to [9] in Bull. (N.S.) Amer. Math. Soc. **11**, p. 426 (1984)).

4. The Kato class for the heat semigroup on the half line

Let U denote the C_0-semigroup associated with the heat equation $\frac{\partial}{\partial t}u(t,x) = \frac{1}{2}\frac{\partial^2}{\partial x^2}u(t,x)$ and zero Dirichlet boundary conditions on $L_1(0,\infty)$.

We are going to show that the Kato classes for this semigroup are given by

$$\widehat{\mathbf{K}} = \mathbf{K} = \{V \in L_{1,loc}(0,\infty);\ 1_{[1,\infty)}V \in L_{1,loc,unif}(\mathbb{R}), (x \mapsto xV(x)) \in L_1(0,1)\}.$$

4.1. Remarks. (a) Recall that the Kato classes for the whole line $\mathbb{R}$ are just $\widehat{\mathbf{K}}_1 = \mathbf{K}_1 = L_{1,loc,unif}(\mathbb{R})$ ([9], [12]). The equality given above shows that the zero Dirichlet boundary condition at 0 only influences the property of the Kato class at 0.

(b) We insert an obvious but fundamental observation. If $\Omega \subset \tilde{\Omega} \subset \mathbb{R}^d$ are open subsets and U, $\tilde{U}$ are the semigroups associated with the heat equation $u_t = \frac{1}{2}\Delta u$ and zero Dirichlet boundary conditions on Ω, $\tilde{\Omega}$, respectively, then $U(t) \leq \tilde{U}(t)$ $(t \geq 0)$(see, e.g., [1], [SV], [5]), and therefore $\widehat{\mathbf{K}}_{\tilde{\Omega}} \subset \widehat{\mathbf{K}}_{\Omega}$, $\mathbf{K}_{\tilde{\Omega}} \subset \mathbf{K}_{\Omega}$, in the sense that $V \in \widehat{\mathbf{K}}_{\tilde{\Omega}}$ implies $1_\Omega V \in \widehat{\mathbf{K}}_\Omega$.

For the equalities given above we shall need the resolvent of $T = \frac{1}{2}\frac{d^2}{dx^2}$ at 1, with zero boundary condition at $x = 0$.

The inverse of $1 - T = 1 - \frac{1}{2}\frac{d^2}{dx^2}$ is given by an integral operator with kernel r_1,

$$r_1(x,y) = \frac{1}{\sqrt{2}}(e^{\sqrt{2}|x-y|} - e^{-\sqrt{2}|x+y|}).$$

Let V be such that $1_{[1,\infty)}V \in L_{1,loc,unif}(\mathbb{R})$, $(x \mapsto xV(x) \in L_1(0,1)$. From Remark 4.1 we obtain $1_{[\varepsilon,\infty)}V \in \mathbf{K}$ for all $\varepsilon > 0$. In order to show $V \in \mathbf{K}$ it is now sufficient to show

$$\left\|1_{(0,\varepsilon)}V(1-T)^{-1}\right\|_{1,1} \to 0 \quad (\varepsilon \to 0).$$

Now an easy calculation shows

$$\begin{aligned}\left\|1_{(0,\varepsilon)}V(1-T)^{-1}\right\|_{1,1} &= \sup_x \int_0^\varepsilon r_1(x,y)\frac{1}{y}y|V(y)|\,dy \\ &\leq (\sup_{x,y}\frac{1}{y}r_1(x,y))\int_0^\varepsilon y|V(y)|\,dy.\end{aligned}$$

Thus the assertion reduces to showing $\sup_{x,y}\frac{1}{y}r_1(x,y) < \infty$. This in turn is obtained by elementary calculations from the explicit form of the kernel.

On the other hand, let $V \in \widehat{\mathbf{K}}$. Let k_t denote the integral kernel for the operator $U(t)$ $(t > 0)$. Recall the inequalities

$$k_t(x,y) \leq \frac{1}{(2\pi t)^{1/2}}e^{-\frac{|x-y|^2}{2t}}(1 \wedge \frac{xy}{t}) \leq ck_t(x,y),$$

where $c > 1$ is such that

$$\frac{1}{c}(1 \wedge \eta) \leq 1 - e^{-\eta}\,(\leq 1 \wedge \eta) \quad (\eta > 0)$$

(cf. [2; Example 4.1.1]). For $0 < t \leq 1$, $x,y \geq 1$ one concludes

$$\frac{1}{(2\pi t)^{1/2}}e^{-\frac{|x-y|^2}{2t}} \leq ck_t(x,y).$$

For $0 < t \leq 1$ this implies

$$\begin{aligned}&\sup_x \int \frac{1}{(2\pi t)^{1/2}}e^{-\frac{|x-y|^2}{2t}}1_{[1,\infty)}(y)|V(y)|\,dy \\ &= \sup_{x\geq 1}\int \frac{1}{(2\pi t)^{1/2}}e^{-\frac{|x-y|^2}{2t}}1_{[1,\infty)}(y)|V(y)|\,dy \\ &\leq c\sup_x \int k_t(x,y)1_{[1,\infty)}(y)|V(y)|\,dy.\end{aligned}$$

Now the definition of $\widehat{\mathbf{K}}$ implies $1_{[1,\infty)}V \in \widehat{\mathbf{K}}_1$ $(= L_{1,loc,unif}(\mathbb{R}))$.

Further, from the explicit form of the kernel r_1 one easily obtains a constant $c_1 > 0$ such that

$$c_1 y \leq r_1(x,y) \qquad \text{for } 0 \leq y \leq 1 \leq x \leq 2.$$

Therefore

$$c_1\int_0^1 y|V(y)|\,dy \leq \sup_{1\leq x\leq 2}\int_0^\infty r_1(x,y)|V(y)|\,dy \leq \left\|V(1-T)^{-1}\right\|_{1,1} < \infty.$$

References

[1] W. Arendt, C.J.K. Batty: Absorption semigroups and Dirichlet boundary conditions. Math. Ann. **295**, 427–448 (1993).

[2] E. B. Davies: Heat kernels and spectral theory. Cambridge University Press, Cambridge, 1989.

[3] I. Miyadera: On perturbation theory for semi-groups of operators. Tôhoku Math. J. **18**, 299–310 (1966).

[4] I. Miyadera: On perturbation for semigroups of linear operators. Scientific Researches, School of Education, Waseda Univ. **21**, 21–24 (1972) (in Japanese).

[5] E. M. Ouhabaz: Invariance of closed convex sets and domination criteria for semigroups. Preprint.

[6] E.-M. Ouhabaz, P. Stollmann, K.-Th. Sturm, J. Voigt: The Feller property for absorption semigroups. Preprint 1994.

[7] W. Arendt: Gaussian estimates and interpolation of the spectrum in L^p. Preprint.

[8] H.H. Schaefer: Topological vector spaces. Springer-Verlag,New York,1971.

[9] B. Simon: Schrödinger semigroups. Bull. Amer. Math. Soc. (N.S.) **7**, 447–526 (1982).

[10] P. Stollmann, J. Voigt: Perturbation of Dirichlet forms by measures. Potential Analysis, to appear.

[11] J. Voigt: On the perturbation theory for strongly continuous semigroups. Math. Ann. **229**, 163–171 (1977).

[12] J. Voigt: Absorption semigroups, their generators, and Schrödinger semigroups. J. Funct. Anal. **67**, 167–205 (1986).

[13] J. Voigt: Absorption semigroups. J. Operator Theory **20**, 117–131 (1988).

Jürgen Voigt
Fachrichtung Mathematik, Institut für Analysis
Technische Universität Dresden
D-01062 Dresden, Germany

Operator Theory:
Advances and Applications, Vol. 78

Gaussian estimates and analytic semigroups

L. Weis

1. Introduction

Traditionally, the analyticity of semigroups generated by differential operators on $L_p(\Omega, \mu)$ is proven via a-priori estimates and duality ([2], [4], [10] Chap. 7)or by Stein's interpolation theorem (see e.g. [5], theorem 1.4.2).

In recent years, an extensive literature on "Gaussian estimates" on the "Heat kernel" of various differential operators A on a domain Ω of $\mathbb{R}^n$ appeared (see the examples below), They imply that if $T_t = e^{tA}$ is the semigroup generated by A on $L_2(\Omega, \mu)$, there are constants $C < \infty, b < \infty, t_o > 0$ such that for $0 \leq f \in L_2(\Omega, \mu), 0 \leq t \leq t_0$

$$(G) \qquad |T_t f(x)| \leq \frac{c}{(2\pi t)^{n/2}} \int_\Omega e^{-|x-y|^2/bt} f(y)dy \qquad \text{a. e. on } \Omega$$

One may hope now to obtain the analyticity of T_t on the spaces $L_p(\Omega, \mu)$ by this domination from the corresponding properties of the Gaussian semigroup. For selfadjoint operators A this was done by E.M. Ouhabaz in [9] using the methods of [5], sect. 3.4. In this paper we widen the scope of this technique by removing the assumption of selfadjointness. So the theorem below is not only applicable to differential operators in divergence form, but also to more general ones including lower order terms and non-symmetric potentials.
The advantage of the approach of [9] and the theorem below over the traditional techniques quoted above is for one that in some cases it requires less smoothness on the coefficients of A, but mainly that one obtains the maximal angle of analyticity, which is usually not true for the interpolation method.

Examples. The main application we have in mind is to differential operators A on $L_p(\mathbb{R}^n)$ of the form

$$(1) \qquad Af(x) = \sum_{i,j} a_{ij}(x)\partial_{x_i}\partial_{x_j} f(x) + \sum_i b(x)\partial_{x_i} f(x) + d(x) \cdot f(x).$$

where A is uniformly elliptic in the sense that there is a constant $\mu \geq 1$ such that

$$\frac{1}{\mu}|\xi|^2 \leq \Sigma a_{ij}(x)\xi_i\xi_j \leq \mu|\xi|^2 \qquad \text{for } \xi \in \mathbb{R}^n$$

If the coefficients in (1) are Holder continuous and bounded then A generates an analytic semigroup of angle $\frac{\pi}{2}$ on $L_2(\Omega)$ (cf. [10] sect 7.2) and satisfies (G) (cf.

[11] theorem 1.1.) These estimates still hold under weaker condition on a_{ij} (e.g. if a_{ij} belong to certain Dini classes; see the literature quoted in [11], [10]) and on b_i, d (they may be non-continuous and unbounded as long as they belong to certain Kato-classes of potentials, see [11], [13] and the papers quoted there). For differential operators in divergence form see [5], [12] and for Schrödinger operators with singular potentials see also [17], [1] Section 8. □

In all of these cases we can directly apply the following theorem. Let Ω be an open domain in $\mathbb{R}^n$ with Lebesgue measure and consider kernel the operator

$$T(z)f(x) = \int K(z,x,y)f(y)dy \tag{1}$$

Theorem. *Assume that* (1) *defines an analytic semigroup on* $L_q(\Omega), 1 < q < \infty$, *of angle* Θ *with a Gaussian estimate*

$$|K(t,x,y)| \leq Mt^{-\frac{n}{2}} \exp[-b|x-y|^2/t] \qquad \text{for } t > 0 \tag{2}$$

Then $T(z)$ *extends to an analytic semigroup on all spaces* $L_p(\Omega), 1 \leq p < \infty$, *of the same angle* Θ. *If in addition*

$$\{T(t)f : f \in C_0(\bar{\Omega}) \cap L_q(\mathbb{R}^n), t > 0\} \subset C_0(\bar{\Omega}) \tag{3}$$

is dense in the $||\cdot||_\infty$ *norm, then* $T(z)$ *also defines an analytic semigroup of angle* Θ *on* $C_0(\bar{\Omega})$.

Remarks. a) The theorem remains true if we replace $L_p(\Omega)$ by a Banach function space E on Ω with order continuous norm (see [3] for such spaces).
b) In the last part of the statement one can replace $C_0(\bar{\Omega})$ by any closed subspace E of $L_\infty(\Omega)$ for which $E \cap L_q(\Omega)$ is $||\cdot||_\infty$ dense in E and the corresponding condition (3) is fullfilled.
c) If (2) holds only for $0 < t \leq T$ one still obtains that $e^{-\omega t}T(t)$ extends to a semigroup of angle Θ on $L_p(\Omega)$ for an approbriate $\omega > 0$. (see, Proposition 2.1 of [9]).

2. Proof of the Theorem.

For the proof of the theorem we need two lemmas which are extensions of the argument in [5], p. 102–104.

Lemma 1. *Let* $T(z), \in \Sigma(\Theta)$, *be an analytic semigroup on* $L_q(\Omega,\mu)$ *for some* $1 < q < \infty, 0 < \Theta \leq \frac{\pi}{2}$. *Assume also that there are constants* C, D *and* $\alpha > 0$ *with*

1) $||T(t)||_{L_p \to L_p} \leq C$ for $t \geq 0$ and $1 \leq p \leq \infty$

2) $||T(t)||_{L_1 \to L_\infty} \leq Dt^{-\alpha}$ for $t > 0$.

Then for all $0 < \Theta_1 < \Theta$ and $z \in \Sigma(\Theta_1)$ we have

$$\|T(z)\|_{L_1 \to L_\infty} \leq 2^\alpha CD \cdot M(\Theta) \left(1 - \frac{\tan \Theta_1}{\tan \Theta}\right)^{-\alpha} (\text{ Re} z\ )^{-\alpha}$$

$$M(\varphi) = \sup\{\|T(z)\|_{L_q \to L_q} : z \in \Sigma(\varphi)\}.$$

Proof. By the Riesz interpolation theorem we get for $t > 0$

$$\|T(t)\|_{L_q \to L_\infty} \leq \|T(t)\|_{L_1 \to L_\infty}^{\frac{1}{q}} \cdot \|T(t)\|_{L_\infty \to L_\infty}^{\frac{1}{q'}} \leq D^{\frac{1}{q}}\ C^{\frac{1}{q'}}\ t^{\frac{-\alpha}{q}}$$

$$\|T(t)\|_{L_1 \to L_q} \leq \|T(t)\|_{L_1 \to L_\infty}^{\frac{1}{q'}} \cdot \|T(t)\|_{L_1 \to L_1}^{\frac{1}{q}} \leq D^{\frac{1}{q'}}\ C^{\frac{1}{q}}\ t^{\frac{-\alpha}{q'}}$$

Given Θ_1 with $0 < \Theta_1 < \Theta$ choose now

$$\delta = \frac{\tan \Theta_1}{\tan \Theta}, \quad \varepsilon = \frac{1}{2}(1 - \delta).$$

Then $0 < \delta < 1, \delta + 2\varepsilon = 1$ and $z = \alpha + i\beta \in \Sigma(\Theta_1)$ implies that $z' = \delta\alpha + i\beta \in \Sigma(\Theta)$. For $f \in L_1(\Omega) \cap L_q(\Omega)$ the semigroup property gives $T(z)f = T(\varepsilon\alpha)T(\delta\alpha + i\beta)T(\varepsilon\alpha)f$. By the above norm estimates

$$\begin{aligned} \|T(z)\|_{L_1 \to L_\infty} &\leq \|T(\varepsilon\alpha)\|_{L_1 \to L_q} \|T(z')\|_{L_q \to L_q} \|T(\varepsilon\alpha)\|_{L_q \to L_\infty} \\ &\leq C \cdot D(\varepsilon\alpha)^{-\alpha} M(\Theta), \end{aligned}$$

where $\alpha =$ Re z and $\varepsilon = \frac{1}{2}\left(1 - \frac{\tan \Theta_1}{\tan \Theta}\right)$. □

Remark. It is well known that every bounded operator $T : L_1(\Omega, \mu) \to L_\infty(\Omega, \mu)$ can be represented by a kernel operator

$$Tf(x) = \int_\Omega K(x,y) f(y) d\mu(y)$$

with $\|T\|_{L_1 \to L_\infty} = \|K\|_{L_\infty(\Omega \times \Omega)}$. Hence Lemma 1 gives in essence a bound on the kernel of T_z for $Imz \neq 0$ which can be improved in presence of a Gaussian estimate for T_z.

Lemma 2. *Let $T(z), z \in \Sigma(\Theta)$, be a bounded analytic semigroup of kernel operators*

$$T_z f(x) = \int K(z, x, y) f(y) dy$$

on $L_q(\mathbb{R}^n)$ for some $1 < q < \infty, 0 < \Theta \leq \frac{\pi}{2}$. Assume that we have a Gaussian estimate with constants b, M :

(*G*) $$|K(t, x, y)| \leq Mt^{\frac{-n}{2}} \exp[-b|x - y|^2/t] \quad \text{for } t > 0$$

then for all $0 < \Theta_1 < \Theta$ and $z \in \Sigma(\Theta_1)$

$$|K(z,x,y)| \leq M(\Theta_1)(\text{ Re}z)^{-\alpha} \exp[-b(\Theta_1)|x-y|^2 \text{ Re }(z^{-1})]$$

where $M(\Theta_1) \sim (\Theta - \Theta_1)^{-\alpha}, b(\Theta_1) \sim (\Theta - \Theta_1)$ for $\Theta_1 \to \Theta$ and fixed Θ. If $\Theta = \frac{\pi}{2}$ we can choose $\Theta_1 = \frac{\pi}{2}$ and $C(\Theta_1) \sim \frac{1}{\varepsilon^\alpha}, b(\Theta_1) = b(1-\varepsilon)$ for all $0 < \varepsilon < 1$.

Proof. If follows from (G) and the corresponding properties of the Gaussian semigroup that 1) and 2) of Lemma 1 are fullfilled for $\alpha = \frac{N}{2}$. Choose $0 < \Theta_1 < \Theta_2 < \Theta$. Lemma 1 gives

(1) $\|T_z\|_{L_1 \to L_\infty} \leq \tilde{C}(\text{ Re}z)^{-\alpha}, \; \tilde{C} = 2^\alpha CDM(\Theta)[1 - \frac{\tan\Theta_2}{\tan\Theta}]^{-\alpha}$ for $z \in \Sigma(\Theta_2)$

Now we can proceed in the same way as in [5] 3.4.8, p. 103/104. Put $D = \{z \in \mathbb{C} : 0 < \arg(z) < \Theta_2\}$ and consider for $z = re^{i\varphi} \in D$

$$g(z) = x^{-\alpha} K(z^{-1},x,y) \exp\{b|x-y|^2 e^{i(\pi/2-\Theta_2)} z / \sin\Theta_2\}$$

then

$$|g(z)| \leq \max(M\tilde{C})(\text{ sec }\Theta_2)^\alpha \tag{2}$$

For $z = re^{i\Theta_2}, r > 0$, this follows from (1) and for $z = r, r > 0$, from 3) and one can apply the Phragmén- Lindelöf theorem for cones to get (+) for all $z \in D$. (2) implies for $z = re^{i\varphi}, 0 < |\varphi| \leq \Theta_1$:

$$\begin{aligned} |K(z,x,y)| &\leq \max(M,\tilde{C}) \left(\frac{\cos\varphi}{\cos\Theta_2}\right)^\alpha (\text{ Re } z)^{-\alpha} \\ &\quad \exp\left\{-b|x-y|^2 \text{Re }(z^{-1}) \frac{\sin(\Theta_2 - |\varphi|)}{\sin\Theta_2 \cos\varphi}\right\} \\ &\leq M(\Theta_1)(\text{ Re}z)^{-\alpha} \exp\{-b(\Theta_1)|x-y|^2 \text{ Re}(z^{-1})\} \end{aligned}$$

where

$$M(\Theta_1) = \max(M, 2^\alpha CDM(\Theta))[1 - \frac{\tan\Theta_2}{\tan\Theta}]^{-\alpha} (\cos\Theta_2)^{-\alpha}$$

$$b(\Theta_1) = (\sin\Theta_2)^{-1} \sin(\Theta_2 - \Theta_1) \tag{3}$$

To have these constants independent of Θ_2, choose e.g. $\Theta_2 = \frac{1}{2}(\Theta + \Theta_1)$. If $\Theta_1 \to \Theta$, note that $\Theta_2 \to \Theta$ and

$$1 - \frac{\tan\Theta_2}{\tan\Theta} \sim \Theta - \Theta_2, \qquad \sin(\Theta_2 - \Theta_1) \sim \Theta_2 - \Theta_1$$

Assume now that $\|T(z)\|_{L_q}$ is bounded on $\{l : \text{ Re } l > 0\}$. For fixed $\varepsilon > 0$ and every $z = re^{i\varphi}, 0 < |\varphi| < \frac{\pi}{2}$, choose $\Theta_1 = |\varphi|$ and Θ_2, Θ in such a way that

$\Theta_1 < \Theta_2 < \Theta < \frac{\pi}{2}, \Theta_2 = (1-\varepsilon)\frac{\pi}{2} + \varepsilon\Theta_1, \tan\Theta = 2\tan\Theta_2$. Since

$$\begin{aligned}\sin(\Theta_2 - |\varphi|) &= \sin\big((1-\varepsilon)(\frac{\pi}{2} - \Theta_1)\big) \geq (1-\varepsilon)\cos\Theta_1, \\ \varepsilon\cos(\Theta_1) &\leq \cos\big((1-\varepsilon)\frac{\pi}{2} + \varepsilon\Theta_1\big) = \cos(\Theta_2)\end{aligned}$$

and

$$1 - \frac{\tan\Theta_2}{\tan\Theta} = \frac{1}{2}$$

estimate (3) takes now the form

$$|K(z,x,y)| \leq \max(M, 2^{2\alpha}C\cdot D(M(\Theta))\varepsilon^{-\alpha}(\text{ Re} z)^{-\alpha}\exp\{-b\,|x-y|^2(1-\varepsilon)\text{Re}(z^{-1})\}$$

and the last claim follows. □

With similar arguments as is [9], we can now prove our main result.

Proof of the Theorem. a) Lemma 2 shows that for a fixed $\Theta_1 < \Theta \quad T(z), z \in \Sigma(\Theta_1)$, is dominated by the convolution operator

$$(K(z)f)(x) = \int g(z, x-y)f(y)dy$$

with

$$g(z,x) = M_1(\text{ Re } z)^{\frac{-n}{2}} \exp[-b_1 x^2 \text{ Re } (z^{-1})]$$

and

$$\begin{aligned}\|g(z,\cdot)\|_{L_1} &= M_1(\text{ Re} z \text{ Re } (z^{-1})b2\pi)^{\frac{-n}{2}} \\ &\leq M_1(b_1 2\pi)^{\frac{-n}{2}}(\cos\Theta_1)^{-n}\end{aligned} \tag{4}$$

for all $z \in \Sigma(\Theta_1)$. By Young's inequality

$$\sup\{\|T(z)\|_{L_{p(\Omega)}} : z \in \Sigma(\Theta_1\} < \infty \qquad \text{for all } 1 < p < \infty. \tag{5}$$

To show the analycity of $T(z)$ in a space $L_p(\Omega), 1 \leq p < \infty$, pick $f \in L_p(\Omega), g \in L_{p'}(\Omega)$ and $f_n \in L_p(\Omega) \cap L_q(\Omega), g_n \in L_{p'}(\Omega) \cap L_{q'}(\Omega)$ with $\|f_n - f\|_{L_p} \to 0, \|g_n - g\|_{L_{p}'} \to 0$. Then

$$z \in \Sigma(\Theta_1) \to \langle g, T(z)f\rangle = \lim_{n\to\infty} \langle g_n, T(z)f_n\rangle$$

is analytic as the limit of a uniformly bounded sequence of analytic funtions and $z \in \Sigma(\Theta_1) \to T(z) \in B(L_p)$ is analytic by [7] theorem 3.10.1. If $f \in L_p(\Omega) \cap L_q(\Omega)$ then $T(z)f \to f$ in L_q and $|T(z)f| \leq K(z)f \to f$ in L_p for $\Sigma(\Theta_1) \ni z \to 0$. This implies $T(z)f \to f$ in L_q. □

b) Let E now be a closed subspace of $L_\infty(\Omega)$ such that $\{T(t)f : f \in L_q \cap E, t > 0\}$ is a dense subset of E with respect to $\|\cdot\|_\infty$. Since $L_q \cap E$ is dense in E (5) implies that $T(t), t > 0$, is uniformly bounded in $B(E)$. Furthermore, for $f \in L_q(\Omega)$ there are $f_n \in E \cap L_q$ with $\|f_n - f\|_{L_q} \to 0$. Since $T(t)$ is bounded from $L_q(\Omega)$ into $L_\infty(\Omega)$ we have $\|T_t f_n - T_t f\|_{L_\infty} \to 0$ and $T(t)f_n \in E$ implies $T(t)f \in E$. Hence $T(t)$ maps $L_q(\Omega)$ into E for all $t > 0$. Given $z \in \Sigma(\Theta_1)$ choose $\varepsilon > 0$ with $z - \varepsilon \in \Sigma(\Theta_1)$ then $T(z)f = T(\varepsilon)T(z-\varepsilon)f \in E$ for all $f \in E \cap L_q(\Omega)$. Hence $z \in \Sigma(\Theta_1) \to T(z) \in B(E)$ is a uniformly bounded, analytic function and it follows now from (3) and [7], theorem 10.6.6, that $T(z), z \in \Sigma(\Theta_1)$, is an analytic semigroup in E. □

References

[1] S. Albeverio, Ph. Blanchard, Z. Ma: Feynman-Kac semigroups in terms of signed smooth measures, preprint.

[2] H. Amann: Dual semigroups and second order linear elliptic boundary value problems, Israel J. Math. 45 (1983), 225–254.

[3] C. Bennett and R. Sharpley: "Interpolation of Operators", Academic Press, Boston, 1988.

[4] P. Cannarsa and V. Vespri, Generation of anayltic semigroups in the L_p topology by elliptic operators in $\mathbb{R}^n$. Israel J. Math. Vol. 61 (1988) 235–255.

[5] E.B. Davies, "Heat Kernels and Spectral Theory," Cambridge University Press, Cambridge 1989.

[6] J.A. Goldstein: "Semigroups of Linear Operators and Applications", Oxford Univ. Press 1985.

[7] E. Hille and R.S. Phillips: "Functional Analysis and Semigroups", Amer. Math. Soc. Coll. Publ. Providence, R.I. 1957.

[8] Y. Katznelson: "An Introduction to Harmonic Analysis", Dover Pub., New York, 1976.

[9] E.M. Ouhabaz: Gaussian estimates and holomorphy of semigroups, preprint.

[10] A. Pazy: Semigroups of Linear Operators and Applications to Partial Differential Equations, Springer, New York – Berlin, 1983.

[11] F.O. Porper and S.D. Eidelman: Two-sided estimates of fundamental solutions of second order parabolic equations and some applications. Russian Math. Surveys 39 (1984), 119–178.

[12] D.W. Robinson: "Elliptic Operators and Lie Groups", Oxford Univ. Press, Oxford 1991.

[13] Y. A. Semenov: On the theory of the operators $L_t = \frac{\partial}{\partial t} - \Delta + b \cdot \nabla + v$, preprint.

[14] B. Simon: Schrödinger semigroups. Bull. Amer. Math. Soc. 7 (1982) 447–526.

[15] H.B. Stewart: Generation of analytic semigroups by strongly elliptic operators. Trans. Amer. Soc. **199** (1974) 141–162.

[16] H.B. Stewart: Generation of analytic semigroups by strongly elliptic operators under general boundary conditions. Trans. Amer. Soc. **259** (1980) 299–310.

[17] J. Voigt, Absorption semigroups, their generators and Schrödinger semigroups. J. Funct. Anal. **67** (1986) 167–205.

Lutz Weis
Department of Mathematics
Louisiana State University
Baton Rouge, La 70803, USA.

Operator Theory:
Advances and Applications, Vol. 78

New Channels of Scattering for Long-Range Potentials

D. Yafaev[1]

Abstract

We consider a system of three one-dimensional particles with one of pair potentials $V^\alpha(x^\alpha)$ decaying at infinity as $|x^\alpha|^{-\rho}$, $0 < \rho < 1/2$. It is shown that such a system can possess channels of scattering not included in the usual list of channels called the asymptotic completeness. Similar channels arise in a two-body system with a long-range potential if a traditional condition on its derivatives is relaxed.

1 Introduction

An aim of the scattering theory is to find the asymptotics as $t \to \infty$ of the solution $u(t) = \exp(-iHt)f$ of the time-dependent Schrödinger equation with a Hamiltonian $H = -2^{-1}\Delta + V(x)$, $V(x) = \overline{V(x)}$, in the space $\mathcal{H} = L_2(\mathbb{R}^d)$. If f is an eigenvector of H, i.e. $Hf = \lambda f$, then obviously $u(t) = \exp(-i\lambda t)f$. Suppose now that f is orthogonal to the subspace $\mathcal{H}^{(p)}$ spanned by all eigenvectors. In the two-body short-range case when $V(x) = O(|x|^{-\rho})$, $\rho > 1$, the asymptotics of $u(t)$ is the same as that for the free system, that is

$$\exp(-iHt)f = \exp(-iH_0t)f_0 + o(1) \tag{1.1}$$

for some $f_0 \in \mathcal{H}$ and $H_0 = -2^{-1}\Delta$. The symbol $o(1)$ means a function such that its norm in the space $\mathcal{H}$ tends to zero as $t \to \infty$. One can rewrite (1.1) in an equivalent way as

$$(\exp(-iHt)f)(x) = \exp(i\Phi_0(x,t))t^{-d/2}g(x/t) + o(1), \tag{1.2}$$

where $\Phi_0(x,t) = x^2(2t)^{-1}$, $g = \exp(i\pi d/4)\hat{f}_0$ and $\hat{f}_0$ is the Fourier transform of f_0.

The relation (1.2) holds true (see e.g.[1]) also for long-range potentials satisfying the condition

$$|D^\kappa V(x)| \le C(1+|x|)^{-\rho-|\kappa|}, \quad \rho > 0, \tag{1.3}$$

for $|\kappa| = 0,1,2$. In this case the phase function $\Phi_0(x,t)$ is a (perhaps, approximate) solution of the eikonal equation

$$\partial\Phi_0/\partial t + 2^{-1}|\nabla\Phi_0|^2 + V = 0. \tag{1.4}$$

[1]On leave from Math. Inst. of St. Petersburg, Russia

The asymptotics (1.2) shows that, if f belongs to the absolutely continuous subspace $\mathcal{H}^{(ac)} = \mathcal{H} \ominus \mathcal{H}^{(p)}$ of the operator H, then the solution $(\exp(-iHt)f)(x)$ "lives" in the region where $|x| \sim t$.

The situation is more complicated in the many-body case when

$$H = -2^{-1}\Delta + \sum_{\alpha} V^{\alpha}(x^{\alpha}), \quad 1 \leq \alpha \leq \alpha_0 < \infty,$$

and x^{α} are orthogonal projections of $x \in \mathbb{R}^d$ on some given subspaces $X^{\alpha} \subset \mathbb{R}^d$. Set $X_{\alpha} = \mathbb{R}^d \ominus X^{\alpha}$. The three-body case is distinguished by the assumption $X_{\alpha} \cap X_{\beta} = \{0\}$ for $\alpha \neq \beta$. Suppose first that all pair potentials V^{α} are short-range, i.e. that they satisfy the condition $V^{\alpha}(x^{\alpha}) = O(|x^{\alpha}|^{-\rho})$ for some $\rho > 1$. Thc relation (1.1) (or (1.2) with $\Phi_0(x,t) = x^2(2t)^{-1}$) is fulfilled for f from some subspace $\mathcal{H}_0 \subset \mathcal{H}^{(ac)}$. In general, $\mathcal{H}_0 \neq \mathcal{H}^{(ac)}$ and the orthogonal complement $\mathcal{H}^{(ac)} \ominus \mathcal{H}_0$ is determined by the point spectra of pair Hamiltonians $H^{\alpha} = -2^{-1}\Delta_{x^{\alpha}} + V^{\alpha}$ acting in the spaces $\mathcal{H}^{\alpha} = L_2(X^{\alpha})$. Let us introduce their eigenvalues $\lambda^{\alpha,k}$ and eigenvectors $\psi^{\alpha,k}$. For every couple $\{\alpha,k\}$ there exists a subspace $\mathcal{H}_{\alpha,k} \subset \mathcal{H}^{(ac)}$ such that for any $f \in \mathcal{H}_{\alpha,k}$

$$(\exp(-iHt)f)(x) \sim \psi^{\alpha,k}(x^{\alpha}) \exp(i\Phi_{\alpha,k}(x_{\alpha},t))t^{-d_{\alpha}/2}g(x_{\alpha}/t) + o(1) \qquad (1.5)$$

for the function $\Phi_{\alpha,k}(x_{\alpha},t) = x_{\alpha}^2(2t)^{-1} - \lambda^{\alpha,k}t$, $d_{\alpha} = \dim X_{\alpha}$ and some $g \in L_2(X_{\alpha})$. Note that the subspaces $\mathcal{H}_0$ and $\mathcal{H}_{\alpha,k}$ are constructed as images of the corresponding wave operators $W_0 : L_2(\mathbb{R}^d) \to \mathcal{H}$ and $W_{\alpha,k} : L_2(X_{\alpha}) \to \mathcal{H}$. The basic result of the scattering theory, called the asymptotic completeness, is that the sum of all subspaces $\mathcal{H}_0$ and $\mathcal{H}_{\alpha,k}$ exhausts the absolutely continuous subspace $\mathcal{H}^{(ac)}$ of the operator H. Actually, $\mathcal{H}^{(ac)}$ can be decomposed into the orthogonal sum

$$\mathcal{H}^{(ac)} = \mathcal{H}_0 \oplus (\bigoplus_{\alpha,k} \mathcal{H}_{\alpha,k}). \qquad (1.6)$$

This result was first obtained by L. D. Faddeev [2] under some additional assumptions. The optimal formulation is due to V. Enß [3].

For three-body systems with long-range pair potentials V^{α} the answer is almost the same if $V^{\alpha}(x^{\alpha})$ satisfy the condition (1.3) with some $\rho > \sqrt{3} - 1$ as functions of x^{α}. As shown by V. Enss [4, 5], in this case again the asymptotic completeness (1.6) holds. For $f \in \mathcal{H}_0$ the asymptotics (1.2) and for $f \in \mathcal{H}_{\alpha,k}$ the asymptotics (1.5) with suitable functions $\Phi_0(x,t)$ and $\Phi_{\alpha,k}(x_{\alpha},t)$ are fulfilled. We emphasize that for $f \in \mathcal{H}_{\alpha,k}$ the solution $\exp(-iHt)f$ is localized in the variable x^{α} and, on the contrary, $|x_{\alpha}| \sim t$. Thus such solutions play an intermediary role between those for $f \in \mathcal{H}^{(p)}$ and $f \in \mathcal{H}_0$. Note also that in the case $\rho > 1/2$ the asymptotic completeness (1.6) was established in the papers [7, 8] under some additional assumptions.

We do not discuss here results on N body systems for $N > 3$. Let us mention only that the condition $\rho > \sqrt{3} - 1$ reappeared in the paper by J. Dereziński

[11], where asymptotic completeness was considered for arbitrary N by a method different from [4, 5].

Our goal is to show that for some three-body systems with sufficiently slowly decreasing pair potentials there exist channels of scattering different from (1.2) and (1.5). Actually, we consider one-dimensional particles with pair potentials satisfying (1.3) for $\rho \in (0,1/2)$. We pick up some α, say $\alpha = 1$, and construct a subspace $\mathfrak{H} \subset \mathcal{H}^{(ac)}$ of initial data f such that for $f \in \mathfrak{H}$ the solution $u(t) = \exp(-iHt)f$ of the Schrödinger equation "lives" in the region where $|x_1| \sim t$ and $|x^1| \sim t^\sigma$ for $\sigma = (\rho+1)/3 \in (1/3,1/2)$. Thus solutions $u(t)$ for $f \in \mathfrak{H}$ are intermediary between those for $f \in \mathcal{H}_0$ and $f \in \mathcal{H}_{1,k}$. The subspace $\mathfrak{H}$ is orthogonal to $\mathcal{H}_0$ and $\mathcal{H}_{\alpha,k}$ for all α and k. The restriction of the operator H on $\mathfrak{H}$ has the absolutely continuous spectrum which coincides with $\mathbb{R}_+$. This contradicts, of course, the asymptotic completeness (1.6). We require that one of pair potentials $V^\alpha(x^\alpha) = \pm v_\alpha |x^\alpha|^{-\rho}$, $\alpha \neq 1$, $0 < \rho < 1/2$, $v_\alpha > 0$, for large positive values of $\mp x^\alpha$. Other pair potentials can be short-range. Thus we consider two cases " + " and " − " which differ by the sign of the long-range interaction. They are parametrized by upper and lower indices, respectively.

From the point of view of quantum mechanics we construct channels of scattering where particles of some chosen pair are relatively close to one another and the third particle is far away. This pair is bound by a potential depending on the position of the third particle but its bound state is evanescent as $t \to \infty$. The third particle moves freely in some effective long-range potential created by the pair of the first two particles. This means that its distance from the center of mass of this pair grows as t.

One can imagine that, for example, the mass of the first particle is infinite and it is fixed at the origin. Positions of the second and third particles are given by coordinates x^1 and x_1, respectively. The second particle is retained on the positive half-axis by a potential barrier on the negative half-axis due to its interaction $V^1(x^1)$ with the first particle. A potential $V^2(x^2) = V^2(x^1 - x_1)$ of interaction between the second and third particles is long-range. In the case " + ", V^2 is repulsive and the third particle moves to plus infinity. In the case " − ", V^2 is attactive and the third particle moves to minus infinity. In both cases $|x_1|$ grows as t for $t \to \infty$. The second particle is jammed to the first one by an interaction with the third particle. In the long run the second particle escapes to plus infinity but x^1 behaves as t^σ, $\sigma \in (1/3,1/2)$, for large t.

Note that existence of new channels for three-particle systems automatically implies the same phenomena for systems of more than three particles. It suffices to take a system where all particles but three are free and the system of these three distinguished particles possesses a described channel.

New channels of scattering arise also for a two-body system with a long-range potential $V(x)$ if one tries to relax the condition (1.3) on its derivatives. A typical example of such a potential is the following. Let

$$\mathbb{R}^d = X_1 \oplus X^1, \quad \dim X_1 = d_1,\ \dim X^1 = d^1,\ d_1 + d^1 = d, \tag{1.7}$$

be some decomposition of $\mathbb{R}^d$ into an orthogonal sum and let

$$V(x_1,x^1) = -v(< x_1 >^q + < x^1 >^q)^{-\rho/q}, \quad \rho \in (0,1),\ q \in (0,2),\ v > 0, \tag{1.8}$$

where we use the notation $< y >= (1 + |y|^2)^{1/2}$. It turns out that under the assumption $q < 2(1-\rho)$ there exist solutions of the time-dependent Schrödinger equation "living" in a parabolic neighbourhood of the subspace X_1. In this case $\mathcal{H}_0 \neq \mathcal{H}^{(ac)}$.

Our proof of the existence of new channels of scattering relies on the following general construction which is similar to that of [9]. We consider a decomposition (1.7) but do not make any special assumptions about a potential $V(x) = V(x_1,x^1)$. Let us introduce an operator

$$H^1(x_1) = -2^{-1}\Delta_{x^1} + V(x_1,x^1) \tag{1.9}$$

acting in the space $L_2(X^1)$. Suppose that the operator $H^1(x_1)$ has an eigenvalue $\lambda(x_1)$ and denote by $\psi(x_1)$ the corresponding normalized eigenfunction. In the particular case when $V(x)$ does not depend on x_1 the operator (1.9) describes a three-body system with only one non-trivial pair interaction. In this case both channels (1.2) and (1.5) (where $\alpha = 1$) exist. We are looking for a generalization of (1.5) in the case when $\lambda(x_1) \to 0$ as $|x_1| \to \infty$. In interesting situations the function $\lambda(x_1)$ decreases slower than $|x_1|^{-1}$. Let us consider it as an "effective" potential energy and associate to the long-range potential $\lambda(x_1)$ the phase function $\Phi(x_1,t)$. We will prove under some assumptions that for every $g \in L_2(X_1)$ there exists an element $f \in \mathcal{H}^{(ac)}$ such that

$$(\exp(-iHt)f)(x) = \psi(x_1,x^1)\exp(i\Phi(x_1,t))t^{-d_1/2}g(x_1/t) + o(1). \tag{1.10}$$

A set of these elements f is a subspace $\mathfrak{H} \subset \mathcal{H}^{(ac)}$. It is constructed as the range of the corresponding wave operator $\mathcal{W} : L_2(X_1) \to \mathcal{H}$. The subspace $\mathfrak{H}$ is orthogonal to the subspace $\mathcal{H}_0$ of elements f for which the relation (1.2) holds.

The existence of solutions of the time-dependent Schrödinger equation with the asymptotics (1.10) requires rather special assumptions which are naturally formulated in terms of eigenfunctions $\psi(x_1,x^1)$ of the operator $H^1(x_1)$. It turns out that typically the asymptotics of $\psi(x_1,x^1)$ as $\lambda(x_1) \to 0$ has a certain self-similarity:

$$\psi(x_1,x^1) \sim |x_1|^{-\sigma d^1/2}\Psi(|x_1|^{-\sigma}x^1) \tag{1.11}$$

for some $\Psi \in L_2(X^1)$ and $\sigma > 0$. For instance, in the case (1.8) this is true for $\sigma = (\rho+q)(2+q)^{-1}$. We prove the asymptotics (1.10) if the relation (1.11) is fulfilled for $\sigma < 1/2$. This gives the condition $q < 2(1-\rho)$ in the example (1.8).

On the other hand, the relation (1.11) for $\sigma > 1/2$ does not ensure the existence of solutions with the asymptotics (1.10). Indeed, consider again the potential (1.8) for $q = 2$ when $\sigma = 1/2 + \rho/4$. This potential is radial so that the asymptotics (1.2) is fulfilled for all $f \in \mathcal{H}^{(ac)}$.

The study of the asymptotics of an eigenfunction $\psi(x_1)$ as the corresponding eigenvalue $\lambda(x_1)$ tends to zero could be a cumbersome problem. In the three particle case an additional difficulty is to distinguish $\psi(x_1)$ generating channels (1.5) and (1.10). Therefore it is important that $\psi(x_1)$ can be chosen as an *approximate* solution of the equation $H^1(x_1)\psi(x_1) - \lambda(x_1)\psi(x_1) = 0$. We construct $\psi(x_1)$ as a bound state in some effective potential well which is getting more shallow but wider as $x_1 \to \pm\infty$. This well has a triangular shape so that $\psi(x_1)$ is more concentrated around its bottom than for a rectangular well. We emphasize that channels (1.10) may arise even if the operator (1.9) has no eigenvalues. In this case $\psi(x_1)$ is generated by a *positive* minimum (in the variable x^1) of the potential $V(x_1,x^1)$ and hence $\psi(x_1)$ corresponds to a resonant state of the operator $H^1(x_1)$. In applications to three-body systems this situation is realized in the case " + ". In both cases "$\pm$" the function Ψ intervening in the asymptotics (1.11) is an eigenfunction of an auxiliary operator $K = -2^{-1}(d/dx^1)^2 + v_\alpha \rho x^1$ in the space $L_2(\mathbb{R}_+)$ with the Dirichlet boundary condition at $x^1 = 0$. Thus Ψ is the Airy function. Note that the channel (1.5) can be constructed in terms of every eigenfunction $\Psi = \Psi_n$ of the operator K.

A construction of channels (1.10) is given in Section 2. Concrete examples of two- and three-body systems for which solutions of the type (1.10) exist are discussed in Sections 3 and 4, respectively.

2 A general construction

Let us, first, recall some simple results of the two-particle long-range scattering theory. We discuss here only the existence of wave operators and, following [10], define them in the coordinate representation. Compared to a tradional approach [11, 12, 1], where a free dynamics is defined in the momentum representation, this allows us to dispense with the stationary phase method. Let us introduce a unitary operator $U_0(t)$,

$$(U_0(t)f)(x) = \exp(i\Phi_0(x,t))t^{-d/2}f(x/t),$$

in the space $\mathcal{H} = L_2(\mathbb{R}^d)$ corresponding to the right-hand side of (1.2). We define $\Phi_0(x,t)$ as an approximate solution of the eikonal equation (1.4).

Proposition 2.1 Let the condition (1.3) be satisfied. Set $R = \partial\Phi_0/\partial t + \frac{1}{2}|\nabla\Phi_0|^2$. Then there exists a function $\Phi_0(x,t)$ such that for some $\varepsilon > 0$ and any constants $c_2 > c_1 > 0$

$$\sup_{c_1 t \le |x| \le c_2 t} |R(x,t) + V(x)| \le C t^{-1-\varepsilon}. \tag{2.1}$$

Actually, in [1] the existence of an exact solution of the equation (1.4) was proven. In [12, 10] an explicit construction of a function $\Phi_0(x,t)$ satisfying (2.1) was given. This requires, however, the condition (1.3) for all $|\kappa| \leq [\rho^{-1}] + 1$. One can show additionally that

$$\sup_{c_1 t \leq |x| \leq c_2 t} |D^\kappa(\Phi_0(x,t) - x^2(2t)^{-1})| \leq C t^{1-|\kappa|-\varepsilon}, \quad |\kappa| = 1,2. \tag{2.2}$$

Clearly, a function Φ_0 depends on a long-range potential V. Denote by Γ the corresponding mapping $\Gamma : V \mapsto \Phi_0$. Set $H_0 = -2^{-1}\Delta$, $H = -2^{-1}\Delta + V$. Let $\mathbf{X}$ be the operator of multiplication by $x^2/2$ in the space $L_2(\mathbb{R}^d)$. The following assertion can be checked by an explicit calculation.

Proposition 2.2 Let a function $\Phi_0(x,t)$ satisfy the conditions (2.1) and (2.2). Then for any $f \in C_0^\infty(\mathbb{R}^d \setminus \{0\})$

$$\|(i\partial/\partial t + 2^{-1}\Delta - V)U_0(t)f\| = O(t^{-1-\varepsilon}), \quad t \to \infty.$$

Corollary 2.3 The wave operator

$$W_0 = s - \lim_{t\to\infty} \exp(iHt)U_0(t) \tag{2.3}$$

exists, it is isometric and the intertwining property $HW_0 = W_0\mathbf{X}$ holds.

We emphasize that W_0 equals the product of the usual (perhaps, modified) wave operator relating H_0 and H and of the Fourier transform.

Our aim now is to establish the asymptotics (1.10). Let us formulate precise assumptions on a function $\psi(x_1,x^1)$. We suppose that $\psi(x_1,\cdot)$ is an *approximate* eigenfunction of the operator (1.9), i.e.

$$-2^{-1}\Delta_{x^1}\psi(x_1,x^1) + V(x_1,x^1)\psi(x_1,x^1) = \lambda(x_1)\psi(x_1,x^1) + Y(x_1,x^1), \tag{2.4}$$

where

$$\|\psi(x_1,\cdot)\|_{L_2(X^1)} \to 1, \quad \|Y(x_1,\cdot)\|_{L_2(X^1)} = O(|x_1|^{-1-\varepsilon}), \quad |x_1| \to \infty, \tag{2.5}$$

for some $\varepsilon > 0$. Approximate "eigenvalues" $\lambda(x_1)$ should satisfy a usual condition of long-range scattering:

$$|D^\kappa \lambda(x_1)| \leq C(1+|x_1|)^{-\rho-|\kappa|}, \quad \rho > 0, \quad |\kappa| \leq [\rho^{-1}] + 1. \tag{2.6}$$

With respect to $\psi(x_1,x^1)$ itself we suppose that

$$\|\nabla_{x_1}\psi(x_1,\cdot)\|_{L_2(X^1)} = O(|x_1|^{-1}), \quad \|\Delta_{x_1}\psi(x_1,\cdot)\|_{L_2(X^1)} = O(|x_1|^{-1-\varepsilon}) \tag{2.7}$$

as $|x_1| \to \infty$. Furthermore, we require that auxiliary functions

$$\tilde{\psi}(x_1,x^1) = (x^1)^2\psi(x_1,x^1), \tag{2.8}$$

$$\xi_\sigma(x_1,x^1) = \sigma\langle\nabla_{x^1}\psi,x^1\rangle + 2^{-1}\sigma d^1\psi + \langle\nabla_{x_1}\psi,x_1\rangle \tag{2.9}$$

satisfy the bounds

$$\|\tilde{\psi}(x_1,\cdot)\|_{L_2(X^1)} = O(|x_1|^{1-\varepsilon}), \quad \|\xi_\sigma(x_1,\cdot)\|_{L_2(X^1)} = O(|x_1|^{-\varepsilon}). \tag{2.10}$$

We emphasize that all assumptions on functions $\psi(x_1,x^1)$, $Y(x_1,x^1)$ and $\lambda(x_1)$ are used for sufficiently large $|x_1|$ only. These conditions are well adapted to treat functions $\psi(x_1,x^1)$ with the asymptotics (1.11). Actually, suppose for a moment that for some $\sigma \in \mathbb{R}$ there is the precise equality in (1.11) and that the function $(x^1)^2\Psi(x^1)$ belongs to $L_2(X^1)$. Then $\xi_\sigma = 0$, both conditions (2.7) are fulfilled and $\tilde{\psi}$ satisfies (2.10) with $1-\varepsilon = 2\sigma$. Thus the conditions $\varepsilon > 0$ and $\sigma < 1/2$ are equivalent.

Let us reformulate the asymptotics (1.10) in terms of the corresponding wave operator. Define the function $\Phi(x_1,t)$ by the construction of Proposition 2.1 which we apply to a function $\lambda(x_1)$ (in place of $V(x)$) in the variable x_1 (in place of x), i.e. we set $\Phi = \Gamma\lambda$. Let $U_1(t)$ be the operator of a modified free evolution in the variable x_1,

$$(U_1(t)f)(x_1) = \exp(i\Phi(x_1,t))t^{-d_1/2}f(x_1/t). \tag{2.11}$$

According to Proposition 2.2 for any $f \in C_0^\infty(\mathbb{R}^{d_1}\setminus\{0\})$ there is a bound

$$\|(i\partial/\partial t + 2^{-1}\Delta_{x_1} - \lambda)U_1(t)f\|_{L_2(X_1)} = O(t^{-1-\varepsilon}), \quad t\to\infty.$$

Let an isometric operator $J : L_2(X_1) \to \mathcal{H}$ be defined by the equality

$$(Jf)(x_1,x^1) = \psi(x_1,x^1)f(x_1), \quad J = J(\psi). \tag{2.12}$$

We prove the existence of the wave operator

$$\mathcal{W} = s - \lim_{t\to\infty} \exp(iHt)JU_1(t). \tag{2.13}$$

To that end it would suffice to verify that a function $u_1(t) = JU_1(t)f$, $f \in C_0^\infty(\mathbb{R}^{d_1}\setminus\{0\})$, is a "good" approximation to a solution of the time-dependent Schrödinger equation, i.e.

$$\|\frac{d}{dt}(\exp(iHt)u_1(t))\| = \|i\partial u_1/\partial t - Hu_1\| = O(t^{-1-\varepsilon}), \quad \varepsilon > 0, \quad t\to\infty. \tag{2.14}$$

Unfortunately, for the function $\psi(x_1,x^1)$ with the asymptotics (1.11) the estimate (2.14) can be satisfied for $\varepsilon = 0$ only.

It turns out that a better approximation to a solution of the Schrödinger equation is given by the formula

$$u(x,t) = \exp(i\sigma\gamma(x^1,t))u_1(x,t), \quad \gamma(x^1,t) = (x^1)^2(2t)^{-1}, \quad u_1(t) = JU_1(t)f. \tag{2.15}$$

The following assertion can be verified by an explicit calculation.

Proposition 2.4 Let a function $u(t)$ be defined by the equalities (2.15) where $f \in C_0^\infty(\mathbb{R}^{d_1} \setminus \{0\})$. Suppose that a function $\psi(x_1,x^1)$ satisfies for some $\sigma \in \mathbb{R}$ the conditions (2.7) - (2.10). Let functions $Y(x_1,x^1)$ and $\lambda(x_1)$, defined by (2.4), obey (2.5), (2.6). Then

$$\|i\partial u/\partial t - Hu\| = O(t^{-1-\varepsilon}), \quad \varepsilon > 0, \quad t \to \infty.$$

This assertion implies the main result of this section. Below $\mathbf{X}_1$ is multiplication by $x_1^2/2$ in the space $L_2(X_1)$.

Theorem 2.5 Let functions $\psi(x_1,x^1)$, $\lambda(x_1)$ and $Y(x_1,x^1)$ satisfy the assumptions of Proposition 2.4. Define the operator $U_1(t)$ by the equality (2.11), where $\Phi = \Gamma\lambda$. Then the limit (2.13) exists and the wave operator $\mathcal{W} : L_2(X_1) \to \mathcal{H}$ is isometric. The intertwining property $H\mathcal{W} = \mathcal{W}\mathbf{X}_1$ holds. In particular, the restriction of H on the range $R(\mathcal{W})$ of $\mathcal{W}$ has the absolutely continuous spectrum which coincides with $\mathbb{R}_+$.

Solutions with the asymptotics (1.1) and (1.10) "live" in different regions of $\mathbb{R}^d$. Let us give a precise formulation of this statement.

Theorem 2.6 Suppose that both wave operators (2.13) and (2.3) exist, for some functions Φ and Φ_0, respectively. Assume that the function (2.8) satisfies the bound

$$\|\tilde{\psi}(x_1,\cdot)\|_{L_2(X^1)} = O(|x_1|^{2-\varepsilon}), \quad \varepsilon > 0.$$

Then the ranges of the operators $\mathcal{W}$ and W_0 are orthogonal.

3 Two-body potentials

We construct here a concrete class of potentials $V(x)$ for which all the assumptions of the previous section are satisfied. These potentials decay at infinity and thus correspond to the two-body case.

Let (1.7) be some decomposition of $\mathbb{R}^d$ and let the potential $V(x_1,x^1)$ be defined by (1.8). This function is infinitely differentiable and $V(x) = O(|x|^{-\rho})$ as $|x| \to \infty$. Outside of any conical neighbourhood of the planes X_1 and X^1 the bound (1.3) is fulfilled for arbitrary κ. This suffices for the existence of the wave operator (2.3). If $q = 2$, then $V(x_1,x^1)$ is a radial function so that this wave operator is, of course, complete. If $1 \leq q < 2$, then the bound (1.3) is satisfied (uniformly in directions of x) for $|\kappa| = 1$ but is violated for $|\kappa| = 2$. If $0 < q < 1$, then (1.3) is violated already for $|\kappa| = 1$. We shall prove that under the assumption

$$1 - \rho < q < 2(1 - \rho) \tag{3.1}$$

the wave operator (2.13) exists. This ensures that W_0 fails to be complete.

Let us construct an approximate eigenfunction $\psi(x_1,x^1)$ of the operator (1.9). On the subspace of functions admitting a factorization

$$\psi(x^1) = r^{-(d^1-1)/2}\tilde{\psi}(r)\mathcal{Y}_l(\omega), \quad r = |x^1|, \quad \omega = x^1 r^{-1},$$

where $\mathcal{Y}_l$, $l = 0,1,2,...$, is a spherical function, $H(x_1)$ reduces to the operator

$$-2^{-1}d^2/dr^2 + \nu r^{-2} - v(a^q + <r>^q)^{-\rho/q}, \quad a = <x_1>, \tag{3.2}$$

$$\nu = 4^{-1}(d^1-1)(d^1-3) + l(l+d^1-2) \geq -1/4, \quad d^1 > 1,$$

in the space $L_2(\mathbb{R}_+)$. It is complemented by the boundary condition $\tilde{\psi}(0) = 0$ if necessary. If $d^1 = 1$, then $\nu = 0$ and the boundary condition is either $\tilde{\psi}(0) = 0$ or $\tilde{\psi}'(0) = 0$. We construct $\psi(x_1,x^1)$ as an exact eigenfunction of a Schrödinger operator with an increasing potential. To that end we replace in (3.2) $<r>$ by r and the potential $(a^q + r^q)^{-\rho/q}$ by the first two terms of its Taylor expansion at the point $r = 0$. Finally, changing in the equation

$$-2^{-1}\tilde{\psi}'' + \nu r^{-2}\tilde{\psi} + v_0 a^{-\rho-q} r^q \tilde{\psi} = (\lambda + va^{-\rho})\tilde{\psi}, \quad v_0 = v\rho q^{-1},$$

the variable

$$R = a^{-\sigma} r, \quad \sigma = (\rho+q)(2+q)^{-1}, \quad \Psi(R) = \tilde{\psi}(r),$$

we obtain the operator

$$K_0 = -2^{-1}d^2/dR^2 + \nu R^{-2} + v_0 R^q \tag{3.3}$$

with the discrete spectrum.

Let us choose a normalized eigenfunction $\Psi(R) = \Psi_n(R)$ corresponding to one of its eigenvalues $\Lambda = \Lambda_n$. The function $\Psi(R)$ decays exponentially as $R \to \infty$ and equals the Airy function if $q = 1$, $\nu = 0$. We set now

$$\psi(x_1,x^1) = a^{-\sigma/2} r^{-(d^1-1)/2} \Psi(a^{-\sigma} r)\mathcal{Y}_l(\omega) \tag{3.4}$$

and

$$\lambda(x_1) = -va^{-\rho} + \Lambda a^{-2\sigma}, \quad 2\sigma > \rho. \tag{3.5}$$

The "potential" (3.5) satisfies the assumptions (2.6) and "eigenfunctions" (3.4) satisfy the conditions (2.7) - (2.10) if $\sigma < 1/2$. The inequality $\sigma < 1/2$ is equivalent to $q < 2(1-\rho)$.

It remains to check the condition (2.5). The function $Y(a,r)$ (see (2.4)) corresponding to (3.4), (3.5) equals

$$Y(a,x^1) = -vy(a,r)a^{-\sigma/2} r^{-(d^1-1)/2} \Psi(a^{-\sigma} r)\mathcal{Y}_l(\omega), \tag{3.6}$$

where

$$y(a,r) = (a^q + <r>^q)^{-\rho/q} - a^{-\rho} + \rho q^{-1} a^{-\rho-q} r^q.$$

Note that

$$|y(a,r)| \leq a^{-q} < r >^{q} |$$
$$C(a^{-\rho-2q} < r >^{2q} + a^{-\rho-q}).$$

According to (3.6), it follows that

$$\|Y(a,\cdot)\| \leq C(a^{-\rho-2q+2q\sigma} + a^{-\rho-q}).$$

The term $a^{-\rho-q}$ satisfies (2.5) under the left assumption (3.1). The inequality $\rho+2q-2q\sigma > 1$ is equivalent to $q > 2(3-\rho)^{-1}(1-\rho)$ and is fulfilled automatically whenever $q > 1-\rho$.

Applying now Theorems 2.5 and 2.6 we arrive at

Theorem 3.1 Let a potential V be defined by (1.8) where ρ and q satisfy (3.1). Let Λ be any eigenvalue and Ψ a corresponding eigenfunction of the operator (3.3). Define the function $\psi(x_1,x^1)$ and the "potential" $\lambda(x_1)$ by the equalities (3.4), (3.5), where $a =< x_1 >, r = |x^1|$, and set $\Phi = \Gamma\lambda$. Then the wave operator (2.13) exists. It is isometric and $H\mathcal{W} = \mathcal{W}\mathbf{X}_1$. The range of $\mathcal{W}$ is orthogonal to that of the wave operator (2.3).

Notice that the above construction does not work in the case $q = 2$. Actually, everything goes through but the corresponding approximate eigenfunction satisfies the equality (3.4) with $\sigma = (2+\rho)/4 > 1/2$. Thus the first condition (2.10) is not fulfilled. Of course, non-existence of the limit (2.13) should have been expected, since the wave operator (2.3) is complete now.

We emphasize that for potentials (1.8) there exists a countable set of wave operators $\mathcal{W}_n$ corresponding to each eigenvalue Λ_n of the operator K_0. The ranges of these operators are, obviously, orthogonal to each other. Furthermore, one can interchange the roles of variables x_1 and x^1. This gives us a new set of wave operators $\mathcal{W}^m$ whose ranges are orthogonal to those of $\mathcal{W}_n$ (and, of course, to that of (2.3)).

Note finally that the first assumption (3.1), i.e. $1-\rho < q$, is of technical nature and can be omitted (see [13], for details).

4 The three-body case

Let us now consider the three-particle Hamiltonian $H = 2^{-1}\Delta + V(x)$ where $V(x) = V^1(x^1) + V^2(x^2)$ and x^α are orthogonal projections of $x \in \mathbb{R}^d$ on given subspaces $X^\alpha, \alpha = 1,2, X^1 \neq X^2$. We suppose that $d = 2$, $\dim X^\alpha = 1$ and that X^2 is not orthogonal to X^1. Then

$$x^2 = l(x^1 - mx_1), \quad l \neq 0, \quad m \neq 0, \tag{4.1}$$

where x_1 is the orthogonal projection of x on $X_1 = \mathbb{R}^d \ominus X^1$. The Hamiltonian H describes three one-dimensional particles of finite masses; one of three pair interactions is zero. With respect to other pair potentials we make

Assumption 4.1 A bounded function $V^1 \geq 0$ and $V^1(x^1) = 0$ for $x^1 \geq 0$. There exists $p_1 < 0$ such that $V^1(x^1)$ is twice differentiable for $x^1 \leq p_1 < 0$ and

$$V^1(x^1) \geq v_1|x^1|^{-r}, v_1 > 0, |D^\kappa V^1(x^1)| \leq C|x^1|^{-r-\kappa}, \kappa = 0,1,2, 0 < r < 2.$$

Assumption 4.2 A bounded function V^2 equals $V^2(x^2) = \pm v_2|x^2|^{-\rho}, \rho \in (0, 1/2)$,
$v_2 > 0$, for $\mp x^2 \geq p_2$ and some p_2 .

Suppose for definiteness $l = 1$, $m > 0$. Let us show that under these assumptions all conditions of Section 2 are satisfied for the potential

$$V(x_1,x^1) = V^1(x^1) + V^2(x^1 - mx_1).$$

Let us consider the "approximate equation" (2.4) with an error term $Y(x_1,x^1)$. Our construction for large *positive* values of $\pm x_1$ of $\psi(x_1,x^1)$ is different for $x^1 > 0$ and $x^1 < 0$. For $x^1 > 0$ we construct ψ as an L_2-solution of the Schrödinger equation with some simple potential. Note that

$$V(x_1,x^1) = \pm v_2(\mp x^1 + a)^{-\rho}, \quad a = \pm m x_1 > 0,$$

for $x^1 < a - p_2$ in the case " + " and for all x^1 in the case " − ". It turns out that it suffices to replace this function by the first two terms of its Taylor expansion at the point $x^1 = 0$. This gives the function

$$V_0(x_1,x^1) = \pm v_2 a^{-\rho} + v_2 \rho a^{-\rho-1} x^1.$$

Let parameters μ and Λ be related by the equality

$$\mu(x_1, \Lambda) = \pm v_2 a^{-\rho} + 2^{-1} v^2 \Lambda a^{-2\sigma}, \quad v = (2v_2\rho)^{1/3}, \quad \sigma = (\rho+1)/3, \qquad (4.2)$$

and let $\mathrm{Ai}(y)$ be the (exponentially decreasing as $y \to \infty$) Airy function. Then for any Λ

$$\psi_+(x^1,x_1;\Lambda) = \mathrm{Ai}(va^{-\sigma}x^1 - \Lambda), \quad x^1 \geq 0, \qquad (4.3)$$

satisfies the equation

$$-2^{-1}\psi_+'' + V_0(x_1,x^1)\psi_+ = \mu(x_1,\Lambda)\psi_+.$$

On the half-axis $x^1 < 0$ we again define ψ as a solution of the Schrödinger equation with some simplified potential but we allow that it depends on the parameter Λ. Here we keep track of a barrier due to V^1 and try to get rid of the potential V^2. The simplest possibility is to neglect V^2, as well as μ, for large negative x^1

and to replace $V^2(x^1 - mx_1)$ in an "essential" region by its value at $x^1 = 0$. More precisely, we introduce a function $\zeta \in C^\infty$ such that $\zeta(x^1) = 1$ for $x^1 \in (-1,0)$ and $\zeta(x^1) = 0$ for $x^1 \leq -2$ and set

$$Q(x^1,a,\Lambda) = 2V^1(x^1) - v^2\Lambda a^{-2\sigma}\zeta(x^1/A), \quad A = a^\delta, \quad \delta < 2\sigma r^{-1}.$$

Note that $Q(x^1,a,\Lambda)$ is close to $2V^1(x^1)$ for large a. Let us define $\psi_-(x^1,a,\Lambda)$ as a solution of the equation

$$-\psi_-'' + Q(x^1,a,\Lambda)\psi_- = 0, \quad x^1 < 0,$$

with the asymptotics

$$\psi_-(x^1,a,\Lambda) \sim Q^{-1/4}(x^1,a,\Lambda)\exp\Big(-\int_{x^1}^p Q^{1/2}(y,a,\Lambda)dy\Big) \tag{4.4}$$

as $x^1 \to -\infty$.

We need rather a detailed information on functions $\psi_-(x^1,a,\Lambda)$ although their values do not intervene into a final result. A construction of $\psi_-(x^1,a,\Lambda)$ reduces to a consideration of a Volterra integral equation (cf. [14]). These functions satisfy the bounds

$$|(\partial/\partial x^1)^\kappa(\partial/\partial a)^\alpha(\partial/\partial\Lambda)^\beta\psi_-(x^1,a,\Lambda)| \leq Ca^{-\alpha}\exp(-c|x^1|^{1-r/2}) \tag{4.5}$$

for $c > 0, \kappa = 0,1$ and arbitrary α and β. Furthermore,

$$\lim_{a\to\infty} \psi_-^{(k)}(x^1,a,\Lambda) = \psi_1^{(k)}(x^1) \tag{4.6}$$

for any fixed $x^1 \leq 0$ (in particular, for $x^1 = 0$) uniformly for bounded Λ. Here ψ_1 is a solution of the equation $-\psi_1'' + 2V^1(x^1)\psi_1 = 0$ with the asymptotics (4.4) where $Q = 2V^1$. This solution, as well as its derivative, does not vanish:

$$\psi_1(x^1) \neq 0, \quad \psi_1'(x^1) \neq 0, \quad x^1 \leq 0. \tag{4.7}$$

We consider (cf. [15]) the matching condition for ψ_+ and ψ_- at $x^1 = 0$ as an equation for $\Lambda = \Lambda(a)$. By virtue of (4.3) it has the form

$$va^{-\sigma}\psi_-(0,a,\Lambda(a))\psi_-'(0,a,\Lambda(a))^{-1} = \mathrm{Ai}(-\Lambda(a))\mathrm{Ai}'(-\Lambda(a))^{-1}. \tag{4.8}$$

Choose any zero $-\Lambda_n$ of the Airy function: $\mathrm{Ai}(-\Lambda_n) = 0$. According to (4.5)–(4.7), the implicit function theorem shows that the equation (4.8) has a solution such that $\Lambda(a) \to \Lambda_n$ as $a \to \infty$. Furthemore, $\Lambda(a) = \Lambda_n + O(a^{-\sigma})$. Now we define "eigenfunctions" $\psi(x_1,x^1)$ by the equalities

$$\begin{aligned}\psi(x_1,x^1) &= \gamma_+(a)\mathrm{Ai}(va^{-\sigma}x^1 - \Lambda(a)), \quad x^1 \geq 0,\\ \psi(x_1,x^1) &= \gamma_-(a)\psi_-(x^1,a,\Lambda(a)), \quad x^1 \leq 0,\end{aligned}$$

where $a = \pm m x_1$. The constants $\gamma_+(a)$ and $\gamma_-(a)$ are determined by the continuity of the function $\psi(x_1,x^1)$ at $x^1 = 0$ and by the normalization $\|\psi(x_1,\cdot)\| = 1$. The first of them yields that $\gamma_-(a) \sim a^{-\sigma}\gamma_+(a)$ and the second one shows that $\gamma_+(a) \sim a^{-\sigma/2}$.

The potential $\lambda(x_1) = \mu(x_1,\Lambda_n)$, where μ is defined by (4.2), obeys, of course, the estimate (2.6). A verification of conditions (2.7) - (2.10) for $\psi(x_1,x^1)$ is different for $x^1 \geq 0$ and $x^1 \leq 0$. On the half-axis $x^1 \geq 0$ it suffices to use, as in the two-body case, the self-similarity of the function (4.3). The condition $\sigma < 1/2$ is equivalent to $\rho < 1/2$. The contribution of the half-axis $x^1 \leq 0$ is negligible because $\gamma_-(a) \sim a^{-3\sigma/2}$ and the function $\psi_-(x^1,a,\Lambda)$ decays exponentially as $x^1 \to -\infty$ uniformly in a and Λ.

Let us discuss the condition (2.5) for $Y(x^1,x_1)$. On the half-line $x^1 > 0$

$$Y = y\psi + (\mu - \lambda)\psi, \quad y = V - V_0, \quad \mu = \mu(x_1,\Lambda(\pm m x_1)).$$

In the case " $-$ "

$$|y(x_1,x^1)| \leq C a^{-\rho-2}(x^1)^2 \tag{4.9}$$

so that the norm of $y\psi$ is estimated by $Ca^{-1-\sigma}$. In the case " $+$ " the bound (4.9) is fulfilled for $x^1 \leq a/2$ which allows us to estimate by $Ca^{-1-\sigma}$ the norm of $y\psi$ in the space $L_2(0,a/2)$. Its norm in the space $L_2(a/2,\infty)$ decays faster than any power of a^{-1} due to the exponential decay of the Airy function. The term $(\mu-\lambda)\psi$ satisfies (2.5) since, by (4.2),

$$|\mu(x_1,\Lambda_n) - \mu(x_1,\Lambda(\pm m x_1))| \leq |x_1|^{-1-\rho}.$$

On the half-axis $x^1 < 0$ we have that $Y = y\psi$, where

$$y(x_1,x^1) = V^2(x^1 - m x_1) - \lambda(x_1) + 2^{-1}v^2\Lambda(\pm m x_1)|m x_1|^{-2\sigma}\zeta(x^1/A).$$

Since $|y(x_1,x^1)| \leq C$, it follows from (4.5) that the norm of the function $Y(x_1,\cdot)$ in the space $L_2(-\infty,-A)$ decays faster than any power of a^{-1} as $a = \pm m x_1 \to \infty$. If $x^1 \in (-A,0)$, then according to Assumption 4.2 and the definition of $\lambda(x_1)$

$$y(x_1,x^1) = \pm v_2(|a \pm x^1|^{-\rho} - a^{-\rho}) + 2^{-1}v^2(\Lambda(\pm m x_1) - \Lambda_n)|m x_1|^{-2\sigma}.$$

Here the first term is bounded by $a^{-\rho-1}|x^1|$ and the second - by $a^{-3\sigma}$. This implies that the norm of $Y(x_1,\cdot)$ in $L_2(-A,0)$ does not exceed $Ca^{-3(\rho+1)/2}$.

Let $\Phi(x_1,t) = \Phi_n(x_1,t)$ be a phase function constructed for $\pm x_1 \geq 0$ with respect to the long-range potential $\lambda_n(x_1)$ and let an identification $J = J(\psi_n) : L_2(\mathbb{R}_\pm) \to L_2(\mathbb{R}^2)$ be defined by the equality (2.12). According to Theorem 2.5 the wave operator (2.13) exists. Actually, one can replace ψ_n by a simpler function $\psi_n^{(0)}$:

$$\left.\begin{aligned} \psi_n^{(0)}(x_1,x^1) &= \gamma_n|x_1|^{-\sigma/2}\mathrm{Ai}(vm^{-\sigma}|x_1|^{-\sigma}x^1 - \Lambda_n), \quad x^1 \geq 0, \\ \psi_n^{(0)}(x_1,x^1) &= 0, \quad x^1 \leq 0; \end{aligned}\right\} \tag{4.10}$$

the constant γ_n is chosen in such a way that $\|\psi_n^{(0)}(x_1,\cdot)\| = 1$. We stress that $\psi_n^{(0)}(x_1,x^1)$ is continuous at $x^1 = 0$ but its derivative has a jump at this point. Clearly,

$$\lim_{x_1\to\pm\infty} \|\psi_n(x_1,\cdot) - \psi_n^{(0)}(x_1,\cdot)\| = 0$$

and hence $(J(\psi_n) - J(\psi_n^{(0)}))U_0(t)f \to 0$ as $t \to \infty$. Thus the wave operator $\mathcal{W}_n$ defined by the equality (2.13) with respect to $J(\psi_n^{(0)})$ exists and equals that for $J(\psi_n)$. Now we are able to formulate our main result.

Theorem 4.3 Let Assumptions 4.1 and 4.2 hold. Let $\mathrm{Ai}(-\Lambda_n) = 0$. Define a long-range potential $\lambda_n(x_1)$, $\pm x_1 > 0$, by the equality (4.2), where $\Lambda = \Lambda_n$, and set $\Phi_n = \Gamma\lambda_n$. Then for the function (4.10) and $J = J(\psi_n^{(0)})$ the limit $\mathcal{W}_n$ defined by the equality (2.13) exists. The wave operator $\mathcal{W}_n : L_2(\mathbb{R}_\pm) \to L_2(\mathbb{R}^2)$ is isometric and $H\mathcal{W}_n = \mathcal{W}_n\mathbf{X}_1$, where $\mathbf{X}_1$ is multiplication by $x_1^2/2$ in the space $L_2(\mathbb{R}_\pm)$.

Assumptions 4.1 and 4.2 almost guarantee the existence of the wave operators W_0 and $W_{2,k}$ corresponding (see (1.5)) to scattering on bound states $\psi^{2,k}(x^2)$ of the Hamiltonian H^2. According to Theorem 2.6 the image $R(\mathcal{W}_n)$ of the wave operator $\mathcal{W}_n$ is orthogonal to that of W_0 whenever W_0 exists. The scalar product of the right-hand sides in (1.5) (where $\alpha = 2$) and in (1.10) tends to zero as $t \to \infty$. Therefore $R(\mathcal{W}_n)$ and $R(W_{2,k})$ are also orthogonal. Clearly, the ranges of the operators $\mathcal{W}_n$ are orthogonal for different n to each other.

Let us finally mention that if the subspaces X^1 and X^2 are orthogonal then due to the separation of variables the asymptotic completeness holds. In this case our construction does not work because $x^2 = x_1$ so that the equality (4.1) fails.

References

[1] L. Hörmander, The analysis of linear partial differential operators IV, Springer-Verlag, 1985.

[2] L. D. Faddeev, Mathematical aspects of the three body problem in quantum scattering theory, Trudy MIAN **69**, 1963. (Russian)

[3] V. Enss, Completeness of three-body quantum scattering, in: Dynamics and processes, P. Blanchard and L. Streit, eds., Springer Lecture Notes in Math. **1031** (1983), 62–88.

[4] V. Enß, Quantum scattering theory of two-body and three-body systems with potentials of short and long range, in: Schrödinger operators, S. Graffi, ed., Springer Lecture Notes in Math. **1159** (1985), 39–176.

[5] V. Enß, Long-range scattering of two- and three-body quantum systems, in: Journées EDP, Saint Jean de Monts (1989), 1–31.

[6] J. Dereziński, Asymptotic completeness of long-range quantum systems, Ann. Math. **138** (1993), 427–473.

[7] X. P. Wang, On the three body long range scattering problems, Reports Math. Phys. **25**(1992), 267–276.

[8] C. Gérard, Asymptotic completeness of 3-particle long-range systems, Invent. Math. **114** (1993), 333–397.

[9] D. R. Yafaev, On the break-down of completeness of wave operators in potential scattering, Comm. Math. Phys. **65** (1979), 167–179.

[10] D. R. Yafaev, Wave operators for the Schrödinger equation, Theor. Math. Phys. **45** (1980), 224–234. (Russian)

[11] J. Dollard, Asymptotic convergence and Coulomb interaction, J. Math. Phys. **5** (1964), 723–738.

[12] V. S. Buslaev, V. B. Matveev, Wave operators for the Schrödinger equation with a slowly decreasing potential, Theor. Math. Phys. **2** (1970), 266–274.

[13] D. R. Yafaev, New channels in two-body long-range scattering, Proc. Int. Conf. on Schrödinger operators, 1993, St. Petersbourg, Gordon and Breach (to be published).

[14] D. R. Yafaev, The low energy scattering for slowly decreasing potentials, Comm. Math. Phys. **85** (1982), 177–196.

[15] M. S. Ashbaugh, E. M. Harrel, Perturbation theory for shape resonances and large barrier potentials, Comm. Math. Phys. **83** (1982), 151–170.

Dimitri Yafaev
IRMAR, Univerité de Rennes 1, Campus de Beaulieu,
35042 Rennes Cedex, France

Participants

Lew Aizenberg: Bar-Ilan University, Department of Mathematics, 52900 Ramat-Gan, Israel.

Sergio Albeverio: Ruhr-Universität, Institut für Mathematik, 44780 Bochum, Germany.

Wolfgang Arendt: Université de Franche-Comté, Équipe de Mathématiques 16, Route de Gray, 25030 Besançon Cedex, France.

Michael Baranowski: TU Berlin, SFB 288, Straße des 17. Juni 136, 10623 Berlin, Germany.

Sebastian Behm: Max-Planck-Arbeitsgruppe "Partielle Differentialgleichungen und Komplexe Analysis", Universität Potsdam, Postfach 601553, 14415 Potsdam, Germany.

Stefan Benvegnu: Ruhr-Universität, Institut für Mathematik, 44780 Bochum, Germany.

Yuri Berezansky: Ukrainian Academy of Sciences, Institute of Mathematics, Tereshchenkivska street 3, 252601 Kiev, Ukraine.

Wulf Böttger: RWTH Aachen, Institut für Theoretische Physik, 52056 Aachen, Germany.

Paolo Boggiatto: Universita di Torino, Dipartimento di Matematica, Via Carlo Alberto 10, 10123 Torino, Italia.

Louis Boutet de Monvel: Université de Paris VI, Mathématique, 4, place Jussieu, 75232 Paris Cedex, France.

Antonio Bove: Universita di Bologna, Dipartimento di Matematica, Piazza di Porta San Donato 5, 40127 Bologna, Italia.

Thilo Buchholz: Universität Potsdam, Institut für Mathematik, Postfach 601553, 14415 Potsdam, Germany.

Vladimir Buslaev: Academy of Sciences of Russia, Euler International Mathematical Institute, Pesochnaya 10, St. Petersburg 197022, Russia.

Jan van Casteren: University of Antwerp (UIA), Department of Mathematics and Computer Sciences, Universitetplein 1, 2610 Wilrijk, Belgium.

Alexandre Chlapounov: Scuola Normale Superiore, Piazza dei Cavalieri 7, Pisa 56126, Italy.

Ludwig Dabrowski: Ruhr-Universität, Institut für Mathematik, 44780 Bochum, Germany.

Michael Demuth: Technische Universität Clausthal, Institut für Mathematik, Erzstraße 1, 38678 Clausthal-Zellerfeld, Germany.

Jan Dereziński: Warsaw University, Division of Mathematical Methods in Physics, Hoza 74, PL-00-682 Warsaw, Poland.

Christoph Dorschfeldt: Max-Planck-Arbeitsgruppe "Partielle Differentialgleichungen und Komplexe Analysis", Institut für Mathematik, Universität Potsdam, Postfach 601553, 14415 Potsdam, Germany.

Pierre Duclos: C.N.R.S., Centre Physique Théorique, Luminy-Case 907, F-13288 Marseille Cedex 9, France.

Yuri Egorov: Université Paul Sabatier, U.F.R., M.I.G., 118, route de Narbonne, 31062 Toulouse Cedex, France.

Volker Enß: RWTH Aachen, Institut für Reine und Angewandte Mathematik, 52056 Aachen, Germany.

László Erdös: Princeton University, Department of Mathematics, Princeton, NJ 08544, USA.

Boris Fedosov: Moscow Institute of Physics and Technology, Department of Mathematics, Dolgoprudny, Moscow region 141700, Russia.

Daisuke Fujiwara: Gakushuin University, Department of Mathematics, Faculty of Science, 1-5-1, Mejiro, Toshima-ku, Tokyo, 171, Japan.

Gianluca Garello: Universita di Torino, Dipartimento di Matematica, Via Carlo Alberto 10, 10123 Torino, Italy.

Peter Gilkey: University of Oregon, Department of Mathematics, Eugene, OR 97403-1222, USA.

Jean Ginibre: Université de Paris-Sud, Laboratoire de Physique Théorique et Hautes Energies, 91405 Orsay Cedex, France.

Ulrich Grieme: Max-Planck-Arbeitsgruppe "Partielle Differentialgleichungen und Komplexe Analysis", Universität Potsdam, Posfach 601553, 14415 Potsdam, Germany.

Bernard Helffer: Ecole Normale Supérieure, Laboratoire de Mathématiques, 45, rue d'Ulm, 75230 Paris Cedex 05, France.

Rainer Hempel: Institut für Analysis, Technische Universität Braunschweig, Pockelstraße 14, 38106 Braunschweig, Germany.

Matthias Hieber: Universität Zürich, Mathematisches Institut, Winterthurerstraße 190, 8001 Zürich, Switzerland.

Tankred Hirschmann: Max-Planck-Arbeitsgruppe "Partielle Differentialgleichungen und Komplexe Analysis", Universität Potsdam, Posfach 601553, 14415 Potsdam, Germany.

Peter Hislop: Mathematics Department, University of Kentucky, Lexington KY 40506-0027, USA.

Thomas Hoffmann-Ostenhof: The Erwin Schrödinger Institute, Institute of Mathematical Physics, Pasteurgasse 6/7, A-1090 Wien, Austria.

Wolf Hofmann: Universität Hamburg, Institut für Angewandte Mathematik, Bundesstr. 55, 20146 Hamburg, Germany.

Matthias Holschneider: C.N.R.S., Centre Physique Théorique, Luminy Case 907, 13288 Marseille Cedex 09, France.

Takashi Ichinose: Kanazawa University, Department of Mathematics, Faculty of Science, Kakuma-machi, 920-11 Kanazawa, Japan.

Victor Ivrii: University of Toronto, Department of Mathematics, 100 St. George Street, Toronto, Ontario, M5S 12A1, Canada.

Chisato Iwasaki: Himeji Institute for Technology, Department of Mathematics, Shosha, Himeji, 671-22-01, Japan.

Werner Kirsch: Ruhr-Universität Bochum, Institut für Mathematik, 44780 Bochum, Germany.

Vladimir Kondratev: Moscow State University, Department of Mechanics and Mathematics, Lenin Hills, 119899 Moscow, Russia.

Michael Korey: Max-Planck-Arbeitsgruppe "Partielle Differentialgleichungen und Komplexe Analysis", Universität Potsdam, Postfach 601553, 14415 Potsdam, Germany.

Ognyan Kounchev: Institute of Mathematics, Bulgarian Academy of Sciences, Acad. G. Bonchev 8, 1113 Sofia, Bulgaria.

Maddaly Krishna: The Institute of Mathematical Sciences, CIT Campus, Tharamani, Madras-600-113, India.

Shige Toshi Kuroda: Gakushuin University, Department of Mathematics, Faculty of Sciences, 1-5-1 Mejiro, Toshima-ku, Tokyo 171, Japan.

Ari Laptev: KTH, Mat. Institutionen, S-10044 Stockholm, Sweden.

Otto Liess: Universita di Bologna, Dipartimento Matematica, Piazza Porta San Donato 5, 40127 Bologna, Italy.

Günter Lumer: Université de Mons-Hainaut, Institute de Mathematiques et d'Informatique, Avenue Maistriau 15, 7000 Mons, Belgium.

Zhi-Ming Ma: Academica Sinica, Institute of Applied Mathematics, P.O. Box 2734, Bejing 1000080, China.

Konstantin Makarov: Max-Planck-Arbeitsgruppe "Partielle Differentialgleichungen und Komplexe Analysis", Universität Potsdam, Postfach 601553, 14415 Potsdam, Germany.

André Martinez: Université Paris Nord, Département de Mathématiques, Avenue J. B. Clement, 93430 Villetaneuse, France.

Victor Maslov: Institute for Problems in Mechanics, Academy of Sciences, Prosp. Vernadskogo, 101, 117526 Moscow, Russia.

Ivor McGillivray: Technische Universität Clausthal, Institut für Mathematik, Erzstraße 1, 38678 Clausthal-Zellerfeld, Germany.

Vladimir Mikhailets: Ukrainian Academy of Sciences, Institute of Mathematics, ul. Repina 3, 252601 Kiev, Ukraine.

Tosinobu Muramatu: University of Tsukuba, Institute of Mathematics, Tsukuba-Shi, Ibaraki 305, Japan.

Shu Nakamura: University of Tokyo, Komaba, Meguro-ku, Tokyo 153, Japan.

Olga Oleinik: Moscow State University, Department of Mechanics and Mathematics, Lenin Hills, 119899 Moscow, Russia.

Markus Pflaum: Universität München, Mathematisches Institut, Theresienstraße 39, 80333 München, Germany.

Boris Plamenevski: Academy of Sciences of Russia, Euler International Mathematical Institute, Pesochnaya 10, 197022 St. Petersburg, Russia.

Radu Purice: Institute of Mathematics, Romanian Academy, P.O. Box 1-764, RO-70700 Bucharest, Romania.

Vladimir Rabinovich: Univ. Rostov-on-Don, Department of Mathematics and Mechancis, Zorge str. 5, 344104 Rostov-na-Donu, Russia.

Michael Reissig: Institut für Angewandte Mathematik I, Technische Universität Bergakademie Freiberg, 09599 Freiberg, Germany.

Luigi Rodino: Universita di Torino, Dipartiment di Matematica, via Carlo Alberto 10, 10123 Torino, Italy.

Jakov Roitberg: Pedagogical Institute, Department of Mathematics, ul. Sverdlova 53, 250038 Chernigov, Ukraine.

Inna Roitberg: Pedagogical Institute, Department of Mathematics, ul. Sverdlova 53, 250038 Chernigov, Ukraine.

Michael Röckner: Fakultät für Mathematik, Universität Bielefeld, Postfach 100131, 33501 Bielefeld, Germany.

Yoshimi Saito: University of Alabama, Department of Mathematics, Birmingham, 35294 Birmingham, Alabama, USA.

Pierre Schapira: Université de Paris VI, Mathématiques, 4 place Jussieu, 75252 Paris Cedex 05, France.

Wolfgang Scherer: Technische Universität Clausthal, Institut für Theoretische Physik A, Leibnizstraße 10, 38678 Clausthal-Zellerfeld.

Elmar Schrohe: Max-Planck-Arbeitsgruppe "Partielle Differentialgleichungen und Komplexe Analysis", Institut für Mathematik, Universität Potsdam, Posfach 601553, 14415 Potsdam, Germany.

Bert-Wolfgang Schulze: Max-Planck-Arbeitsgruppe "Partielle Differentialgleichungen und Komplexe Analysis", Institut für Mathematik, Universität Potsdam, Postfach 601553, 14415 Potsdam, Germany.

Alexander Shamaev: Moscow State University, Department of Mathematics and Mechanics, Lenin Hills, 119899 Moscow, Russia.

Victor Shatalov: Moscow State University, Department of Calculating Mathematics and Mechancis, Lenin Hills, 119899 Moscow, Russia.

Zinovi Sheftel: Pedagogical Institute, Mathematical Department, ul. Sverdlova, 250038 Chernigov, Ukraine.

Alexander Sobolev: MAPS, University of Sussex, Falmer, Brighton, BM 19QH, United Kingdom.

Victoria Steblovskaya: Polytechnical Institute, Department of Mathematical Methods of System Analysis, Prosp. Pobedy, 252056 Kiev, Ukraine.

Boris Sternin: Moscow State University, Department of Calculating Mathematics and Mechanics, Lenin Hills, 119899 Moscow, Russia.

Peter Stollmann: Universität Frankfurt, Fachbereich Mathematik, Robert-Mayer-Straße 6-10, Postfach 111932, 60054 Frankfurt, Germany.

Günter Stolz: Department of Mathematics, University of Alambama, Birmingham, 35294 Birmingham, Alabama, USA.

Nikolai Tarkhanov: Max-Planck-Arbeitsgruppe "Partielle Differentialgleichungen und Komplexe Analysis", Institut für Mathematik, Universität Potsdam, Postfach 601553, 14415 Potsdam, Germany.

Dimitri Vassiliev: Sussex University, Division Mathematics, Falmer, Brighton BN 19QH, UK.

Mark Vishik: Moscow State University, Department of Mechanics and Mathematics, Lenin Hills, 119899 Moscow, Russia.

Jürgen Voigt: Technische Universität Dresden, Abteilung Mathematik, Institut für Analysis, Mommsenstraße 13, 01062 Dresden, Germany.

Lutz Weis: Universität Kiel, Mathematisches Seminar, Ludewig-Meyn-Straße 4, 24098 Kiel, Germany.

Harald Widom: University of California, Department of Mathematics 34-135, Santa Cruz, California 95064, USA.

Ingo Witt: Max-Planck-Arbeitsgruppe "Partielle Differentialgleichungen und Komplexe Analysis", Institut für Mathematik, Universität Potsdam, Posfach 601553, 14415 Potsdam, Germany.

Dimitri Yafaev: IRMAR, Université Rennes 1, Campus de Beaulieu, 35042 Rennes Cedex, France.

Karen Yagdjian: Armenian Academy of Sciences, Institute of Mathematics, Marshal Bagramian Av. 24B, 375019 Yerevan, Armenia.

Kenji Yajima: University of Tokyo, Department of Mathematics, Coll. of Arts and Sci., 3-8-1 Komaba, Meguro-ku, Tokyo 153, Japan.

Talks

L. Aizenberg: The Cauchy problem for the Cauchy-Riemann system

S. Albeverio: Some new developments in stochastic partial differential equations

W. Arendt: p-independence of the spectrum of elliptic operators in $L^p(\Omega)$

Yu. M. Berezansky: The spectral approach to the integration of nonlinear differential-difference equations

L. Boutet de Monvel: Real analogue of the Bergman kernel

V. Buslaev: On a dynamical system generated by Hofstadter model

J. van Casteren: Some compactness properties of differences of Feynman-Kac semigroups

A. Chlapounov, N. Tarkhanov: Bases with double orthogonality

M. Demuth: Finite capacities in spectral theory

J. Dereziński: N-body classical scattering

P. Duclos: Quantum Fermi accelerators with pure point quasi spectrum

Yu. Egorov: On the moments of the negative eigenvalues of an elliptic operator

V. Enß: Inverse two- and N-body scattering by geometrical methods

L. Erdös: Spectral estimates of the Schrödinger operator

B. Fedosov: On the trace density in deformation quantization

D. Fujiwara: A remainder estimate of stationary phase method as the dimension of the space goes to ∞

G. Garello: Microlocal properties for general classes of pseudo-differential operators

P. Gilkey: Inequivalent metrics of positive scalar curvature, equivariant spin bordism, and the eta invariant

B. Helffer: Transfer matrix, Schrödinger operators and applications

R. Hempel: On the absolute continuity of Schrödinger operators with periodic vector potentials

M. Hieber: Gaussian estimates and imaginary powers of elliptic operators on L^p-spaces

P. Hislop: Localization for 2-dimensional random Schrödinger operators with magnetic fields and application to the quantum Hall effect

Th. Hoffmann-Ostenhof: Interior Hölder estimates for solutions of Schrödinger equations and their regularity for their nodal sets

M. Holschneider: Spectral wavelet dimensions and time-evolution

T. Ichinose: Path integral for the relativistic Schrödinger semigroup

V. Ivrii: Multiparticle quantum theory and semiclassical asymptotics

Ch. Iwasaki: The symbolic construction of the fundamental solution for parabolic mixed problems and the Gauss-Bonnet-Chern theorem

W. Kirsch: Lifshitz tails with magnetic fields

V. Kondratev: On behaviour of the solutions of non-linear elliptic equations in cylindrical domains

M. Krishna: Inverse theory for random Jacobi matrices

S. T. Kuroda: Product formulas and some error estimates

A. Laptev: On the Dirichlet eigenvalue problem

O. Liess: Decay estimates for Fourier transforms of densities on surfaces with singularities

G. Lumer: Singular problems, generalized solutions, and topology

Z. Ma: Malliavin calculus in Ito's framework

A. Martinez: A reduction scheme for adiabatic transition probabilities

V. P. Maslov: Hyperbolic and elliptic equations with jumping nonlinearity in classical and quantum statistics and nuclear physics

I. McGillivray: A recurrence condition for some subordinated strongly local Dirichlet forms

S. Nakamura: On an example of phase space tunneling

O. Oleinik: Homogenization problems in partial differential equations

M. Reissig, K. Yagdjian: Hypoellipticity of some operators with degeneration of infinite order

B. Plamenevski: On the compositional series in algebras of pseudodifferential operators with discontinuous symbols

R. Purice: A distinguished self-adjoint realization of the Dirac-Hamiltonian with strong local singularities and arbitrary behaviour at infinity

V. Rabinovich: Boundary value problems in domain with non-standard singularities

M. Röckner: Essential self-adjointness of generalized Schrödinger operators and applications

L. Rodino: Multi-quasi-elliptic operators in $\mathbb{R}^n$

I. Roitberg: On approximation of the solutions of elliptic boundary value problems for Petrovskii elliptic system by linear combinations of the fundamental solutions

J. Roitberg: Elliptic and parabolic value problems in non-smooth domains in complete scales of Banach spaces

Y. Saito: The reduced wave equation for two media with cylindrical discontinuity

P. Schapira: Leray's quantization of projective duality

E. Schrohe: Conical singularities and boundary value problems

B.-W. Schulze: Ellipticity on singular spaces with components of different dimensions

Z. Sheftel: On approximation by solutions of non-local elliptic problems

A. Sobolev: Discrete spectrum of the Schrödinger operator with singular potential

V. Steblovskaya: On some problems of infinite-dimensional calculus of variations

B. Sternin, V. Shatalov: The Lefschetz fixed point theorem in symplectic geometry

P. Stollmann: Continuity of Dirichlet boundary problems with respect to varying domains

G. Stolz: Localization for the Poisson model

D. Vassiliev: The Berry conjecture for "fractal drums" and the renewal theorem

M. S. Vishik: Attractors of nonautonomous partial differential equations and their dimensions

J. Voigt: The Feller property for absorption semigroups

L. Weis: Stability of positive semigroups on L^p

H. Widom: Fredholm determinants and systems of partial differential equations

D. Yafaev: New channels in two-body long-range scattering

K. Yajima: The L^p-boundedness of wave operators for Schrödinger operators

Titles previously published in the series

OPERATOR THEORY: ADVANCES AND APPLICATIONS

BIRKHÄUSER VERLAG

65. **M.M. Djrbashian:** Harmonic Analysis and Boundary Value Problems in the Complex Domain, 1993, (3-7643-2855-X)
66. **V. Khatskevich, D. Shoiykhet:** Differentiable Operators and Nonlinear Equations, 1993, (3-7643-2929-7)
67. **N.V. Govorov †:** Riemann's Boundary Problem with Infinite Index, 1994, (3-7643-2999-8)
68. **A. Halanay, V. Ionescu:** Time-Varying Discrete Linear Systems Input-Output Operators. Riccati Equations. Disturbance Attenuation, 1994, (3-7643-5012-1)
69. **A. Ashyralyev, P.E. Sobolevskii:** Well-Posedness of Parabolic Difference Equations, 1994, (3-7643-5024-5)
70. **M. Demuth, P. Exner, G. Neidhardt, V. Zagrebnov** (Eds): Mathematical Results in Quantum Mechanics. International Conference in Blossin (Germany), May 17-21, 1993, 1994, (3-7643-5025-3)
71. **E.L. Basor, I. Gohberg** (Eds): Toeplitz Operators and Related Topics. The Harold Widom Anniversary Volume. Workshop on Toeplitz and Wiener-Hopf Operators, Santa Cruz, California, September 20–22, 1992, 1994 (3-7643-5068-7)
72. **I. Gohberg, L.A. Sakhnovich** (Eds): Matrix and Operator Valued Functions. The Vladimir Petrovich Potapov Memorial Volume, (3-7643-5091-1)
73. **A. Feintuch, I. Gohberg** (Eds): Nonselfadjoint Operators and Related Topics. Workshop on Operator Theory and Its Applications, Beersheva, February 24–28, 1994, (3-7643-5097-0)
74. **R. Hagen, S. Roch, B. Silbermann:** Spectral Theory of Approximation Methods for Convolution Equations, 1994, (3-7643-5112-8)
75. **C.B. Huijsmans, M.A. Kaashoek, B. de Pagter**: Operator Theory in Function Spaces and Banach Lattices. The A.C. Zaanen Anniversary Volume, 1994 (ISBN 3-7643-5146-2)
76. **A.M. Krasnosellskii**: Asymptotics of Nonlinearities and Operator Equations, 1995 (ISBN 3-7643-5175-6)
77. **J. Lindenstrauss, V.D. Milman** (Eds): Geometric Aspects of Functional Analysis Israel Seminar GAFA 1992-94, 1995 (ISBN 3-7643-5207-8)
78. **M. Demuth, B.-W. Schulze** (Eds): Partial Differential Operators and Mathematical Physics: International Conference in Holzhau (Germany), July 3-9, 1994, 1995 (ISBN 3-7643-5208-6)

MIX
Papier aus verantwortungsvollen Quellen
Paper from responsible sources
FSC® C105338

If you have any concerns about our products,
you can contact us on
ProductSafety@springernature.com

In case Publisher is established outside the EU,
the EU authorized representative is:
Springer Nature Customer Service Center GmbH
Europaplatz 3, 69115 Heidelberg, Germany

Printed by Libri Plureos GmbH
in Hamburg, Germany